The Story of Telecommunications

The Story of
Telecommunications

by
George P. Oslin

MERCER UNIVERSITY PRESS
MACON, GEORGIA
· 1992 ·

ISBN 0-86554-418-2   MUP/H322

The paper used in this publication meets
the minimum requirements of American National Standard
for Information Sciences—Permanence of Paper
for Printed Library Materials, ANSI Z39.48-1984.

Library of Congress Cataloging-in-Publication Data

Oslin, George P., 1899–
The story of telecommunications / by George P. Oslin.
xii+507pp. 7x10" (18x25.5cm.)
Includes bibliographical references and index.
ISBN 0-86554-418-2 (alk. paper)
1. Telecommunications—History.
2. Telecommunications—United States—History.
I. Title.
HE7631.082   1992
384—dc20   92-38503
                                                                    CIP

Contents

List of Illustrations

Preface

This book is the result of an enormous amount of research of journals, diaries, and letters of the pioneers who invented, built, and organized what is now the communications industry's global network. It attempts to provide the vicarious experience of being there during centuries of development, and provide knowledge readers need to adjust to our complex modern information age and have successful and happy work and private lives.

Our way of life would be impossible without telephone, telegraph, television, and radio communications flashing through satellites, microwaves, and cables crisscrossing the nation and the world. They are required to meet our business, bank, news, government, national defense, food, and travel needs.

This is the story of our communications advances from semaphore to Morse telegraphy to high-speed computer/data transmission, and from hand cranked telephones to television and microwaves.

Each great advance has been achieved by using past achievements and adding the new. We have a blending of our heritage and modern developments. The last chapters describe modern wonders, predict future trends and urge a national policy that allows the industry to continue its progress instead of tearing to pieces the world's finest communications systems—symbols of American free enterprise.

This gathering of communications history began to answer requests by government officials, writers, and educators. No book had been published on the industry since James D. Reid's *The Telegraph in America* in 1886. American histories did not even mention the transcontinental telegraph which was vital in preserving the Union in the Civil War, or Western Union's Russian-American Expedition that caused the purchase of Alaska.

Among hundreds who told me about their experiences were Thomas A. Edison, William Henry Jackson (Civil War soldier, pioneer photographer, and covered wagon bullwhacker, then nearly 100 years old), William Campbell (last surviving Pony Express rider, then 95), and Martin Cahoon (on the Great Eastern when the first transatlantic cable was laid). Gradually file drawers were filled with diaries, letters, clippings, and interviews.

For thirty-five years, I wrote communications history as it was being made; I planned and staged the centennial and other national communications celebrations; and I publicized daily industry developments. During the 1960s, 1970s, and 1980s, I traveled and researched an enormous mass of telecommunications documents in historical libraries. This book was written because a unique store of our nation's history would otherwise be lost forever.

Facts from original sources, company documents, thousands of old newspapers, magazines, and books, and more than 100,000 letters and diaries of the pioneers were pieced together and condensed into the real story of their struggles, strategy, and success. One discovery was the manuscript for a book—missing for 88 years—by Morse's partner F. O. J. Smith.

This book was written without sponsorship, subsidy, or obligation to any company. It is the true story of colorful heroes, pirates, and villains, and men like Morse and Bell, famous for inventions largely produced by others. It rescues unsung heroes from undeserved limbo.

Hundreds of people helped. Among them were Leila Livingston Morse, granddaughter of the inventor; Harper Sibley, Jr., great grandson of Hiram Sibley; Wilmer R. Leech and Arthur J. Breton, New York Historical Society Library; Mrs. Donald Kissil, Morristown (New Jersey) Library; Dr. Blake McKelvey, Rochester City Historian; John Keating and Mrs. Thelma Jefferies, Local History Division, Rochester Library; Mrs. Edith M. Fox, curator and archivist, Cornell University; Miss Elizabeth Ring, vice president and director, Maine Historical Society; M. R. Marsh, former head of Western Union Patent Department; Mrs. Alene Lowe White, librarian, Western Reserve Historical Society, Cleveland; Miss Isabella Sims, gifts and exchanges librarian, University of Washington; Mary M. Johnson, research consultant, National Archives; Fred W. Henck, editor, *Telecommunications Reports*; Norman R. Speiden, Edison National Historic Site; John O'Brien, RCA; Joseph Johnson, Western Union International, Inc.; George A. Shawy, RCA Communications; Edward J. Felesina, ITT World Communications; C. A. Delhomme, ITT; Matthew Gordon, COMSAT; Joseph E. Baudino, Westinghouse Electric Corporation; Warren R. Bechtel, corporate communications director of Western Union; Norma H. McCormick, AT&T historical photograph specialist; and Carl L. Bolton, New York Stock Exchange official.

Mary F. Toomey, retired English teacher, read copy. The final typing, recording the text onto computer diskettes, was by my daughter Jeanne Fernald; her husband Olaf, a computer engineering consultant, gave helpful advice.

My greatest debt is to my wife Susan, who made this book possible by taking notes on thousands of documents, letters, and clippings in libraries, handling voluminous correspondence, typing and retyping condensations of the manuscript, and making helpful suggestions.

—*George P. Oslin*
July 1992

Man's Rebellion against Time and Space

This is the story of man's rebellion against the barriers of time and space, and his success in overcoming them.[1] It tells how inventors created equipment with which people could communicate at a distance, and relates the pioneers' own stories, from their diaries, of how they built telegraph lines that had a vital role in expanding, building, and preserving the United States.

It also tells how the lines were extended around the world and even reached remote parts of the universe. This is the first history of one of the largest industries in the world, with thousands of companies and millions of employees. It is designed to provide readers with an understanding of the wonderful, complex telecommunications world in which we live. Skillful use of telecommunications services will help everyone achieve greater success and happiness in business and private life.

Our story begins with prehistoric cavemen shouting warnings and challenges, or building fires to signal others. Fire, smoke, and flag signals aided Egyptian, Athenian, and Roman armies. A legend of the Trojan

War 3,000 years ago tells of King Agamemnon's son returning victorious after helping to free Helen of Troy. As he sailed into the harbor he forgot to lower the black sails of death. Seeing that signal, the king, watching from his castle on the cliff, believed his son had been killed, and plunged to his death on the rocks below.

Sentinels of Julius Caesar, stationed in a series of towers, relayed messages by shouting. In 1588, a chain of signal fires in a series of towers warned Queen Elizabeth I that King Philip of Spain had sent a Spanish Armada to conquer Protestant England and make it a Catholic country. The Armada arrived with 13,773 soldiers and sailors, 2,000 slaves, and 180 priests on 130 ships. The alerted British Navy destroyed the Armada. In the jungles of Africa, native drums "talked" with distant villages. American Indians used smoke signals by day and shot flaming arrows into the sky at night.

General Aneas, a Greek contemporary of Aristotle, prescribed a way to send sentences using water in jugs on two hills. At a signal the water would be run out of both jugs until

1.1 INDIANS SENT SMOKE SIGNALS BY DAY AND FLAMING ARROWS AT NIGHT TO CONVEY MESSAGES AT A DISTANCE.

a flag was waved to stop. Each water level indicated a prearranged sentence. Later, Cleoxenus of Greece placed two to five torches on a wall in successive combinations representing letters of the alphabet. That method, described by Polybius, a Greek historian about 200 B.C., became the basic idea of modern printing telegraphy.

As late as 1825 it required about eighty minutes for cannons five miles apart to relay the signal of the Erie Canal opening from Buffalo to New York City.

Early communications were between governments, kings, military leaders, or tribal chieftains. Messengers were sent with written documents, but could go no faster than the horses drawing the imperial chariot. For centuries the post rider and stagecoach remained the standard speed.

Harnessing the Lightning

Electricity is an elemental force, used in almost every aspect of our daily lives. Without it our information age could not exist. But how did man harness that elemental force to serve him? The ancient Chinese discovered and used the lodestone, a natural magnet, and knowledge of it traveled westward. Roger Bacon suggested the use of electricity for communication as early as 1267. The Franciscan Order, of which he was a member, accused him of dealing in "black magic," and Bacon spent the next twenty years in prison. Dr. William Gilbert, Queen Elizabeth's physician, announced in 1600 that some substances, such as glass, have the power of attraction, which he called "electricity," because *ēlektron* is the Greek word for amber.

The first electricity-producing device was a sulphur ball that Otto von Guericke, burgomaster of Magdeburg, Germany, rubbed with his hands around 1650. In 1675 Sir Issac Newton of England, who discovered the laws of gravity, substituted a glass globe for the sulphur ball.

Stephen Gray of England described the basic principles of telegraphy in 1730, and his partner, Wheeler, suggested that the wire be suspended by a silk thread. With that insulation, electricity was sent through 886 feet of wire. At one end of the wire they rubbed a glass globe, and at the other end a feather moved.

Seeking a way to store electricity, Pieter van Musschenbroek, a professor at Leyden, Holland, learned in January 1746 that bottles coated with tinfoil would store static electricity that could be conveyed from the jar by a wire. They were called Leyden jars.

The next year Winckler, Nollet, and Lemonnier experimented with static electrical discharges through a wire with a water and earth return for two miles. In 1747 Dr. William Watson sent current across Westminster Bridge in London, dipping an iron rod in the Thames River to complete the circuit.

Hearing of the Leyden jar, Benjamin Franklin discharged electricity through a wire across the Schuylkill River at Philadelphia in 1748. His famous kite experiment in 1752 proved that lightning and electricity were identical. Franklin then invented the lightning rod to protect buildings, and some French people wore lightning rods on their hats in stormy weather.

The First Telegraphs

Many have the mistaken idea that the telegraph, telephone, radio, TV, satellite, or computer each sprang from the mind of one person. Actually thousands paved the way for the final step in each invention.

The first practical suggestion for telegraphy was by "C. M."[2] whose letter in *Scots's Magazine* in February 1753, proposed the use of a wire for each letter of the alphabet. At the receiving end, he said, "Let a ball be suspended from every wire, and about one-sixth to one-eighth of an inch below the ball, place the letters of the alphabet, marked on bits of paper. . . ." As the "charge" would be sent on each wire, it would pick up the paper beneath it and indicate that letter. He also suggested using bells of different sizes and tones instead of paper. Using static electricity, Georg-Louis Le Sage, a French professor, tried in 1774 to carry out "C.M.'s" idea in Geneva, attaching a pith ball to each wire.

In 1787, Lomond, a French mechanic, used only one wire to send words in code to his wife in another room. The impulses produced deflections of a small pith ball.

In 1794 Reusser of Geneva and Salva of Barcelona operated visual telegraphs by interrupting electric circuits on the desired wire and causing sparks to appear. Humboldt reported that a telegraph using only one wire was established by Salva, twenty-six miles from Madrid to Aranjuez.

Homing pigeons were used for centuries because they could fly hundreds of miles at speeds up to eighty miles an hour. The

Rothschild brothers used pigeons to speed news between European capitals.[3]

Semaphore Systems

On May 21, 1684, long before "C.M.'s" telegraph suggestion, Dr. Robert Hooke of England, inventor of the balance wheel in watches, proposed, before the Royal Society of England, a semaphore[4] system. He wanted to use triangles and circles by day, and lights in different arrangements at night. The telescope (invented in Holland about 1608) was to be used. The Liverpool Town Council erected towers at Bidston to signal the arrival of ships.

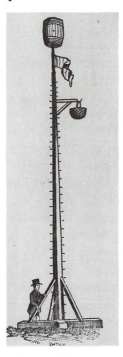

1.2 Semaphore system used by Washington's army during the American Revolution.

During the American Revolution, George Washington's forces used semaphore signals, moving a barrel at the top of a mast, a flag below the barrel, and a basket on a crossarm

to various positions.

During the French Revolution, Claude Chappe operated three different semaphore

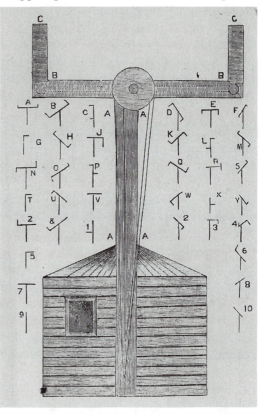

1.3 A German semaphore system of 1798, based on Chappe's.

systems. The first depended on two clocks working synchronously at sending and receiving towers. The sending operator struck a gong. The receiving operator heard the sound, noted the number indicated by the clock, and got the meaning of the number from a code book. Revolutionary fanatics thought the system was being used to communicate with King Louis XVI, a prisoner in the Temple nearby, and wrecked it in September 1792.

Chappe then built a "shutter" system, with a rectangular frame at each station, with five shutters which could be opened and closed. The mob wrecked that also.

Undaunted, Chappe and two of his brothers constructed a third system, this time with the government's financial aid and protection. Each station had a tall pole with a long arm across the top. At each end of the crossarm was a wing, drawn by pulleys into various positions to convey messages, as Boy Scouts send wigwag signals with a flag in each hand.

1.4 Shutter semaphore system used in England until 1816.

By 1794 Chappe had fifteen stations linking Paris with Lille. The first message, on August 15, 1794, reported that Quesnoy had been taken from Austria, then at war with France. The semaphore was extended to Brest in 1796, to Dunkirk in 1798, to Brussels in 1803, and to Milan in 1805. Napoleon Bonaparte ordered the Paris-Milan Line established. By 1844 France had more than

500 towers; a message from Paris to Calais required four minutes; to Strasbourg, six; and to Brest, seven.

Until 1816 a shutter system, invented by Lord George Murray, was operated by turning six circular shutters to various positions to indicate letters of the alphabet. It linked London with Yarmouth, and Portsmouth with Plymouth.

1.5 A Russian semaphore system of 1858.

Government semaphore systems were used in nearly all European countries between 1795 and 1840. One connected Berlin with Cologne and Coblenz in 1833. Another, with 1,320 operators and 220 towers five and six miles apart, was inaugurated by Czar Nicholas I in 1838 to link St. Petersburg, Moscow, and Warsaw.

In eighteenth-century America, itinerant merchants, travelers, stagecoaches, and special couriers carried mail over trails and poor roads. Benjamin Franklin, our first postmaster general, used post riders, but it required a week to get mail from New York to Boston. News of the adoption of the Declaration of Independence July 4, 1776 was taken by swift express riders to the thirteen states, but did not reach New York until July 9; Worcester, Massachusetts, July 15; Newport,

Rhode Island, and Williamsburg, Virginia, July 20; Halifax, North Carolina, July 22; and Savannah, Georgia, August 8.[5]

The first commercial semaphore system in America was built in 1801 by Jonathan Grout, Jr., seventy-two miles from Martha's Vineyard across the base of Cape Cod and north through Cohasset, Weymouth, and Dorchester to Boston. For six winters, from October to May, men with telescopes noted the positions of the arms of the next tower in the line of sixteen, and reported the arrival and departure of ships. The line was not a financial success and was abandoned, but the name "telegraph" is still applied to the hills that were used, and to a street in Boston. Grout moved to Philadelphia, where he began operating a semaphore ship-reporting system on November 9, 1809, from Reedy Island at the head of Delaware Bay.

The Boston line was revived by Samuel Topliff in 1820 and extended to Long Island Head, where the light keeper operated it to supply ship news to the merchants and captains who gathered in Topliff's commercial news room in Merchants' Hall, Congress, and Water Streets. Topliff's reporters then boarded the incoming ships as they entered the harbor.[6]

Even before Topliff, a semaphore line was operated by James M. Elford, using black and white balls, and later seven blue and white flags in combinations representing words, on the cupola of his "observatory" at 149 East Bay Street, Charleston, South Carolina. The cupola was replaced in 1822 by a higher one. Ships adopted his code to "talk" with Charleston and other ports they approached.[7]

John R. Parker established a semaphore system from Boston to Topliff's old station on Long Island, using Elford's flag signals. Three years later he moved his Boston Light

station across the harbor to Point Allerton, and in 1827 to a hill in Hull. He also relayed messages from incoming ships to Boston. His successors substituted the magnetic telegraph in 1853. Similar ship-news reporting was provided by New Orleans shipping companies in the 1850s.

In 1807, in Portland, Maine, Captain Lemuel Moody, who had been a water boy in the American Revolution, built an eighty-two-foot, octagonal tower on Munjoy Hill, 227 feet above the blue waters of Casco Bay. Using flags, he signalled the approach of craft to owners and merchants. On 4 September 1813, the battle of the U. S. Brig Enterprise and H. M. Brig Boxer was watched by Moody, who relayed the events to an excited crowd below like a modern sports broadcaster.[8]

In 1812 a semaphore line was built by Christopher Colles from the New York Customs House to Staten Island, and south to Sandy Hook on the New Jersey coast. "On January 22, 1813, British war vessels were seen off Sandy Hook, and furnaces for heating cannon balls were made ready. The telegraph, consisting of a number of white and black balls or kegs, hoisted in a preconcerted manner, gave signals easily seen from New York."[9] The Colles line earned little and fell into disrepair, but another was built to Staten Island by Samuel C. Reid in 1821.

An eighty-year-old sea captain, John Green, set up a line to Sandy Hook and Highlands in 1829. The New York office, on the south side of Castle Garden, now the Battery, had a pole with a movable arm operated by rope and pulley. John F. Myers was a semaphore and telegraph operator on that route for sixty-five years, from 1829 to 1894.[10]

New York merchants and ship owners built a new semaphore line to Sandy Hook in 1837, placing the New York station atop the Broadway Stephen Holt Hotel, later named United States Hotel. Ships were boarded, and foreign news was sent to the tower at Sandy Hook by carrier pigeon for transmission by semaphore to New York hours ahead of the vessel. At first nine newspapers joined in sending a boat to board ships; then David Hale's *Journal of Commerce* bucked the combine with his own boat, and had a horseman bring the news from the outer tip of Staten Island.

From 1840 to 1846, a semaphore line from New York was used by William C. Bridges and other Philadelphia brokers to speed news of stocks, produce, and lottery numbers. That and the New York-Sandy Hook semaphore were replaced later by telegraph lines.[11]

The Electric Battery

In 1790 the wife of Luigi Galvani, an Italian physician, noticed a convulsive muscular movement in a dead frog when its leg touched a scalpel charged with electricity. Galvani hung some frogs by a copper wire to an iron balcony and they twitched when they touched the iron.

What Galvani found was that current could be produced by bringing two dissimilar metals into contact with a moist substance, such as a frog's leg. In 1800 Allesandro Volta separated alternate discs of dissimilar metals by wet cloth or cardboard, and invented the electric battery. He shaped the discs like cups and put a salt solution in them.

Salva built an electrolytic telegraph in 1805; the signals were bubbles of hydrogen rising from the end of a wire in water. Samuel Soemmering, a Munich surgeon, had a similar system in 1809, using a wire for each letter and numeral.

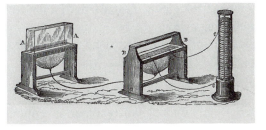

1.6 SOEMMERING'S TELEGRAPH OF 1809.

Francis Ronalds, an English merchant, invented a crude telegraph in 1823. A disc at each end of a wire revolved slowly in unison with the other. A signal, sent over the wire when the desired letter appeared on the sending disc, caused pith balls at the receiving disc to spring apart and show the same letter.

History has neglected an American, Harrison Gray Dyar, who built and operated a telegraph line several miles long at a Long Island race track in 1828. He recorded the signals in red marks on litmus paper, which he moved by hand against the end of the wire.

Dyar asked New Jersey for permission to build a telegraph line from New York to Philadelphia, but left the country when a spiteful employee, unable to obtain stock in the enterprise, informed the government that Dyar was conspiring to send "secret information" over the wires. Since he obviously used a code to convey information, Dyar, then twenty-one, was on the threshold of inventing the telegraph in America.

Speaking at Dyar's funeral at Rhinebeck, New York, on 3 September 1875, Rev. Charles S. Harrower charged, "Mr. Charles Walker, the brother-in-law of Mr. S. F. B. Morse, was Mr. Dyar's counsel, as well as intimate friend. After having absented himself from New York for a few months, Mr.

Dyar returned and consulted with Mr. Walker, who thought that, however groundless such a charge might be, it might give his client infinite trouble to stand suit. Accordingly the enterprise was then and there abandoned."[12]

Dyar lived for nearly thirty years in Paris and London, where other inventions made him a fortune, and returned to New York for the final six years of his life.

The Secret of Electromagnetism

A lucky accident enabled Hans Christian Oersted to learn, in 1819, the secret of electromagnetism. While teaching a science class at the University of Copenhagen, he noticed the deflection of a compass needle when he moved a neighboring wire carrying an electric current. He realized that a wire carrying a current exerts magnetic force. The deflection of the needle was to the right or left, according to the direction of the current. Andre Marie Ampere of Lyons, France, determined the unit of measurement of electric current, which was named for him, *ampere*.

While he was attached to the Russian embassy, Baron Paul Ludovitch Schilling saw a telegraph demonstration by Soemmering in Munich in 1809. Returning home, Schilling sent signals across rivers, and in 1820 produced a telegraph with five magnetic needles. He reduced the number of needles to one, and devised a code of letters and numbers indicated by movements of the needle. In 1835 the inventor exhibited it to a congress of physicists at Bonn. In 1837 Czar Alexander I ordered him to build a telegraph line, but Schilling died on June 25 of that year.

Johann S. C. Schweigger of Halle, Germany, found in 1820 that the deflection of

the needle could be increased by surrounding

1.7 ALEXANDER'S TELEGRAPH OF 1837.

it with separated turns of naked wire, and produced the galvanometer, which measures currents of electricity. William Sturgeon in England put that discovery to use in 1825 by inventing an electromagnet, consisting of iron in the shape of a horseshoe wound with wire, which would attract a small iron bar when current was passed through it.

It was Joseph Henry,[13] a teacher at the Albany (New York) Academy, who divided electromagnets into two classes: "quantity" and "intensity." In 1833 he had a 100 pound quantity magnet that would lift 3,500 pounds.[14]

Henry's intensity magnet was vital to telegraph development because it could be operated with small currents. He used a soft iron core bent in the shape of a horseshoe because it lost its magnetism the instant the electrical current was turned off. In 1830, he operated an electromagnetic telegraph over more than a mile of copper wire in his school building at Albany. At the receiving end, he used a small magnet, pivoted to

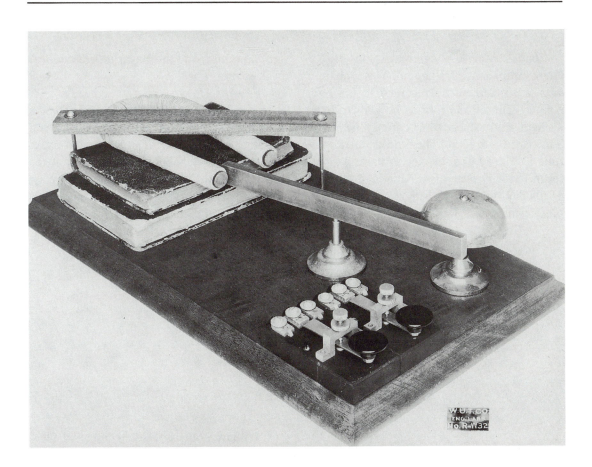

1.8 MODEL OF HENRY'S ELECTRIC BELL. (THE MORSE KEYS ARE MODERN, USED ONLY TO ILLUSTRATE OPERATION.)

swing like a compass needle, to tap a bell when reversals of current were sent through the line. In 1831, when Morse was studying art in Europe, Henry wrote that the way was then clear for the commercial electromagnetic telegraph.

In 1833 Henry was appointed professor of natural philosophy at Princeton University. Two years later he produced the electromagnetic relay to send electrical impulses on when they became weak from long travel.

Henry's own story is in the Princeton Library: "I think the first actual line of telegraph using the earth as a conductor was made in the beginning of 1836. A wire was extended across the front campus from the upper story of the college library to the Philosophical Hall. Through this wire signals were sent from time to time from my house to my laboratory." The operator at the other end usually was Henry's wife.[15]

Henry took the first step toward inventing wireless telegraphy and radio in 1838 when he found that variations in the current or oscillatory discharges in one wire would induce currents in another separate and parallel wire. He demonstrated this with wires hundreds of feet apart across the Princeton campus. He said every spark of electricity in motion exerts "these inductive effects at distances indefinitely great."[16] In effect, he had a wireless transmitter and receiver.

James Macie, illegitimate son of Sir Hugh Smithson, of England, was unhappy with the treatment he received in his own country. He left the fortune he had made as a chemist to America—a country he never had visited—to found The Smithsonian Institution.[17] Henry was selected as its first director in 1846. The foremost American scientist of his day, he died 13 May 1878, at eighty-one years of age.

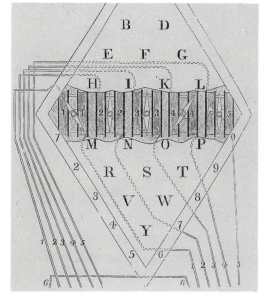

1.9 COOKE AND WHEATSTONE'S FIVE-NEEDLE TELEGRAPH OF 1837.

A number of magnetic needle telegraphs were developed. Wilhelm E. Weber and Karl F. Gauss set up an electromagnetic line in 1833, and sent messages more than a half mile at the University of Gottingen.[18] Karl August Steinheil of Munich, who invented the electric clock, used two wires and two magnetic needles in 1836. When deflected, they would sound two bells with different tones, to send messages by sound. Placing two ink markers on the apparatus, he also recorded dots on a moving paper tape. Later he used only one wire, with the earth as a

ground return, and the Bavarian government built several lines.

Edward Davy exhibited a needle telegraph in London in 1837 and 1838. Twelve keys were depressed to send positive and negative signals over eight wires and cause letters of the alphabet to be exposed at the receiving end. Davy also invented a relay, which he called a "renewer," to boost the strength of signals and send them on, and patented a chemical telegraph in 1838.

1.10 WHEATSTONE'S ABC TELEGRAPH. SENDING DIAL IS AT LEFT, RECEIVING DIAL AT RIGHT.

William Fothergill Cooke attended a lecture and telegraph demonstration by Schilling at Heidelberg University in March 1836. He then constructed a telegraph, needed technical help, and went to Sir Charles Wheatstone, a professor at Kings College, London. The two became partners, and in July 1836 operated a telegraph using five needles and five wires. They obtained a patent on June 12, 1837, and in 1839 installed a two-needle telegraph on the Great Western Railway from London to West Drayton.[19]

Wheatstone brought out in 1839 the first "ABC" or "step-by-step" telegraph, employing a principle later used in stock tickers. The clock-shaped dial receiver had a hand to indicate the letters of the alphabet. A weight caused the hand to rotate, and a spring escapement holding the weight was controlled by a magnet. Each time the magnet received an impulse over the wire, it caused the spring to release the weight and allow the hand to move to the next letter. At the sending end was a similar letter dial, with a hand to be turned to the desired letter, causing the desired number of impulses to pass over the wire.

Wheatstone and Cooke quarreled over credit for the invention as their system was spread over England by the Electric Telegraph Company they formed in 1846. Wheatstone, a great inventor, is still best known as the creator of the Wheatstone bridge, which he did not invent. He gave full credit for it to Samuel Hunter Christie, who in 1833 described the bridge as a "differential resistant measurer." The Wheatstone bridge is the father of our modern devices for measuring resistances, inductances, and capacities.

For example, when a landline or ocean cable is broken, the exact spot must be found to repair it. One terminal of the bridge is attached to the broken line, and the other to a series of resistance coils. The coils are one arch of the bridge; the broken line is the other. Resistance coils are added until the galvanometer needle shows no deflection, either to right or left. The resistance to the break and distance to it is obtained by counting the coils. That is how a cable ship can find a break in a cable buried in the soft ooze, perpetual cold, and darkness of the ocean bottom far below.

That 1833 bridge was a forerunner of modern radar which bounces microwaves off distant objects and measures the time the microwaves take to return.

Notes

[1]General sources: my article on "Telegraph" in *Encyclopaedia Britannica,* privately printed booklets, pamphlets, and other publications long out of print, found in historical society libraries. Nothing regarding the American inventors mentioned in this chapter was found in history textbooks.

[2]S. I. Prime, *Life of S. F. B. Morse* (New York, 1875) 259, p. 259, said that "C. M." was believed to be "Charles Marshall, of Paisley, who was at the time at Reffrew, from which place the letter was written." Others have said "C. M." was Charles Morrison.

[3]It was widely believed that Nathan Rothschild received the news of Wellington's defeat of Napoleon at Waterloo by pigeon, enabling him to make a fortune. However, Joseph Wechsbert, in *The Merchant Bankers* (Boston: Little, Brown, 1966), stated it was not a pigeon but a courier, bringing a Dutch newspaper to Rothschild a day before the official dispatches arrived.

[4]The word "semaphore" comes from the Greek *sema,* which means "a sign," and *phero,* "I bear," just as "telegraph" comes from words meaning "far off" and "write."

[5]Charles D. Deshler, in *Harper's New Monthly Magazine* (July 1892). The Continental Congress decided on July 5 to send copies of the Declaration to all states. Some couriers did not start from Philadelphia until July 6 and the dates given are those on which they "appear to have reached" some places.

[6]Victor Rosewater, *History of Cooperative News Gathering in the United States* (New York: D. Apelton and Co., 1930); Oliver Gramling, *A.P. The Story of News* (Boston: Farrar and Rinehart, 1940); and *The Columbia Sentinel* (Boston, November 20, 1911). Topliff took charge of the "news books" in Samuel Gilbert's Exchange Coffee House reading room in Boston

on November 20, 1811, gathering and writing marine and other commercial news in two "diary" books for merchants. Around the same time Aaron Smith Willington, proprietor of the Charleston (S.C.) *Courier,* boarded ships as far as 100 miles at sea to get the news. One of his employees was James Gordon Bennett, who later became a famous New York editor.

[7]Katherine Drayton Simons, *Stories of Charleston Harbor* (Charleston Historical Society, 1930). The 1706 Palace of Arms replaced the 1680 public building near that spot.

[8]*Portland Press Herald* (July 13, 1951); *Portland Evening Express* (July 21, 1965); William Willis, *The History of Portland from 1632 to 1864* (Portland: Baily and Hoyes, 1865); and *Portland City Guide* (1940). The tower still stands as a monument to a bygone era.

[9]C. W. Long and W. T. Davis, *Staten Island and Its People* (New York, 1930) 1:218-19. Also *Staten Island Historian* (April 1939).

[10]*Telegraph Age* (July 1, 1894): 279.

[11]Fear that the observation towers at Sandy Hook would serve as landmarks and aid German submarines to attack New York in World War II caused the military to order them razed. Ship reporting thereafter was from a tower at Quarrantine, on Staten Island, and the pilot ship at Ambrose Light.

[12]"Address in Memoriam at the Obsequies of Harrison Gray Dyar," by Rev. Charles S. Harrower (New York: C. H. Jones & Co., 114 Fulton Street, 1875).

[13]Born in Albany, New York in 1797, of Scottish ancestry, Henry left school at fourteen and was apprenticed to a watchmaker. A book on science spurred his ambition at sixteen, and he went to Albany Academy. Later he was appointed professor of mathematics and natural philosophy at the Academy.

[14]That "father" of magnets is at Princeton University.

[15]*Princeton Architecture,* by Greiff, Gibbons, and Menzies (Princeton NJ: Princeton University Press, 1967). Wires also were strung to Henry's home on the campus, which is still used as a home by a university official. The home, erected in 1837, was moved three times, and in 1968 was at Nassau Street, northeast of Nassau Hall.

[16]Paper read before the American Philosophical Society, June 17, 1842.

[17]*James Smithson and the Smithsonian Story,* by Leonard Carmichael and J. C. Long (New York: Putnam, 1966).

[18]Weber worked out the law of forces between electrical charges, basing his studies on Fechner's theory that positive and negative impulses move over a wire with equal but opposite velocities. Gauss proposed a system of absolute units, based on length, mass, and time, and the unit of magnetic field is named *gauss.*

[19]The International Telecommunications Union's anniversary book stated (p. 57) that this system was used in England for many years, and about 1,000 boy and girl operators, about fifteen years old, were employed in London alone. In the code used, *a* was one left deflection of the left needle, *b* was two, *e* was one left deflection of the left needle and two right deflections of the right needle, and so forth.

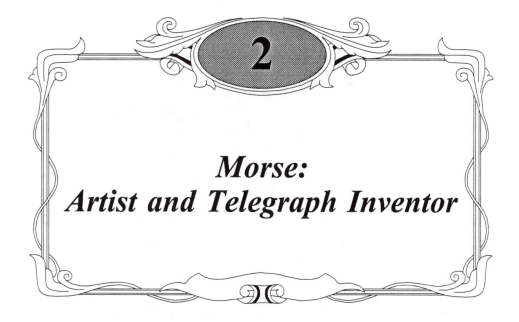

Morse:
Artist and Telegraph Inventor

Ask almost any American who invented the telegraph, and the answer will be "Morse," but he did not create the dot-and-dash Morse code, the Morse key, or the stylus recorder. One of Samuel Finley Breese Morse's associates produced the first telegraph equipment, practical and simple enough to attain general use in America. The invention of Morse's telegraph—like Bell's telephone—could have been credited to others if all of the facts had been known.

Probably no invention which has profoundly affected the world's progress has been the sole product of one man. Since each was made possible by earlier discoveries, many great inventions have been announced almost simultaneously by two or more persons.

Plato once said no sensation that enters the soul is ever lost; the world is a repository of all that ever has been. Perhaps our thoughts, our actions, our efforts to advance the world's understanding, our influence on our children and others, is our "hereafter." This also is the philosophy of invention: nothing is lost; the present comprehends all of the past and spans eternity.

Morse's heritage and education strongly influenced his telegraph activities. He was born April 27, 1791 in a parsonage at the foot of Breed's Hill (where the battle of Bunker Hill was fought in 1775) in Charlestown, Massachusetts (now part of Boston). His father was Jedediah Morse, D.D., Congregational pastor, author of the first geography printed in America, and a friend of George Washington and Daniel Webster. The Morses were Federalists, who believed the government should be guided by men of intellect and wealth.

Morse's mother was Elizabeth Ann Breese, daughter of Judge Samuel Breese, who founded Shrewsbury, New Jersey, and the granddaughter of Dr. Samuel Finley, Calvinist president of the college that is now Princeton University.[1]

Morse devoted the first forty-one of his eighty-one years to art. Many do not know he was a great artist because his renown as telegraph inventor overshadowed that of Morse the painter.[2] As a student at Yale, in 1808, 1809, and 1810, Morse attended lec-

2.1 Samuel F(inley) B(reese) Morse (1791–1872).

tures on electricity by Benjamin Silliman, professor of chemistry, and Jeremiah Day, professor of natural philosophy. Volta's battery was used to produce electricity, and Morse spent a vacation assisting with electrical experiments.

2.2 PAINTING BY MORSE OF GEN. LAFAYETTE, IN MEETING ROOM OF NEW YORK CITY HALL.

Morse went to England in 1810 to study art, and became a friend of such immortals as Lamb and Coleridge. When the War of 1812 with England began, Morse's letters were in the vein of Henry Clay's fiery patriotic speeches, but his family opposed the war. Two days before America declared war, Great Britain withdrew her restriction on commerce, but, unaware of the Treaty of Ghent on Christmas Eve 1814, Andrew Jackson's troops, Indians, and backwoodsmen

killed 2,000 English regulars advancing on New Orleans on January 8, 1815. Morse deplored the lack of fast communications which would have prevented the needless bloodshed.

Because his father could not continue to support him abroad, Morse returned home after four years. He was disappointed when New England people did not want historical paintings, and would only buy a few portraits for fifteen dollars. In 1818 he went to Charleston, South Carolina, a gathering place of Southern aristocracy. Through the influence of an uncle, he was received by the best families, and did fifty-three portraits at sixty to three hundred dollars apiece. The city of Charleston had Morse paint a full-length portrait of President Monroe for $750. He earned $9,000 above expenses in his second year there which enabled him to marry Lucretia Pickering Walker of Concord, New Hampshire.

Returning North in 1821, Morse did portraits of his neighbors Eli Whitney and Noah Webster, and a large painting of the House of Representatives, which included small portraits of eighty-eight members. He exhibited it at Boston and New York, but had to sacrifice it in 1829 for about $1,000. It is now in the Corcoran Gallery of Art in Washington, D.C.

In 1824, Morse opened a studio at 96 Broadway in New York, sleeping on the floor for lack of a bed, and his luck turned. The city commissioned him to paint a heroic-size portrait of General Lafayette, then visiting the White House. That portrait, now in New York City Hall, is a reminder of the greatest tragedy of Morse's life.

On February 10, 1825, Morse happily wrote to tell his wife at New Haven of his friendship with Lafayette and his progress on the portrait, not knowing his wife had died two days before, following the birth of a second son. Word of her death reached him seven days after the funeral. His three children went to live with relatives.

Morse attended lectures on electricity at Columbia, and was a founder and first president of the National Academy of the Arts of Design in 1826. He returned to Europe in 1829 to continue his art education.

James Fenimore Cooper, American novelist and Morse's roommate in Paris, and R. W. Habersham of Augusta, Georgia recalled later that Morse was interested in Chappe's semaphore system, Benjamin Franklin's electrical experiments, and scientific progress in Europe, which must have included the telegraph advances described in chapter 1.

Sailing home from Le Havre, France, on October 6, 1832, on the Packet Ship *Sully,* Morse heard Dr. Charles T. Jackson of Boston talk of Oersted's discovery of electromagnetism in 1819 and of Sturgeon's electromagnet (1825). Jackson remarked that electricity would pass through miles of wire as fast as lightning. "If this be so, and the presence of electricity can be made visible in any desired part of the circuit," Morse exclaimed, "I see no reason why intelligence might not be instantaneously transmitted by electricity to any distance."[3]

Morse went on deck, pacing up and down. On one of the little sketch books he always carried, he jotted down a rough telegraph code. Instead of dots and dashes to indicate letters of the alphabet, he jotted down numbers corresponding to words. Thus, the numbers 252 might mean "England," and 403, "Wednesday." To spell names, he assigned a number to each letter. His drawings show the wires encased in clay tubes.[4] He also sketched a recording instrument with a paper tape over two rollers, passing beneath the point of a pendulum.

By the time the ship neared New York, Morse was so confident of success he told Captain Pell that if he should hear of the telegraph, one of these days, to remember the discovery was made on the good ship *Sully.*

Arriving at New York on November 15, 1832, Morse informed his brothers Richard and Sidney of his "discovery." Years later, Sidney remembered, Morse rigged up "a kind of cogged or sawtoothed type, the object of which, I understand, was to regulate the interruptions of the electric current, so as to enable him to make dots, and regulate the length of spaces on the paper upon which the information transmitted by his telegraph was to be recorded." Richard's wife Louise saw Morse melting lead over the parlor fire and casting the type, but was most unimpressed when he spilled hot lead on her carpet and a chair.

His funds exhausted by his stay in Europe, his three little children living with relatives, Morse had to continue painting. He exhibited his large painting *Gallery of the Louvre,* including copies of thirty masterpieces in miniature. Few came to see his *Gallery,* and he finally sold it for $1,200. In 1982, Syracuse University sold it for $3.25 million to an Evanston, Illinois museum.

In 1836 four panels remained to be filled with paintings in the rotunda under the great dome of the Capitol. Morse was a logical choice to do one. John Quincy Adams, former president and member of the committee to select the artists, said there were no Americans who could do the work. This drew an indignant article in the *New York Evening Post* by an anonymous writer. The committee believed it was Morse and rejected his name. It was learned too late that James Fenimore Cooper wrote the article.[5]

2.3 Replica of Morse's "portrule" (1835), a telegraph sending device never used commercially. (Also see 2.4.)

Of the four artists chosen, only one was trained, like Morse, as a historical painter. Indignant artists employed Morse for $3,000 in 1837 to paint a historical picture,[6] but Morse returned the money; he had turned his mind to invention.

President Franklin D. Roosevelt told me it was too bad the public did not realize Morse was "one of the nation's four greatest artists." On March 28, 1932, the Metropolitan Museum of Art exhibited a majority of Morse's 174 portraits, thirty-three landscapes, and other paintings. Cass Gilbert, president of the National Academy, presented a President's Medal "in posthumous recognition" of Morse, to his granddaughter, Miss Leila Livingston Morse.

Morse had been appointed Professor of the Literature of the Arts of Design of New York University on October 2, 1832, before he returned from Europe. He could not begin work, however, until the fall of 1835 when the main building of the university neared completion on the east side of the Washington Parade Ground, now Washington Square.

Morse had a large classroom on the third floor. In it he ate, slept, taught, painted, and experimented. He received no salary and depended on fees from his students and the occasional sale of a portrait. His students, and other artists gave Greenwich Village, in which Washington Square is located, its start as an artists' colony.

In his classroom Morse built his first telegraph apparatus. His lengthy description years later said the recorder was an old canvas painting frame, a pencil suspended on a pendulum, wheels of an old wooden clock, a strip of paper wound around a wooden drum, a crank to raise a weight to pull the paper, and an armature to lower the pencil. The transmitter had a portrule in which metal teeth were set to make and break the current. The pencil made a succession of *V*s "as it passed to and fro" across the paper.

"With this apparatus, rude as it was, and completed before the first of the year 1836," Morse concluded, "I was enabled to and did mark down telegraphic intelligible signs, and

2.4 Picture frame telegraph apparatus Morse made at New York in 1835. On it a pencil was suspended to be moved by a magnet and make long and short marks on a moving paper tape. The marks were called dots and dashes and the magnet was moved in accordance with interruptions made in the current by the teeth of a "portrule" being moved under a sending device.

to make and did make distinguishable sounds for telegraphing; and having arrived at that point, I exhibited it to some of my friends early in that year, and among others to Professor Leonard D. Gale."[7]

The scientific knowledge of Gale, professor of chemistry and geology, was of great value to Morse. Although Gale told Morse about his friend Joseph Henry's method of winding strong magnets, building a battery for telegraph use, and invention of the relay, Morse later claimed the invention of the relay.

When he first saw Morse's instrument, Gale said, his "use of the single cup battery, were to me, on the first look of the instrument, obvious marks of defect; and I accordingly suggested to the Professor . . . that a battery of many pairs should be substituted for that of a single pair, and that the coil on each arm of the magnet be increased to many hundred turns each." When his suggestions were tried, Gale said, the distance the instrument would send the electric current was increased from fifteen or forty feet to as many hundred.

2.5 Alfred Vail, who created the Morse telegraph key and sounder and telegraph code at Morristown, N.J. while Morse was in New York devising a number for each word commonly used. Morse's idea was to transmit numbers instead of words to send messages.

Who invented the first "practical" telegraph? An opinion by Justice Woodbury, in the case of Smith vs. Downing, stated: "Among about sixty-two competitors to the

discovery of the electric telegraph by 1838, Morse alone, in 1837, seems to have reached the most perfect result desirable for public and practical use." In 1854 the U.S. Supreme Court decided, in the case of O'Reilly vs. Morse, that Morse's invention "preceded the three European inventions."

As noted earlier, however, Henry had operated a mile-long electromagnetic telegraph at Albany in 1830, and Schilling exhibited his telegraph using a code of letters before a congress of physicists at Bonn in 1835. Morse filed his caveat, a declaration of intention to apply for a patent, in the United States in October 1837. On December 23, 1836, Steinheil published that he had installed a seven and one-half mile experimental line; and in July 1838, that he was establishing a line from Nuremburg to Furth.

Announcement that two Frenchmen, Gonon and Servell, had invented a telegraph, and news of Wheatstone and Cooke's telegraph worried Morse.[8] He feared "other nations will take the hint and rob us both of the credit and the profit."

Gale assisted Morse in improving the equipment, and on September 2, 1837, a few friends witnessed the sending of messages over 1,700 feet of wire in Morse's room. This demonstration won Morse the enthusiastic aid of twenty-nine-year-old Alfred Vail,[9] a member of the Mechanics' Institute of which Gale was secretary. Alfred Vail was a son of Judge Stephen Vail, and nephew of George Vail, owners of the Speedwell Iron Works at Morristown, New Jersey, which had furnished artillery to the Colonial forces during the Revolutionary War, and built many of the first American locomotives when the firm's name was Baldwin and Vail.[10]

Vail took Morse to Morristown to meet his father, a Quaker whose word was law.

2.6 BUILDING AT OLD SPEEDWELL IRON WORKS AT MORRISTOWN, N.J. OWNED BY JUDGE STEPHEN VAIL AND HIS BROTHER GEORGE. IN THIS BUILDING, ALFRED VAIL AND HIS ASSISTANT PRODUCED THE FIRST SUCCESSFUL TELEGRAPH INSTRUMENT OVER WHICH VAIL AND MORSE DEMONSTRATED TO ALFRED'S FATHER, JUDGE VAIL, THAT HIS FINANCIAL BACKING HAD RESULTED IN AN INSTRUMENT THAT WOULD SEND LETTERS OF THE ALPHABET OVER A WIRE.

Judge Vail provided financial aid, and on September 23, 1837, Alfred signed a contract to build instruments and pay the costs of securing patents. Alfred received a twenty-five percent interest in the invention, but assigned half of it to his brother George.

A long second-floor room in a building near the house was assigned to the project. William Baxter, a young but expert mechanic designated to work with Alfred, recalled later the difficulties he and Alfred had. They appropriated the local supply of milliner's wire, used in making womens' bonnets, and

found it would work.[11]

While Vail created the instruments, Morse went to Washington. On October 3, 1837, Morse filed a caveat on the invention with Commissioner of Patents H. L. Ellsworth, once his classmate at Yale. (The 6,732-word application followed on April 1, 1838, resulting in the first Morse patent, June 20, 1840.)

Vail's father advanced $2,000. With Vail at Morristown, constructing instruments, Gale conducting experiments with stronger batteries and greater lengths of wire, and

Morse at New York writing a dictionary of numbers, the work progressed. Having worked in the machine shop for a dozen years and having invented stenographic printing machines, drawing machines for artists, fountain pens, and other things, Vail was well prepared for his task.

On October 18, 1837 Morse wrote to Vail: "I long to see the machine you have been making and the one you have been maturing in the studio of your brain."[12] Later Vail invited Morse to Morristown, where the artist realized his cumbersome picture-frame equipment was to be superseded by the practical and simple Vail instruments. Morse was so upset, Baxter said, that he became ill and was in bed for some weeks at the Vail home. Then he recovered and painted portraits of Judge and Mrs. Vail for which he received payment from Judge Vail.

"If we were meeting with difficulties," Baxter said, "he [Morse] succumbed to the blues and left us to work our way out as best we might. When we had scored a success, however, his elation was beyond all bounds, and he would boast gleefully of the wonderful achievement and tell everyone what *he* had accomplished."

Judge Vail threatened to withdraw his support when he saw no results. The worried inventors locked themselves in their room and worked feverishly. They avoided the judge for about six weeks, eating at the nearby home of Alfred's brother-in-law, Dr. Cutler. At last, on January 6, 1838, the instruments worked, and Alfred invited the judge to see the telegraph working.

"There was a long bench on one side of the room and one of the instruments was placed at each end of it," Baxter wrote, "with three miles of wire intervening." After hearing a short explanation, Judge Vail wrote a message that he handed to Alfred to send. He was delighted when Morse deciphered the message, "A patient waiter is no loser," with the aid of his dictionary of numbers.

The *Morristown Jerseyman* reported that on January 11 hundreds of people were shown how the telegraph worked: "The communication, which was made through a distance of two miles, was the following sentence: 'Railroad cars arrived—345 passengers.'"

A Morristown woman remembered another demonstration in a large upstairs room at the plant, and that Morse and Vail also spoke before a local audience in the second floor dining room of O'Hara's Tavern, on the town square.[13]

When Vail's new transmitter and a register that made long and short marks on paper tape were ready, invitations were issued for an exhibition on January 24, 1838, in Gale's Geological Cabinet room at New York University.[14]

One of the stories Morse apparently told his biographer S. I. Prime was that someone present gave him a facetious military command to send. It was: "Attention, the Universe! By Kingdoms, Right Wheel!" That story was not true, however. The tape Morse preserved was in a form not used until later, and we can only surmise that the person who wrote it was lampooning Morse's grandiose posturing.

On February 8, 1838, Morse and Vail exhibited the invention before the Science and Arts Committee of the Franklin Institute, Philadelphia, and obtained its endorsement. Next was the House Committee on Commerce in the Capitol. Committee Chairman Francis Ormond Jonathan Smith,[15] an unscrupulous lawyer from Portland, Maine,

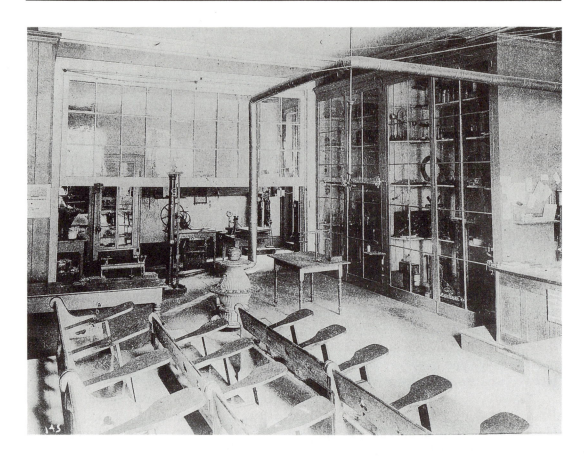

2.7 In this room the first public demonstration of Morse's telegraph took place. This was the Geological Cabinet room and Physics Laboratory at New York University, in which, on January 24, 1838, the historic message was sent: "Attention, the Universe! By Kingdoms, Right Wheel!"

realized a fortune could be made from the invention, and persuaded his committee to make a favorable report. Although he was only thirty, Smith was serving his third term in Congress. A man of great energy, he made a striking appearance with his gray eyes, high forehead, sharp profile, and clean-shaven face, but had a three-inch beard under his chin.

President Martin Van Buren and his cabinet, including Postmaster General Amos Kendall, then saw messages sent over ten miles of wire.[16] Morse asked the Government to finance a 100-mile line, but Congress received the proposal incredulously,

and let it wait. Smith became a partner in the enterprise, obtaining a one-fourth interest. Concealing his personal interest, he submitted an act to Congress on April 6 to build a fifty-mile telegraph line at a cost of $30,000.

Unsuspecting and naive, Morse was a babe in the business world, and was to regret this alliance as long as he lived. He never was able to shake off the relentless clutches of what his son, Edward Lind Morse, called the "malign animosity" of his ill-chosen partner. Patent Commissioner Ellsworth tried to save Morse by pointing out the impropriety of Smith's dual role, but Smith agreed to request a furlough from Congress, then near adjournment.

2.8 F. O. J. "Fog" Smith, one of the greatest developers of our telegraph system, but a stormy petrel whose first name should have been "Fight."

Annoyed because the government did not act, Morse wrote to the Republic of Texas in 1838 offering the patent rights of his invention as a gift. The Texas officials evidently did not think the letter was important enough to answer.

Morse and Smith went to England in June 1838 to patent the invention there, but the *London Mechanics Magazine* of February 1838 had published an article describing the Morse invention, invalidating it for a British patent.

After a demonstration on September 10, 1838, before the Institute of France, with François Arago, Baron Humboldt, and other scientists present, a patent was obtained in France, but it had little value; it would lapse if Morse did not build a line within two years. Smith returned to America while Morse negotiated with Russia and other countries. The telegraph was "growing in favor, testimonials of approbation and compliments multiply," Morse wrote to Smith from Paris on February 2, 1839, but on February 13, "I am at a loss how to act. . . . As for me, I feel that I am a child in business matters."

Morse returned to the United States without a cent after wasting eleven months abroad, while Wheatstone and Cooke were building telegraph lines in England and Belgium, as was Steinheil in Bavaria. From New York on May 24, 1839, Morse wrote to Smith: "Instead of finding my associates ready to sustain me with counsel and means, I find them all dispersed, leaving me without either the opportunity to consult or a cent of means, and consequently bringing everything in relation to the telegraph to a dead stand."

Vail Invented Morse Instruments and Code

In November and December 1837, when Vail built the instruments, he visited Louis Vogt, proprietor of a print shop at Morristown, and, over a case of type, learned which letters of the alphabet were used most frequently. He assigned the fewest dots and dashes to those letters.

Morse's caveat of October 3 and his letter to Vail on October 24, 1837, announcing the completion of his dictionary of numbers for words, did not mention a dot-and-dash alphabet. Six years after Vail created the Morse code, Morse wrote to

Smith about the numbers-for-words dictionary he was preparing.

Baxter said the first recorder made at Speedwell used a pencil for marking, but, since the pencil needed frequent sharpening, Vail substituted a "gang" of fountain pens. Later, while Morse was in Europe, Vail built a lever transmitter to make and break the circuit when it was moved up and down. It became known as the Morse key. Baxter said: "For the first time we had a mechanism capable of making dots, dashes and spaces. . . . Vail saw in these new characters the elements of an alphabet which would transmit language in the form of words and sentences, and he set about the construction of such an alphabet. . . . The letter 'e' was used oftener . . . and he gave it the shortest combination—a single dot (1). 'J' which occurs infrequently is expressed by dash, dot, dash, dot."[17]

Baxter described the origin of the steel stylus recorder, and said Morse saw the instrument for the first time when he unpacked the box in Paris, where it had been sent by Vail.

The *Engineering News* of April 14, 1886, stated that credit for "the alphabet, ground circuit, and other important features of the Morse system belongs not to Morse at all, but to Alfred Vail, a name that should ever be held in remembrance and honor." Numerous letters indicate that Morse tried to keep Vail's name submerged. Vail wrote that when he sought to go with Morse to see the president and other dignitaries, Morse would make excuses, and then go alone. In a letter to his brother George on August 15, 1844, Vail said: "I have repeatedly asked him to have his assignment to me recorded on the patent papers, and he either says that 'there is time enough,' 'it can be done any time,' or that I have got the assignment. I shall

demand it of him or know the reason why it is not."

Morse gave no public recognition to Alfred, but wrote to George Vail on March 10, 1852: "Many are pleased to compare me with Fulton. If Fulton had a Livingston to aid him in his early extremities, I had Vail to aid me in mine." Writing to Amos Kendall on March 11, 1853, Vail said his agreement with Morse provided "that whatever Mr. Smith, Dr. Gale, or myself should invent or discover, going to simplify or improve Morse's telegraph would belong to all jointly. . . . I could not therefore have taken out a patent for this invention." However, Vail said, he did "invent what I stated. . . . The whole value of the telegraph is not enough to induce me to say less or more. . . . Yet this agreement with Morse and others does not refuse me the honor of being the inventor of anything I did invent. . . . What I do desire is truth."

Vail watched Morse gradually eliminate him from credit with mounting astonishment and anger, making no public outcry because Morse, involved in a multiplicity of court battles, required all possible support to preserve the patents. When Morse later referred to Vail and his father merely as "furnishing the means to give the child a decent dress," Vail supporters boiled, and telegraph journals contained many strong words.[18]

When Vail in 1845 wrote a book stating his claims, Morse told him it would destroy any chance of selling his system in Europe. Vail knew that was not true, and published the book.

During the five years before he died (on October 15, 1876), F. O. J. Smith wrote a history of the invention of the telegraph, praising Joseph Henry and Vail, and charging that Morse's claims were a "monstrous fraud."[19]

Called as a witness, Henry supported Morse's invention, but told who had previously invented the essentials in telegraphy. Morse then wrote an article casting "implications" on Professor Henry. He concluded that Henry blamed him because credit due Henry was omitted from Vail's book. However, when Vail had wished to write to Henry and set the matter straight, Morse wrote to Vail on January 15, 1851, "to bear it yet longer in silence."

It is evident that Henry showed how to telegraph, Morse planned a cumbersome system to do it,[20] Gale made valuable contributions, and Vail developed the code and instruments necessary for successful operation. Wheatstone and Cooke wrote to Morse on January 17, 1840, asking him to secure a patent for them in the United States and offering a half share of it.[21] Morse refused, but Wheatstone and Cooke obtained an American patent in June 1840.

Morse returned to Washington in December 1842, demonstrated the telegraph between two committee rooms in the Capitol, and lobbied for his bill. Representative John Pendleton Kennedy of Maryland, chairman of the Committee on Commerce, moved passage of the bill on February 21, 1843. Cave Johnson of Tennessee proposed an amendment to spend half of the appropriation to aid the "science of mesmerism." Representative Houston of Alabama suggested that Millerism, a religious sect, be included. The amendment lost. Two days later Kennedy moved the bill for its third reading, and it passed with a vote of eighty-nine to eighty-three, with seventy members abstaining.[22]

Only a few days remained before the close of Congress. The bill could not be passed that year unless the Senate reached it,

and Morse waited anxiously each day in the Senate gallery. The final day arrived on March 3, 1843 with 140 bills ahead of his. Always dramatizing himself, Morse wrote later that he feared his life's work was falling in ruins and received the sympathy of friends who told him all hope was lost. Morse said he left for his room shortly before midnight—a defeated man with only "a fraction of a dollar" in his pocket. Failure of the bill to pass, he wrote, would leave "little prospect of another attempt to introduce to the world my new invention."[23]

Morse's statement was not true. The *Congressional Globe* reported that the bill was passed at the morning session on March 3, and the *House Journal* said the president's signature of the bill was announced early in the evening. Also, Morse wrote to Smith on 3 March: "Well, my dear sir, the matter is decided. The Senate has just passed my bill without division and with opposition, and it will probably be signed by the president in a few hours."[24]

Morse added to that untrue story twenty-three years later in a letter to Bishop Stevens of Pennsylvania, dated "Paris, November 1866." After repeating the story about leaving the Senate a defeated man, Morse wrote:

> In the morning, as I had just gone into the breakfast room, the servant called me out, announcing that a young lady was in the parlor, wishing to speak with me. I was at once greeted with the smiling face of my young friend,[25] the daughter of my old valued friend and classmate, the Hon. H. L. Ellsworth, the Commissioner of Patents. On expressing my surprise at so early a call, she said, "I have come to congratulate you." "Indeed, for what?" "On the passage of your bill." "Oh, no, my young friend, you are mistaken; I was in the Senate-chamber till after the lamps were lighted, and my senato-

rial friends assured me there was no chance for me." "But," she replied, "it is you that are mistaken. Father was there at the adjournment, at midnight, and saw the president put his name to your bill; and I asked father if I might come and tell you, and he gave me leave. Am I the first to tell you?" The news was so unexpected that for some moments I could not speak. At length I replied: "Yes, Annie, you are the first to inform me; and now I am going to make you a promise; the first dispatch on the completed line from Washington to Baltimore shall be yours." "Well," said she, "I shall hold you to your promise."[26]

That fanciful tale was swallowed by Morse biographers, who did not read his letters. At a later date, Morse did ask Annie Ellsworth to provide the first message. Though he was credited with far more than he did, Morse led an advance in communications that changed and enriched the lives of nearly everyone.[27]

Notes

[1]Finley was the fifth president of Princeton, 1761–1766.

[2]Facts regarding Morse are based on the following: the papers of Smith, O'Reilly, and other pioneers; Morse's papers in the Library of Congress; Alfred Vail's papers, Smithsonian Institution Archives, and in the Morristown, New Jersey library and at the New York Historical Society; Edward Lind Morse's (Morse's son), 2-vol. *Samuel F. B. Morse. His Letters and Journals*; and Carelton Mabee, *The American Leonardo* (New York: Alfred A. Knopf, 1944).

[3]Morse's words as he later reported them.

[4]The sketch books are among the Morse papers in the Library of Congress.

[5]S. I. Prime, *Life of Samuel F. B. Morse*, 290.

[6]From A. B. Durand's original minutes of the New York Historical Society.

[7]Autobiographical article in the Morse papers, Library of Congress.

[8]Volume 12 of Morse's papers in the Library of Congress shows that Morse had obtained detailed drawings of Cooke and Wheatstone's telegraph.

[9]Vail was born September 25, 1807, and died on January 18, 1859.

[10]*The Telegraph and Telephone Age* (April 1, 1937): "The history of the Speedwell Iron Works dates back to 1776, when two Morristown general store proprietors, Captain Jacob Arnold and Captain Thomas Kinney, built a furnace and a slitting mill, a mill for cutting bar iron into small strips for making nails."

After 1800 control of Speedwell Iron Works passed into the hands of Stephen Vail. The "Son" in the name of the firm, Vail & Son, was Alfred's brother George. In 1864, Stephen Vail died, and operations ceased.

[11]A piece of this wire, wound with heavy, white thread, is mounted on a plaque in the Morristown, New Jersey library.

[12]The papers of Alfred Vail, including fifty volumes of letters, notes, diaries, clippings, documents, and drawings, were transferred on October 8, 1890 by his sons Stephen N., James C., and George R. Vail, from the New Jersey Historical Society to the custody of the American Historical Association which moved them to the Smithsonian Institution in Washington. Most of the Vail material in this book is based on these papers, on handwritten transcriptions presented to

the New York Historical Society by Dr. William Penn Vail of Blairstown, New Jersey, and on a 745-page abstract from the Vail papers (by S. Ward Righter of East Orange, New Jersey) in the Morristown library.

[13]The tavern, built about 1750 by Ezekiel Cheever as a home, was used as offices of the Continental Army during the Revolutionary War. It burned on May 5, 1846, and was replaced in 1850 by a building still occupied by a restaurant, the Town House on the Green.

[14]There were two days of demonstrations. On January 23, Vail wrote to his brother George: "Yesterday I wrote by stage informing you of our intention to exhibit today. We have done so with perfect success, through ten miles, to hundreds of ladies and gentlemen. Tomorrow is our last day. I shall box up, and it is decided we make a stop in Philadelphia for a few days. Our audiences were astonished and delighted."

[15]Manuscript on Maine political history by Miss Elizabeth Ring, vice president and director, Maine Historical Society. Smith was elected to the House of Representatives in 1833, 1835, and 1837.

After Smith's path crossed that of Morse in 1838, he blossomed from politician to man of affairs, and erected on the outskirts of Portland an architectural monstrosity, complete with dome, separate servants' quarters, fountains, and sculptured lions. From this singular edifice, named Forest Home, smith modestly rode forth in an imported Stanhope, driven by a coachman in livery, sitting stiffly atop a richly brocaded hammercloth fringed with silver and gold.

[16]Vail letter to his father and brother, February 21, 1838: "The President and his Cabinet have just left and I have the pleasure to inform you that they were highly delighted and entirely satisfied. The President proposed the following sentence, 'The enemy nears' to Prof. Morse silently so that I could not and did not hear it. It was then put in numbers and written on the register."

[17]Morse to Smith, April 8, 1843.

[18]On October 18, 1888, Alfred Vail's widow wrote to H. C. Adams, president of Cornell University, that because of Morse's promises to "write a history of the telegraph and do justice to Mr. Vail," she did not allow anyone to have access to Vail's papers, but that her eyes were opened at the Morse testimonial on June 10, 1871, when she saw Morse telegraph around the world with Vail's original instrument, which she had loaned for this occasion. When she heard Morse "then address that vast audience with a mere allusion to Alfred Vail's financial aid, my whole soul thrilled with indignation."

(What Morse actually said was that his first equipment "found a friend, and an efficient friend in Mr. Alfred Vail of New Jersey, who with his father and brother furnished the means to give the child a decent dress preparatory to its visit to the seat of government.")

Mrs. Vail's letter also said that when an article by Lyman W. Case appeared in 1872, "I am living witness to the effect of this article upon Prof. Morse, when visions opened up to him of receding glory. He sent for me, and on his dying bed, with the forefinger of his left hand raised and moving to give expression to his words, he said: 'The one thing I want to do now is justice to Alfred Vail.'" Morse died on April 2, 1872, almost 81 years old.

[19]That manuscript, which mysteriously disappeared when Smith died, was found eighty-eight years later, by my wife and me, in a basement storage room in the Maine Historical Society Library in Portland.

[20]The Smithsonian Report for 1878 contained William B. Taylor's "A Historical Sketch of Henry's Contribution to the Electro-Magnetic Telegraph with an Account of the Origin and Development of Prof. Morse's Invention" (Government Printing Office, 1879). It quoted a letter by Morse stating that he "was utterly ignorant" that anyone had the idea of the telegraph before him, and had never heard of Henry's experiments.

[21]Smith papers, Maine Historical Society Library.

[22]Morse to Smith, February 23, 1843: "The

long agony (truly agony for me) is over, for you will perceive by the papers of tomorrow that so far as the House is concerned, the matter is decided."

[23]Prime, *Life of Morse,* 464-65.

[24]Morse to Smith, March 3, 1843, Smith papers, Maine Historical Society Library.

[25]Annie Goodrich Ellsworth married Roswell Smith, who became president of *Century Magazine,* and published two powerful anti-Morse, pro-Vail articles in 1888. In April, Professor Franklin Leonard Pope described the Vail instruments, giving him major credit. The other article, in July, by Alfred's son, Stephen, told the Vail story. From 1888 to 1904 Stephen bombarded the press with "Letters to the Editor," ably presenting his father's cause.

[26]Prime, *Life of Morse,* 495-96.

[27]When John Huston rehearsed for his role as Morse in a network TV show in the 1940s, he asked the writer what Morse was like. Based on the opinions of those who knew him best, the reply was that Morse's craving for fame was so strong that he postured, pontificated, tried to convince everyone he was great, and was zealous in defending his claims.

Building the First Telegraph Lines (1844–1847)

Secretary of the Treasury John C. Spencer wrote to Morse on March 14, 1843, authorizing him to construct the experimental line between Washington and Baltimore. He appointed Morse superintendent of telegraphs at a salary of $2,000 a year; Professors Gale and J. C. Fisher, assistants at $1,500 each; and Vail, an assistant at $1,000.

Vail set to work building the apparatus and stations. Fisher superintended the preparation of the wire and its enclosure in lead pipes. Gale assisted Morse in supervising the project. Morse decided to lay the line underground, and Ezra Cornell, a plow salesman of Ithaca, New York, had the job of simultaneously making the trench and laying the line in it, with a special plow he designed. The contractor for the pipe laying, however, was Levi S. Bartlett, a brother-in-law of F. O. J. Smith, who had no scruples about giving contracts to relatives and sharing their profits.

By August 1843, Morse had 160 miles

of wire, manufactured at the Stephens & Thomas plant on Mill Street, Belleville, New Jersey. In April he had appealed to Louis McLane, president of the Baltimore and Ohio Railroad, for permission to use the railroad's right of way, the only one entering the capital. McLane, a former secretary of state, had Morse talk with John H. B. Latrobe, the railroad's counselor, and Latrobe persuaded the directors to approve.[1]

Work began at Baltimore on October 21, 1843. The plowshare, cast in the railroad shops at Mount Clare, made a furrow two inches wide and twenty inches deep, and the pipe was fed through the plow into the trench. Construction had progressed nine miles from Baltimore to a point near the railroad station at Relay, Maryland, when the line failed to operate. Cornell tried to tell Morse what was wrong, but Morse brusquely told him Fisher was a scientific man and knew what he was doing. When the line remained dead, however, Morse became so angry that he discharged Fisher.[2] To play for

3.1 REPLICAS OF SAMUEL F. B. MORSE'S RECORDER, RELAY, AND KEY USED ON THE FIRST COMMERCIAL TELEGRAPH CIRCUIT, FROM WASHINGTON TO BALTIMORE, 1844.

time, he asked Cornell to stop work, but not let the newspapers know about the failure. Cornell did a little acting.

"Hurrah boys! Whip up your mules, we must lay another length of pipe before we stop today," he shouted to his men. The powerful, six-foot Cornell seized the controls and whipped up the team of eight mules. He threw the point of his plow against a rock and broke it.

Things looked black. Gale resigned, and Cornell was appointed assistant superintendent at $1,000 a year.[3] Only $7,000 of the

$40,000 appropriation remained, and Smith demanded payment of $4,000 under the 40-mile cable-burying contract. Morse refused, and he threatened to sue.

Concealing his worries, Morse wrote to the *National Intelligencer* on December 23, 1843:

Although about ten miles of pipe containing the conductor have been laid down, yet the lateness of the season embarrasses any further operation until spring. In an enterprise so entirely new it can be hardly expected that every part can be conducted with that precision and perfection which is gained only by experience. Unforeseen delays will be encountered and are to be overcome. . . . There are no intrinsic ones, as yet, of a nature to check the confidence of the most sanguine in the final triumph of the enterprise.

Morse's real feelings were recorded in his diary on January 8, 1844:

> I doubt if the experiment is ever tried, and am at a loss to decide whether or not to remain in the employ of the government. There is much inefficiency in the chief superintendence of it, much indecision, and economy ill-devised. Mr. Smith, I understand, carries his case to the Senate tomorrow. I fear if the appropriation is spent without a trial, that utter disgrace will fall on all concerned; so far, I can conscientiously say, I am not in any way implicated.

To his brother Sidney, Morse wrote of Smith: "Where I expected to find a friend, I find a fiend. . . . He seems perfectly reckless and acts like a madman."

To a proposal by Morse on May 2, 1844, to ask Congress to extend the telegraph to Philadelphia or New York, Smith replied he "would upon no terms yield . . . until the government shall have honorably and justly redeemed its faith, plighted through you in the Bartlett contract." Morse wrote to Smith that he had "lost sight of no means to discourage, disable, and harass me. . . . That I have succeeded in rescuing it from ruin is owing to no kindness on your part, but in spite of all your efforts to destroy it."

Vail read that Wheatstone and Cooke had shifted from underground to pole line in England and decided to build above ground. With plans for insulating the wire, drawn separately by Vail and Cornell, Morse selected Vail's plan to insulate each wire and to bunch the wires together at the poles. Going to New York to have the fixtures made, Morse stopped at Princeton, where Professor Henry told him the Vail plan would not work, but to select Cornell's plan to put each wire between two glass plates on the pole, with each wire separated from any other.

Cornell extracted most of the copper wire from the pipe and sold the pipe to salvage badly needed money. Abner Cloud Shoemaker, whose land is now a part of Rock Creek Park in Washington, D.C., contracted to provide 500 chestnut poles. Construction of the line from Washington, begun on April 1, 1844, progressed rapidly with Cornell in charge. The poles, twenty-four feet high, were erected 200 feet apart. Number 16 copper wire was used, insulated with cotton thread treated with shellac and a mixture of beeswax, resin, linseed oil, and asphalt. Eighty acid cells produced the current of about eighty volts.

The First Telegraph News Dispatch

The line was operating to Annapolis Junction, twenty-two miles toward Baltimore, when the National Whig Convention opened in that city on May 1, 1844. While Cornell built on toward Baltimore, Vail remained at Annapolis Junction, sitting on a platform made of railroad crossties. When a train from Baltimore reached the Junction, Vail obtained the names of the men the Whigs had nominated, and wired the news to Morse. At Washington, Morse bent over his apparatus that had a heavy weight, a huge coil, and a primitive battery in pots and jars.

"Late in the afternoon, suddenly, the instrument on the table began to click," John Kirk of the Post Office Department wrote:

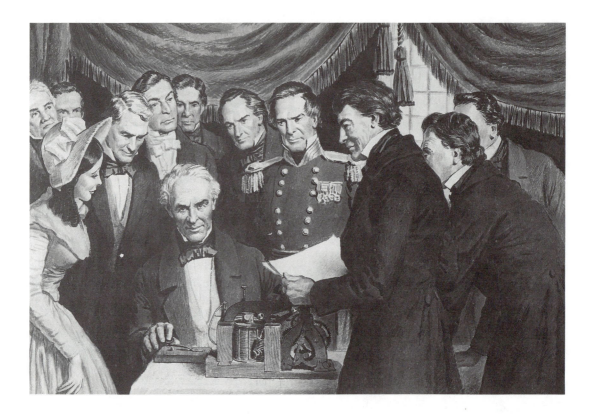

3.2 ARTIST'S PORTRAYAL OF MORSE SENDING THE FIRST TELEGRAPH MESSAGE IN THE SUPREME COURT ROOM OF THE CAPITOL AT WASHINGTON, D.C., MAY 24, 1844. ANNIE ELLSWORTH, LEFT, SUGGESTED THE FIRST MESSAGE: "WHAT HATH GOD WROUGHT!"

Eagerly Professor Morse bent forward over the strip of paper that slowly unrolled from the register. The paper halted, moved ahead, stopped, and moved again in an irregular way, 'till finally Morse arose from his close scrutiny of the paper, stood erect, and looking about him, said proudly: "Mr. Kirk, the convention has adjourned; the train for Washington from Baltimore has just left Annapolis Junction bearing that information, and my assistant has telegraphed me the ticket nominated. The ticket is Henry Clay for president, and Theodore Frelinghuysen for vice president."

Morse passed slips of paper bearing the news to people who were incredulous until the passengers arrived and confirmed it.

The First Public Telegram

When the line was completed, Morse invited prominent people to its formal opening on May 24, 1844 in the Supreme Court Chamber on the ground level of the Capitol. The room was semicircular, its high ceiling supported by fluted columns connected with graceful arches. The wire to Baltimore passed through the center window of five in the room. Henry Clay, candidate for president, was present.

Morse gave Miss Annie Ellsworth the honor of sending the first message. Knowing how religious Morse was, Miss Ellsworth selected a quotation from the Bible—Numbers 23:23: "What Hath God Wrought!"

3.3 Photograph of Morse's first embossing recorder, used on the Washington–Baltimore line in 1844. Now at Cornell University, Ithaca, N.Y.

Morse, at 53, tall, lean, with deep-set eyes, sent the message to Vail in Baltimore who sent it back. With sharp, metallic "clacks" the stylus slowly embossed the dots and dashes on paper tape moving through the machine. A telegraphic conversation followed, concerning the news, time, and weather. Morse, smiling triumphantly, was surrounded by the crowd and congratulated.

News of the historic event was published three days later in the New York Tribune, under a small headline: "The Magnetic telegraph—Its Success." The story began: "The miracle of annihilation of space is at length performed."

On May 26, news bulletins from the National Democratic Convention at Baltimore were received by Morse at the Capitol and read to the Senate. On May 27, James K. Polk was nominated for president, and U.S. Senator Silas Wright of New York for vice president. Vail, on the third floor of the railroad warehouse in Pratt Street, Baltimore, wired the news to Washington. Informed of his nomination, Wright joined Morse at the

telegraph, and the following was sent:

> Important! Mr. Wright is here and says: "Say to the New York delegation that I cannot accept the nomination."

Then,

> Mr. Wright is here, he will support Mr. Polk cheerfully, but cannot accept nomination for vice president.

Several convention officials with Vail replied, urging that Wright reconsider. The reply was:

> Under no circumstances can Mr. Wright accept the nomination. He thanks the convention and refers to his two former answers.[4]

The convention adjourned for the day, and a committee was sent to Washington to confirm the report. The committee returned the next day with confirmation, and a telegraph conference followed, with Senator Wright beside Morse still refusing the nomination. George M. Dallas was then nominated, and Polk and Dallas were elected that fall.

On May 29, 1844, Senator W. P. Mangum of North Carolina wrote:

> Yesterday evening from 4 to 7 o'clock more than a thousand people were in attendance at the window, at which placards in large letters were exhibited upon the receipt of each item of news. Today from 700 to 900 were attending when the news came that Polk was unanimously nominated.

Morse's news bulletins convinced many that the telegraph was a reality, but there was little comprehension of its possible practical uses.

Morse sent a "memorial" to Congress on June 3, 1844 offering "to transfer the exclu-

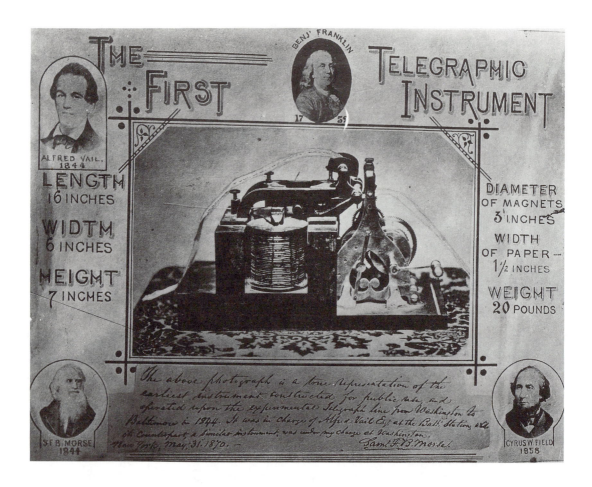

3.4 A contemporary picture and description, signed by Morse, illustrating the receiving apparatus used on the Washington–Baltimore line in 1844. (Also see 3.3.)

sive use and control of it from Washington City to the City of New York to the United States . . . if Congress shall proceed to cause its construction, and pay the proprietors what it is worth." Morse repeated the proposal in December, but Congress took no action.

The telegraph was exhibited to the public without charge until February 15, 1845, when the appropriation was exhausted. On March 3, Congress provided $8,000 to operate the line, and shifted its supervision to Postmaster General Cave Johnson, who had proposed that part of the telegraph funds be spent for mesmerism! Morse was reappointed superintendent, with Vail and Henry Rogers as operators. The messages were to be placed "in the hands of the penny post for delivery." The first public telegraph office was opened on April 1, 1845, in a second floor front room of a two-story house on Seventh Street near E Street, N.W., Washington, D.C., with an outside stairway to the office.[5] Before that, the telegraph was exhibited in a second floor room of the post office.

One cent was charged for every four

characters. An office seeker was the first patron on April 4. He had only one cent and a twenty dollar bill. The operator accepted the cent, and wired Baltimore "4" which meant "What time is it?" Baltimore replied "1" meaning one o'clock. On April 5, the revenue was twelve and one-half cents; on the seventh, sixty cents; on the eighth, $1.32; on the ninth, $1.04. Income for the first six months was $413, and expenses $3,274.

Practical use began almost at once. A person in Baltimore wired the Bank of Washington to know if a man presenting a check had money in that bank, and the reply said he did. A family in Washington learned to its relief that one of its members had not died.

On Christmas day 1844, Vail at Washington wrote Judge Vail:

> Last week they at Baltimore ordered the train here by telegraph to proceed through to Baltimore without stopping at the halfway turnout where they always met, as the Baltimore train would delay starting for some time. So the Washington train passed through without stopping *and there was no collision.*

Still another use, on June 1, 1845, was the first telegraph money order. Vail's diary recorded that he paid twenty dollars at Washington to Mr. L. W. Lilly on an order by telegraph from Baltimore. After October 29, newspapers were charged half rates for each 100 words after the first 100.

Postmaster General Johnson reported to Congress, however, that

> The operation of the telegraph between this city and Baltimore has not satisfied me that, under any rate of postage that could be adopted, its revenues can be made equal to its expenditures. Its importance to the public does not consist in any probable income that can ever be derived from it.

That was all the opponents of government ownership needed to prevent further appropriations. Thus, to the great good fortune of America, the telegraph was turned back to free enterprise, which produced far greater progress and service than those under government ownership in other countries.

On December 1, 1846, the Post Office Department gave the use of the line to Vail and Rogers during the transitional period.

Private Capital Extends the Telegraph

On March 10, 1845, Morse, Vail, and Gale placed the telegraph business—for a ten percent commission—in the experienced hands of thin-lipped, hatchet-faced Amos Kendall, lawyer, former editor of the Frankfort, Kentucky *Argus*, politician, and former postmaster general.[6]

Weary of his feud with Smith, Morse was glad to turn his problems over to Kendall. He continued to leap into the fight, however, when anyone challenged his honors. Refusing to accept Kendall as his agent, Smith wrote to Morse on March 10, 1845 about "the inequity of advantages and disadvantages that have accrued to the different proprietors hitherto, and which operate as an insuperable objection in my mind to future joint operations."

The Magnetic Telegraph Company was formed on May 15, 1845 to extend the Washington-Baltimore Line to New York. Smith, Cornell, and Cornell's brother-in-law Orrin S. Wood went to Boston and New York to raise money. Capitalists looked at the straggling wire put up for exhibition across rooftops and the hungry exhibitors, and said, "No!" Cornell, future millionaire

3.5 AMOS KENDALL (1789–1869).

founder of Cornell University, slept on chairs. The quarters the public paid to see the telegraph work in an upstairs room at 563 Broadway did not pay expenses.

Cornell wrote that he and Wood had worked six months and received nothing. Kendall and Smith then employed Cornell at $1,500 a year. On July 5, 1845, Smith offered Morse $100,000, to be paid over an extended period, for his patent rights, and his offer was refused.

Finally, $15,000 was subscribed for stock to build the New York-Philadelphia section. William W. Corcoran, the Washington banker who later founded the Corcoran Art Gallery, put up $1,000 and became the first subscriber. F. O. J. Smith, with $2,750, was the largest. Eliphalet Case,

Cincinnati Enquirer editor and Smith's brother-in-law, took $1,000. Kendall and Cornell each bought twenty shares for $500, Cornell's payments to be deducted from his salary. Since each stockholder was given two shares for one, $30,000 in stock was issued. A like amount to the patent owners made the total of $60,000. Kendall was president.

Construction was under the direction of Cornell from New York to Somerville, New Jersey, and under the direction of Dr. A. C. Goell from Somerville to Philadelphia. The poles were small, 200 feet apart; glass "bureau knobs" were used for insulation, and number 14 copper wire, unannealed. The railroad having refused the use of its right of way, the route was roundabout: from the Merchants' Exchange in Philadelphia, seventy-four miles through Norristown and Doylestown, across the Delaware River at New Hope, and through Lambertville, New Jersey to Somerville.[7] Cornell built from Fort Lee to Newark, then followed the Morris Canal, on which the reels of wire were sent to Somerville.[8] The line was opened from Philadelphia to Norristown in November 1845 and to Fort Lee on January 20, 1846.

On January 26, 1846, the company issued a circular stating the hours the New York office at 10 Wall Street would send, and the hours it would receive telegrams from Philadelphia. The rate for ten words between New York and Philadelphia was twenty-five cents.

Charles T. "Tap" Smith, manager at Fort Lee, invented the lineman's climbers, had Asa Vandergrift of Jersey City make the first pair, and let Dr. J. Craven of Newark, use them first. He lost a fortune by failing to patent them.

In later years James D. Reid[9] recalled how the glass insulators glistened in the sun and provided targets for boys with rifles. Extreme cold also had disastrous effects on the copper wire. Unable to obtain copper wire for an emergency repair, Reid substituted iron and found that it would serve. After that, iron wire came into general use.

At first there was one operator at Washington, three at Jersey City, one clerk and four boys at New York, three operators and a boy at Baltimore, and one operator at Wilmington. The first three months showed a profit of $294.17 at New York and $223.50 at Philadelphia.

To cross the mile-wide Hudson River, the *Scientific American* (November 20, 1845) proposed supporting the wires with balloons. Cornell, however, laid two lead pipes—the first submarine cable in America—each containing two wires covered with India rubber and cotton saturated with pitch, across the river from Fort Lee, where the majestic George Washington Memorial Bridge now stands. The pipes worked, but ice swept them away that winter.[10]

Early in 1848 two wires were stretched across the river between tall iron towers, but it was necessary to lower the wire into the water whenever a ship with a tall mast approached. The *Scientific American* reported on April 15, 1848 that the wires were torn away by the "fly of a sloop" when they sagged too low.

Boatmen then carried messages across the river to the home of John James Audubon, which served as the telegraph office. From this home, named "Minniesland" for Audubon's wife, the messages were telegraphed to downtown New York.

The company then built a line up both sides of the Hudson to a narrow point south of the West Point Military Academy, where a wire was stretched across the river, but poor construction necessitated constant

repairs, and it was abandoned. The company then had Reid build the line direct from Newark to Jersey City, and messengers crossed the Hudson by ferry to deliver telegrams in New York.

The O'Reilly Contract

The imaginations of enterprising Rochester and Utica men were fired by news of the telegraph. Among them were John J. Butterfield and Theodore S. Faxton of Utica, stagecoach line pioneers, who went to Washington to see the first line placed in service. They obtained a license on May 30, 1845 to build a line between Buffalo, Utica, Albany, and Springfield, Massachusetts. Returning from Washington, Butterfield met Henry O'Reilly[11] of Rochester on the night boat going up the Hudson to Albany. That started O'Reilly on a stormy career as the greatest of all pioneer line builders.

On June 13, O'Reilly obtained a contract from his friend, Amos Kendall. It gave him authority to "raise capital for the construction of" a Morse line from Philadelphia to Harrisburg, Pittsburgh, Wheeling, Cincinnati, and "such other towns and cities as the said O'Reilly and his associates may elect, to St. Louis, and also the principal towns on the Lakes." The patent owners were to receive one-fourth of the capital stock and not "connect any Western cities or towns with each other, which may have been already connected by said O'Reilly." It required O'Reilly to build the line to Harrisburg and provide capital for its extension to Pittsburgh within six months, or the agreement would be null and void.

This ambiguously worded contract plagued the patent owners for years. Kendall thought it only gave O'Reilly the right to build some lines from Philadelphia west for his company.[12] However, O'Reilly regarded it as authority to organize, build, and manage lines for numerous companies, and establish his own telegraph empire, with the growing Midwest as his exclusive territory. By taking stock in the lines he built, he expected to control them.

The Philadelphia-Baltimore line was constructed during the first half of 1846 by O'Reilly, who provided $4,000 of the $10,000 initially subscribed; William M. Swain, owner of the *Philadelphia Public Ledger,* paid $3,500.

Reid said the wires were covered with tar for "insulation." A newly landed Scotsman was hired, who "with a tar bucket slung to his side, and a monster sponge in his hand, tarred the wire as far as Wilmington, Delaware. The tar proved too much for him. He went to sleep and never woke." Reid recalled with feeling that he completed the job, and O'Reilly made a bonfire of his stinking garments.

When he completed the Philadelphia-Baltimore line on June 2,[13] O'Reilly attached it to the one to Washington. The line was broken constantly; there were many errors, and receipts from January to July 1846 were only $4,228.77. The line was improved with good joints and a regime of efficiency was instituted in 1850 when William M. Swain became president. Receipts from July 1, 1851 to July 1, 1852 were $103,860.

O'Reilly,[14] an enthusiastic, dynamic Irishman, with a short moustache and beard, and a round, rosy face, became the mercurial firebrand of the telegraph pioneers, the popular darling of press and public, but poison to the Morse patentees. Weighing only 124 pounds, O'Reilly was referred to by Vail's friend George Wood as "busy as ever playing still the part of the Napoleon of the telegraph."

Although he was Kendall's protege, O'Reilly's itemized bill for constructing the Philadelphia-Baltimore line ended their friendship. Delays by the company in obtaining rights of way, and changes in plans authorized by Kendall, added several thousand dollars to O'Reilly's costs, but he was coldly notified in a letter on August 13, 1846 that a resolution of the board of directors allowed him only $13,252.82.

The stern and aging Kendall, facetiously referred to as "Amos the Pious," was not a man to let friendship swerve him from duty. He wrote to O'Reilly that since he had not accepted the company's offer, it was nullified and withdrawn, and the company would "Concede nothing to extravagant, pretentious, and passionate threats." He said the company would remedy defects in the line and charge the cost to O'Reilly.

O'Reilly offered to compromise for $14,000, and Kendall replied that a committee would "adjust your amount." O'Reilly noted in his letter that after he had "spent so much time and effort to make their line . . . I was nearly all summer in 1846 trying to get pay for the work done."

The Second Telegraph Line

With the financial backing of Hervey Ely and his nephews Elisha D. and Herman B. Ely, lawyer brothers Samuel L. and Henry R. Selden, George Dawson, and Alvah Strong, all of Rochester, New York, Judge Skinner of Batavia, and M. Brooks of Livingston County,[15] the Atlantic, Lake, and Mississippi Telegraph Company was organized on September 14, 1845, to finance and build lines under the O'Reilly contract. Henry Selden was elected president, O'Reilly secretary, and Dawson treasurer.[16]

The company gave O'Reilly a contract

for $4,200 to build a forty-mile line between Lancaster and Harrisburg. With his brother Captain John O'Reilly and Bernard O'Connor in charge of construction, James D. Reid as superintendent, and Anson Stager, David Brooks, and Henry Hepburn, all of Rochester, O'Reilly had it built by November 24, but, because of a blizzard, line breaks, and other problems, it was not placed in operation until January 8, 1846.

Lancaster was selected because Kendall could not use the right-of-way of the Philadelphia, Wilmington, and Baltimore Railroad. He wrote to O'Reilly on July 12, 1845 that the line probably would have to go from Philadelphia to Baltimore via Lancaster as the connecting point between the lines. On November 11, Kendall wrote to O'Reilly that he could extend his line from Lancaster to Philadelphia.

Another delay resulted when "Morse" equipment arrived only three days before the December 13 completion date under his contract. Relay magnets developed by Cornell had been substituted, and O'Reilly's signature of a receipt was demanded to protect Cornell's rights to obtain a patent. O'Reilly refused to accept the instruments, and held up operation of the line until January 8, he said, to protect the Morse patentees.[17]

When the line began operating, O'Reilly issued a circular bidding for press and public support:

> Dispatches not exceeding fifteen words, including address and signature, sent for twelve and a half cents. Newspaper editors half this rate, and a larger reduction when much intelligence is sent. Visitors desirous of seeing the operation of the telegraph between Harrisburg and Lancaster, may have their names sent and returned, 72 miles, for

six and one-fourth cents.

The Philadelphia-Harrisburg line was the second in the United States. When it was completed, Reid at Harrisburg sent to Brooks at Lancaster: "Why don't you write, you rascals?" People flocked to see the "wonder of the age," but the only revenue was from sending names of visitors.

Weird musical sounds were produced by the wind and copper wire, to the great discomfort of the rustics. People walked long distances to avoid passing under it, and a woman built a fence around a pole to prevent her cow from rubbing against it for fear it would spoil the milk. A Pennsylvania legislator solemnly concluded, "It will do well enough for carrying letters and small packages, but it will never do for carrying large bundles and bale boxes."

David Lechler, proprietor of the North American House, in which the Lancaster office was located, would gather a crowd of country folks in the barroom and explain in "Pennsylvania Dutch" the wonders of the invention. Then he would hurriedly enter the telegraph office and bring out a pair of hose, a handkerchief or newspaper, which he had previously punctured with holes, and gravely announce, "I received these in just forty seconds from Philadelphia."

As little revenue developed, and the cold constantly contracted and broke the wires, O'Reilly had the line taken down.

The Atlantic and Ohio Line

O'Reilly then rebuilt the Lancaster-Harrisburg line and continued to Pittsburgh. In an effort to block him, the Morse patentees published an advertisement on December 12, 1846 and distributed copies along the route, threatening to prosecute any person assisting O'Reilly.

West of Harrisburg, the Atlantic and Ohio line was built through Carlisle, Chambersburg, McConnellsburg, and Bedford. When it reached Pittsburgh on December 26, 1846, a crowd gathered to witness the sending of the first message. It was from Adjutant General G. W. Bowman to President Polk, stating that the Second Pennsylvania Regiment would be ready to leave on January 6 for the Mexican War front.[18]

Hugh Downing was elected president, O'Reilly secretary, and Reid superintendent of the line. The $300,000 capital stock was divided as $75,000 for the patentees, $75,000 for O'Reilly, and $150,000 for the subscribers. Since the patentees refused their stock, the company held it pending settlement of the dispute. The company was a success and paid dividends ranging from three and one-eighth to eight percent every few months.

The Third Line

On July 16, 1845, the Springfield, Albany and Buffalo Telegraph Company was formed in Utica with Theodore S. Faxton as president. The line was to connect at Springfield with one F. O. J. Smith was to build from New York to Boston. Because of difficulty in obtaining rights of way, Faxton wisely had Cornell complete the line from Albany to New York, instead of Springfield, on September 9, 1846, less than a month after planting the first pole at the corner of Wall Street and Broadway.[19]

Cornell wrote to J. J. Speed, Jr. on October 9, 1846 that he built the section from Albany to New York at $125 a mile, earning a profit of $7,000, and was appointed superintendent of the 500-mile New York-Buffalo line at $1,750 a year.

When the line reached Rochester on June 1, 1846, the telegraph office was opened in the basement of the Congress Hall Hotel. As soon as the apparatus was connected, a message began coming in from Albany, and it asked, "Do you hear me?"

"To be sure I do," Rochester replied.

"Ha, ha! Dr. Tichener, give me your hand," said Albany.

Shortly afterwards, the telegraph office was moved to Reynolds Arcade[20] in the business center of Rochester, and the Arcade started on its road to fame as the cradle of the telegraph industry.

Faxton proposed a federation of papers to supply news by telegraph, and the press began sending daily reports on 1 January 1847. The line was profitable, and Cornell put up a second wire.

New York and Boston Linked

After many delays, largely due to "Fog" Smith's refusal to use glass insulators and crossarms, and penny-pinching on labor, the line was completed from Boston to New York. It reached Lowell on February 21, Springfield on March 22, Hartford on March 26, New Haven about May 1, and New York on June 27, 1846, through Harlem, Third Avenue, and the Bowery. Ira Berry was manager of the Boston office.

Smith had contracted to build the line for $160 a mile, or $40,000. The subscribers were to pay $40,000 for $60,000 in stock and the patent owners were to receive $60,000 in stock, making the total issue $120,000. Due to poor construction, the line was out of order much of the time. Stock subscribers would not pay up, and Smith had to take over the line and repair it with his own funds.

From the office at 26 Washington Street,

Boston, Smith issued a circular announcing rates ranging from ten to twenty-five cents for ten words or less, and two cents for each extra word. Newspaper rates were the same, but fifty percent was added for each additional paper receiving the same matter. An extra charge of twenty-five percent was made for transmission between 9:00 p.m. and 8:00 a.m. Those rates and poor service angered the press.

"Fog" Smith was such a czar in running the line that newspapers caused the erection of two competing lines in 1849 and 1850. One was operated with an automatic telegraph printer invented by Royal E. House, and the other with a chemical process of Alexander Bain.

Smith's New York and Boston Magnetic Telegraph Association was, of course, a Morse line. The Bain line, built by O'Reilly, was the New York and New England, also known as the Merchant's Telegraph. It was headed by Marshall Lefferts, importer of galvanized wire and iron. The House line, the New York and Boston Telegraph Company, later known as the Commercial Telegraph Company, was built by Hugh Downing, a Philadelphia wire manufacturer.[21]

The first woman telegrapher, Miss Sarah G. Bagley, was appointed operator at the Lowell Depot when the line reached there. At that time few business jobs were open to women.

When the competing House line was built, Smith announced in the *Boston Transcript*:

> I will deposit $1,000 against a like sum that I have a Durham Bull, whose weight exceeds 2,500 pounds, who will travel from Boston to New York City with a message of 1,000 words, in less time than the whole telegraph system patented to House can convey the same message.

Downing replied that he would accept and make the bet $10,000 if Smith would substitute Morse instruments for the bull. He offered to race both the House method and the bull if Smith would ride the bull.

His ridicule having boomeranged against him, Smith charged his competitors with piracy, claiming he had the sole right to the telegraph in that area. The *Boston Daily Mail* replied on March 8, 1849 that Smith might as well say the heirs of Robert Fulton were entitled "to a monopoly of all the navigable waters." Smith said the Bain system was as worthless as "a horseshoe without nails, or a shoe without a horse." When the two other lines began competing with Smith's, a rate war cut the cost between New York and Boston as low as one cent a word. The press loved that.

Notes

[1]McLane letter to Morse, April 7, 1843. Smith papers, Maine Historical Society Library.

[2]Morse letter to Secretary of the Treasury J. C. Spencer, January 27, 1843.

[3]Morse letter to Cornell, December 27, 1843, giving Cornell the management of preparing the pipe with the wire enclosed.

[4]From May 25, 1844 to February 15, 1845, Vail maintained a "Journal of the Telegraph" in which he recorded all messages sent over the line. A copy of this "conversation" appeared in the *New York Herald* on June 4, 1844.

[5]"History of the First Telegraph Office in the United States," a pamphlet by James Cummings Vail, son of Alfred. Donated to Morristown, New Jersey library by James Vail in 1920. Also *Washington Evening Star,* September 5, 1945 and Januarry 9, 1949.

[6]Kendall was one of President Jackson's chief advisers. He also founded and provided the original property for Gallaudet College, where the Columbia Institution for the Deaf at Washington, D.C., and the Deafness Research Foundation were located.

[7]Kendall wrote Smith on October 15, 1846 that the Camden and Amboy Railroad had finally consented to allow the use of its right of way, and the route was being changed.

[8]Cornell papers. Kendall to Cornell, September 9, 1845; Stephens and Thomas to Cornell, October 31, 1845.

[9]Reid, a Scotch immigrant, was employed by Henry O'Reilly, then postmaster at Rochester. When O'Reilly began building lines, Reid joined him. He was hired later by Kendall for the Magnetic Telegraph Company. His book *The Telegraph in America* was published in 1886.

[10]New York Historical Society library. Cornell to Peter Cooper, president of the American Telegraph Company, December 1, 1858; and *The Telegrapher,* December 26, 1864.

[11]Sources included thousands of letters and documents in the O'Reilly papers, New York Historical Society library, the Rochester Historical Society, and the article "Henry O'Reilly" by Dexter Perkins, in *Rochester History* (January 1945), a quarterly published by the Rochester Public Library.

[12]"Morse's Telegraph and the O'Reilly Contracting," a pamphlet by Kendall (Louisville, 1848).

[13]The *Baltimore Sun,* June 3, 1846, said the first connection was made to that city at 4:00 p.m. the day before, but the only news received was of the *Hibernia*'s arrival at Boston. More time was needed "to arrange the registers in proper working order."

[14]O'Reilly was born in Carrickmacross, County Monaghan, Ulster, Ireland on February 6, 1806. In 1816, he came to America with his mother and sister while his father was in a debtor's prison. He began work for the *New York Columbian,* a newspaper, and at age seventeen became assistant editor of the *New York Patriot,* organ of the People's Party. At age 20, he became the first editor of the Rochester, New York *Daily Advertiser.* In 1838, O'Reilly was appointed postmaster of Rochester by Kendall, then Postmaster General, but he lost the job when the Whigs won the national elections in 1842. O'Reilly wrote a book entitled *Sketches of Rochester.*

[15]Hervey Ely owned a flour mill; Dawson was the editor and Strong the owner of the *Rochester Daily Democrat.* James D. Reid and Anson Stager quit the *Daily Democrat* to work for O'Reilly.

Strong's autobiography (provided to the writer by Strong's grandson, Alvah G. Strong of Alton, New York) states that O'Reilly labored hard at Philadelphia, New York, Albany, and Utica "to secure the attention and subscriptions of leading businessmen to the new company, and not until he had laid his plans before the citizens of Rochester could he secure the confidence of men to invest a single dollar. It looked to them too commercial. Hervey Ely was the first to listen and to say: 'It is just what the world wanted.'"

[16]*Albany Argus,* September 19, 1845.

[17]O'Reilly to Cornell, December 30, 1845.

[18]*Pittsburgh Gazette and Advertiser*, December 30, 1846: "A good deal of business was awaiting the opening, and it kept the operators busy until late at night." The paper also carried news from Washington, Philadelphia, and Baltimore under the headline "By Telegraph."

[19]Kendall to O'Reilly, August 1, 1845.

[20]Reynolds Arcade, built in 1828 by Abelard Reynolds, saddler, postmaster, and tavern keeper, was referred to as "Reynold's Folly" because Rochester was a small town. Chicago then had a population of less than 100 outside the fort, and New York City boasted 200,000. The Arcade housed the post office, but later contained the offices of companies that became national institutions. Rochester men were leaders in establishing the telegraph industry, and the Arcade was headquarters for their companies.

[21]While forming the Atlantic and Ohio Telegraph Company to build from Philadelphia to Pittsburgh, O'Reilly, Samuel L. Selden, and Hugh Downing obtained an agreement on November 16, 1846 from House, giving them the right to use the House patent. However, O'Reilly used Morse equipment on that line and Bain on his Boston line, while Downing used House on his Boston line. Downing and Selden assigned their interest in the House agreement to O'Reilly on May 20 an 22, 1847.

Progress South and West
(1845–1851)

"Fog" Smith watched with cold, calculating, and greedy eyes the westward progress of O'Reilly, whom he called the "Irish Aztec." As enthusiastic citizens subscribed to finance O'Reilly lines, Smith saw a fortune at stake, and tried to elbow O'Reilly out, and himself into control.

He also maneuvered constantly to gain at the expense of his partners Morse, Vail, and Gale. When O'Reilly blocked the sale of equipment invented by Cornell, it strengthened Cornell's alliance with Smith. Cornell and Vail disliked each other from the start.

Kendall knew he had to get along with Smith somehow, or all progress would halt. He tried the role of mediator, but it was impossible to keep both Smith and O'Reilly in line. He was well aware of Smith's avariciousness and cunning, and O'Reilly's exuberance and overconfidence, but could not avert a head-on collision.

Kendall wrote to O'Reilly that Smith's brother-in-law Eliphalet Case proposed to

build a line from Cincinnati to meet O'Reilly's line at Pittsburgh, saying Smith was confident the Irishman would accept.[1] On September 20, Kendall took O'Reilly to task for not replying to Case's proposal. He said Case would support the project in his newspaper, and raise funds for it. "It is due to the patentees who have suffered you to proceed when your contract has expired, to inform them what arrangements justify you in refusing assistance which promises to be so efficient."

Case wrote that he could raise $50,000 for the line if his local group could organize and control it. O'Reilly was especially annoyed by a letter from Smith on October 2, 1846, saying, "You will perceive it [your contract] has long since expired in its limitations—I hope, however, you have the section to Harrisburg so nicely completed as to have that section disposed of under your contract. Its retardation has greatly delayed the progress further west."

O'Reilly finally offered to compromise

by giving the patent owners three-eighths of all stock in his lines west of Pittsburgh, and one-fourth east. Morse and the others wanted to accept, but Smith refused.

4.1 HENRY O'REILLY, GREATEST OF THE PIONEER LINE BUILDERS.

Kendall wrote that he had learned "a considerable amount of stock" had been distributed by O'Reilly in Philadelphia and Baltimore.[2] "What position would you be in," he asked, "were it announced to the public that their line is totally without authority . . . and that those certificates represent absolutely nothing?"

O'Reilly replied that he had issued a few certificates to influential editors, and as receipts for payments. To further threats O'Reilly replied that his backers included Judge Selden, and "Lt. Governor Gardiner,

president of our court of last resort, in whose judgment the highest tribunals of our state would place full confidence, considered the contract valid," and "the rights and duties under my contract (which is their contract also) have been and will be most sedulously considered for they esteem that contract quite as sacred as the patent."[3]

Going to Cincinnati, Judge Selden found that Case, believing the O'Reilly contract void, was organizing a company to build from that city to Pittsburgh, but city leaders were eager to join in the O'Reilly line. O'Reilly believed he had blocked Case, but he was mistaken. Hand in glove with his brother-in-law Smith, Case organized the Western Telegraph Company on January 16, 1847, with Morse patent rights.

The Morse patentees then applied for an injunction against O'Reilly in the Circuit Court in Pennsylvania. On February 18, 1847, Judge Kane decided O'Reilly had built the line as required in his contract, and that since the patentees had "formally conveyed to Eliphalet Case all their right of constructing and using the Magnetic Telegraph" in that area, they had no right to an injunction.

The patentees charged they were not notified about the "new line." A man who saved all correspondence, O'Reilly had a letter Kendall addressed to him on November 11, 1845, as "Agent of the Atlantic, Lake, and Mississippi Telegraph Co." Underscoring those words, O'Reilly wrote below them:

> Mr. Amos Kendall afterwards swore he has never known that I organized any such company!!! Although I sent him every printed circular on such subjects up to the time he swore so.

After the court decision, Smith and Case offered to compromise. O'Reilly went to

Cincinnati and reached an agreement with the Western Telegraph group for each party to build different routes, but the patentees refused that plan, and a second one also. O'Reilly proposed arbitration, but the patentees demanded that he stop all building.[4]

Morse wrote that Smith's desire for half of the stock on O'Reilly's western lines caused him to insist on annulling O'Reilly's contract, though he and Kendall did not wish to do so.[5]

Weary of the controversy, Kendall wrote to Smith on May 10, 1847, proposing a division of interests. The result was a series of contracts, signed on June 22, 1847. Smith received the Philadelphia, Pittsburgh, Cincinnati, Louisville, St. Louis route, the Great Lakes line, unsold routes in the New England States, New York, Wisconsin, and Michigan, and most of Illinois, Indiana, and Ohio. The other patentees received the rest of the country.

Smith agreed to settle the O'Reilly battle, at his expense, but he was determined to take everything. He immediately asked the Cincinnati group to surrender their patent rights. They refused, but finally accepted $500. That cleared the way for Smith's scheme to build a line to Pittsburgh, Cincinnati, and New Orleans, link that with his northern empire, and put O'Reilly out of business.

At the annual meeting of O'Reilly's Atlantic and Ohio Company, July 3 and 4, 1848, a demand from Smith was read for double the amount of stock required in the contract. The answer was "no."

Smith put Case to work on his projects. Protesting that he had raised about half the stock for a line from Boston to Lowell, but had not been paid for that or earlier work, Case wrote to Smith from Cincinnati on October 11, 1847:

I came here last night on the Ben Franklin. It was crowded to overflowing; for besides its regular customers, it had an infernal circus on board, horses and all. I got a bed in the captain's room, and he did not lock the door. And the consequences were that some villain took my pantaloons out from under my pillow, and stole from the pockets about $150—perhaps a little more—perhaps a little less. This is another chapter in my telegraphic history, which has been all bad and no good.[6]

Expansion South and Westward

Kendall recognized the need for a line through the southeastern states to New Orleans and asked the government to construct it as a defense measure in the Mexican War.[7] Congress, Secretary of War William L. Marcy, and Postmaster General Cave Johnson turned him away. He then asked the press to help, but got nowhere.

Then F. O. J. Smith and Dr. A. Sidney Doane, Treasurer of the Magnetic company, offered to build the line and give one-fourth of the stock to the patentees. Kendall demanded one-half and finally had his way. Under the contract of November 2, 1846, the patentees would also participate in the contractor's profits. To gain Smith's approval, the contract was awarded to John J. Haley, one of Smith's available cousins. Haley had baked good pies in his restaurant on Nassau Street, New York City, but erected a poor line.

In those days of poor roads, slow mail, and old news, Southerners along the New Orleans route readily subscribed $280,850 to finance the Washington and New Orleans Telegraph Company, incorporated on March 20, 1847. David Griffin of Columbus, Georgia was elected president, but soon was succeeded by Col. Elam Alexander of Macon.

The line was built via Petersburg, Richmond,[8] Wilmington, Columbia, Charleston, Augusta, Savannah,[9] Macon,[10] Columbus, Montgomery, and Mobile. It reached New Orleans on July 13, 1848, too late to speed Mexican War news arriving by boat from the front. While the line was progressing southward, however, newspapers operated horse relays to take the war dispatches from New Orleans north to the nearest point the telegraph had reached.

Kendall wrote that fewer poles would reduce the number of points at which the insulation could be bad, and save "not far from $25 per mile." The result was poles too far apart, the use of many live trees, much maintenance and trouble. Charles S. Bulkley was in charge of construction.[11]

The line, the longest that had been built, was constantly out of order, but flooded with more business than its two wires could handle at rush hours. Newspapers charged gross mismanagement and hinted at bribery by competitors to get their stories sent first. The *Charleston Courier* said the line had operated properly only one day in nineteen. Manual retransmission at points along the route resulted in frequent errors until Bulkley invented and installed automatic repeaters.

Something else wrong was the deal between Smith and Kendall to share the construction profits, which Vail denounced in a letter to Kendall on March 11, 1853. Vail said the line cost the company $235,950. A profit of $95,875, Vail said, was divided: $49,937.50 to Haley, $23,968.75 to Smith, and $23,968.75 to Kendall in stock. Kendall gave half of his share to Morse and the Vails, charging them a commission; Kendall received $13,495.30, while Alfred and his brother George got only $1,250.

Subscribers to a line should pay only actual costs, plus a reasonable profit to the contractor, instead of $150 a mile for a line costing $89, Vail said, and the subscribers' money should have gone into building a good line earning profits for all. He and George gave their $1,250 to charity.

On March 14, Kendall replied to Vail's voluminous letter in a few words, that "more important duties will not allow me at present if ever to give it such an answer as its terms would justify." He said Vail's letter "abounds in errors."

Progress Westward

Relying on Judge Kane's decision, O'Reilly also built west and north, fighting Smith's forces on both fronts. To avoid confusion, we will consider west and north separately instead of chronologically.

As O'Reilly extended his Philadelphia-Pittsburgh line to Cincinnati and Louisville, Smith refused to allow his New York-Boston line to send or receive messages with it. O'Reilly quickly "memorialized" the press, business, and legislatures, charging the "monopoly" with depriving the public and business of the right to send messages where they wished. A deluge of editorial criticism followed: "Robbery," said the Troy, New York *Daily Budget*; "monstrous pretensions," the *Richmond Whig*; "construct a parallel line, with House's machines," the Rochester, New York *Daily Advertiser*; "high-handed usurpations of public and private right," the *New York Tribune*; "F. O. J. Smith's grinding monopoly," the New Haven *Daily Palladium*; "Kendall & Co., if they continue the course they are pursuing, will find themselves at last in the condition of the dog in the fable, that, in grasping at the shadow, lost the substance," *Chicago Daily Tribune.* Smith finally rescinded his order.

One of O'Reilly's Pittsburgh, Cincinnati, and Louisville Telegraph Company messengers at Pittsburgh was a sixteen-year-old Scotch immigrant, a towheaded lad named Andrew Carnegie, who worked long hours and became a skilled operator who could receive by "sound," without reading the tape.[12] Impressed by his ability, Thomas A. Scott of the Pennsylvania Railroad hired Carnegie, who became expert in train dispatching by telegraph, superintendent of the Pittsburgh Division, and an operator in the War Department telegraph office in the Civil War.

The route of O'Reilly's line from Pittsburgh was through Wheeling over the National Pike to Dayton, then Zanesville, Columbus, and Springfield. On August 21, 1847, the *Cincinnati Gazette* hailed its arrival: "Cincinnati and Pittsburgh shook hands yesterday and exchanged compliments by means of Mr. O'Reilly's telegraph."

Cincinnati in 1847 was the gateway to the West. Chicago's population was 17,000; Detroit's, 18,000; Cleveland's, less than 13,000; the frontier settlement of St. Paul, 100. San Francisco had 450 American residents.

From Cincinnati the line was extended down the west side of the Ohio River in Indiana through Cleves, Lawrenceburg, and Madison, reaching a point across the river from Louisville, Kentucky on September 23, 1847. Messages were taken across the river by boat until wires were strung on masts to Towhead Island and Louisville.

While O'Reilly was building to Cincinnati, his Ohio and Mississippi Telegraph Company line was being strung from that city to St. Louis via Lawrenceburg, North Vernon, Seymour, New Albany, and Vincennes, Indiana. Against the advice of his associates, he plunged into that project without raising subscriptions to finance it. When Congress met on December 7, 1847, the line had reached Vincennes. President Polk's 18,000-word message was telegraphed simultaneously to Pittsburgh, Cincinnati, Louisville, and Vincennes in eleven hours. It was carried from Vincennes to the *Missouri Republican* at St. Louis in twenty-six hours in the rain by a senior editor in a coach, using relays of horses.

With spectacular speed, the line reached the Mississippi, and on December 21 the *St. Louis Union* declared the linking of the Mississippi with the Atlantic "marked a new era in the history of St. Louis." To cross the river, a wire was suspended between tall masts. The masts were blown over in a storm, and a submarine cable was laid in October 1850. There were no bridges over large rivers in the midwest, and tall masts were made by binding poles together.

Hard Lives of Pioneer Builders

On January 30, 1848, seventy-four leading citizens and firms of St. Louis "begged leave to tender" O'Reilly a public dinner. He had more than a normal vanity, but replied that no formal testimonial could impress him more deeply than the hospitality and manifestations of confidence already shown. Perhaps so much adulation overbalanced his judgment. From St. Louis he built a half dozen lines in all directions without waiting for sound financing.

O'Reilly was the greatest pioneer builder, but, like Morse, was complex, full of contradictions, inconsistencies, and dreams of fame. His character would tax the credulity of a reader if portrayed in fiction. He inspired intense loyalty in men who put everything into his projects and were rewarded with massive ingratitude. He turned savagely and vindictively on his closest friends and supporters, sometimes upon hearing a baseless rumor that something derogatory had been said about him. He kept changing his alliances, when he thought he could gain, until long-time friends became his bitter enemies, and then some became his associates again.

Some went into debt heavily to pay O'Reilly's pressing bills only to have him dishonor their claims. The chaotic early years of the telegraph had a fascinating array of heroes and villains, in ever-changing combinations, but O'Reilly, a scoundrel one day and a knight in shining armor the next, was the most unpredictable.

Ubiquitous Reid, who worked at different times with the Morse group and O'Reilly, had the best opportunity to judge the pioneers, but his opinions changed as rapidly as his jobs. One of his letters to O'Reilly said,

> I have often wept the live-long night that I was drawing from your meagre stores the weekly stipend of life . . . we will name our first boy after you. . . . Mr. Morse and his associates have rolled themselves in filth and must abide by their own stench. I am sick of the bare mention of them.

Later, however, Reid wrote of the "beauty of Mr. Morse's character, the versatility of his intellect, the sweetness of the life of a man as modest as he was great."

He also wrote to O'Reilly:

> When you are in anger, you are insane. This opinion is held by your best friends. . . . The kindness of your better nature which I have shared in for many years at times when my star shone unclouded to you is only equaled by the hyena-like ferocity of your present malice.

From reading many thousands of letters of the pioneers, a picture of their lives emerges: Many were away from their families six months or a year, and then had only a brief visit home before going away again for a rough life in which they often slept on the dirt floor in log cabins of wilderness settlers whose food they shared. Often their food was corn or winter wheat, pounded to a pulp and boiled.

They had the courage to bet their lives on a dream and the fortitude and strength to make it come true. With few exceptions, such as Smith and Vail, each was so poor that for years his wife and children had to depend largely on the generosity of relatives and friends for food and clothing. Each knew he was making history, and went to great lengths to save voluminous correspondence in which he stated again and again his high principles and purposes, defended his actions and denounced the "infamy" and "pretensions" of others. So many thousands of O'Reilly's papers are in the New York Historical Society Library,[13] and elsewhere, that one wonders how he could have done anything except write.

The Race to New Orleans

The Morse group obtained the passage of a Kentucky Law on February 27, 1847, authorizing lines along public roads. While Case raised subscriptions and built from Pittsburgh to Lexington, Kentucky for Smith, who he claimed had never provided a dollar, in February 1848 the Kendall group began building a 1,500-mile line down the Mississippi River. O'Reilly recklessly decided to race it to New Orleans.

Announcement of O'Reilly's plans produced a controversy. On January 15, 1848, the *Louisville Journal* printed a letter by Henry Pirtle, a prominent lawyer, saying nothing in O'Reilly's contract gave him the right to construct a line to New Orleans. Thirteen leading citizens of Tennessee, headed by Return Jonathan Meigs, issued a statement agreeing. The *Louisville Morning Courier* came to O'Reilly's aid, saying he "may afford to be blackguarded and abused by Amos Kendall and F. O. J. Smith, as long as he has almost the entire newspaper press of the United States in his favor."

Kendall issued a "Notice to the Public," which stated that

> O'Reilly and his hands have no more right to use Morse's telegraph on this line, than they have to kill the farmers' horses and hogs along the road . . . a patent right is as much private property as a farmer's land, house and stock. . . . If they proceed, in defiance of all this authority, they will be stopped by the strong arm of the law. . . . They have no right of way in Tennessee or Kentucky and their posts are liable to be removed by the state or turnpike authorities.

Disregarding these attacks, O'Reilly organized the People's Telegraph Company to build from Louisville to New Orleans via Nashville, Tuscumbia and Jackson, Mississippi, and Clinton and Baton Rouge, Louisiana. When O'Reilly held a meeting of Louisville citizens on November 9, 1847 to raise subscriptions, Case denounced him, but the *Morning Courier* reported: "O'Reilly triumphantly sustained all he had said by documentary proof, and laid bare a portion of the outrageous villainy that had been practiced towards him."

The *Louisville Journal* said his opponents "were so completely annihilated by Mr. O'Reilly that the audience refused to listen to Mr. Case." Writing to Smith on November 12, Case called this "All a gross falsehood," saying, "O'Reilly is a patron of the newspapers . . . and this liberal expenditure of money has its effect and gets more."

The *New York Tribune*, November 13, 1847, said Smith, Kendall and associates were "utterly void of patriotism, grovelling in the quagmire of litigation, and quarreling and scrambling after enormous gains and close monopoly of a discovery."

The Frankfort, Kentucky *Yeoman* took the opposite view, saying of the Morse patentees: "On ascertaining the virtual abandonment of his contract by O'Reilly, they vested the patent right in trustees in Cincinnati." William Tanner, editor of the paper, was biased. He and a Louisville man had a contract to build the Morse line from Lexington to Nashville.

In the *Louisville Journal,* December 22, 1847, Kendall accused O'Reilly of attempting to set up a gigantic monopoly and take about half of the stock "without paying anything for it." On 3,000 miles that O'Reilly claimed the right to build, Kendall said, a profit of $100 a mile would have given O'Reilly $300,000 in cash and $300,000 in stock.

Walter N. Halderman, editor of the *Courier,* replied the next morning that if Smith had not tried to break the O'Reilly contract, to get fifty percent instead of twenty-five, "perfect harmony would have subsisted today." Supporting O'Reilly, the *New York Journal of Commerce* hoped he "will establish a telegraph line alongside of every principal line owned by other patentees throughout the country."

Saying it would be "a terror to evildoers only," Kendall got a Tennessee law passed requiring proof, before a line could be built, that it would not infringe anyone's patent rights, but O'Reilly's employee, F. S. Rutherford, persuaded the House to repeal it.

A furious race began with rival line gangs sometimes in sight of, and shouting insults at, each other. The O'Reilly line was opened to Nashville on March 6, 1848, and had poles erected to New Orleans in June, but no wire. O'Reilly charged Col. Charles Doane, his agent in New Orleans, with failing to raise funds and provide the wire.

The route of the Morse line was from the end of the Western Telegraph line at Wheeling through Nashville, Waynesboro, Pontotoc, Grenada, Vicksburg, Natchez, and Baton Rouge, to New Orleans.

O'Reilly told the merchants and editors of New Orleans he would ask for no stock subscriptions until his line was built, and they were quite willing to let him take all of the risk. He was building several other lines and soon was pressed for payment by Downing and other suppliers, and by employees for wages. That forced him to "borrow from Peter to pay Paul," hoping to get money to pay Peter later.

His Rochester associates began backing away. Facing bankruptcy, O'Reilly made Joshua Hanna, of the banking firm Hussey, Hanna, and Company of Pittsburgh, Sanford Smith, and John I. Roggen trustees of his affairs. They worked out a deal with creditors, promising to undertake no new lines, complete those under construction, and pay creditors on a pro rata basis as income was received.[14]

Struggling to free himself from Kendall's lawsuits, O'Reilly announced he no longer needed the Morse patent license because he would use "Columbian" instruments. The instruments, devised by O'Reilly employees Edmund F. Barnes and Samuel F. Zook, were described by Reid as "the veriest plagiarism" and merely "Morse working backwards."

The "Columbian" instruments were adjudged by Judge T. B. Monroe in the U.S. District Court of Kentucky at Franfort on September 7, 1848 to be an infringement of the Morse patents. The *Scientific American* (October 14), the Albany, New York *Daily News* (October 13), and others printed editorials, partly identical, calling the decision "unjust as it is ridiculous—the exclusive monopoly of electromagnetism is held to be secured to Prof. Morse for telegraphic purposes."

O'Reilly claimed the injunction only applied on intrastate traffic in Kentucky, closed his offices in the state, and sent interstate messages across the state from Indiana to Tennessee. The court then had the line seized and broken by the U.S. Marshal.

This was disastrous for O'Reilly, but Horace Greeley paid his respects to Kendall and Smith in the *New York Tribune*:

> Age and avarice go together, and here we have the sickening spectacle of a couple of pharisees of Democracy, one an ex-member of Congress, the other an ex-Postmaster General, holding on to a monopoly most injurious to the press and the public with the grip of a dying miser to his money-bags.

The Albany, New York *Knickerbocker* (November 21, 1848):

> Should this be true, the judge who could make such a decision must be a superanuated old ignoramus, an ass clothed in ermine, or a hired and paid agent of Fog Smith, Kendall & Co.

O'Reilly issued circulars saying he had completed thousands of miles of telegraph, "extending between the Atlantic seaboard and the Mississippi and from the Canadian Frontier to the Mexican Gulf, including the Lake country and Ohio Valley," and had contracts to build more thousands.

On January 15, 1849, Senator Underwood of Kentucky presented a memorial from O'Reilly to Congress urging an investigation of special laws he said the Morse patentees had smuggled through Congress. The memorial also said six suits were pending in different states—all with the same plaintiffs and defendants and on the same matter—but the patentees refused to combine them in one trial. Morse, Vail, and Kendall denied the charges and said O'Reilly's attacks were echoed by a "bought press."

Seeking a system to use on the New Orleans line, O'Reilly suspected the trouble Alexander Bain was having in patenting his electrochemical telegraph was caused by Morse's old associate Professor Gale and Charles G. Page, now examiners in the Patent Office, who had subscribed to stock in the Magnetic company. Gale had been recommended for the job by Morse.

Under his pen name "Morion," O'Reilly charged in the *New York Tribune* that Bain went to the Patent Office and put his specifications in the hands of Page and Gale, who told Bain a patent would be issued. O'Reilly said Page, Gale, and a chief clerk spent the evening at Bain's hotel, but the next morning Page told Bain that Morse had filed for a similar invention shortly before Bain arrived at the Patent Office.

The *New York Express* said the Commissioner of Patents should compare Morse's application with Bain's. "He (Bain) is about to bring his family to this country," the *Express* said, "though he may be restrained from doing so by the fear, as said in the *Louisville Courier,* that Amos Kendall will claim his wife and children as inventions of Prof. Morse, or perhaps of Amos Kendall."

On March 12, 1849, Chief Justice W. Cranch of the U.S. District Court, District of Columbia, reversed the patent commissioner's decision, ruling that both Morse and Bain were entitled to patents. The elements of telegraphy were not new, his opinion stated, pointing out that Davy obtained a patent in England in 1838. "Two persons may use the same principle and produce the same effect—by different means—without interference or infringement, and each would be entitled to a patent for his own invention."

The Kentucky part of the line was then returned to O'Reilly. However, a hundred miles of wire were missing and the line required rebuilding. O'Reilly tried to placate employees and other creditors by sending them a few shares of stock. A typical response was from F. S. Rutherford at Tuscumbia, on June 23, that he couldn't get fifty cents for the stock, the line could not be repaired without funds, and he was ill and didn't have ten cents to buy pills.

The anger of stockholders, unpaid employees, and creditors boiled over when O'Reilly chose that time to announce plans to build west from St. Louis to Leavenworth as the first part of a line to California. If he had money to start a new line, they asked, why not get the lines already being built into operation and pay some of his debts?

R. H. Woolfolk at Nashville wrote on August 3, 1849 that the Morse office had attached the wire at Gallatin, claiming $70 was due, and he had just paid another attachment and shipped new wire along the line, but "the Morse people sent the sheriff after it the day after the wire left Nashville."

His affairs getting worse, O'Reilly announced the end of the trusteeship. His Rochester backers, Selden and Ely, told him they had no more funds to invest through him. The Bain lines O'Reilly was building from New York to Buffalo, Toledo, and

Chicago were hostile, of course, to those House printing telegraph lines. Downing also was annoyed at O'Reilly building a Bain line to Boston to compete with him. All of O'Reilly's notes and drafts on Smith at St. Louis and Doane at New Orleans were returned unpaid.

Just in time to avert a collapse in O'Reilly's finances, Walter N. Halderman of the *Louisville Courier* and Dr. T. S. Bell became surety for a loan variously reported as $6,000 and $10,000, and repairs on the New Orleans line were resumed. On September 19, Charles Doane reported he had paid all of O'Reilly's large debts, using stock to do so. Agents of the line, however, continued to draw on O'Reilly, and he was happy to turn it over to the company in November 1849.

On November 22, O'Reilly reported that the line was working from Louisville to Baton Rouge, and the company was organized with Hamilton Smith as president. Of the $700,000 capital stock, Doane received $130,000, Bain $105,000, and John Reilly, Barnes, and O'Connor various amounts.

Still in trouble, the line was leased by Reid on July 1, 1850 for $13,000 a year. He pledged his salary and borrowed $5,000 to swing the deal. Reid was operating two other O'Reilly lines—the Atlantic and Ohio, and the Pittsburgh, Cincinnati, and Louisville—and needed the connecting New Orleans route. With those connections the line was loaded with business.

The Morse line to New Orleans was finally completed in January 1851. It was consolidated with O'Reilly's line on May 13, 1853 and became the United New Orleans-Ohio People's line. In financial straits the next year, the merged company was reorganized as the New Orleans and Ohio Telegraph Lessees, with Dr. Norvin Green of Louisville as president, George L. Douglass as vice president, and Reid as lessee. It was rebuilt and became profitable in 1855.

The Western Telegraph Company

Licensed by Kendall, the Western Telegraph Company was organized to bring the New Orleans traffic east from Pittsburgh. The route was from Baltimore to Frederick, Harper's Ferry, Cumberland, Uniontown, Brownsville, Washington, Pennsylvania, and Wheeling. A branch from Brownsville to Pittsburgh connected with the New Orleans line and another was from Frederick to Washington, D.C. Western was designed to connect with the Magnetic at Baltimore, but at the last moment Magnetic refused. It was afraid O'Reilly's busy Atlantic and Ohio line would retaliate by giving its business to Hugh Downing's new House-printer line to New York. After years of loss, wrangling, and reorganizing, agreements were reached to apportion the business.

Downing's line had been built along the turnpike from Philadelphia to Fort Lee in 1848, with piano wire suspended between tall masts over the Delaware, Raritan, and Hudson Rivers. Little business developed, and Downing sold the line in 1851 to Johnston Livingston, Francis Morris, and R. W. Russell who extended it to Washington, D.C.

One of O'Reilly's self-defeating practices was to delegate authority to someone to raise subscriptions, build and organize a line, and then hire others to handle parts of the job. When he did that to Marshall Lefferts in 1849, he was flatly told to stop hiring people for the New York-Boston and New York-Buffalo lines Lefferts had financed and was building. If O'Reilly wanted the management, Lefferts would give it up, he wrote, but "it had better be done by one or the other."

O'Reilly had a variety of odd "hangers-on." One was William Lyon MacKenzie, a *New York Tribune* writer. The *New York Courier and Inquirer* said:

> Hired writers have been employed to attack Morse and the Patent Office. . . . One of these writers was William L. MacKenzie, the Canadian renegade, who for a long time filled the columns of the *New York Tribune* with ribald tirades against not only the Patent Office and its officers, but also against Mr. Buchanan, Secretary of State.

O'Reilly had gained MacKenzie's loyalty in 1838 by gathering petitions to the President that obtained the Canadian's release from prison. MacKenzie had been convicted of violating the neutrality law when upstate New York people sent volunteer troops, arms, and provisions to aid an insurrection in Canada.[15] President Van Buren's message to Congress on December 2, 1839 attributed the insurrection to "emigrants from the Provinces who have sought refuge here," and deplored the hostility to the United States it had aroused in Canada.

Notes

[1]Kendall to O'Reilly, September 2, 1846. Smith papers, New York Public Library.

[2]It was customary for the contractor to raise funds to meet some or all costs by issuing certificates to subscribers. The certificates were replaced by corporate stock after the company was organized and the completed line was turned over to it. The amount was then deducted from what the company owed the contractor.

[3]O'Reilly to Kendall, October 24, 1846.

[4]Crafts J. Wright, letter in Maysville, Kentucky *Flag,* January 1848. Wright's father, Judge John C. Wright, was one of the trustees of the Cincinnati group and editor of the *Cincinnati Gazette,* an O'Reilly supporter.

[5]Morse to Smith, December 25, 1849.

[6]Smith papers, New York Public Library.

[7]Kendall to Smith, May 12, 1846. On June 8, Kendall added that Senator Calhoun had attacked the proposal for the government to build on constitutional grounds, and it had been withdrawn.

[8]Vail wrote to his wife from Richmond on July 25, 1847: "Yesterday the telegraph commenced operations with Washington in 30 min-

utes after the wires reached the office. I have just taken upon myself the whole responsibility of the Washington and Petersburg line."

[9]The Savannah office was opened on March 22, 1848.

[10]An office was opened in Macon on April 8, 1848. According to a mimeographed booklet, *Atlanta's Telegraph History* by Fred L. Hester, "The Savannah—Macon line was extended north along the Macon and Western (now Central of Georgia) Railroad to Atlanta, and the first office was opened in the M & W depot," in May 1849. "Western Union's first office in Atlanta," Mr. Hester continued, "was opened in 1856 at the corner of Wall Street and Central Avenue with a telegrapher named David U. Sloan. His messenger was an alert youngster named Evan P. Howell, who later became editor of the *Atlanta Constitution.*"

[11]The line was leased by Magnetic on July 7, 1856. Between Washington, D.C. and Petersburg, Virginia is a monument inscribed: "Telegraph Road. One of the first telegraph lines in the world." In 1957 another historical marker for the

line was placed three miles north of Columbus, Georgia by the Daughters of the American Revolution.

[12]Burton J. Hendrick, *The Life of Andrew Carnegie* (Garden City NY: Doubleday, Doran, 1932) said Carnegie's uncle, Andrew Hogan, played checkers with David Brooks, the telegraph manager in Pittsburgh. In 1850 Brooks needed a messenger, and Carnegie got the job at $2.50 a week. He brought in four other Scotch lads as messengers and the line superintendent James D. Reid was so impressed that he had a tailor make a uniform with dark green jackets and knickerbockers.

In the *Autobiography of Andrew Carnegie* (Boston and New York: Houghton Mifflin, 1924) Carnegie said his life as a messenger was in every respect a happy one and laid the foundation of his closest friendships. Delivering telegrams to Edwin M. Stanton, e.g., he came to know the man who became Lincoln's Secretary of War.

Investing in express, sleeping car, bridge, locomotive, and other companies, Carnegie assembled a steel empire that was sold for $492,000,000 in 1901 in forming U.S. Steel. J. P. Morgan congratulated Carnegie on being the richest man in the world. Carnegie gave about $300,000,000 to charities and schools, and built 2,500 libraries.

[13]Librarians there said no one else had researched the huge O'Reilly collection. Thousands of O'Reilly letters also were found at the Rochester Public Library, Cornell University, Portland Historical Society, and other places.

[14]Gleaned from letters between O'Reilly, the trustees, and others during May, June, and July 1848. O'Reilly Papers, New York Historical Society.

[15]Congress papers, document no. 74, 25th Congress.

Fight to Control Midwest (1847–1854)

When "Fog" Smith gained control of Morse patent rights in the north in 1847, he made Ezra Cornell and John J. Speed, Jr., an Ithaca, New York merchant, his agents in Ohio, Indiana, Illinois, Michigan, and Wisconsin.[1] He ordered them to build lines with all possible speed, seize routes, and halt O'Reilly's progress.

Cornell and Speed then built the Erie and Michigan Telegraph Company line from Buffalo to Erie, Cleveland, Milan, Sandusky, Toledo, Monroe, Detroit, Ypsilanti, Ann Arbor, Jackson, Albion, Marshall, Battle Creek, Kalamazoo, Niles, South Bend, Michigan City, Chicago, Southport, Racine, and Milwaukee. It was called the Speed Line because Speed was its president.[2]

Smith provided no money, and investors wanted to see who won the Morse–O'Reilly battle before deciding whom to back. People came to see demonstrations by the promoters and their agent, D. T. Tillotson, however, and finally $10,000 was subscribed in Buffalo, $2,000 in Detroit, $2,100, Ann Arbor;

$2,150, Jackson; $1,000, Battle Creek; $2,000, Niles; $10,000, Southport, Racine, and Milwaukee combined; and $2,000, South Bend.[3] O'Reilly and H. B. Ely issued circulars along the line claiming that O'Reilly had the exclusive patent right and warning against fraudulent promoters.

Convinced that money could not be collected "until the O'Reilly matter was arranged," Speed wrote Cornell April 26, 1847 that he had promised not to ask subscribers to pay until it was settled. On May 2, 1847, Speed wrote Cornell that, not having heard from him, Tillotson had subscribed $10,000 for Cornell "so that if we should be forced into court, we could show that we complied with the contract" and protect their exclusive right to the route. Thus was Cornell involuntarily committed to put everything he had earned on other lines and all he could borrow into this one, but it proved to be the basis of his fortune. He literally had wealth thrust upon him.

Livingston and Wells, A New York and

Buffalo express firm, was given the contract to build the Buffalo-Detroit section, but when they delayed, waiting for the O'Reilly dispute to be settled, Cornell and Speed took over the contract. Speed constructed the line from Detroit to Milwaukee,[4] getting Jeptha H. Wade, later a dominant telegraph figure,[5] to build the section from Detroit along the Michigan Central Railroad to Jackson, Michigan, and a part of the line from Detroit east. The *Detroit Free Press* on November 30, 1847 reported that a member of its staff had exchanged messages with the operator at Ypsilanti.

5.1 JOHN J. SPEED, JR.

Dr. S. D. Cushman of Chicago, one of the line's builders, said farmers were so pleased they provided free poles. At night he gave exhibitions and solicited a bonus for including towns in the route.

The first telegraph office in Chicago was in the Saloon Building at Lake and Clark Streets, and Chicago received its first message from Milwaukee January 15, 1848. When the line was completed April 6, Detroit sent the following message:

> To Milwaukee, Racine, South Portland, Chicago: We hail you by lightning as fair sisters, as bright stars of the West. Time has been annihilated. Let no element of discord divide us. May your prosperity, as heretofore, be onward. What Morse has devised and Speed joined let no man put asunder.

One branch was built to Pittsburgh, Wheeling, Columbus, and Cincinnati. It was an invasion of territory O'Reilly had built and ruined Morse's and Vail's prospects of collecting from O'Reilly, while Smith enjoyed the profits of his new lines.

While in the midst of the Erie and Michigan project in 1847, Cornell built a line from Montreal to Troy, New York, connecting with one from Saratoga. He was bombarded with pleas by his men for funds to buy poles and supplies. The poorly constructed line was repaired with wire on high masts over the St. Lawrence River on July 3, 1848.

By borrowing, pinching, working day and night, and sometimes going a week without taking off his clothes, Cornell established a trunk line—the New York and Erie—to New York City from Fredonia, through Pike, Nunda, Dansville, and Binghamton, New York; Montrose and Honesdale, Pennsylvania; and Middletown, Goshen, Newburgh, Peekskill and White Plains, New York.

Kendall protested March 1, 1850, that it was built "in bad faith," because it competed with Faxton's New York-Buffalo Morse line. Although it was designed to divert business from the other patentees for Smith's profit, when it lost money, he tried to force them to share his losses. He constantly sought, without success, to drive a wedge between Morse and Kendall. Without Kendall's shrewdness, Morse would have been at Smith's mercy.

Smith intended the New York and Erie to be the eastern outlet for the Erie and Michigan, and he expected to control it. He instructed Cornell to stay behind the scenes, but Cornell was elected president. The company then refused to pay Smith, as originally agreed, fifty dollars a mile for the patent rights and half of the profit on construction. He had not settled with O'Reilly, the company said, and could not give clear patent rights. Since Smith had refused to settle with O'Reilly until he could set up competing lines and ruin him, his own lieutenants were now using that to his loss, while Kendall was demanding that Morse and Vail share the patent rights on the new lines. Smith angrily severed relations with Cornell and Speed.

Kendall objected strongly to Smith's licensing Cornell and Speed, and them passing the Morse rights on to Wade, Martin B. Wood of Albion, Michigan, and John J. S. Lee of Cleveland to build lines into the southern part of the mid-west.[6]

The New York and Erie line had bad insulation, was constantly out of operation, and was sold for debt in 1852. It was bought by Cornell for $7,000. With creditors hounding him, Cornell gave the Erie Railroad the right to use Morse on a line it had built along its tracks, in return for $5,000 and permission to add the New York and Erie wires on the railroad's poles.

This enabled Cornell to save the line, and his own threadbare shirt. He changed its name to the New York and Western Union Telegraph Company, because it provided his Erie and Michigan Line (the west) a direct connection with New York and the Atlantic seaboard. In 1853 he leased it for $2,000 a year to the New York Albany and Buffalo Telegraph Company.

The Lake Erie Telegraph Company

When Cornell and Speed began the Erie and Michigan line, O'Reilly met the challenge by extending his Pittsburgh line to Cleveland via Lisbon, Massillon, Canton, Akron, and Cuyahoga Falls, naming it the Lake Erie Telegraph Company.

The Rochester group issued a circular in August 1847 asserting O'Reilly had exclusive patent rights and Cornell and Speed would not be permitted in that area "until we cease to have laws to enforce the observance of contracts." Signers of the circular were Henry R. and Samuel L. Selden, Alvah Strong, Elijah D. Ely, Jonathan Child, and George Dawson.[7]

When the O'Reilly line was completed to Cleveland August 20, the Smith group was shocked. "I think the Lake line is the fall of Waterloo," Speed wrote to Cornell, "and if we yield that, we may as well retire from the contest." O'Reilly's next move was a Detroit-Buffalo section. Ely, the contractor, built simultaneously from east and west—from Buffalo through Cleveland, and Detroit through Toledo so speedily that it was opened March 1, 1848.

While building the O'Reilly line to Detroit, Don Mann's progress was blocked by superstitious mixed French and Indian people at "French Settlement." Then Wade arrived, building the opposing Erie and Michigan line east from Detroit. He soon found who the "head devil" was, and won him over with "plenty of soft soap and some

5.2 SOME OF THE EARLIEST TELEGRAPH BLANKS. FROM THE COLLECTION OF SIGMUND ROTHSCHILD.

whisky." He attended a squirrel dinner at the half-breed's home, through which muddy pigs roamed. The next morning Wade's half-breed friend ordered his neighbors: "On west side all right, lief em be. On east side give em hell." To Mann's amazement, Wade's line got through without trouble. Later Wade relented and persuaded the settlement leader to let Mann's party through.[8]

The *Toledo Blade* issued a one-column "Extra" on February 12. It said the O'Reilly line had commenced operations there, and printed the latest news from the New York and Boston markets. On March 2, a *Cincinnati Daily Times* story, headed "First Flash from Detroit," said the line totaled nearly 600 miles. Detroit received its first telegram from New York over O'Reilly's line more than a month before Smith's Erie and Michigan line was opened to Buffalo.

O'Reilly wrote to the Rochester group August 10, 1849, that he was being hounded for money while they owed him about $25,000.

> I see you declared a four percent dividend on July 1; is it right that I should be insulted and outraged without some effort to raise the means advanced by me for constructing that line, for which, payment was promised by you last January? I am now going home on one of my half-yearly visits, without a dollar to pay even the three-fourths quarters rent for the house. It required all of my energies to complete and pay for lines. I am in debt for my board for the last three months. In one case I gave an order on H. B. Ely for a debt to J. H. Jones, an employee of yours at Cleveland, and it was not paid. For God's sake pay this man from what is due me.

Gillet, O'Reilly's trustee, wrote to the Lake Erie stockholders December 24, 1850, that the line had paid its debts, contracted by H. B. Ely, but not for material provided by O'Reilly. He asked that justice be done at the annual meeting, but before his letter could reach many stockholders, the Rochester group suddenly switched the meeting date.

Upset by the opening of the Lake Erie line Buffalo office at 8 Exchange Street, Smith notified Kendall he had employed astute counsel to get an injunction. George Dawson proposed that O'Reilly compromise, but the Irishman replied on March 4, 1848:

> With that faithless, double-dyed scoundrel, Fog Smith, I can have no intercourse, as you know how he has acted respecting former compromises. The dastardly villain! He and Kendall tried to ruin me every way, by slander, law suits and special legislation.

Smith did institute the suit, but without success.

The Ohio, Indiana, and Illinois Line

To connect the Lake Erie with his Ohio Valley lines, O'Reilly had the Ohio, Indiana, and Illinois Telegraph line built. Work began at Chicago before the end of 1847. It was built east to meet the Lake Erie line at Toledo, and southwest to meet the Pittsburgh, Cincinnati, and Louisville line at Dayton. Crawfordsville, Indiana was reached before O'Reilly ran out of funds.

William J. Delano of Dayton, dependable, honest and industrious, was sent to take over the project. He was fought at every town by Cornell and Speed men, also seeking subscriptions, and his letters to O'Reilly in 1848, 1849, and 1850 were a tale of woe, as his progress was repeatedly delayed by the lack of money. The line did not reach Toledo and the Lake Erie connection until the summer of 1850. Delano then wrote to the unhappy stockholders that they had paid only $37,000 of their subscriptions, forcing O'Reilly to provide $43,000 to complete the line.

Facing backruptcy, O'Reilly wrote to H. R. Selden March 15, 1851, to ask where he could find labor "whereby to earn the daily bread of my family." Delano, faced with demands for payment, borrowed $5,000 for two months, pledging $50,000 of the stock. In an effort to rescue him, O'Reilly confessed judgment to Delano for nearly $15,000, and pledged stock.[9]

The line fell so heavily into debt that O'Reilly had to sell $100,000 of its stock to Hiram Sibley for $2,000. Cornell bought some of the stock, to get friendly directors elected, and leased the line for five years.[10] O'Reilly then had his Pittsburgh, Cincinnati and Louisville line make a sealed bid of only one half of one percent of its capital to buy it. Somehow, Cornell learned the amount, bid $100 more on behalf of his Erie and Michigan, won the line and made it profitable.

The Illinois and Mississippi Line

In 1848 O'Reilly built a line north from St. Louis, crossing the Mississippi just above St. Louis, and proceeding at breakneck speed through Alton, Jacksonville, Springfield, Peoria, Peru and Ottawa to Chicago. A branch west from Jacksonville to Quincy, Warsaw, Keokuk, and Burlington also had branches to Dixon, Galena, Dubuque, Rock Island, Davenport, Iowa City, and from Quincy to Hannibal.

Charles G. Oslere, in charge of construction, did his utmost, often with gangs of unpaid workers and no wire. He wrote September 22, 1848, that he could collect no further subscriptions until the line was built. He was in debt $800 and had borrowed $500 from friends, but somehow he completed the 750 miles of line early in 1849. The stockholders met at Peoria April 10 to organize the company, and elected William Hempstead of Galena president, and Sanford J. Smith treasurer.

5.3 From *The Palimpsest* 6/11 (November 1925) published by the Historical Society of Iowa.

Business was at a standstill that summer because of spring floods and an epidemic of cholera. Oslere died of it June 19 at Keokuk. In two years the company was $17,000 in debt, and subscribers met to abandon the line. Associate Justice J. D. Caton of the

Illinois Supreme Court[11] happened to enter the room, became interested, and was elected a director. He opposed abandoning the line, and was elected president.

Caton secured legislation enabling him to assess stockholders and sell their stock for unpaid assessments. A levy of $2.50 a share was imposed, and O'Reilly wrote repeatedly, protesting the forfeiture of his stock. Many did not pay, and Cornell and Wade bought their shares at $2.50. Raising money in that way and using his own funds, Justice Caton reconstructed the 1,086-mile line and made it pay.

O'Reilly and Associates

O'Reilly had a telegraph blank for each part of his "Telegraph Range": (1) Atlantic and Ohio, (2) Pittsburgh, Cincinnati, and Louisville, (3) Ohio and Mississippi, (4) Ohio, Indiana, and Illinois, (5) Lake Erie, and (6) Illinois and Mississippi.

Sometimes O'Reilly's associates gave him as many problems as his enemies. On one of his papers, he wrote "Gratitude!" above Talleyrand's remark: "For every appointment I have had, I have gained ninety-nine enemies and one ingrate." O'Reilly added that he had appointed many agents, and "found some choice specimens—take such specimens as Ls, Ja, Rd, and Sanford Sh."

Since the Morse patentees considered the contract void and would not accept stock and dividends, O'Reilly accumulated them in a fund to be paid when the dispute was settled. On March 18, 1850, he made Marshall Lefferts trustee of the fund and provided for it in his will.

Two of his agents, Sanford J. Smith and George W. Olney, a former attorney general of Illinois, quarreled. After receiving letters from Olney about "Smith's failures to fulfill promises," O'Reilly wrote in August 1848 of Smith's "utter inattention to the business." At the same time, Smith wrote to O'Reilly that people were afraid to pay stock subscriptions, because men who drank and gambled gathered in Olney's room. In amusing letters, however, Olney told O'Reilly how he entertained prospective subscribers. His ability to get subscriptions was a major asset in O'Reilly's frantic race.

In one letter Olney told of swimming a team across a swollen, icy stream where the bridge had washed out. Horses and wagon parted, and the horses went on, but Olney swam out and saved the wagon. Walking to a farmhouse miles away, he obtained permission to thaw his frozen clothes and sleep on the earth floor in front of the fire. It was a world of no electricity, gas, plumbing, refrigeration or water, and often no tables, chairs, beds and utensils.

In 1849 Olney tried to obtain pay for his services, saying he was in debt, wearing threadbare clothes, and too embarrassed to go home to St. Louis. Two years later, he wrote that he had raised $63,000, had received only $200 aside from expenses, was broke and needed $2,500. Replying on August 1, 1850, O'Reilly told Olney to raise subscriptions immediately at Logansport and Lafayette "and block Fog Smith and Kendall who are busy in that area." Seventeen days later, Olney died of cholera, and was penniless.

To a plea for funds from J. B. Perkins, constructing the Missouri and Nebraska line, O'Reilly sent his typical pep letter:

> Stand firm. You have seen that I always accomplish what I undertake; sooner or later, yea, far more than I promised. . . . You may be sure that 200 or 300 miles in Missouri

will not stop me. . . . I will ere long cross the Rocky Mountains and reach the Pacific.

Perkins replied that he had a force of men but no money. "Shaffner has told the people out here so much about our insolvency that we can get no credit." His next letter said, "(James) Creighton has just come down from the boys and says 'they have dismissed all of their hands, unpaid, of course.'" O'Reilly replied that he was sending O'Connor, to which Perkins answered, "Fine. But what we need is money." Perkins next reported: "We were compelled to stop right at Jefferson, and the men came back to St. Louis by boat, while Shaffner put on a force of men and went ahead." At this time O'Reilly notes as far back as seven years were unpaid.

Two More New York-Buffalo Lines

When O'Reilly had Lefferts build the Bain-operated Merchant's State Telegraph line from New York to Buffalo, Faxton's Morse line over the same route cut rates and placed an advertisement in the March 23, 1849 Rochester, Utica, and other papers, headed "Telegraph Speculators Beware!"

The papers, however, welcomed competition. The *Albany Daily Messenger* (on September 6, 1849) said:

> We will not use you a cents worth, if we can help it, O'Reilly is coming. He it is that has driven you to reduction and now we don't thank you for it. Charge us $5 and furnish Buffalo for $4! Out upon you. Three cheers for O'Reilly.

Backing the House patent, the Selden group announced late in 1849 that they also would build to Buffalo. They organized the

New York State Printing Telegraph Company, with Selden's brother-in-law, Levi A. Ward, as president, and Isaac R. Elwood, secretary. Now O'Reilly's original backers at Rochester sent agents along the line, battling him for subscriptions. Don Mann built the western; and H. C. Hepburn, the eastern section of the O'Reilly line.

Trying to free the completed line from O'Reilly's control, Lefferts, its president, claimed it was not in good working order, while O'Reilly blamed Lefferts for hiring inexperienced operators. O'Reilly notified Lefferts to stop issuing stock and interfering with the business until he was ready to turn it over to the company.[12]

O'Reilly refused to recognize stock the company issued,[13] and charged that Lefferts issued $17,000 of it to himself. At a suddenly-called board meeting, the directors authorized the stock to Lefferts and $21,300 to Bain. O'Reilly then charged that Lefferts prematurely issued the stock so that he could buy it from Bain at a low price.

The dispute was submitted to Charles O'Connor, who advised that transfer of the line should proceed, and O'Reilly could sue Lefferts if he wished. Acting for his New York wire firm, Morehouse & Co., Lefferts attached O'Reilly's interest in the Boston–Portland and New Orleans lines. That forced O'Reilly into a settlement. Lefferts received certificates for $6,200 in the Boston-Portland line and $17,900 in the New York State line in exchange for O'Reilly's debt to the wire firm. His written settlement with Lefferts, September 27, 1850, made O'Reilly so angry that he gnawed away a part of it. He wrote on the margin: "I have a bad habit of biting and chewing paper when much enraged. I did so with the above memo—but the torn words are restored in the copy annexed."

Faxton's line bought the Merchant's State line on June 7, 1852 and leased the House-operated New York State Printing line in 1856.

O'Reilly regarded the telegraph as a great "cause," and his enthusiasm inspired many, but none more than his brother, John, who had worked faithfully for him since he helped build the original Lancaster-Harrisburg line. O'Reilly turned against John when he asked for pay in February, 1851. O'Reilly's counsel and trustees Gillet and Mann, replied to John that O'Reilly had not had sufficient money for two or three years to pay his family's living expenses.

Smith excoriated Kendall's stand that the division agreement did not give him the right to sign Morse's and Vail's names to contracts. Morse and Vail disavowed any such use of their names. Smith sarcastically wrote to Kendall, "I think you had better buy a dog's muzzle for him (Vail) and keep him in his place."

Taking advantage of O'Reilly's financial distress, after five years of vehement enmity, Smith suddenly reached a "settlement" with O'Reilly on December 2, 1851, giving control of the Morse patents in Ohio and all states northwest of the Ohio River to the greatest enemy of the patentees. The "Patentees" were to receive one-fourth of the stock on O'Reilly's lines; one fifth of the balance to O'Reilly; and four fifths of the balance to Smith.[14]

Bankrupt and weary, O'Reilly tried to consolidate the Speed, Cornell and Wade lines with his,[15] but failed because he no longer had much influence.

Wade and Thomas T. Eckert agreed March 19, 1853 to build along the Ohio and Pennsylvania Railroad through Rochester and New Brighton, Pennsylvania, Columbiana, Salem, Alliance, Canton, Massillon, Wooster, Loudonville, and Mansfield, to Crestline, Ohio. Later that year they contracted with the Ohio and Indiana Railroad to build from Crestline to Bucyrus, Upper Sandusky, Lima, Delphos and Van Wert, Ohio, and Fort Wayne, Indiana. Another deal by Wade June 11, 1853, was with the Bellefontaine and Indiana Railroad to build from Galion to Union. These activities convinced Kendall that Wade should be cultivated, and he wrote to Cornell:

> I regret that you and Mr. Wade do not get along amicably. These quarrels kill the telegraph, and no sooner is one quieted than another springs up.

Cornell sent a friendly letter to Wade, but Wade's lengthy reply was a masterpiece of sarcasm. He said Cornell was "forever growling like a dog with a sore head" and compared Cornell's New York and Erie line, which lost $6,800, with his prosperous Cleveland-St. Louis line. Wade's letter ended:

> You showed your ingenuity by claiming to be the inventor of the brimstone insulator—and showed your firmness by sticking to them manfully two years after everybody else had found them a failure and threw them away.

Proud of this scathing letter, Wade sent a copy to Kendall, who replied that Morse had a hearty laugh over it.

Some lines cooperated with others. Wade, president of the Cincinnati and St. Louis line; Veitch, representing the St. Louis and Missouri River line; and Shaffner, the St. Louis and New Orleans line agreed on December 6, 1851 to share a St. Louis office. Wade and Shaffner agreed to divide equally the income from messages for transmission over either line. Also, the Pittsburgh,

Cincinnati and Louisville and the Ohio and Mississippi lines agreed on July 6, 1852, to divide their net earnings in the proportion of $135 and $45.

Most early lines were poorly constructed. Some had three-inch saplings as poles, wire with poor joints, and insulators like small bureau-drawer knobs. Extensive repair or rebuilding was necessary, and the press and public were exasperated by poor service.

Endless bickering resulted when a promoter could not complete a line and another finished it, as Wade did with J. S. S. Lee's Cincinnati-St. Louis line in 1851. After making repeated demands for the money he had put in the line, Lee wrote Wade on May 17, 1852: "Why in hell don't you send me that money?" In four blistering pages Wade replied that when Lee's letter came, he used the funds elsewhere. "Drop this cross firing, low abuse and insult, and wait quietly," Wade said. Otherwise, no payment would be made until a court forced him to.

The role of the telegraph in gathering election returns was forecast by a circular from A. Jones & Co. of the New York Merchants Exchange on behalf of the New York newspapers October 5, 1848, to telegraph correspondents, reporters and operators. It prescribed the form and information to be wired from all states and counties.

Working conditions were crude, but were improved at a few points by the introduction of women operators. Miss Sarah Bagley, on the New York—Boston line in 1846 was first. Miss Emma A. Hunter, later Mrs. Thomas T. Smith, was on a branch line to Westchester, Pennsylvania in 1851; and Miss Ellen A. Laughton, at 14, was operator-manager at Dover, New Hampshire in 1852. Women operators wore huge skirts and had wasp waists.

Maine Telegraph Company

Fog Smith extended his New York-Boston line to Portland early in 1848, followed by the Maine Telegraph Company line 275 miles from Portland to Calais, at the Canadian border. Reid wrote that its promoters thought it "occupied in telegraph history a place similar to Palestine in connection with sacred learning, and influence out of all proportion to its limited extent."[16]

James Eddy obtained the right to use the Morse patent from Fog Smith, and built the line from Portland to Brunswick, Bath, Rockland, Wiscasset, Rockport, Camden, Belfast, Winterport, Bangor, Machias, Eastport, and Pembroke, reaching Calais, Maine in March, 1849. Eddy was superintendent of the line and secretary of the company. Prospering, the company used its surplus earnings in September 1853 to buy Smith's Portland-Boston line, thus obtaining a continuous route from Boston to Calais.

Reception By Sound

Some operators learned to read messages by listening to the short and long intervals between clicks of the stylus receiver. James F. Leonard, 15, at Frankfort, Ky. apparently was first. Moving to the Louisville office in 1848, he "read" Morse signals, transmitted by James Fisher from the Nashville office, at fifty-five words a minute.[17] The great showman, P. T. Barnum, offered Leonard "a considerable salary" to exhibit his skill in New York. Leonard did not wish to be shown as a freak and refused, but his fame drew many visitors to the Louisville office.

Soon other cities had sound operators. Smith's New York-Boston line dismissed copyists and ordered operators to learn to "read" the dots and dashes and write messages directly on paper.

Gradually the register and tape were eliminated. The receiving instrument, the sounder, was an electromagnet that was energized by the current and attracted a small iron lever. When the sending key was closed or opened, the sounder lever struck an anvil. The Morse operator distinguished between a dot and dash by the short or long interval between the two clicks. Sounders were placed in boxes known as resonators, with the open side toward the operator's ear, to amplify the sound.

Operators, writing rapidly in longhand, developed artistic styles of penmanship known as "telegraphers' fists." Every Morse operator had his own distinctive style of sending, and friendships grew up between operators who knew one another only by individual style and the letters assigned for use in "signing" the messages sent—known as the operator's "sine."

Prior to 1852, a message sent by telegraph was called a telegraphic dispatch or communication. E. Peshine Smith, a Rochester lawyer who was for five years legal adviser to the Mikado of Japan and a grandfather of Mrs. Rudyard Kipling, introduced the word "telegram" into the English language in 1852. The first issue of the *American Telegraph Magazine,* October 1852, edited by Donald Mann, announced the coinage of the word.

Train Dispatching by Telegraph

When the Erie built the first long railroad in the United States from New York to Lake Erie, Charles Minot, its general superintendent, persuaded the company to erect a telegraph line along the right of way. The first two Erie trains in May 1851, carried President Millard Filmore and 300 other notables. Secretary of State Daniel Webster insisted on riding in a rocking chair strapped on a flat car so he wouldn't miss the scenery. One engine had trouble, and Minot telegraphed ahead from Middletown to Port Jervis to have another engine ready at that point.

In the fall of 1851, Minot was on the westbound express at Turner, fifty-seven miles west of New York waiting for an eastbound train. He telegraphed Goshen, fourteen miles west, learned the eastbound train had not arrived there, and ordered Goshen: "Hold the train for further orders." He handed an order to Conductor W. H. Stewart: "Run to Goshen regardless of opposing train."

"I took the order," Mr. Stewart said later, "showed it to the engineer, Isaac Lewis. The surprised engineer handed it back and exclaimed: 'Do you take me for a damned fool? I won't run by that thing!'"

"I reported to Superintendent Minot, who used his verbal authority on the engineer, but without effect. Minot then climbed on the engine and took charge of it himself. Engineer Lewis jumped off and got in the rear seat of the last car. The superintendent ran the train to Goshen. The eastbound train had not yet arrived." Minot ran the train to Middletown, then to Port Jervis, and arrived there just as the other train came in. "An hour and more in time had been saved to the westbound train," he said, "and the question of running trains on the Erie by telegraph was at once and forever settled."

D. C. McCallum, who succeeded Minot, said in his annual report:

> I would rather have a road of single track with the electric telegraph to manage the movement of its trains, than a double track

without it.

Telegraphic dispatching spread to other railroads, and lines were built along the right of way of railroads when the tracks were laid. Morse equipment was used because it was easy for station agents to learn to operate and repair. Experts said the telegraph practically quadrupled the capacity of the single-track railroad.

When the engine of a midwestern train broke down in 1858, Anson Stager, a frock-coated, impatient passenger, asked the conductor, "Will you order an engine from the next station if I telegraph for you?" When the answer was "yes," Stager climbed a telegraph pole and lowered a wire to the ground. He thrust into the ground an iron poker from the coal stove in the coach, and tapped the end of the wire against it to order an engine. He asked that the message be acknowledged by sending "OK" slowly three times. Stager then stuck out his tongue, placed the wire upon it and received the electrical impulses.

When the Telegraph Came to Texas

After O'Reilly's and Kendall's lines to New Orleans were consolidated in 1853, a line was built from New Orleans west to Morgan City and northwest to Alexandria, Louisiana. Organized January 5, 1854, the Texas and Red River Telegraph Company branched east from Alexandria to connect with the O'Reilly line at Natchez.

The main line continued northwest to Shreveport and west to Marshall, Texas. Most of its 1,189 residents gathered in the town square of Marshall on February 14, 1854. Six oxen, straining at their yokes, slowly pulled a heavy wagon loaded with

poles and reels of wire to the square. Two workmen set up a small pole; linemen unwound wire to the telegraph office and connected a Morse key. The metallic chattering of dots and dashes followed. It was announced that Shreveport and New Orleans had replied, and a shout arose from the crowd.

"The magnetic telegraph is at length in operation between Marshall and New Orleans," proclaimed the weekly *Texas Republican* at Marshall. "We are no longer cut off from the balance of the world by low water and slow mails." The paper printed the first telegraph news in Texas on February 18, about the collision of two ships on the Alabama River in which several lives were lost, the New Orleans cotton market, and the war between Russia and Turkey.

From Marshall the line was built through Henderson, Rusk, Palestine, Crockett, Huntsville, and Houston, which had 2,396 residents. By the end of 1854 it reached Galveston, with a population of 4,177. Unfortunately, the line was built with small green poles and was constantly out of order. On Christmas Day 1855, it was seized by the operators for wages due them, and sold by the sheriff.

The Bain Lines

Alexander Bain of Edinburgh, Scotland, invented the first printing telegraph to print letters of the alphabet on tape in 1840. His electrochemical telegraph was widely used in America and Europe. Electrical contacts, made through holes in perforated tape, sent pulses over the wire. At the receiving end, an iron pen connected with the line rested on a moving paper tape treated with potassium prussiate. When the pulses passed through the pen, they decomposed the chemical and made corresponding marks on the tape. A speed of 1,000 words per minute was claimed, but it required time to perforate the sending tape and write the message received.

In 1849 the North American Telegraph Company was organized to operate under the Bain patents between New York and Washington; O'Reilly built a Bain line in 1849 from New York to Boston, followed by a line from Boston to Burlington in 1850. The Morse group finally defeated the Bain patent in U.S. District Court, Eastern Pennsylvania, on November 3, 1851, and chemical telegraphy came to an end.

The six-story Astor House, built in 1834 by John Jacob Astor at Broadway and Vesey Street, was a center of commercial activity in New York. In September 1853 Gustavus A. Swan resigned as superintendent of the merged New York-Boston line, and built lines from his office in the hotel to the telegraph companies willing to pay him a commission on the messages.

Until the Astor House was torn down in 1913, famous men dined there on sixty-cent steaks and fancy fifteen-cent desserts. It also was a telegraph landmark. On telegrams he sent there, Morse drew a skull and cross bones to indicate they were to be sent "dead head," or free.

House and His Printer

Royal E. House was one of the most remarkable men in communications history. Born in Vermont, he moved with his parents to tiny Choconut, Susquehanna County, Pennsylvania. Without drawings, written plans, or models of any kind, House, who was so poor that he slept under his lathe for lack of a bed, performed the astounding mental feat of planning in his mind all of the complicated mechanical details of a printing

telegraph system.

He then went to New York, had parts of his machine built at several different shops to preserve his secret, assembled the parts and exhibited his printer at the Fair of the Mechanics Institute in the basement of the City Hall in the fall of 1844, a few months after the first line was built between Washington and Baltimore. On November 16, 1846, Hugh Downing, O'Reilly, and Judge Samuel L. Selden bought the right to use the House patent on the Atlantic and Ohio line.

5.4 ROYAL E. HOUSE, INVENTOR OF THE FIRST PRACTICAL PRINTING TELEGRAPH.

O'Reilly sent the first telegram by House printer to his brother John in Louisville September 22, 1847. In the Cincinnati Gazette October 4, 1847, House offered to provide his "Lightning Letter Printer" to telegraph lines. Downing built his House line from New York to Boston in 1847, and to Philadelphia in 1848.

Contemporary enthusiasm over House's

5.5 MODEL OF HOUSE'S PRINTING TELEGRAPH (1852), AN IMPROVED VERSION OF HIS 1848 ORIGINAL.

invention was reflected in a book by Dr. Lawrence Turnbull:

> To make the cold, dull, inanimate steel speak to us in our own tongue, surpasses the mythological narratives of ancient Greece and Rome, throws into the shade the fabulous myths of superstitious Arabia, and sinks into insignificance the time honored traditions of the Oriental World.[18]

The House sending machine looked like a small piano with twenty-eight keys. The black keys corresponded to the letters A to N and the white keys to the letters O to Z, the period and hyphen. Under the keyboard was a revolving cylinder. When the operator depressed a key, it would catch a corresponding tooth in the cylinder and hold it while other parts revolved in alphabetical order until the desired letter was reached. Magnets in the receiving machine moved an equal number of times and when the desired letter arrived on the type wheel, a paper tape was pressed against it, printing the letter. Rufus B. Bullock, later Governor of Georgia, and Theodore Fullon could read messages from the time between sounds of the House printer.

Late in 1847 Downing obtained the House rights for the Atlantic seaboard; and Selden did for New York State, with an option for the rest of the country that he exercised later. House and his associates, William Ballard and John B. Richards, were to receive one-fourth of the stock in House lines.

Advocates of printing telegraphy said it would eliminate Morse operators' errors. One Morse message: "Sarah and little one doing well," was recorded as "Sarah and litter all doing well." The recipient wired back, "For Heaven's sake, how many has she got?" The message, "See the judge at once and get excused. I cannot send a man in your place," was delivered as "See the judge at once and get executed. I can send a man in your place."

The speed of the House machine was announced as 2,600 words an hour. In 1852, House was used on four main lines, and another was being built from Buffalo to St. Louis, connecting with lines to Cleveland, Cincinnati, and Louisville. Rapid progress indeed.

Notes

[1]*Ithaca Daily Chronicle,* June 19, 1847.

[2]R. B. Ross and G. S. Catlin, *Landmarks of Wayne County and Detroit* (Detroit: Evening News Association, 1898) 556-58. Also, C. M. Burton and others, *The City of Detroit, Michigan*, vol. 1 (Detroit: Clarke, 1922) 389-90.

[3]Speed to Cornell, March 21; Daniel T. Tillotson to Cornell, March 20; and Schuyler Colfax to Speed, April 3, 1847. The letters on which this section is based are in the Cornell Papers at Cornell University, O'Reilly Papers at the New York Historical Society, Rochester Public Library, Rochester Historical Society, and Wade Papers at Cleveland.

[4]Speed wrote Cornell on November 6, 1847, that he had contracted to build from Detroit to Milwaukee with iron wire at $125 per mile, 25 poles per mile, and give Kendall and Smith half the profits.

[5]Wade Papers, Western Reserve Historical Society, Cleveland, box 1. Other material is from Wade's autobiographical handwritten letter to his grandson Homer, July 1, 1889; *Directory of American Bibliography*, vol. 19 (New York: Scribners); William Ganson Rose, *Cleveland—The Making of a City* (Cleveland and New York: World Publishing Co., 1950); Elroy McKendree Avery, *Cleveland and Its Environs*, vol. 2 (Chicago and New York: Lewis Publishing Co., 1918) 510.

Jeptha Homer Wade (August 11, 1811– August 9, 1890) was born at Romulus, Seneca County, N.Y. His father, Jeptha, died in 1813 and the family moved to Seneca Falls. At 12 Wade worked as a shoemaker, at 13 a brick maker, at 18 a carpenter and joiner in Pennsylvania when coal was found there, and at 20 foreman of a sash, blind, and door factory at Seneca Falls.

Wade began building telegraph lines in 1846, became the cleverest strategist in the industry, president of Western Union, and a multimillionaire. More than six feet tall, slender but powerfully built, bearded but with shaven upper lip, Wade was an important national figure. To Cleveland he gave 75-acre Wade Park for Western Reserve University, and the Cleveland Museum of Art.

[6]Wade Papers, Western Reserve Historical Society, box 1. Wade letters to Kendall, Sept. 20 and Nov. 30, 1849. Kendall to Wade, Sept. 25, Dec. 4 and 22, 1849. Cornell license to Wade April 20, 1849 for Cleveland, Columbus, Cincinnati line. Speed license to Wade May 2 for same route. Speed assignment of contract to Wade Sept. 20 for Cleveland-Cincinnati line.

[7]Dawson later was an editor of the *Albany Evening Journal*, Child was mayor of Rochester, and Strong publisher of the *Rochester Democrat*.

[8]Wade's autobiographical letter to his grand-

son.

⁹Letters between O'Reilly and Delano, April and March 1851.

¹⁰Cornell to F. O. J. Smith, March 4 and April 17, 1853. Smith Papers, New York Public Library and Maine Historical Society Library.

¹¹Associate Justice 1842–1855, Chief Justice 1855–1864. Caton's papers in the Library of Congress contain 9,000 items in 30 boxes. More Caton correspondence is in papers of other pioneers.

¹²Numerous letters between O'Reilly and Lefferts. O'Reilly Papers, New York Historical Society Library.

¹³Jerome to O'Reilly, August 21, 1850: "Donald Mann and I have made ourselves liable for a large amount. . . . This we have done relying upon the stock being available to meet our paper." O'Reilly had forbidden the issuance of the stock, he said, and "it must follow that we are ruined."

¹⁴Article by J. W. Orr, "The Lightning World," in *The American Odd Fellow Magazine* (1867). O'Reilly Papers, Rochester Public Library; Wade Papers, Western Reserve Historical Society.

¹⁵O'Reilly to Wade, May 13 and 25, 1853.

¹⁶Reid, *Telegraph in America*, 394.

¹⁷*The Telegrapher* (November 28, 1864).

¹⁸*The Electro-Magnetic Telegraph* (Philadelphia: A. Hart, 1853).

Birth of Our Telegraph System (1851–1861)

In 1851 the eastern half of the United States was thinly covered with more than fifty short telegraph lines. Terminal offices of eleven lines were in New York City alone. Thirteen companies operated in the five states north of the Ohio River.

All were suffering the disadvantages of multiplied and inharmonious management; high cost to the public when messages passed from one line to another, and each collected its full rates; slow and unreliable service; hastily built short lines, constantly out of repair; and contradictory rules and methods of doing business. A telegram might require three days to go from Boston to St. Louis. Most companies were barely self-sustaining, and could not afford to replace the poor poles, three-strand iron wire, and bad insulation. Public confidence was at a low ebb, and the business was suffering.

Kendall and O'Reilly tried to organize their groups of lines and get them to adopt uniform rules and practices. Delegates from a number of companies met and discussed mutual problems, but little improvement resulted. Each line continued to make a deal with another whenever it saw a competitive advantage, regardless of the effect on service.

The poor condition of communications was viewed with concern by public-spirited men in the rich and fertile Genesee Valley, of which Rochester, New York was the center. Two men decided to remedy the situation—Judge Samuel L. Selden and Hiram Sibley.[1] For years Sibley had spent winter months in Washington, D.C. where he knew Presidents Pierce and Harrison. He had aided Morse in getting Congress to pay for the first line, watched it built, and seen the first message sent. Convinced that the telegraph would be one of the "wonders of the age," Sibley became a director of the New York State Printing and Lake Erie companies, and bought stock in others.

Judge Selden discussed the need for reliable telegraph service with Sibley in

6.1 PICTURED IS THE OLD REYNOLDS ARCADE BUILDING, ROCHESTER, N.Y., IN WHICH WESTERN UNION'S FIRST EXECUTIVE OFFICE WAS LOCATED. THE COMPANY, FOUNDED APRIL 1, 1851, WAS CALLED THE NEW YORK AND MISSISSIPPI VALLEY PRINTING TELEGRAPH COMPANY. THE NAME WAS CHANGED TO WESTERN UNION IN 1856. THE TWO MEN WHO LED THE FORMATION OF THE FIRST COMPANY WERE JUDGE SAMUEL L. SELDEN AND HIRAM SIBLEY. SIBLEY BECAME THE SECOND WESTERN UNION PRESIDENT IN 1856, SUCCEEDING HENRY S. POTTER, THE FIRST PRESIDENT.

1850 and proposed a House line westward from Buffalo, but Sibley saw no value in adding another company to the many already failing or just managing to survive. He suggested that a new company be organized to buy all of the weak lines in the west and unite them in a single network. He would be interested, he said, if he had a substantial interest in the House patent rights owned by

Selden and Freeman M. Edson, and in the profits and management. Selden agreed, and they developed a dream of combining all companies in one great telegraph system to serve the social and commercial needs of the nation.

To start the project, Selden and Edson began constructing a line from Buffalo to St. Louis. The contractors, Isaac Butts,[2] Sanford

Smith, and Charles L. Shepard agreed to complete 100 miles of the line by January 1, 1851.

On December 9, 1850, Sibley and Butts bought an interest in House's patent, as trustees for the group. This proved valuable because article 8 of the company's Articles of Association and Incorporation, four months later, gave the patent owners half of the capital stock.

To finance the enterprise, Sibley and Selden assembled a group of Rochester's wealthiest citizens at the Elon Huntington mansion at 1100 St. Paul Street. At first, the group objected to buying nearly bankrupt companies and assuming their debts. Sibley's story in the *New York Evening Post* (February 27, 1885) and Jane Marsh Parker's article *How Men of Rochester Saved the Telegraph* (Rochester Historical Society Publication Fund Series, 1926) said they called it "Sibley's crazy scheme."

Finally the group met in Sibley's office in Reynold's Arcade, and he threatened to go to another city to find the money. Aristarchus Champion, a millionaire, was asked to sign first, and inquired, "You admit, Mr. Sibley, that the telegraph is a failure?" Upon receiving an affirmative answer, Mr. Champion asked, "And you further admit that each company is a failure?" Again the answer was "Yes."

"Then how is the consolidation of failures to escape failure?" Mr. Champion continued. "Would collecting all the paupers—the social failures of Monroe County—into one organization, composed entirely of paupers, insure their success and make them men of fortune?"

Judge Addison Gardiner asked, "Admitting that this organization . . . may reap a certain and increasing success, is it at all probable that you or I, or any here present,

will live long enough to see your prophecy fulfilled, to reap the benefit of faith in your seership?"

Sibley handed the form to Judge Gardiner who signed for $10,000, followed by George H. Mumford, for $10,000. Don Alonzo Watson signed for $5,000 the next morning. Isaac Butts promised $10,000, and paid it in stock of other lines.

"The $90,000 subscribed was all the money ever paid," Sibley said later. "The balance was money loaned on bonds of the company, and individual loans." He said the property of its successor, the Western Union Telegraph Company, soon exceeded the whole assessed value of the property, real and personal, in the city of Rochester.[3]

The New York and Mississippi Valley Printing Telegraph Company was incorporated April 1, 1851, with capital of $360,000 authorized, but Sibley and Selden had difficulty collecting what was subscribed, and offered a fifty percent stock bonus to those who would pay. Some objected to the advantages held by the House patent owners, who then agreed to transfer stock to the company. Even then, only thirteen subscribers paid $83,000, which permitted construction from Buffalo through Cleveland, Columbus, Dayton, Cincinnati and Frankfort to Louisville but not to St. Louis.

Henry Sayre Potter, a wealthy Rochester merchant, paid $10,000 and was elected president.[4] Medbery was vice president and Isaac R. Elwood, secretary-treasurer. The other directors were: Selden, Sibley, Butts, Judge Addison Gardiner, Freeman Clarke,[5] Gideon W. Burbank, Joseph Hall, George H. Mumford, E. Darwin Smith, Isaac Hills, Samuel Medary, James Chapman, Rufus Keeler, Freeman M. Edson, Sanford J. Smith and Royal E. House.

The company ran $15,000 into debt in

6.2 The room in Reynolds Arcade Building, Rochester, N.Y., where Western Union's first headquarters office was located. Many years later the room was moved, intact, with stovepipe iron hat cuspidor and papers, to the Rochester Museum to preserve it as a memento of the company's great role in American history.

three years in spite of every economy. Its shaky condition is illustrated by the minutes of its executive committee June 26, 1852: "Resolved, that the secretary be authorized to order or purchase two instruments for this line of J. B. Richards if he will furnish them after an explanation of the condition of the line and the financial affairs of the company." However, the directors showed their determination to carry on in January 1854 by filing new articles of association, reducing the capital stock to $170,000, reducing the

patent owners' share to $42,500, and authorizing $15,000 in bonds to pay debts.

By skillful negotiation, Sibley and Anson Stager, superintendent of the line, obtained a contract February 7, 1854, for the Cleveland & Toledo, Michigan Southern, and Northern Indiana railroads to build telegraph lines along their rights of way and link them with the Sibley line at Detroit, Chicago and other places without the payment of any cash. The company was to give $125 of its stock to the railroads for each mile, provide free telegraph service for railroad business, and maintain the lines. The railroads were to transport telegraph people and materials to repair lines.

This railroad contract gave fresh courage to Sibley. He carried a bag filled with all of the cash he could raise, and bought stock in lines he believed the company would acquire, wherever he could find a discouraged owner. He got some for only two cents on the dollar.

One New York and Mississippi Valley stockholder wrote, begging for the return of his subscription money: "If nothing I can say will touch your heart, have pity upon my wife and children."[6] Most of the original subscribers became afraid and sold, but Sibley, Potter, Medbery and Watson bought more.

The company leased the Lake Erie line on March 30, 1854. This T-shaped line, from Buffalo to Detroit and Cleveland to Pittsburgh, had experienced stiff competition from the Erie and Michigan and was bankrupt.[7] The company agreed to pay the dividends on $50,000 of the Lake Erie stock.

Sibley also obtained an exclusive connection to New York, again without money. His group controlled the New York State Printing Telegraph Company, which had tried repeatedly to buy or lease Faxton's strong New York, Albany and Buffalo line. This plan seemed doomed when the American Telegraph Company offered a high percentage on capital to lease Faxton's line. Undaunted, Sibley persuaded Faxton's line on February 15, 1856, to lease the New York State Printing line for ten years at seven percent a year on its $200,000 capital, to agree to buy it on March 1, 1866 for $200,000 in stock, and to give the New York and Mississippi Valley an exclusive connection to New York.

Of the dozen other lines in the Midwest,

two were sold for debt, one assessed its stockholders to the point of confiscation, and others were in such poor condition that their owners offered to sell to the Rochester group. Five lines in which J. J. Speed was involved were abandoned.[8] Consolidations followed rapidly, financed by bond and stock issues for which the New York and Mississippi directors and their friends pledged their personal fortunes and borrowed money.

The strongest opposition was from the Erie and Michigan's 900-mile line from Buffalo to Milwaukee. Controlled by Cornell, Speed, and Wade, it had important connections with lines to the east, and Sibley's Rochester-House patent group proposed to unite with it.

Cornell underrated the House group, believing they would fail, and he could control a consolidation of lines. When the group met with him and O'Reilly at New York to discuss consolidation, Cornell left for Indianapolis, saying he "had not time to remain in New York while their ideas ripened into practical form."[9] His partner, Speed, had lost hope of much gain in the telegraph business and wished to leave it. Cornell tried without success to get an accounting of the finances of Wade's Cincinnati and St. Louis line, in which he and Speed had a large interest.

Hearing of their bad feeling, Sibley asked Speed and Wade to name the price for their interests in the Erie and Michigan, Cleveland & Cincinnati, Cincinnati & St. Louis, and Ohio Telegraph Company lines, and the rights to use Morse methods in six states. They named a price of $50,000.[10] Sibley surprised them by accepting at once, and the agreement was signed on April 29, 1854.[11] Sibley hired Wade at $3,000 a year, and gave him $10,000 in New York and Mississippi Valley stock. Speed received an extra $4,000 for his patent interests.

When Cornell learned of this deal, he was furious, declaring it the "foulest piece of treachery." Writing to Justice Caton, May 5, 1854, Cornell said he had gone to New York and was surprised to find Speed and Wade there. He asked what was going on, and they assured him nothing was.

> Yesterday, I was thunderstruck by information from Speed [he wrote] that he and Wade had sold out their telegraph interest to the House folks. In the operation Speed sold all the Morse patents, my interest as well as his, thus giving the House folks our own tools with which to defeat us. . . . I asked Speed mildly why he sold my interest in the patent. His answer was "To make money, by God!" This is the foulest piece of treachery towards me that I have ever known . . . all parties to the fraud refuse even to let me see the contract.[12]

He said that when he had invited Speed to enter the telegraph business Speed was in debt $10,000. He had furnished expenses and $26,000 to Speed to build the Erie and Michigan line, and had invested as much more in it. To be sold out in this manner now, Cornell wrote, "almost destroys my faith in man."[13]

Cornell fought back with notices to other companies, claiming the House line could not succeed, and urging them not to be trapped into selling. He obtained control of the strategic Southern Michigan line by leasing it. Reid wrote to Cornell June 22, 1854, that when he told the House group he had sold his interest in that line to Cornell, they warned that Cornell would be bankrupt within 12 months, and a judgment for $1,600 against Cornell could be bought for $1,200.

The New York and Mississippi Valley Company issued a circular: "Telegraph Consolidation. Union of House, Morse, Speed and Wade Lines." It said it had purchased "the undisputed and exclusive control of all patents pertaining to telegraphing in the states of Ohio, Indiana, Illinois, Michigan, Wisconsin, Iowa, and the Territory of Minnesota, including connections with Pittsburgh and St. Louis."

The circular blamed the "great confusion," poor service, and low earnings that existed, on the conflicting claims to the Morse patent, competition between House and Morse systems, and "the short, disconnected and poorly built lines, acting without concert and without responsibility beyond their respective limits." This was the first time, it said, that the different patents, interests and lines "could be sufficiently got under one control to enable the proprietors to meet the public's wants." It invited railroads to arrange with its "principal agent," Wade, at Columbus for telegraph lines.

Cornell replied, claiming exclusive rights to the Morse patents in the area.[14] Now on Sibley's team, Wade bought up the claims of Cornell's creditors and began court actions to get Erie and Michigan controlling stock pledged as security for the debts.

Cornell managed to retain control of the Erie and Michigan, but the Sibley group harassed him with a court injunction, rumors about his finances and bribery of his employees. Writing to Caton November 23, 1854, he said he had dismissed Emory Cobb, Eastern Division superintendent, who allowed a suit to go to judgment without even informing him, and had leased the offices of the Erie and Michigan, and Ohio, Indiana and Illinois lines to the House company. He said James Haviland, Western Division superintendent of the Erie and Michigan, and Cobb accepted bribes from Sibley and were trying to ruin their own company.

Worried and ill, Cornell finally agreed to talk in 1855. After months of hard and sometimes angry negotiations, an agreement was reached giving Sibley's company the Erie and Michigan, and Cornell's interest in the Morse licenses for the Midwest. Cornell gained wealth.[15]

The New York and Mississippi Valley increased its capital stock to $500,000, with $350,000 of it going to the Sibley group and $150,000 to Cornell and the other Erie and Michigan stockholders. The agreement, effective November 1, 1855, named the new company The Western Union Telegraph Company.

Ezra Cornell

Descended on his mother's side from the first settlers of Nantucket Island, Massachusetts, and Thomas Cornell, an associate of Roger Williams who came from England to Boston in 1638,[16] Ezra was born in Brooklyn, New York, January 11, 1807. He was the eldest of eleven children of Elijah Cornell, a Quaker pottery maker. In 1818 his family moved to a farm at De Ruyter, forty-three miles northeast of Ithaca, New York.

Ezra attended winter school, cleared heavily timbered land for planting, built split-rail fences, and plowed with oxen. He built a two-story frame house for his father, worked in Syracuse as a carpenter, at Homer, New York making wool carding machinery, and at Ithaca as a carpenter, mechanic, manager of a plaster mill and a flour mill, and as a farmer.[17]

Cornell's entrance into the telegraph industry was pure luck. In 1842 he bought sales rights to a patented plow and to promote it went to see F. O. J. Smith, editor of *The Maine Farmer,* in 1843. Smith hired him to provide the plow and bury the first

line. Often working all day to repair a line and spending the night on a train coach to reach another point, he engaged in several telegraph projects at the same time. In 1849 he said he had built a third of the telegraph lines in the United States.

6.3   Ezra Cornell, a great early telegraph line builder, who merged his company with Western Union, gave the company its name, and made a fortune on W.U. stock with which he founded Cornell University.

Cornell wrote to Kendall in 1851 that Smith never paid a dollar for his services, that he and his family were dependent on the charity of relatives and friends, that he was $15,000 in debt, and that his wife wanted him to give up the telegraph for something that would earn a living.[18] In the early 1850s when a friend asked how he was getting along. Cornell took a quarter from his pocket and said, "There, that will buy my dinner, but where the next is coming from, I don't know."

Stern-faced, tall, thin, and bearded, Cornell was said to bear a striking resemblance to Abraham Lincoln. Beneath the surface was an iron will. In 1854 he was

$40,000 in debt. He had put every cent he could into Erie and Michigan and other telegraph stock. When Western Union needed them, Cornell's holdings had a high value; in 1865 it issued $2,000,000 in stock to Cornell for stock that had cost him about $50,000.[19]

The value of the stock grew as Western Union developed huge earning power. He was the company's largest stockholder, and attended directors' meetings wearing a broadcloth frock coat, a high stiff collar with a black satin stock around it, and a silk hat filled with important papers.

Like Carnegie, Cornell had little education and wanted to aid others to obtain what he had missed. He contributed more than $1,000,000 to establish and build up Cornell University, about $65,000 to establish a public library at Ithaca, and large sums to other causes.

When Cornell University was established, the president of the University of Rochester attacked it because it was nonsectarian. That caused Sibley to go to Cornell's aid. He accepted a trusteeship and gave $150,000 to establish the Sibley College of Mechanical Engineering at Cornell.[20]

The Western Union Telegraph Company

Ezra Cornell insisted that the name of the united company be The Western Union Telegraph Company, to indicate the union of the western lines in one system. He made this the first item in the consolidation agreement of November 1, 1855. The formal consolidation came after an enabling act was passed by the New York legislature April 4, 1856. The incorporators and first directors were as follows: President Potter, Samuel L. Selden, Sibley, Medbery, Butts, Mumford, Clarke, Elwood, Cornell, J. M. Howard, and Alvah Strong.

At the annual meeting July 30, Hiram Sibley's brilliant work was recognized by his election as president. He started at once to rebuild poorly constructed lines, connect them into one network, and consolidate duplicating offices of merged companies. The right to use the less expensive and more adaptable Morse method of operation made rapid expansion easier. The company also used the 1856 invention by A. F. Woodman and Moses G. Farmer of an automatic repeater that strengthened signals and sent them, in either direction, over the next line. Before that, operators had to reverse the instruments at relay points to send in the opposite direction.

In the consolidation, the company gained control of the telegraph in the "West" except the remaining O'Reilly lines. Then with one blow on February 13, 1856, it severed a major link in O'Reilly's chain of National Lines,[21] by subleasing the Ohio and Mississippi line from Louisville to St. Louis, from Joshua N. Alvord, the superintendent. Alvord had called on Western Union in January, and assigned the lease to that company only six days after getting it, for $2,700 annually, plus a job and salary.[22] Alvord's associates felt they had been tricked, but other lines saw the handwriting on the wall, and the Ohio, Indiana and Illinois was leased to Western Union September 22, 1856.

Three held out: The New Orleans and Ohio Telegraph Lessees; Pittsburgh, Cincinnati, and Louisville; and Atlantic and Ohio. The first had reorganized the financially distressed line to New Orleans and, under the leadership of Dr. Norvin Green of

6.4 An example of the first Western Union telegram blank. (Note the "consolidation of" line.)

Louisville, rebuilt it and made it pay.[23] Edward Creighton traveled along the New Orleans route in 1856, wiring messages over the line to Sibley about the route and prices of poles. This convinced the lessees that Western Union was about to build a competing line, and Dr. Green quickly accepted a request by Sibley for a chance to compete for the haul east of Louisville.

When Dr. Green notified the Pittsburgh, Cincinnati, and Louisville of the termination of their exclusive contract, it was evident that they had lost the business, and they leased their line to Western Union June 1, 1856. This left the New Orleans and Ohio company completely at the mercy of Western Union, and Reid sold his lease of the line June 1, 1856, with two and a half years to run, for $200 a month. It was reorganized January 6, 1860 as Southwestern Telegraph Company, with Dr. Green continuing as president.

Now only the Atlantic and Ohio stood firm, and the clever strategy of Wade was needed. The line paralleled the Pennsylvania Railroad from Philadelphia to Pittsburgh, and on February 12, 1856, Wade quietly made a deal with the railroad, which had installed a pole line along its right of way to handle railroad business. He offered to install House printers free if he could add two wires on the railroad's poles. His offer was readily accepted, and the Atlantic and Ohio soon found a competing line operating through the heart of its territory and taking over much of the business. Francis Morris and Robert W. Russell realized their situation was hopeless, but they drove a hard bargain.

Wade had confided his plans to D. B. Brooks of A. and O., but Brooks had "tried to betray us," Wade wrote to Sibley. He was confident he could get Brooks fired for "his base, villainous and foolish treachery."[24] Sibley replied that he had a "message from Morris (written by Russell)" saying the time had passed for an exchange traffic deal. "That old fool Russell," Sibley wrote, "would take advantage of the time to break a contract if the bread of his wife and children was at stake on the fulfillment of it."[25] Sibley's letters indicated he and Morris had reached an agreement which Russell upset. After much maneuvering, Western Union obtained control through an agreement January 9, 1857, merging A. and O. with the Pennsylvania Telegraph Company which Wade had formed to operate the line along the railroad.

This gave Western Union a line of its own to the Atlantic seaboard. The company had depended for its outlet to the east on its connection with the New York, Albany, and Buffalo line, with which it had an exclusive agreement. As Sibley had planned, Western Union acquired the New York, Albany, and Buffalo Company with $600,000 in stock on December 23, 1863.

Caton found it expedient to switch his alliance to the powerful Sibley group. This resulted in three contracts giving Caton the patent rights to build lines along railroads in his Midwest area, and assuring exclusive connections with Western Union.[26]

These deals amazed and angered the Morse patentees. Writing to Kendall, Smith denounced Judge Caton: "It is a reproach to the public morality of his state that such an unprincipled scoundrel should be on the Judicial bench." He threatened suits against Caton and Wade.[27]

From the day Wade and Speed "sold out" Cornell in 1854, Wade was Sibley's right arm, a fellow plotter and schemer, with a knack for using other men. In many letters, Sibley kept Wade informed of his plans and

his estimate of the intelligence and character of others. He asked Wade's advice and aid on many matters.

Each would ask the other to further some scheme, or ask what reply should be made to a letter. Often the two bought stock together, or agreed on plans to trick others. For example, Sibley wrote Wade September 29, 1856 that he was going to make a "confidant" of Reid and tell him Western Union proposed to take all of the business it could in the Atlantic and Ohio area, so that the information would get back to the A. and O. people and alarm them. Soon Sibley had Reid advising his own A. and O. group to make a deal.

6.5 Jeptha H. Wade, pioneer and strategist in forming the early telegraph companies that spread their lines across the Midwest.

Rebuilding and extending lines required all of Western Union's earnings until December 1857, when it paid its first dividend.

Railroad mileage jumped from less than 10,000 to nearly 25,000 in a decade. In the next few decades Western Union built more than 200,000 miles of pole lines on railroad rights of way.

People in small towns did not have to wait until their telegraph business justified an office; railroad stations had a bay window, telegraph set and railroad agent with green eyeshade and black sleeve covering that became synonymous with Western Union service in thousands of small towns. People began to realize that the telegraph shot the red blood of outside trade and commerce into local business veins, and growth in manufacturing and other business followed.

In 1856 the 35,000 miles of telegraph line in the United States carried 17,136,000 messages for about $7,000,000. Union of the Midwestern lines under single management resulted in unprecedented speed and reliability of service. As the public gained confidence in the ability of the telegraph to send messages and get quick answers, business grew.

A Final Word about Henry O'Reilly

When O'Reilly's alliance of lines was smashed, he was left penniless. As our defeated Napoleon leaves the stage, it should be remembered that he was the greatest pioneer line builder. His temperament got him into bitter conflicts, but he did a herculean job. As O'Reilly wrote to Hugh Downing about the Morse patentees in June 1847:

What a world of trouble would have been saved, if those gentlemen had been willing to settle even thus much "before they began" to make all this trouble.

He once summarized his work:

I more than fulfilled all that I originally proposed, by constructing and organizing the Great Telegraph Range of about 8,000 miles whereby the different sections of the United States were connected with electric intercourse—the longest and widest Telegraph Range then in the world—at a time when not a solitary capitalist in this great commercial metropolis, not even the rich kindred of Professor Morse, would risk a single dollar in extending the lightning from Baltimore.

For all of this, O'Reilly testified in July 1879, he never received enough "to pay me the wages of a responsible clerk."

In one case (F. O. J. Smith vs. Ezra Cornell) E. W. Chester, counsel for Cornell, said,

A better temper and wise counseling would have saved millions to the Patentees and to those engaged in building telegraph lines.

In another case in 1853 Chester said,

Before the 13th day of June 1845, the owners of the Morse patent rights had so far failed to secure the public confidence, that they were unable to raise the funds for any important line. The patent was dead property in their hands. Not a dollar of profit had been realized from it, and the future promised as little as the past had realized.

On that day, Henry O'Reilly made a contract with the patentees. . . . O'Reilly . . . gave the first effective impulse to telegraphing to this country. . . . From the moment that O'Reilly had a prospect of success—from the day when he had made the telegraph to be felt as a commercial and social necessity in the country, these patentees commenced a war upon him. They demanded, instead of one-fourth, one-half the capital—one-half the stocks—one-half the profits. . . . They left no plan untried to break the contract, and to deprive him of his rights under it. He was persecuted in the courts, pursued through the newspapers, met at every step by threatening letters.

Chester said Cornell, Speed, Wade, and others "had been used by the heartless (Morse) patentees as mere tools" and the patentees were as ready to "strip these, their own overzealous partisans of all that was left them in the fight, as they had been to wrest the very bread from the mouth of O'Reilly."

O'Reilly was acclaimed by the press as a champion of the people, fighting to protect the common man from the "special interests." What the press really backed was competition, hoping it would lower their costs. O'Reilly's empire was built on financial sand—increasing debt as each line was constructed with money borrowed on preceding ones—and it fell in ruins.

In 1869 Alonzo B. Cornell, son of Ezra and Collector of the Port of New York, hired O'Reilly as a storekeeper. The *New York Mail,* January 9, 1879, said

> Among the Custom House Storekeepers recently dismissed was Henry O'Reilly, a man of 70 years [an error—he was born February 6, 1806], who at one time seemingly had in his grasp the control of the telegraphs of the country.

In 1884 O'Reilly returned to Rochester, where he died on August 17, 1886.

It was a dog-eat-dog period in which each promoter fought for survival. O'Reilly, Cornell, Sibley and others fought to accomplish what they did, and helped to build a nation.

The American Telegraph Company

On May 8, 1854, Cyrus W. Field, Peter Cooper, Moses Taylor, Marshall O. Roberts, and Chandler White organized the New York, Newfoundland & London Electric Telegraph Company to lay a cable across the Atlantic.

Morse was made honorary electrician to use the prestige of his name. To the amazement of his patent partners, Morse gave free use of his share in the Morse patent for a landline to New York, subscribed $10,000 of the stock, and asked the telegraph companies to handle cable traffic at half their usual rates. Deciding that Morse wanted them to cut their rates for his own profit in the cable company, the companies refused.

Since a landline to Newfoundland would be necessary, the Field group organized the American Telegraph Company on November 1, 1855, to establish it. Peter Cooper was president; Hiram O. Alden, vice president; and James Eddy, treasurer and general superintendent. Offers were made to lease the principal seaboard lines as far south as New Orleans. Morse and Kendall asked the companies to accept, and some agreed. Then Field withdrew the offers because, he said, they had not been accepted gladly. He blamed Smith.[28]

Morse believed in Field, little suspecting that behind the scenes was Daniel H. Craig of the Associated Press, who had located another telegraph inventor and was preparing a series of attacks on Morse in the press. It was announced on the day the American company was organized that the Field group had bought a printing telegraph system being developed by David E. Hughes, a music teacher in Kentucky. Hughes did not use the House printer's step-by-step method, but a continuously revolving type wheel that printed 130 letters a minute. Spring governors, tuned to the same number of vibrations per second, ran synchronously, and the impulses they transmitted caused a type wheel to print the letters that were sent.

Field and Cooper assured Morse they agreed to buy the Hughes patent just to keep it out of the hands of the House group. Morse swallowed that, writing Kendall that all was well, but the "old Fox" said the Field group would hold Hughes "in terrorem" over their heads to reduce the price of their lines.[29] That was what happened.

In 1856 American leased the New Brunswick Electric Telegraph Company line to Calais, the Maine company to Boston, the House line to New York, and several branch lines. It built a line from New York to Philadelphia, competing with the Morse and House lines, which refused reduced offers from the Field group. Field then threatened to build on to Washington, and the Magnetic company offered to lease itself for 25 years at eight percent, but when this offer was refused, Magnetic itself leased the Washington-New Orleans line on July 7, 1856.

Upset by American's ambitious plans, the companies met July 1, 1857, to offer the Field group an agreement to interchange traffic. Kendall, as chairman, presented an agreement under which participating companies would connect exclusively with the others, and work to merge all lines into one company. Field demanded that the other companies first join American in buying the Hughes patent. Kendall, of course, replied that Magnetic had no use for Hughes. The meeting was then adjourned until those present could consult with their companies.[30]

Without letting Kendall know, Sibley met with representatives of the other lines.[31] This group drew up an agreement, signed August 10, 1857, that became known as the "Treaty of Six Nations." Ignoring Kendall's plan, they pledged to join American in paying $56,000 for exclusive use of the Hughes patent.

Under the "treaty," American received the Atlantic seaboard from Newfoundland to Florida, and a line to New Orleans; and Western Union, the states north of the Ohio River and parts of Iowa, Minnesota, Missouri and Kansas. The New York, Albany, and Buffalo, got New York State; the Atlantic and Ohio received Pennsylvania; and the Illinois and Mississippi, the upper Mississippi Valley (parts of Illinois, Iowa and Missouri). The New Orleans and Ohio Lessees were given the lower Mississippi Valley and the Southwest. Each was to exchange traffic exclusively with the others, shutting out all competition.

The "treaty" also provided for getting Morse, the "Old Fox" and "Fog" out of the telegraph business. Kendall's Magnetic, and Smith's New York and New England Union were left out and declared war. Smith extended his line from Boston to Portland, and even threatened to build through Nova Scotia and back a rival cable project via Greenland, Iceland and the Faroe Islands to Europe. Kendall leased the Western Telegraph Company and extended that line along the Ohio River Valley to Cincinnati. To punish Dr. Norvin Green and his New Orleans and Ohio Lessees for what he considered his double cross in the Six Nations deal, Kendall also planned a line from Mobile to Memphis, to connect with a line to Illinois.

The agressive steps by Kendall and Smith proved to leaders of the Six Nations that they could not force the original Morse companies out of business. They held another meeting on October 20, 1858, and changed their name to North American Telegraph Association.

Purchase of the Morse and Smith interests presented a tough problem. Every man hated some of the others. Lengthy meetings followed in which Smith held out for $300,000. The others called it "blackmail" and "exorbitant." Sibley wrote from the Astor House to Wade December 16, 1858:

> Smith has been here for two or three days cracking his whip, and he has succeeded in getting the parties here willing to pay his price $300,000 if he will sell all his property, but in making out the papers he reserves so large a portion that they broke up. . . . Smith left this evening in the boat mad because he could not sell his telegraph interests and keep them too. He will return better natured on a little reflection and close the contract unless he is a bigger fool than I think he is.

It was nearly a year before an agreement was reached on October 12, 1859. Smith sold his telegraph stock and patent rights for $301,108.50.[32] Morse, Kendall, and their partners received $107,000 in American stock for their patent rights.[33]

To finance its rapid expansion, American reorganized in 1859, with $740,000 capital stock and the right to increase it to $2 million. It exchanged $400,000 of its stock for $369,300 of Magnetic stock, and issued stock for that still outstanding in Smith's New York and New England. It took over Magnetic's lease of the New Orleans line, leased Western Telegraph, and leased or absorbed other lines as far south as Georgia.

In 1860, the American Telegraph Company controlled the entire Atlantic seaboard. On its board were Kendall, Morse, Field, Abram S. Hewitt, Wilson C. Hunt, Zenas Barnum, Swain, Russell, Morris, and other leaders of the industry.

Notes

[1]From a thesis submitted by Harper Sibley, Jr., a great-grandson of Hiram Sibley, to the History Department of Princeton University, in January 1949, and loaned to me by him.

Hiram Sibley, son of Benjamin, a farmer, and Zilpha Davis Sibley of North Adams, Mass., left school at age ten and did odd jobs to augment the family income. At 16 Hiram started walking west. He worked in a machine shop at Lima, and as a wool carder at Sparta and Mount Morris, New York. At the latter place he worked in the same shop with Millard Fillmore, later the 13th President of the United States. In partnership with Don Alonzo Watson, a machinist, he set up a shop making machinery for woolen mills at Mendon, New York, and was so successful that the area was named Sibleyville in 1833. Sibley also started a flour mill and farm, and was elected sheriff in 1843. He and Watson sold out and moved to Rochester in 1844.

[2]The *Rochester Union American,* November 29, 1874, said Butts was born in Dutchess County, January 11, 1816, and started in the newspaper business in 1844 on the *Rochester Democrat American.* At one time and another he owned or had a managing interest in that paper and the *Union* until his retirement in 1864. By age 30 he also owned the *Daily Advertiser.* The *Rochester Daily Democrat* of April 27, 1866 quoted an article from the *Buffalo Courier* saying Butts was a dealer in patent medicines and sold out for $25,000, which he invested in Western Union and made $1,500,000. He died at Rochester November 20, 1874.

[3]Mrs. Jane Marsh Parker, *How Men of Rochester Saved the Telegraph* (Rochester Historical Society Publication Fund Series, 1926) vol. 5.

[4]Potter's father was the first settler at Seneca Falls in 1801. Potter was born at Galway, Saratoga County, N.Y., February 11, 1798.

[5]Clarke was born at Troy, N.Y., March 22, 1809. He moved to Rochester in 1845 and organized and was president of the Rochester Bank.

In 1857 he organized and was president of the Monroe County Bank, subsequently the Clarke National Bank. President Lincoln appointed him Comptroller of the Currency in 1865, and during his incumbency legislation was enacted for the organization of national banks. He also served two terms in Congress.

[6]Parker, *How Men of Rochester Saved the Telegraph.*

[7]In his autobiography, Alvah Strong said: "The Lake Erie telegraph line went into bankruptcy and S. P. Ely and myself bid the line in at sheriff's sale."

[8]A deposition by Speed in 1856.

[9]Cornell letter to Speed, April 17, 1853.

[10]In his autobiographical letter Wade indicated that he had promoted the deal: "To make this sale was a long tedious job. To get them to put $60,000 into the business where they had already put so much and were so heartily sick of it, was no easy job, but it was finally accomplished, and was the first step in the way of consolidating telegraph lines, and formed the nucleus around which was built the mammoth Western Union Telegraph Company."

[11]The agreement is among the O'Reilly Papers, New York Historical Society Library.

[12]Cornell Papers, Cornell University Archives.

[13]Caton Papers, June 1854. Library of Congress.

[14]Cornell circular in Caton Papers also.

[15]In his manuscript autobiography, Alvah Strong said, "A bitter fight was kept up until both ourselves and Cornell were nearly exhausted. Yet Cornell, by his persistent bluff and pluck, beat us in the end."

[16]*Pre-Cornell and Early Cornell*, a genealogical study by Albert Hazen Wright, 1965.

[17]Morris Bishop, *A History of Cornell* (Ithaca NY: Cornell University Press, 1962).

[18]Carl Becker, *Cornell University: Founders and the Founding Fathers* (Ithaca NY: Cornell

University Press, 1943).

[19]James D. Reid, *The Telegraph In America* (New York: John Polhemus, 1886) 470.

[20]A large, domed building at Cornell University today bears the inscription: "Sibley College. In honor of Hiram Sibley. Trustee of Cornell University from 1865 to 1888. By whose generosity this college was founded and very largely equipped. Erected 1870." Nearby, on the campus heights overlooking the blue waters of Cayuga Lake, is a bronze statue of Cornell with a bronze replica of the instrument with which Morse sent the first telegram in 1844. The historic instrument itself is a treasured possession of the university, whose library provided access to Cornell's papers.

[21]Several O'Reilly lines had James D. Reid as their superintendent, exchanged business with one another, and made agreements with other lines. This loose federation was called "National Lines."

[22]Letter from Elwood to G. K. McGunnigle, president of the Ohio and Mississippi line.

[23]See above, chap. 4.

[24]Wade Papers. Wade to Sibley, September 18, 1856.

[25]Wade Papers. Sibley to Wade September 20, 1856.

[26]O'Reilly Papers, 1855 and 1856. New York Historical Society Library.

[27]Smith to Kendall, December 4, 1857. Smith Papers, New York Public Library.

[28]Morse to Kendall, September 25, 1855.

[29]Morse Papers. Morse to Kendall, December 7, 1855. Kendall to Morse, same date.

[30]Smith Papers. Proceedings, June 30 and July 1, 1857. Cornell Papers. George Curtiss to Cornell, July 4, 1857.

[31]Wade Papers. Sibley to Wade, June 20, 1857.

[32]American's report to stockholders August 20, 1860, said the other companies paid $129,000 of it.

[33]Proceedings of the North American Telegraph Association, August 31, to October 12, 1859.

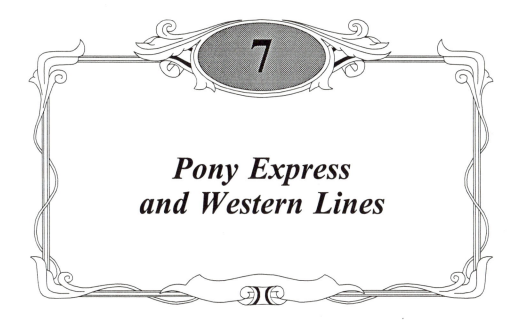

Pony Express and Western Lines

Three names—Covered Wagon, Pony Express, and Transcontinental Telegraph—symbolize two thrilling decades of American history, 1841 to 1861. It was a period of great expansion in which the West was joined to the nation.

In 1830, the first caravan of covered wagons left a little frontier town—named St. Louis for a French king—and was guided to the Rocky Mountains by W. L. Sublette.

In 1840 the nation's population was 17,069,453, but only 150 Americans were in the Northwest. In the next six years more parties went to Oregon and California. They told tales of hardship and horror, thirst in the desert, starvation in the mountains, and attacks by Indians.

The Oregon area, extending to the present boundary of Alaska, (54°40′ latitude, north) was in dispute between Great Britain and the United States. The slogan "Fifty-four forty or fight," swept James K. Polk, the great expansionist president, into office. Because of the rapid movement of settlers

into the area, the British agreed in 1846 to set the boundary at the 49th parallel. The covered wagon had helped to win what is now Oregon, Washington, and Idaho for the United States. In 1848 Mexico ceded the areas that now are California, New Mexico, Nevada, Utah, most of Arizona, and part of Colorado and Wyoming, for $15,000,000.

Most pioneers gathered in tent cities along the Missouri River in winter. They waited until the grass was high enough to feed the animals, at least part of the way, before setting out with their oxen and their Conestoga wagons. Usually with six or eight yoke of oxen to a wagon, they formed parties, not so large as to need too much grass, but large enough to fight off Indians. One wagon would be suicide. Strong parties made friends of the Indians with gifts of clothing, mirrors, and sugar. The Indians loved sugar, and enjoyed dressing in high fashion, with the seats of trousers they received cut out for comfort. They kept umbrellas safely under blankets when it rained.

The procession of white-topped prairie schooners stretched out hundreds of miles, and a pall of dust hung over the deeply cut trail. Weary men and women walked beside plodding oxen. Furniture was abandoned to make the going easier. Oxen died, wagons broke down, Asiatic cholera and mountain fever took a sudden and deadly toll, and relatives could only pause long enough for brief services at shallow graves beside the trail. Water and food supplies did not permit lingering; they had to "live off of the country." Yet, in spite of all hardships, they carried on, through sunshine and rain, singing and praying, dancing around campfires with rifles ready a few yards away. Night guards were posted. Marriages were performed and babies were born. From 10,000 to 20,000 died from the hardships of the trail and Indian arrows, and their graves marked the way.

In California the pioneers found pastoral scenes in which small colonies of Indians and Spaniards tilled patches of ground and lived in quiet, unhurried peace. Then came the hustling people from the East, hungry for letters from loved ones and friends back home. Mail arrived monthly, by ship to the Isthmus of Panama, across the Isthmus on muleback and up the coast by ship again, or by ship around Cape Horn. When the mail reached San Francisco, a village of 700 people, months-old newspapers were bought eagerly for a dollar.

Gold was discovered in California on January 24, 1848 by James W. Marshall, constructing a saw mill for John Sutter at Coloma on the American River. The cry of "Gold!" sparked a mad rush by men who left farms half-planted, houses half-built, and stores closed.

To show the value of the lands wrested from Mexico, President Polk confirmed the rich gold find in his annual message to Congress December 5, 1848, and the spring of 1849 saw 20,000 persons on the trail. The population of California jumped to 300,000 in 1850.

In 1849 Alexander H. Todd found miners so anxious for mail that he obtained their letters at San Francisco and collected four dollars a letter on delivery. He also sold old New York newspapers at eight dollars apiece and carried gold to San Francisco for deposit. Soon express companies were organized to haul gold, and many were absorbed by Wells, Fargo & Co., incorporated in 1852 by Henry Wells, Johnston Livingston, W. G. Fargo, A. Reynolds, and E. B. Morgan.

Suddenly San Francisco became one of the world's busiest ports. One Forty-Niner wrote, "Outside of hell, it is the dirtiest, meanest city in the world." In the rainy season the streets were deep in mud. One sign read, "This street is impassable, not even jackassable." About half of the houses were of canvas. A twenty-year-old New Yorker named Levi Strauss arrived at San Francisco in 1850 and began making the rugged blue denim pants miners called "Levis."

Ships entering the Golden Gate were seen from Telegraph Hill, where a semaphore tower was built in September 1849. To see ships approaching sooner, however, a semaphore tower was built on Point Lobos in 1852, and its signals were duplicated on Telegraph Hill, eight miles away. When the semaphore arms signalled the approach of a mail ship, people rushed to the post office and stood in line.[1] Because of frequent fogs, a telegraph line was built from Lobos to Telegraph Hill September 22, 1853—the first line in California. This led to the establishment downtown of the Merchants Exchange, with reading and rest rooms. Then the line was run direct from Lobos to the Exchange.

7.1 One of the Pony Express stations where riders changed to fresh horses. The saddle cover (mochila) being placed on the horse's back had two small pouches in which the telegrams were carried.

Mail to California

Major George Chorpenning, Jr. and Absalom Woodward obtained a $14,000 contract in 1851 to haul mail 750 miles on muleback between Sacramento and Salt Lake City. Woodward and a party of four were killed by Indians in the Humboldt Valley in November.

In the 1850s people were so convinced that the Great American Desert was a second Sahara that Congress voted to purchase seventy camels as mounts for Army couriers and fighters in the southwest. In 1856, thirty-three camels and eleven Arab drivers were imported, landing at Indianola, Texas.

Forty-one camels arrived on a second voyage. The rough trails of the rocky American deserts cut the soft pads of their feet, however, and they were of little use. The War Department abandoned the experiment and sold some in 1864 for a caravan route from salt fields near Columbus to the miners around Virginia City, Nevada. Others wandered off into the desert. People reported seeing camels in the desert years later—and swore off drinking! At Quartzsite, Arizona, a monument was erected in 1935, surmounted by the copper figure of a camel, in memory of the leader of the drivers, Hadji Ali, called "Hi Jolly" by the soldiers.

Because of Southern control of Congress, the government gave the Southern

Overland Mail, organized by Fargo and John Butterfield,[2] a $600,000-a-year contract for a 2,760-mile mail route from St. Louis to Memphis, across Oklahoma and Texas to Los Angeles and San Francisco. The first stagecoach left St. Louis September 15, 1858, and reached San Francisco in twenty-four days and eighteen hours. The first from San Francisco reached St. Louis October 9, 1858, in twenty-three days and four hours.

In 1858 Congress gave Major Chorpenning a subsidy to establish the first central mail route by stage between Salt Lake City and California. That "jackass express" had numerous delays and failures because of snow in winter. Lewis Brady & Co. ran it east of Salt Lake City.

The firm of Russell, Majors, and Waddell had a large overland freighting business, hauling supplies principally to military garrisons in "prairie schooners." It had thousands of men and 75,000 oxen. When gold was found in the Pike's Peak region in 1848, William H. Russell established a stagecoach line from Leavenworth, Kansas, to Denver. It did not pay, so the firm bought the Hockaday and Liggett stage and Mail line, with a government mail contract on the Central Overland route between Fort Kearney and Salt Lake City. Russell established stations with fresh Kentucky mules on the Central route. The trip from Fort Kearney to Salt Lake City was made in ten days, an average of 120 miles a day!

The line was still not profitable, so Russell went to Washington for aid. Senator Gwin of California and other western Senators told Russell that if his firm would establish a fast messenger service to the Pacific Coast, they would use their influence to obtain a government subsidy. Relying on their promises, Russell established the Pony Express,[3] a twenty-four-hour-a-day race of horses and riders in relays, from St. Joseph, Missouri to California.

The Pony Express

The Pony Express had 190 stations, with 500 horses, and eighty daring, light, and wiry young riders."[4] A devil-may-care, proud and swaggering lot they were.

The stage and mail route was followed to Salt Lake City. Then it wound through a vast desert region of sagebrush-covered valleys, low hills and "badlands" in Utah and Nevada, through Cold Springs and Carson City, and on through the Sierra Nevada Mountains. Sacramento was the western end of the run; from there the messages went down the Sacramento River by steamer to San Francisco.

Stations were ten to twenty-five miles apart. The rider would take a fresh horse at each station, and gallop to three or more stations, pausing only to carry to another horse his Spanish mochila (a heavy-leather saddle cover with pockets in the corners in which the messages were locked). Then he would dash on with thundering hoofbeats, eating up his part of the 1,966[5] miles to Sacramento.

The first westbound packet of telegrams left St. Joseph, Missouri on April 3, 1860. St. Joe was in holiday attire; flags floated in the breeze, and the band played. "Mail" from the East for the first run was late reaching Hannibal, and the new railroad set out to make a speed record and win a contract. Engineer "Ad" Clark pulled his throttle on the wood-burning engine wide open, and the swaying little train sped over the frail roadway at a death-defying pace while white-faced passengers prayed. It set a re-

7.2 WESTERN UNION BUILDING, 1015 SECOND STREET, SACRAMENTO, WHICH WAS ORIGINALLY THE WESTERN TERMINUS OF THE PONY EXPRESS RIDERS. PHOTO SHOWS THE UNVEILING OF A PLAQUE PLACED ON THE BUILDING BY THE DAUGHTERS OF THE AMERICAN REVOLUTION, MARKING IT AS THE "SITE OF TERMINAL OF PONY EXPRESS."

cord of 206 miles to St. Joseph in four hours and fifty-one minutes.

When the train arrived, the speeches had been delivered; Billy Richardson,[6] the rider, and his horse were ready. In a minute the rapid beat of the horse's hoofs sounded down the street. A ferry took horse and rider across the Missouri River, and the world's longest horse race was on.

That first run was in rain and storm, yet the news dispatches reached Salt Lake City amid great rejoicing in six days. In spite of a blizzard in the Sierras, they reached Sacramento in four more days. When the Pony Express arrived on April 13, 1860, using 40 riders in a little more than 10 days, all Sacramento turned out to celebrate. A band played, cannons boomed, church bells rang, and floral arches were built. That night, when William Hamilton, a rider, reached San Francisco on the sidewheel steamboat, a band played "See, The Conquering Hero Comes." A parade was formed, with fire apparatus, Hamilton and his pony, and a large part of the population in line.

The race of the first eastward Pony Express also began April 3, when James Randall and his pony left San Francisco with the messages on the river boat, arriving at Sacramento about 2:00 a.m. The rider and pony were taken along just to dramatize the occasion. (After the first trip, the pouch traveled on the boat without horse or rider.) Another rider, Harry Roff,[7] then galloped from Sacramento to Folsom and Placerville at the foot of the Sierras. Others went on through a howling blizzard and snow drifts in the mountains and heavy rains in Nevada and Utah, reaching Salt Lake in four days.

The eastbound Pony Express reached St. Joe at 3:55 p.m. on April 13, covering the entire distance in nine days, fifteen hours, twenty-five minutes—faster than the ten-day westward trip.

7.3 Plaque placed by the Daughters of the American Revolution on the Western Union Building in Sacramento, marking it as the western terminal for the Pony Express.

It is doubtful that history has produced anything to match the glamour of the Pony Express, or any race with more daring riders than "Buffalo Bill" Cody,[8] "Pony Bob"

Haslam, "Wild Bill" Hickok, Bill Fisher, Nick Wilson, "Major" Howard R. Egan,[9] Jack Keetley, Charley Cliff, Bill Streeper, Jim Moore, Jay G. Kelly, William Campbell, John Frye, Kim Beatley, and others. Alexander Majors distributed Bibles to all riders and made them swear "Before the Great and Living God" not to drink, curse, or brawl. With Czar-like powers, Bolivar Roberts, superintendant of the Western Division, held them to their oath.

7.4 James Butler ("Wild Bill") Hickok (1837–1876), Pony Express rider.

"We tried to put it through," said Billy Fisher, "rain or shine, night or day, Injuns or no Injuns."

And they did "put it through," racing continuously for eighteen months over 1,966 miles of plains, deserts, and mountains. The howl of the prairie wolf, the yelp of the coyote, the roar of the mountain lion, or the moan of the panther merely quickened the tempo of the horse's flying hoofs in the night, and the yell of the Indian was the signal for even greater speed.

7.5 WILLIAM CAMPBELL, LAST PONY EXPRESS RIDER SURVIVING IN 1932, WHEN HE WAS 95 YEARS OLD, TOLD THE STORY OF HIS AND OTHER RIDERS' EXPERIENCES.

They commanded the nation's admiration and became a symbol for heroic endeavor and achievement. In the fastest trip of the Pony Express, the news of Abraham Lincoln's election was carried from the telegraph office at St. Joe to the telegraph office at the eastern foot of the Sierras in six days and seventeen hours. When Mark Twain went west by stagecoach, a Pony Express rider galloped past, and he called the pony rider the national hero of 1860.

The last Pony Express rider alive in 1932, William Campbell, ninety-five years old, of Stockton, California, told me:

I was a bullwhacker, hauling provisions and military supplies by wagon train to forts in the West in the spring of 1860, when Russell, Majors, and Waddell decided to establish the Pony Express. Then I was sent to take supplies to stations of the Pony Express, which ran from the end of the Western Union line at St. Joe.

It was December, 1860, before I had my chance to ride. I was six feet tall, weighed 140 pounds, and was too large, but many riders could not stand the grind and more were needed. My relay was between Valley Station, eleven miles east of Fort Kearney, and Box Elder Station, three miles west of Fort McPherson. This was 95 to 100 miles along the Platt River, and my first ride was in a heavy snow storm.

At Fort Kearney, then at the western end of the telegraph line, I stopped to pick up telegrams that traveled the rest of the way west by Pony Express. Richard Ellsworth, operator of the Western Union office at Fort Kearney, usually had coffee on hand. It was just before the Civil War, and he told us the news.

Once I spent 24 hours in the saddle, 120 miles to Fairfield with snow two or three feet deep and the mercury around zero. I could tell where the trail lay only by watching the tall weeds on either side and often had to get off and lead my gray mare. There was no rider to go on at Fort Kearney, so I went on to Fairfield twenty miles away. As the telegraph line was built, the gap the pony ran was shortened. When it was completed, the Pony Express went out of existence.

There were more than 200 riders in all, for few could last long under the strain. "Pony Bob" Haslam rode 380 miles in thirty-six hours during an uprising of Paiute Indians in 1860, in which stations were destroyed, men were killed, and service was disrupted between Carson City and Salt Lake City. "Buffalo Bill" Cody rode 322 miles once when his relief rider was killed. He was chased by Indians on various occasions. Jim Moore rode 280 miles in twenty-two hours when his relief rider was ill.

The Paiutes were the most troublesome in the West; the Sioux, in the East. Only a few soldiers were stationed at Fort Laramie and Fort Bridger, and it was remarkable that a lone rider usually could get through during Indian uprisings.

Although the demand for it was strong, the Pony Express was a financial failure. Its cost was $700,000, including $100,000 to equip it, $30,000 for sixteen months maintenance, and $75,000 lost from Indian raids. Receipts were about $500,000 paid to the Pony Express by the telegraph companies.

Writers called the Pony Express a carrier of the U.S. Mail, but mail was carried in large bags on pack mules or piled on stagecoaches. Because of the cost, only messages of telegraphic importance were sent by Pony Express. Only a few communications, written on tissue paper, could be carried in the two oiled, silk pouches, the size of a writing tablet.

Many people did not realize it, but the Pony Express was, in effect, a part of the telegraph line. It filled the gap between the eastern and western lines, shortened its run as the transcontinental line was built, and ended when the line was completed.

Russell depended on the money-losing Pony Express[10] to gain mail contracts for his firm, but Washington took no action. Repeated appeals to Washington for relief finally resulted in Secretary of War Floyd lending Russell $830,000 in Indian Trust Fund bonds, with Russell leaving as security

acceptances he had received from the War Department. This was discovered after Russell borrowed money on the bonds from a bank, and a national scandal resulted in December 1860. Russell made the amount good at once, disclaimed any intent to defraud, and was not prosecuted; but the affair ruined Russell's chance of getting the contract.

7.6   DASHING BEN HOLLADAY (1819–1887), "NAPOLEON OF THE PLAINS," SHOWMAN OF THE PONY EXPRESS AND STAGECOACH BUSINESS.

Congress finally acted March 2, 1861, but gave the $1,000,000 contract to Butterfield-Wells Fargo, which organized the Overland Mail Company. The contract required the company "until the completion of the overland telegraph, to run a pony express semiweekly." The Pony Express and stagemail was just what Russell had sought so long without success.

7.7 Pony Express rider waving to workers constructing the first transcontinental telegraph line in 1861.

Newspaper and magazine articles told how Ben Holladay, in charge east of Salt Lake City, dashed up and down the route at top speed in a luxurious private coach in which he slept. At one time he operated nearly 5,000 miles of stage lines, 500 freight wagons, and sixteen steamships plying the Pacific.[11] He bought new horses, hired more men, and gave new life to the Pony Express. However, mail was still stuffed into large bags piled with passengers into overloaded stagecoaches, and harassed drivers often "accidentally" left bags behind when coaches turned over or were stuck in mud holes.

Harry Peterson, curator of Sutter's Fort, a pioneer village near Sacramento, said:

The Pony Express was a failure as a business venture, but one service alone justified its existence: James McClatchy, founder of the *Sacramento Bee,* relayed a fast dispatch to Washington that saved California to the Union in the Civil War.

McClatchy learned General Albert Sydney Johnson, later a brilliant Confederate General, planned to turn the Benicia Arsenal over to the rebel cause, if and when war came. He sent a message by Pony Express to Senator E. D. Baker, informing him of the plot. Baker told President Lincoln, and General Johnson was relieved of his post. Had the plot not been exposed, General Johnson and his Southern sympathizers could have controlled California and her gold output, which largely financed the war for the North.

Early California Lines

Between 1853 and 1860 several telegraph lines were built in California. The California Telegraph Company was organized by Oliver E. Allen and Clark Burnham of New York, with a franchise from the state legislature to establish a line from San Francisco to San Jose, Stockton, Sacramento, and Marysville, but its plans went up in smoke in the San Francisco fire of 1852.[12] Since the franchise was to expire unless the line was in operation by November 1, 1853, a new group organized the California State Telegraph Company. John Middleton was president and E. R. Carpentier, secretary.

The story of the race against the expiration date was told in "The Californian Magazine" in 1881 by the line's builder, James Gamble, a newspaperman and Mississippi River "steamboater," formerly with the Illinois and Mississippi Telegraph Company. Gamble built the line with five men in September and October, 1853, bewildering the simple Mexican people at San Jose,[13] and reaching Marysville on October 26, after six hectic weeks.

Connecting with this line at Sacramento was the Alta Telegraph Company, managed by J.E. Strong, which began operating 121 miles in January 1854 into Mother Lode gold-mining country at Mormon Island, Diamond Springs, Placerville, Coloma, Auburn, Grass Valley and Nevada City. Demand for service was so strong that the line was extended to San Francisco, linking Santa Clara, San Jose, Martinez, Benicia, Vallejo, Folsom, and El Dorado. Gamble laid cables for the Alta line across the bay from Oakland to San Francisco, and in the Straits at Benicia. They failed, and he built the line around the bay through San Jose,

trespassing on California State line franchises. After years of litigation the two companies merged.

In 1856 the Northern California Telegraph Company, formed by I. M. Hubbard and J. E. Strong, built north from Marysville to Yreka (Eureka), a mining center half way between San Francisco and Portland.

Frank Bell, later governor of Nevada, built the pioneer line over the Sierras[14] for the Placerville, Humboldt, and Salt Lake Telegraph Co., organized by Frederic A. and Albert W. Bee in 1858. The line, called the "Bee Line," was strung on trees, and teamsters often took the wire to mend the harness of horses and tie loads on wagons.[15]

The California legislature in April 1859 offered a subsidy of $6,000 a year to the first line connecting with one to the East. With that in mind, the Bees extended their line from Genoa to Carson City and Virginia City in 1860.

Henry Comstock, a trapper and fur trader, discovered in 1859 that the "blue rocks" the miners tossed aside were almost pure silver. That caused a rush to Nevada, and in the next 30 years over $300,000,000 in silver and gold came from the Comstock lode. These mining towns were wild, gaudy, ripsnorting, and profane; fortunes were made and lost at saloon gambling tables.[16]

The Placerville and Humboldt wires to the mining towns handled 1,000 messages a day. Many were in simple code from the Ophir, Bonanza, Consolidated Virginia and other mines—but it soon became evident that contents of messages were "leaking" to a stockbroker in San Francisco. Several operators were discharged, but the leak continued.

When a young Canadian operator, Johnny Skae, whose salary was $115 a month, was discharged, the kind Virginia City man-

7.8 This scene from the motion picture *Western Union,* starring Barbara Stanwyck and Joel McCrea, shows the type of equipment used in the early days of the telegraph in the West.

ager, George "Graphy" Senf, offered him stage fare to San Francisco. Johnny thanked the manager, and exhibited his bank book with a balance of $960,000. He went to San Francisco and lost it as a mining speculator.

Jeff Hayes, a pioneer telegrapher, told in his *Tales of the Sierras* of Sun Lee, a Chinese operator at Bodie, and Miss Minnie Lee, the operator at Genoa, who longed for romance. "S. Lee" and "M. Lee," exchanged many poetic phrases when the line was not busy.

Minnie read of a marriage by telegraph and wired the item to Sun, who suggested they do likewise. A minister was called at each end of the line, and the ceremony was performed. The preacher at Bodie told a reporter, and Minnie learned in print that she was the wife of an Oriental. Her brother started for Bodie armed to the teeth, but Sun had gone. No one had heard of Sun since, Mr. Hayes said, but there was a suspicious sign in San Francisco's Chinatown reading: "Telegraph Laundry, Sun Lee, Manager."[17]

O'Reilly, Speed, and Shaffner agreed on February 25, 1858 to pay Morse and Vail $7,500 and Smith $2,500 for the patent rights for a line to Salt Lake City. Senator

Broderick of California then introduced an act to pay $70,000 a year for ten years for a line to his state. His rivalry with Senator Gwin of the same state, and opposition by Southern senators, defeated that bill.[18]

Planning
the Transcontinental Telegraph

Hiram Sibley, president of Western Union, asked his directors in 1857 for approval to build a transcontinental line. They decided, however, that the company was not ready to undertake such a costly and hazardous project.

"Gentlemen, if you won't join hands with me in this thing, I'll go it alone!" Sibley declared.

The board then decided to invite the other lines to join in forming a new company for the purpose, and Sibley presented the plan to the North American Telegraph Association. Not one of the other companies was willing to join, but American and Southwestern Company officials favored a line from New Orleans, west, with the business through New Orleans over their lines. Sibley again declared, "If you won't join me, I'll go it alone."

The Western Union board then authorized Sibley to organize a separate company to construct the transcontinental line, and he spent much of the next three years at the Willard Hotel in Washington, D.C. His struggles with Congress were described in letters to Elwood, who sent them on to Wade. In January 1860 Sibley held meetings with the Bees and James S. Graham of California who advocated different routes.[19] Russell and Alden of The American opposed Sibley, who sent "SOS" messages to Wade, Caton, Green, Elwood and others to come to his aid at Washington, but they did not go.[20] He wrote that three groups had bills before the Senate.[21]

A bill introduced by Gwin would give his constituents of the Placerville, Humboldt, and Salt Lake Telegraph Company $50,000 a year for ten years to provide the transcontinental line. All other companies opposed it. Gwin then proposed that the leaders of all carriers be included, and the Post Office and Post Roads Committee substituted the names of Sibley, Green, Caton, Frederic Bee, and Zenas Barnum of the American.

Russell and Alden made a last-ditch effort to wreck the bill with a resolution opposing it. This embarrassed Barnum and Kendall, who persuaded the American board to repudiate its resolution and authorize them to act for the company. Hewitt resigned the presidency and Barnum succeeded him.[22]

Sibley reported on March 15 that now "We are all acting together." The Senate passed the bill March 26, 1860.

Hoping a pro-South bidder would win, Congressman Henry C. Burnett of Kentucky got an amendment accepted, eliminating all names, and providing for sealed bids. With the aid of a strong speech by Chairman Schuyler Colfax[23] of the House Post Office and Post Roads Committee, the bill passed on May 24,[24] It appropriated a maximum of $400,000 payable in ten yearly installments, to be paid back $40,000 a year in free telegraph service to the government.

The government was to have priority in using the line and could connect lines from military posts to it. A maximum rate from the Missouri River to San Francisco was set at three dollars for ten words. Completion in two years was required, and it could use public lands for right of way and stations for ten years.[25]

At the North American Telegraph Association annual meeting August 29, 1860, the American company, busy with its Atlantic cable project, had two resolutions passed declaring the requirements of the law were objectionable, and calling for members to report at the next annual meeting whatever provisions would make it satisfactory.

Well aware of Russell's motive to delay the project, Isaac Elwood of Western Union presented a resolution that any member of the association could submit a bid. Sibley remarked this would merely enable a member to move, if necessary, to block some outside company. The opposition fell into the Sibley—Elwood trap, and the resolution was passed. Of course Sibley sent his bid to the Secretary of the Treasury at once. His bid was $40,000; Benjamin F. Ficklin's,[26] $33,000; Theodore Adams', $29,000; Harmon and Clark's, $25,000. However, all except Sibley, the high bidder, withdrew, and he was awarded the contract September 22.

The Pacific Telegraph Company was incorporated on January 11, 1861, with $1,000,000 capital, to construct the line from Omaha to Salt Lake City. Wade was elected president; Sibley, vice president; and Elwood, secretary. With them as incorporators were Butts, Selden, and Medbery of Western Union's board; Edward Creighton; Stebbins of the Missouri and Western line; Albert W. Bee and James S. Graham of California; John H. Berryhill of the Atlantic and Ohio; and Thomas R. Walker of the New York, Albany and Buffalo. The incorporators also included those low bidders who had bowed out: Ficklin, Adams and John H. Harmon. Sibley transferred his contract to the new company.[27]

Wade said Sibley made a strong claim for the presidency, "but the board elected me president and Sibley vice president." Western Union took one share more than half of the stock. Wade's first official act was to appoint Edward Creighton superintendent of the line.

Creighton had examined two routes in 1859 and advised against them. In November, 1860, he surveyed what generally was the Pony Express route. Leaving Salt Lake City in midwinter, he was snow-blind when he reached Carson City, but battled on over the snow-covered mountains on muleback to the Pacific. He recommended that route.

The California Telegraph Company had extended its original Sacramento-San Francisco line to Los Angeles by October 8, 1860. Western Union then sent Wade and Albert W. Bee to San Francisco, where they began calling in suppliers and discussing contracts for supplies for a line to Salt Lake City. This pretended plan to invade persuaded the Western lines to negotiate.

Many pages of a large ledger were filled with copies of Wade's gloomy reports to Sibley and Elwood. General Horace W. Carpentier and associates of the California State Telegraph Company were "the hardest of the lot to get along with," he wrote. On January 24 he feared that he had failed, but on January 30 all companies agreed to furnish a statement of their lines, business, and prospects.[28]

California State agreed on March 19, 1861 to absorb the other western lines and build from Carson City to Salt Lake City.[29] The California Legislature appropriated $100,000 to aid the project. Like Western Union, California State formed a separate company—the Overland Telegraph Company, on April 10, 1861, with $1,250,000 capital, and Western Union began buying its stock.

At Washington President Lincoln said to Sibley, "Well, I hear you are the man who is going to build the line across the plains from the Atlantic to the Pacific."

"That is what I plan to do," replied Sibley.

"I think it is a wild scheme," said Lincoln. "It will be next to impossible to get your poles and materials distributed on the plains, and as fast as you build the line, the Indians will cut it down."

Sibley replied that he believed the Indians could be persuaded to leave the wires alone. Getting poles on the plains, he admitted, would be a great task, but with the aid of the Mormons it could be done.

Preparing to build the western end of the transcontinental line, the California company rebuilt its line from Placerville to Carson City in 1861. Work on the Eastern line had started in 1860. The Missouri and Western

Telegraph Company, of which Charles M. Stebbins[31] was president and Robert C. Clowry superintendent, extended its St. Louis-Kansas City line from St. Joseph through Brownsville, Nebraska City, and Platte City, to Omaha and 200 miles west to Fort Kearney, Nebraska.

Stebbins wrote in his autobiography that Western Union provided the wire and funds,[32] Clowry opened the first office in Omaha September 5, 1860,[33] when a message was sent to President Buchanan who replied: "It is another link in the grand chain in telegraphs and railroads which binds the states of the Union together." A banquet and dance was held in honor of the telegraph men, and there Clowry met his future wife, Miss Gussie Estabrook.

Notes

[1]"Postmen of a Century," Society of California Pioneers, vol. 2, p. 140:

> Printed notices were distributed giving a diagram of the semaphore and what each particular posture and signal meant . . . brigs, schooners, sidewheelers. . . . *Hamlet* was being given by a barnstorming company and the little theater was packed. One night "the lead" waved his arms in the approved manner of Hamlets of that time and cried: "What means this, my Lord?" Before he could break pose or "my Lord" answer, a newsboy up in the third tier piped, "Side-wheel Steamer!" The house rose with a howl and *Hamlet* turned into five minutes of comedy.

[2]Butterfield was the stage-line operator who had joined Faxton in building the first telegraph line between New York City and Buffalo in 1845 and 1846.

[3]When the first shots of the American Revolution were fired at Concord in 1775, hard-riding

dispatch bearers carried the news to New York City in four days, and South Carolina in three weeks. In 1833 and later, New York newspapers operated a pony express to get news from Washington, D.C.

In January 1845 the *New York Herald* started a news express from New Orleans because of strained relations with Mexico and Texas. When the Mexican War began in the Spring of 1846, the *New York Herald, Baltimore Sun, Philadelphia Ledger,* and *New Orleans Picayune* operated a pony express with sixty horses from New Orleans to New York in six days. That is how the President and cabinet received news on April 10, 1847 of the capitulation of Vera Cruz.

[4]An advertisement in the San Francisco newspapers read:

> Wanted—young skinny wiry fellows not over eighteen. Must be expert riders willing to risk death daily. Orphans preferred. Wages: $25 a week. Apply Central Overland

Express, Alta Bldg., Montgomery St.

[5]The railroad mileage from St. Joseph to San Francisco is 1,421 miles, but the winding trail was much longer.

[6]One of the riders, Charles Cliff, said he was there and John Frye was the first rider. The St. Joseph newspaper, *The Weekly West*, said lots were drawn and Richardson won the honor. John H. Keetley, the third rider, wrote that Alex Carlyle was first and John Frye second.

[7]Some writers of that period said William Hamilton made this ride, but Alexander Majors of the firm said it was Roff, and added details about riders, distances, stations, and time that leave no doubt of his accuracy.

[8]Cody was then fifteen. His nickname "Buffalo Bill" was earned later when he was hired to kill buffalo to feed construction crews of the Kansas Pacific Railroad. I saw him and Annie Oakley, the famous markswoman, perform in a touring wild west show.

Dr. Howard R. Driggs, head of the Education Dept. of New York University, and a close friend of mine—I was a director of his Pioneer Trails Association—knew several of the riders. He said Nick Wilson never liked to take his hat off even in the house because of a scar left by an arrow shot by a Gosiute (Western Shoshoni) Indian.

[9]At Deep Creek Station, Utah, Egan later founded a general store, reported by Jack Goodman in the *New York Times* on August 7, 1966 to be still in business. He said west of Salt Lake City, at the Fairfield Station, the Carson Inn had been restored. The bronze medallions set in stone monuments at Dugway and other pony stations, showing a racing rider above the inscription plate, have been pried out by souvenir hunters and the inscription plaques defaced by rifle bullets. At forty, Egan had been the oldest rider.

[10]Sending a telegram west cost $16, but only $3 of it was received by the Pony Express, according to a 1964 booklet issued by Waddell F. Smith (great-grandson of the firm's partner William Waddell). When a centennial ride was reenacted on the old route, posts bearing a galloping horse and rider signs were erected, and 1,000 copies of the old revolver were issued. Smith had a Pony Express Art Gallery at San Rafael, California.

[11]William Lightfoot Visscher, *The Pony Express* (Chicago: Rand, McNally, 1908) 21.

[12]Ralph Friedman, "Early Telegraph Days in California." *Westways Magazine* (February 1957).

[13]Hubert Howe Bancroft, in his *Chronicles of the Builders,* 7 vols. (San Francisco: The History Co., 1892), said a Mexican woman in San Jose, seeing the crossarms on the poles, exclaimed, "Well, I believe those Americans are becoming good Catholics!" Others believed the "crosses" were to shoo devils out.

[14]James Scrugham, *Nevada,* 2 vols. (Chicago and New York: American Historical Society, 1935) 2:156-59.

[15]Ralph Friedman, "Early Telegraph Days in California," *Westways Magazine* (February 1957).

[16]Comstock, who had sold out his interest for $10,000 after making the find, realized what a huge fortune he had tossed away, and committed suicide or was murdered. There were 100 saloons in the little town and the first drink in the morning was free, but few managed to visit more than a dozen saloons before noon.

[17]In 1940, William M. Maule of Minden, Nevada, found, near Minnie Lee's old office at Genoa, a square-cut, tapered, red-incense cedar pole from the Sierra Nevada forest, wind-eroded but strong. Attached to it still hung some of the old iron wire, and a hard rubber insulator on a black walnut block from the East.

[18]Smith Papers, New York Public Library. Letters from Speed to Smith, March and April 1858, reported how the three, in debt for their lodgings, lobbied for the bill. Speed's April 13 letter said the Secretary of War wrote a strong letter in favor, but Southern senators were against it, so "our bill is killed."

[19]Wade Papers. Western Reserve Historical Library, Cleveland, Ohio. Sibley to Elwood, January 24, 1860.

[20]January 27, 1860.

[21]February 3, 5, and 7, 1860.

[22]Wade Papers. Sibley to Elwood, March 11,

14, and 15, 1860.

[23]Congressional Proceedings of 1860, Library of Congress. Colfax, a young newspaperman, had been Speed's agent in raising funds for the Erie and Michigan line. Later he was elected vice president of the United States, with President U.S. Grant.

[24]This was sixteen years to the day after the first public telegram was sent.

[25]The Act (Public 49) became law on June 16, 1860.

[26]Ficklin was general superintendent of Russell, Majors, and Waddell.

[27]Sources include company records and notes provided by the late Lewis McKisick, a Western Union official for many years.

[28]Wade Papers. Wade to Sibley and Elwood, December 27, 1860, January 24 and 30, February 11, and March 19, 1861.

[29]Western Union assigned the eastern section of its line contract to Pacific Telegraph on January 10, 1862. Before it became a part of Western Union on May 16, 1867, California State Telegraph Co. included the Alta, Atlantic and Pacific, National, Northern California, Overland, Placer and Humboldt, and Tuolumne telegraph companies. Company records.

[30]As Sibley reported it. Biographical Sketch by his son, Hiram W. Sibley. Sibley Mansion, Rochester, New York.

[31]Stebbins and Isaac Veitch had constructed the line from St. Louis to Kansas City in 1852. Later Stebbins sold his lines to Western Union and started a freighting business on the western plains. Remembering his own bitter childhood, he left $700,000 to establish an orphanage in Denver.

[32]Wade Papers. 1860 correspondence between Stebbins, Sibley, and Wade.

[33]John Wesley Clampitt, *Echoes from the Rocky Mountains* (Chicago and New York: Belford, Clarke & Co., 1889) 63, states that the line to Omaha was completed on September 5, 1860, by Robert C. Clowry. Others said September 10.

First Transcontinental Line

Edward Creighton was placed in charge of construction of the transcontinental line east of Salt Lake City, as James Gamble was for the line west. Each put parties to work at both the eastern and western ends of their lines.

Creighton took personal charge of the forces working westward 700 miles from Fort Kearney; and W. R. Stebbins, a brother of Charles M. Stebbins, of the party working 400 miles eastward from Salt Lake City. Gamble, with about 800 miles to build, had James Street in charge west from Salt Lake City, and I. M. Hubbard, east from Carson City.

Supplies for the eastern line were shipped by steamship up the Missouri River to Omaha early in 1861. There they were checked by Charles H. Brown.[1] A bullwhacker, driving covered wagons to Denver, Brown had been hired at $50 a month as bookkeeper and secretary to Creighton.

Large coils of number 9 galvanized wire, bulky "Wade" insulators, large bottles filled with chemicals for batteries, telegraph instruments, nails, shovels, picks, iron bars, axes, and other materials, arrived in great quantities on boats coming up the Missouri River that spring. On the river bank the supplies were loaded, 4,200 to 7,800 pounds to a wagon, on train after train of wagons. Drivers who could swing axes and laborers who could dig holes and set poles were hired.

Among captains of wagon trains or in other capacities were Edward Creighton's brothers Joseph and John, his cousin James, George Guy, John and David Hazard, James N. Dimmocks, Aaron Hoel, Matthew J. Ragan, and Robert Tate. Ragan's wagon train left first, to cut poles for use west of Julesburg. W. B. Hibbard's train was next, to build 200 miles of line from Fort Kearney to Julesburg. Then Jim Creighton's train left to start building at Julesburg, about 400 miles west of Omaha.

Traveling through muddy areas in which heavily loaded wagons mired deeply, and fording swollen streams, Creighton's army of

400 men, 700 oxen, and 100 mules moved slowly from Omaha to Julesburg, driving 100 cattle with them to provide beef when the camp hunters failed to obtain deer and antelope. Each man was equipped with a rifle and revolver, and one was accidentally shot and killed.

8.1 JAMES GAMBLE, IN CHARGE OF BUILDING THE FIRST TRANSCONTINENTAL TELEGRAPH LINE FROM THE PACIFIC OCEAN TO SALT LAKE CITY.

One young man cutting and hauling poles in Creighton's party was Theodore P. Cook, whose woodcraft and friendship with the Indians was a valuable asset. His family had settled four miles south of what is now Omaha when he was eight. Only one other white settler was in Nebraska—an Indian trader named Peter Sarpy. Cook spent the major part of his boyhood hunting with the Indians.[2]

In 1861 the rush of gold seekers to Nevada and California was in full swing; stagecoaches along the trail were filled with adventurers; and the route was infested with highwaymen. People along the way cooperated gladly with the builders because completion of the line meant greater safety, news of the Civil War, and a means of communication.

At Fort Kearney there was no room in the military quarters for the telegraph, but Moses H. Sydenham, the postmaster, had built a structure of earth and sod, and gave it a part of his space. Along the South Platte River, the party found cedar trees in canyons near Cottonwood Springs, about 100 miles west of Fort Kearney, for use as poles. From there to the Rocky Mountains poles were scarce, and the greatest problem was to find and haul them as much as 200 miles.

Getting the heavy wagons across the Platte River at Julesburg, from one to four feet deep, was a herculean task. It was accomplished by loading wagons less heavily, attaching as many as twelve teams of oxen, and keeping wagons in motion, so they would not sink in the soft river sand. The line was suspended across the river on tall spliced poles. Camps of friendly Sioux and Cheyenne Indians were seen frequently, and Brown decided the Indian was more sinned against than sinning. Mosquitos, "as bloodthirsty as starved tigers," gave the men more trouble than redskins, he said. In Julesburg there was one trading house, one stage station, and four sheds and barns, all built of logs.

Creighton and Brown visited one party after another in a buggy drawn by two mules and they assisted in finding trees in ravines and along streams. When he was not busy with Creighton's correspondence, Brown marked the route, indicating where each pole would be set, twenty-five poles to the mile.

Second parties dug holes. The next cut poles, nailed on brackets, and set them up, and the final group strung wire on the poles.

The cook-wagon went along the line of work with lunch for the men, as it moved to the next night's camp. In the evening the men gathered in a big tent; mock courts were held, and yarns were spun. There was much good-humored kidding among the men, who were given such nicknames as Curley, Dutch, and Greasy.

Creighton's force began construction at Julesburg, July second. Passing Mud Springs, Court House Rock, and Chimney Rock, they reached Scott's Bluff July 29, and Fort Laramie August 5. A station was set up every night to telegraph orders to Omaha for materials and receive the latest news. It was hot, exhausting work, long hours, under primitive conditions, but the line was built at top speed.

The Voice of "The Great Spirit"

Members of the expedition were ordered to treat Indians well; gifts were given to them, and anyone getting into trouble with them was immediately dismissed from the service.

When the line reached Fort Bridger, Utah, Creighton contrived to get Chief Wash-e-ka of the Snake tribe to visit that office, and a leading Sioux chief to visit Horseshoe Station, west of Fort Laramie. These chiefs of two of the most warlike tribes on the plains were friends. With telegraphed questions and answers, Creighton convinced Chief Wash-e-ka that his friend was at the other end of the line. Each chief was informed the telegraph was the voice of Manitou (the Great Spirit) and must never be harmed. Still skeptical, Chief Wash-e-ka asked the other chief to meet him at a point

halfway between the two stations. They met and were convinced. For years their tribes rarely molested telegraph lines.[3]

James Street, Gamble's general agent, made friends with Sho-kup, head chief of the Shoshones, who had influence with the Gosiutes (of the Western Shoshoni) and Paiutes. Chief Sho-kup was persuaded to go by stagecoach to Carson City and talk with General Carpentier who was there inspecting the line. Sho-kup was informed that the telegraph was like an animal; it ate lightning and carried messages with that speed. Chief Sho-kup called the telegraph "We-ente-mo-ke-te-hope," meaning "wire rope express." He was entertained royally, and sent a message to the "telegraph chief" that his Indians would not injure the line.[4]

Once, Gamble said, the wire was so charged with electricity during a thunderstorm that the men had to use buckskin gloves. Some strange Indians came up, and one helped pull on the wire the men were stretching. When his bare hands touched the wire, the electricity discharged through his body and bare feet into the ground and doubled him up. He ran several hundred yards at top speed. Then he gravely motioned his party to him and probably told them Indians had better leave it alone.

On another occasion, Gamble wrote, Indians wrecked a telegraph office, and one took a swig of the nitric acid, used in batteries, thinking it was liquor. He was laid out cold, and respect was aroused for white men who could drink such strong "firewater."

For years buffaloes rubbed their shaggy hides on knotty rough poles until they fell. Spikes were put in many poles to stop that, but the buffaloes loved them, and so smooth poles were used when available.

Gamble faced greater difficulties building 800 miles in the West than Creighton with 1,100 miles in the East. He had to wait for a ship to come around Cape Horn, laden with wire, insulators, and other supplies. Then he, I. M. Hubbard, and fifty others drove 228 oxen, twenty-six wagons, and eighteen mules and horses on rough mountain trails over the Sierra Nevada Mountains to get the supplies to Carson City. Their first pole was set up at Fort Churchill on June 20, 1861. Daily progress for each party was often five to ten miles. Once sixteen miles of line were erected on the plains in an effort to reach a water hole.

Brigham Young Aids

In the spring of 1861 Street made contracts with Mormons to supply poles for several hundred miles west of Salt Lake, but they found they could make no profit and refused to deliver the poles. In his book, *Roughing It,* Mark Twain said Street went to Brigham Young who examined the contracts, made a list of the contractors' names, and declared,

> "Mr. Street, this is all perfectly plain. These contracts are strictly and legally drawn, and are fully signed and certified. These men manifestly entered into them with their eyes open. I see no fault or flaw anywhere!"
>
> Then Mr. Young turned to a man waiting at the other end of the room and said, "Take this list of names to so-and-so and tell him to have these men here at such-and-such an hour."
>
> They were there to the minute—so was I. Mr. Young asked them a number of questions, and their answers made my statement good. Then he said to them, "You signed these contracts and assumed these obligations of your own free will and accord?"
>
> "Yes."

> "Then carry them out to the letter, if it makes paupers of you. Go!"
>
> And they did go, too! They are strung across the deserts now, working like bees. And I never hear a word out of them. There is a batch of governors, and judges, and other officials here, shipped from Washington, and they maintain a semblance of a republican form of government, but the petrified truth is that Utah is an absolute monarchy and Brigham Young is king.

Creighton also had a taste of Mormon justice. He said Brigham Young's son informed the line builders his pole bid had been too low, and was given a new contract at a higher figure. Hearing this, Young summoned Creighton to his home and asked to see the contract. After examining it, Young threw it in the fire, and said, "The poles will be furnished by my son in accordance with the terms of the original contract."[5]

By October 1, the western line was completed except for sixty miles midway between Salt Lake and Carson City. Gamble and Hubbard had to go into the mountains with their men, who feared they would be trapped there all winter by the heavy snow, but they got the poles.

News of the Civil War crowded the wires and was eagerly awaited in the West. Datelines of newspaper stories read like this: "By telegraph from Washington, D.C. to Fort Kearney, thence by Pony Express to Roberts Creek Station, thence by telegraph to San Francisco."

The Deseret News of Salt Lake City, September 11, 1861, said excitement grew as the race was watched in each direction. Creighton's men completed the line to Salt Lake City October 17, 1861, making their final joint at Fort Bridger, Utah, about 100 miles east of Salt Lake City. That night Creighton sent a telegram from Fort Bridger

8.2 SALT LAKE CITY TELEGRAPH OFFICE.

to his wife at Omaha—the first by wire from Utah. The next day Brigham Young sent the first official telegram, congratulating Wade at Cleveland and added:

Utah has not seceded, but is firm for the Constitution and laws of our once happy country, and is warmly interested in such successful enterprises as the one so far completed.

That was important because rumors had circulated that Utah and California had seceded.

With no ties of rapid communication to bind the West with Washington, it was feared that the western states, with many thousands of Southerners, could not be held in the Union.

Replying to a message from Frank Fuller, acting governor of the Territory of Utah on October 20, 1861, President Lincoln said, "The completion of the telegraph to Great Salt Lake City is auspicious of the stability and union of the Republic."

The Atlantic and Pacific Joined

The two lines were joined, and the Atlantic and Pacific linked in communication October 24, 1861. Because of the late hour, however, the offices east of St. Joseph were closed for the night, and the first eastbound messages were held at "St. Joe" overnight.

The first message west from Salt Lake City was tapped out by Street to General H.W. Carpentier, president of the Overland:

Line just completed. Can you come to office?

He then sent the first official message, from Brigham Young to Carpentier.

The first news was from the *Daily Alta California* to New York:

> General Sumner, with 500 troops and 10,000 stand of arms, and $1,000,000, left for the East on the steamer of the 21st inst. [that is, the present or current month].

8.3 MORSE KEY.

A great battle had been won for the Union without a shot being fired. The San Francisco *Evening Bulletin* said:

> California by this operation of endowing her with "a great sympathetic cord" is no longer an extremity, but a vital part of the Union. By virtue of the wire, the remotest possibility of the Western Slope becoming a separate Pacific Republic is quashed utterly.

The first transcontinental message was from Chief Justice Stephen J. Field of California to Abraham Lincoln:

> The people of California desire to congratulate you. . . . They believe that it will be the

means of strengthening the attachment which binds both the East and the West to the Union, and they desire . . . to express their loyalty to the Union and their determination to stand by its Government on this day of trial.

Mayor Henry Teschermacher of San Francisco telegraphed Mayor Fernando Wood of New York City:

> The Pacific to the Atlantic sends greeting, and may both oceans be dry before a foot of all the land that lies between them shall belong to any other then our united country.

People were amazed when the line became a reality in only three months and twenty days. Suddenly a watch instead of a calendar became the measure of time.

The demand for service was so great that editors begrudged the time used for numerous official messages. Plans for a celebration in California were abandoned when the second message from Salt Lake City told of the death of Colonel E. D. Baker, October 21, at Ball's Bluff, Virginia. The line rapidly piled up profits. For a week, $1.00 a word was charged, although Congress had set the maximum west of the Missouri River at $3.00 for ten words. After the first week the charge from San Francisco to St. Louis was $5.00; Chicago, $5.60; New York and Washington, $6.00; and Boston, $7.00. The word rate above ten ranged from 45¢ to 75¢. When a bill was introduced in the House of Representatives to withdraw the subsidy unless the rates were reduced, it was lowered on February 2, 1863, to $3.00 for ten words, and 22¢ for each additional word. Sibley and Wade were delighted with the line's financial success and had additional wires placed on it.

Stations on the transcontinental line between St. Louis and San Francisco were at

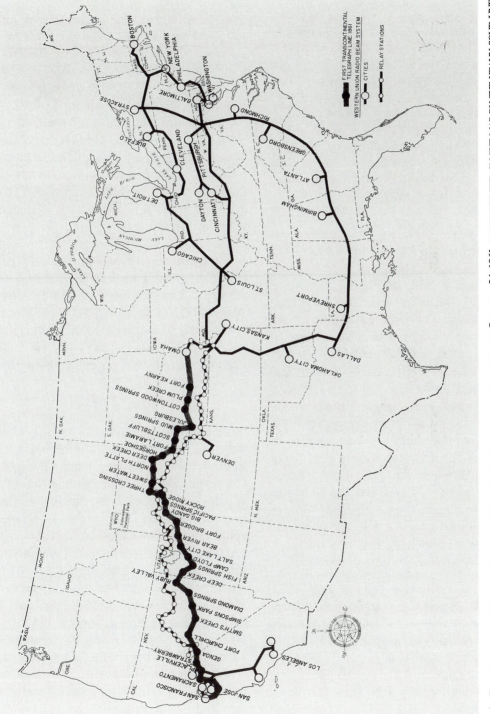

8.4 THE FINAL LINK IN THE FIRST TRANSCONTINENTAL TELEGRAPH LINE, COMPLETED ON OCTOBER 24, 1861, IS REPRESENTED ON THE MAP BY THE HEAVY SOLID LINE BETWEEN OMAHA, NEBRASKA AND SACRAMENTO, CALIFORNIA. A CENTURY LATER, IN 1964, WESTERN UNION BUILT THE FIRST TRANSCONTINENTAL MICROWAVE RADIO BEAM SYSTEM ALONG APPROXIMATELY THE SAME ROUTE.

•ST. LOUIS •BROWNSVILLE •NEBRASKA CITY
•OMAHA •FREMONT •COLUMBUS •GRAND ISLAND
•FORT KEARNEY •WILLOW ISLAND
•COTTONWOOD SPRINGS •ALKALI
•OVERLAND CITY •POLE CREEK •CHIMNEY ROCK
•HORSE CREEK •FORT LARAMIE •HORSE SHOE
•DEER CREEK •SWEET WATER BRIDGE
•ROCKY RIDGE •GREEN RIVER •FORT BRIDGER
•WEBER RIVER •GREAT SALT LAKE CITY
•FORT CRITTENDEN •DEEP CREEK •SHELL CREEK
•RUBY VALLEY •REESE RIVER •MIDDLE GATE
•FORT CHURCHILL •CARSON CITY •GENOA
•STRAWBERRY •PLACERVILLE •FULSOM
•SACRAMENTO •SAN FRANCISCO

When Lincoln's annual message to Congress on Dec. 3, 1862 was transmitted to San Francisco, tolls for the 7,000 words were $600, but letters over the Bufferfield stage route had cost the government more than $60 apiece. The *Alta* proudly printed the message on December 5.

The *San Francisco Bulletin, Alta,* and *Sacramento Union* split the telegraph costs of Associated Press news between them. They formed a state press association, with a $5,000 membership fee to keep smaller, rival papers out. The other papers cried "Foul play!" and when the A.P. news arrived ahead of theirs, they organized the American Press Association. Some charged Western Union with favoring the A.P. and published scorching editorials against the telegraph "monopoly."

Its mission having been accomplished, Pacific Telegraph was merged with Western Union on March 17, 1864. Its $1,000,000 capital stock was exchanged for $3,000,000 of Western Union which issued a 100-percent stock dividend.[6] Those who had bought Pacific stock for very little, then, had Western Union stock worth $6,000,000. In sharp contrast, building the line from Brownsville to Salt Lake City had cost only $150,000,

$34,710 of which was for the Brownsville-Julesburg section. Likewise, California State absorbed Overland. To control California State, Western Union paid $1,652,130.62 for 12,650 shares June 12, 1866, and absorbed it, through two bond issues, for $1,908,500.

The entire line probably cost less than $400,000, but it made fortunes for Sibley, Wade, and Creighton. Creighton had bought $100,000 of it for only 18 cents on the dollar. Western Union made Brigham Young a present of $20,000 of the stock in appreciation for his assistance.

George Hart Mumford, president of California State, became Western Union's general agent at San Francisco in 1869, and vice president and secretary in 1873. Under Wade's skillful handling, the other Pacific Coast lines were merged with Western Union, forming its Pacific Division, and James Gamble succeeded Mumford as its head.

Indian Trouble

Finally realizing that the "talking wires" were used to send military information, the Indians tried to destroy them. When the Cheyennes were on the warpath in Nebraska, Wyoming, and Colorado during the Civil War in 1864 and 1865, the government could provide little protection, and service was frequently at a standstill. Five hundred Indians were reported killed on December 9, 1864, in a battle with Colonel Chevington's force at Sand Creek. They retaliated by burning the telegraph station and destroying other property at Julesburg on January 7, 1865. The few troops at the fort did not dare attack the Indians, who held a war dance and barbecue just out of gun range, and then calmly drove away several thousand cattle.[7] A young operator, Philo Holcomb, stuck an axe in the ground and played the ends of telegraph wires on the iron axehead, to send signals eastward for help. He received the reply by putting the wire in his mouth.[8] The Indians had burned poles for eight miles to the west.

Indians generally did not attack forts. Outrages usually were antics of young bucks, who considered themselves having fun when they scalped men and women and took their horses and goods.

The Telegraph World of October 1937 had a different report:

> The survivors . . . at midnight erected a pole, adorning its top with a tattered flag and an Indian arrow pointing west, and burying at its foot a statement signed by all within the walls of the post:
>
> "This pole is erected by Philo Holcomb and S. R. Smith, operators Pacific Telegraph Line and J. F. Wisely, Surgeon, U.S.A., six days after the bloody conflict of Jan. 7th, 1865, between 1,200 Cheyenne warriors and 40 brave boys of the 7th Iowa Cavalry, under command of Capt. N. C. O'Brien. On this occasion the Telegraph office and Hospital at the Mail Station were totally destroyed. . . . The lives of 15 soldiers and five citizens were lost during this terrible raid, and their remains are interred nearby."

Even after the Civil War, Creighton had only thirty cavalrymen to guard 300 miles of line through hostile Indian territory. They moved up and down the line, fighting from behind a triangular barricade of wagons when attacked. When night came, telegraph men tied pads around the hooves of their horses to muffle sounds and went under cover of darkness to repair the line. When attacked, they scattered in the bushes and returned to camp by circuitous routes.

Where permanent repairs were impossible, Creighton ran a small insulated wire over the tops of the sagebrush from one side

of the break to the other. Sometimes when the Indians camped on the line, he took messages around the camp, and sent them on.

8.5 Edward Creighton, whose fortune founded Creighton University in Omaha, was one of the key figures in the construction of the first transcontinental telegraph line between Omaha and Sacramento in 1861.

Lieutenant Casper Collins, the Indian fighter for whom Casper, Wyoming was named, lost his life going to the rescue of linemen repairing wires Chief Red Cloud's Sioux warriors had torn down. Eight died with Collins, and seven were wounded.

Seventy-five men, women, and children were killed in June and July 1866, on the line's eastern section. Army deserters, highwaymen, and gamblers infested the route. The telegrapher was never without a six-gun and ammunition belt, but station operators had only the company of a horse, cat, and dog. One lone operator telegraphed that Indians were attacking his station:

> I do not know how long I can keep this up.

The red devils are beginning to burn the station. . . .

Then the line was silent. Soldiers found his mutilated body.

When Creighton left the company in 1867, he disregarded the construction of the Union Pacific and Central Pacific railroads and founded a large freighting business, hauling wagon caravans of goods to the West. Noticing that cattle left to roam by the line builders were fat and sleek, he became one of the largest cattle raisers on the plains, and helped to develop Omaha, where he had large real estate and banking interests. Four years after his death in 1874, part of his fortune was used by his widow to establish Creighton University at Omaha.

Ben Holladay also viewed with splendid disregard the railroads spanning the continent. He continued his stage lines until 1866 when he sold out to Wells, Fargo & Company for about $3,000,000. Senator Gwin, who fought to have the transcontinental line built, was the victim of his own good deeds. The line helped to save the Union, but since he believed in the rights of the Confederate states, he had to flee to Mexico, where he received the title of Duke of Sonora from the ill-fated Emperor Maximilian.

Other Western Lines

The Galveston-Houston line was revived in 1859 by E. H. Cushing (proprietor of *The Telegraph,* the Houston newspaper), L. K. Preston, D. L. Davis, and A. C. Burton. In charge of construction, C. C. Cluts had the fifty-mile line ready for operation on January 31, 1860. It crossed Galveston Bay on the railroad bridge, completed a few months before. From Houston it was extended east to Beaumont. The Southwestern Telegraph

Company bought it and extended it to New Orleans in 1868.

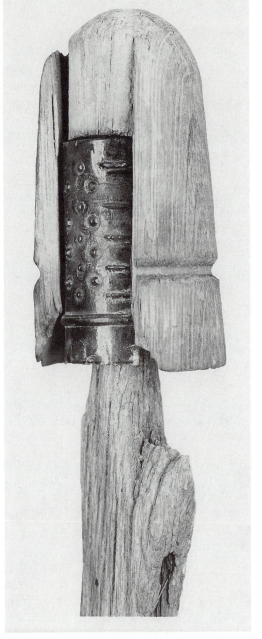

8.6 INSULATOR ON FIRST TRANSCONTINENTAL LINE.

When the Civil War began, there were

about 600 miles of lines in Texas, but supplies were unobtainable. Cow horns were substituted for insulators, and water from Sour Lake for battery acid.

8.7 THE FIRST HEADQUARTERS BUILDING OF THE WESTERN UNION TELEGRAPH COMPANY IN NEW YORK CITY. IN THIS STRUCTURE, AT 145 BROADWAY, WERE LOCATED THE HEADQUARTERS AND NEW YORK OPERATING FORCES OF THE COMPANY FROM 1866 TO 1875.

The first line into Arkansas was built by the Arkansas Telegraph Company in 1860 from Memphis through a swamp to the St. Francis River, Madison, Clarendon, Des Arc, and Little Rock. H. A. Montgomery of Memphis was its builder and superintendent. Charles P. Bertrand of Little Rock was president, and W. E. Woodruff, founder of the *Arkansas Gazette,* a director.

Another line was built into Arkansas in 1860 by the St. Louis and Missouri River Telegraph Company, organized by T. P. Shaffner and Isaac M. Veitch. It was leased to the Missouri and Western, and C. M.

Stebbins extended it that July from St. Louis west to Kansas City and south to Fayetteville and Fort Smith, in the northwestern corner of Arkansas. The Arkansas River was crossed with one wire from a high bluff at Van Buren.

The first office in Denver was opened on October 1, 1863, with a spur line from the transcontinental at Julesburg. The office was in the Nassau Building at 16th and Larimer Streets. The manager was David Moffat, famed later as builder of the Moffat Railroad. A line was built south from Denver in 1868, reaching Pueblo in June, and Santa Fe received its first service July 14.

During the Civil War, so many interruptions to the transcontinental line occurred in Missouri that an alternate line was built for the St. Louis-Omaha section. When it was completed from Chicago to Des Moines to Omaha in January 1862, the *Des Moines Register* said:

> Ever since Adam was an infant, the City of Des Moines has been cut off from the exterior world. . . .

Admission of the "battle-born state" of Nevada to the Union was proclaimed by Lincoln October 31, 1864, after its 15,000-word Constitution was telegraphed from Carson City by the operator Frank Bell,[9] later governor of Nevada. The Nevada votes were urgently needed to pass the abolition resolution ratifying the thirteenth amendment of the Constitution, and to gain its electoral votes for Lincoln's second term as president.[10]

Enterprising citizens of Portland, Oregon, formed the Pacific Telegraph Company in 1854, and built a line from Portland to Corvallis in 1856, but a storm wrecked it. The Oregon Telegraph Company built from Portland in 1863 to Aurora. It reached Salem on April 21, and Eugene City on Valentine's

Day 1864. When it was connected with the San Francisco line at Yreka, California, on March 1, 1864, Portland celebrated with a torchlight parade and speeches, and Governor A. C. Gibbs wired Lincoln, "We want no Pacific Republic."[11]

In 1865 Brigham Young organized the Deseret Telegraph Company, which built a 1,000-mile line connecting all parts of Utah Territory. A new type of insulator was used, with an iron outer shell around the glass substituted for the wooden shell used on the transcontinental line.[12] The Deseret line was extended in 1871 from Kanab, Utah to Pipe Spring, Arizona.

Perhaps galvanized by the success of the transcontinental telegraph, Congress on July 1, 1862 empowered railroads to build "a continuous railroad and telegraph" to the West. Section 19 of the Act said, "the several railroad companies herein named are authorized to enter into an arrangement" with the transcontinental telegraph line so that it would be moved along the railroad "as fast as" it was built.

The Act enabled the Union Pacific to build from Omaha to a point near Ogden, Utah, and secure 12,000,000 acres of public land. It enabled the Central Pacific to construct from Sacramento to meet it, and receive 10,000,000 acres. The two railroads also received $27,000,000 in government bonds. Each rushed construction to gain the largest prizes of land, bond subsidies, and miles of track. When the tracks met at Promontory Point near Ogden on May 10, 1869, the last spike—of gold—was driven with a silver maul, with Leland Stanford of the Central Pacific, and Thomas Durant of the Union Pacific alternating the strokes.[13]

Western Union had rebuilt the transcontinental line along the railroad right of way, and over it each blow on the gold spike was flashed to principal cities. The last blow was by President Stanford, and he missed the spike, but the telegraph operator sent, "Dot, Dot, Dot. Done!" anyway, and the nation celebrated. In New York the chimes of Trinity Church played "Old Hundred," and the last blow was marked by a salvo of cannon. In San Francisco each stroke was repeated by the bell on old City Hall.

The railroads grew rapidly from 30,626 miles of track in 1860 to 52,922 in 1870, 93,262 in 1880, 156,414 in 1890, 193,314 in 1900, and 259,705 in 1916—the maximum reached. Along each mile of track went telegraph lines. As the railroads grew, with leases, purchases, and consolidations, fortunes were made by Vanderbilt, Sage, Gould, Morgan, Belmont, Hill, Harriman, Van Sweringen, and others.

Western Union's 1869 annual report said the transcontinental line was rebuilt with three wires and offices forty to eighty miles apart. The new route was through Ogden—37 miles north of Salt Lake City—because the railroad followed a water-level route through Weber Canyon.

The social and economic results of the transcontinental telegraph line were tremendous. Americans East and West became almost as interconnected as families in the same settlement in Colonial times. The transcontinental telegraph line was a major contribution to the nation, and a proud part of our heritage.[14]

Notes

[1]Brown's diary was made available by his niece, Mrs. Margaret Brown Burgess Ward.

[2]When the transcontinental line was completed, Cook became an operator and worked at stations as far west as Sweetwater Pass, Wyoming. He later became general manager of Western Union at Chicago. His son Morris T. Cook was the highly respected general manager of the Pacific Division, with headquarters at San Francisco, until the 1940s.

[3]*The Telegrapher*, March 1, 1867. The Indians also learned that the telegraph men did not want to take their land.

[4]Gamble, "When the Telegraph Came to California," *Telegraph Age* (November 16 and December 1, 1902).

[5]B. H. Wilson, "Across the Prairies of Iowa," *The Palimpsest* (August 1926).

[6]Western Union's Annual Report, October 1, 1865, said its directors owned $6,000,000 of its $21,355,100 stock outstanding. It said the purpose of the stock dividend was to lower the price and obtain "a more general distribution."

[7]Eugene F. Ware, *Indian War of 1864* (Topeka KS: Crane & Co., 1911).

[8]Until he was 75, around 1912, Holcomb worked a fast wire in the office of the *Atlanta Journal*. His giant-size son, also named Philo, was for many years on the staff of the traffic vice president of Western Union, and invented the Varioplex, important in modern telegraphy. Holcomb's other sons, Andrew and Egbert, also were operators.

[9]The *Nevada State Journal*'s obituary of Bell (February 14, 1927) said he learned to telegraph at 15 in his home town in Michigan, went to San Francisco and aided the Bees in building a line from Placerville to Genoa, and to Carson City in 1859 and Virginia City in 1860. He then worked with Gamble on the transcontinental line.

[10]Effie Mona Mack, *Story of Nevada* (Glendale CA: Arthur H. Clark Co., 1946) 264. Statutes of Nevada, 1864, p. 187.

[11]Fred Lockley in the *Oregon Journal*, August 22, 1926.

[12]In 1964 Dr. Alvin C. Hull, Jr. of Logan, Utah found one of these insulators on the route of an extinct line between Paris and Franklin, Idaho, just across the Utah border.

[13]Personal conversation with 100-year-old William H. Jackson, bullwhacker on covered wagon trains who photographed the scene. He had been a combat artist in the Civil War, and painted scenes of the transcontinental telegraph line when it was being built in 1861.

[14]For the celebration of the first transcontinental line, see appendix 1, "Making History Live," below.

Another source for chaps. 7 and 8 was Dr. Howard R. Driggs, former Education Department head of New York University, native of Utah and historian of the West, who knew personally some Pony Express riders.

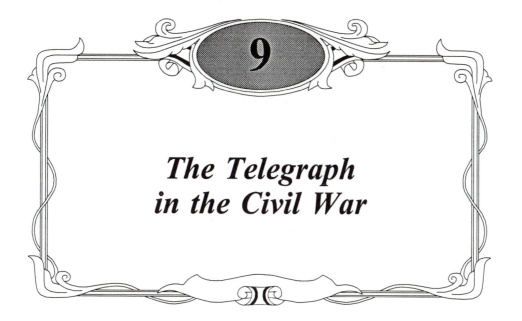

The Telegraph in the Civil War

When Fort Sumter fell, and President Lincoln telegraphed his call for 75,000 troops on April 15, 1861, to states supporting the Union, the telegraph had the greatest test since its birth. Everyone wanted service at once, and telegraph people were up to their ears in work and problems.

The government, preparing for war, needed it most of all. Lincoln's call for troops was followed by telegrams from Secretary of War Simon Cameron giving governors quotas. Their answers promised troops or reported when regiments of state militia started for Washington.

People sent telegrams to relatives and friends about their plans. Businessmen closed out old arrangements, made new deals, and shifted people, money and goods north or south. Newspapers for the first time ordered columns and even pages of news telegraphed daily. The public's appetite for war news spurred the growth of newspapers and started the publication of Sunday papers.

The Government seized the American Telegraph office in Washington, cutting off communication with the South three days after Lincoln wired for troops. American's northern and southern officials met on the Potomac River bridge on April 21, but could do nothing to resume the flow of messages. Authorities in the North and South used the telegraph to organize, order supplies and prepare for war. Western Union's lines, largely in northern and western states, were used extensively by the government to carry on the war.

After American was split at Washington, the northern half, under the presidency of Edward S. Sanford, built lines needed by the government. The other half was named Southern Telegraph Company, with Dr. William S. Morris as president. It was a powerful aid to the Confederate cause.

Southwestern was split south of Louisville, with its president, Dr. Norvin Green, managing its lines from Kentucky north. George L. Douglass, its treasurer, and then John Van Horne, general superintendent, served as president of its southern lines.

Many Northerners regarded abolitionists as crackpots, and rioted to protest Army drafts. President Lincoln advocated returning $400 of the price Southerners paid for the slaves to Yankee slave ship owners, and sending them back to Africa. Representative Daniel E. Sickles of New York threatened secession by New York City, and its mayor, Fernando Wood, proposed on January 7, 1861,that the city secede and become a "free city." Leslie's *History of the Greater New York* said, "The Common Council adopted his suggestion with wild enthusiasm."[1]

9.1 ANDREW CARNEGIE, A MESSENGER BOY AND LATER A TELEGRAPHER IN LINCOLN'S TELEGRAPH OFFICE DURING THE CIVIL WAR, BECAME ONE OF THE NATION'S GREAT STEEL INDUSTRY LEADERS.

There was grave danger of Washington, D.C. falling into the hands of the South, and emergency calls for help went out over the telegraph lines. Southern sympathizers destroyed telegraph and railroad lines in the vicinity of Baltimore, isolating Washington. Secessionist mobs in Pennsylvania and Massachusetts attacked troops en route to Washington with bricks, stones and bottles. William Bender Wilson, operating a telegraph line in the Pennsylvania Governor's office, received news of the escape of the Pottsville, Pa. National Light Infantry en route to Washington from a mob of 10,000 in Baltimore. Maryland was a slave state, and one line of its state song "the despot's heel is on thy shore," referred to Lincoln.

Railroads were taken over, and Colonel Thomas A. Scott of the Pennsylvania Railroad was placed in charge of them. He immediately summoned Andrew Carnegie and other telegraph men to restore communications lines. Carnegie and other telegraphers skilled in train dispatching worked at Washington guiding trains loaded with supplies and troops. When Scott was appointed Assistant Secretary of War, he placed Carnegie in charge of military railroads and telegraphs.

The vital importance of the telegraph in military operations was demonstrated in the Civil War. Lincoln spent much of his time in the War Department Telegraph office, anxiously receiving reports from the front and sending orders to his generals. Sometimes he remained there all night. He called the operators by first names or nicknames.[2]

Military telegraphers David Strouse and David Homer Bates wrote the book *Lincoln in the Telegraph Office.* One of the first jobs of the military telegraph was to build lines between the War Department, the Navy Yard and arsenal, and between the main telegraph office and the President's mansion.

Military telegrapher David Strouse's health was ruined by the strain, and he died on November 17. He was succeeded by nineteen-year-old James R. Gilmore, who had built a line from Washington through Confederate territory to the Shenandoah Valley headquarters of General Banks. As troops poured into Washington, other lines were built to their camps in Chain Bridge, Arlington and Alexandria. Telegraphers were recruited to operate these lines and accompany the armies.

If the telegraph had gone with General Patterson, who was to divert General Johnston while McDowell attacked General Beauregard's Army at Bull Run, the rout of the Union Army in July 1861 might have been avoided. However, when Patterson discovered that Johnston had slipped away, he had no telegraph to inform McDowell, who was at Fairfax Court House receiving reports by courier from his army ten miles away at Bull Run.

So supremely confident were Washington officials that the Confederates would be routed that they did not bother to send to McDowell the news of Johnston's disappearance. Lincoln and his Cabinet assembled in the telegraph office to receive the victory news. Other officials and members of Congress drove out in carriages with their wives to watch the "fun," and newspaper editors set up their largest headlines to proclaim "Victory" across Page One.

During that afternoon McDowell reported success in pushing the Confederates back, and war correspondents sent glowing stories. Such songs as "Old Man Tucker," "Oh Susanna," "John Brown's Body," and "Tramp, Tramp, Tramp, the Boys are Marching" resounded in hundreds of towns and cities as news of a great victory spread.

Then a long silence followed; no couriers arrived from the front. At last Bates, in the War Department telegraph office, received a message: "Our army is in full retreat." This was followed by descriptions of troops and carriages of dignitaries battling with one another to be first to escape over the small bridges and roads. All news of the army's defeat was supressed by ordering the telegraph companies not to send the stories.

9.2 ANSON STAGER, A WESTERN UNION TELEGRAPHER, ORGANIZED THE MILITARY TELEGRAPH CORPS WITH WHICH LINCOLN DIRECTED UNION FORCES DURING THE CIVIL WAR. LATER STAGER WAS A LEADING WESTERN UNION OFFICIAL.

The vital need for a centrally organized and directed military telegraph corps was demonstrated by Bull Run, and Scott sent for Anson Stager, general superintendent of Western Union.[3] The day Fort Sumter was fired on, April 12, 1861, Governor Denison of Ohio had asked Stager to take charge of the military telegraphs and provide assistance to Major General George B. McClellan, commander of the army in the Department of Ohio.

Stager's men won the praise of McClell-an and his officers, who had at first opposed his appointment, by building a telegraph line—the first time this had been done—as the army moved that summer to the victori-ous battle of Rich Mountain. With the aid of Stager and his men, McClellan was able to direct his armies in five states and keep informed of their movements. That is why, when cocky, little McClellan became com-mander of all Union armies in the fall of 1861, he made Stager a colonel, and asked him to organize the Military Telegraph Corps. Stager and Wade were Clevelanders and close friends, and he had Wade look after his affairs in his absence from Western Union. In a lengthy letter to Wade on War Department stationery, he described the Union strategy.[4]

The situation required action of all Union armies in concert. Each army was ordered to report its position and actions and all available information concerning the enemy daily by telegraph. In turn, the War Department wired orders to each army at frequent intervals and provided information concerning the enemy that had been received from other army units and other sources.

In organizing the Corps, Stager appoint-ed experienced telegraph men to take charge in the various departments, or areas, of the war. Colonel Robert C. Clowry, who had been active in Western Union's early expan-sion, was placed in charge of the military lines in Missouri, Arkansas and Kansas.[5]

Major Thomas T. Eckert,[6] in charge of the Department of Virginia and North Caro-lina, was appointed telegraph manager and aide-de-camp to General McClellan on April 7, 1862. When ill health from overwork forced Stager to move his headquarters to Cleveland in April 1863, Eckert was placed in charge of operations, but Stager continued

to direct it.

When Stanton replaced Cameron as Sec-retary of War, he was irked to find that Ec-kert was delivering Army messages only to General McClellan. He tried to get rid of Ec-kert, but Lincoln would not permit it. Stan-ton then moved Eckert from McClellan's headquarters to the War Department, cre-ating a long-lasting feud with that general.

9.3   THOMAS J. ECKERT, WHO HEADED LINCOLN'S TELE-GRAPH OFFICE IN THE CIVIL WAR, WAS JAY GOULD'S TOOL IN ROBBING WESTERN UNION. ECKERT TRICKED EDISON INTO AGREEING TO SELL INVENTIONS TO COMPET-ITORS UNDER GOULD'S CONTROL. WHEN GOULD GOT CONTROL OF WESTERN UNION, HE MADE ECKERT ITS PRESIDENT.

On some occasions Eckert had the nerve and intelligence to withhold messages. In 1864 Grant ordered him to send a message removing a man he disliked—General George H. Thomas—who was opposing General Hood in Tennessee. No report had been received from Thomas because the lines had been cut. Eckert delayed sending the unjust order, and eventually a message arrived that night telling of a smashing victory by Thomas. He then rushed to tell Stanton and Lincoln the news. Grant's order, of course, was not sent. In 1864 Eckert became a Brigadier General and Assistant Secretary of War.[7]

The Corps built about 15,000 miles of telegraph lines, and generals were delighted when, at the end of a day's march, they had the telegraph to communicate with one another and the War Department. However, the regular telegraph lines were used when available.

No War with England

The sympathies of England were so strongly with the South that it accorded full recognition to the Confederate Government in 1861, and seemed about to go to war as an ally of the Confederacy. In the North feeling ran high, and a strong demand developed to declare war on Great Britain.

The telegraph had been extended from London to Galway, Ireland, and across Newfoundland to St. John's, through 400 miles of wilderness. Telegrams from London were carried by packet ship from Galway to St. John's in about two weeks, and then transmitted to New York and Washington. In June 1861, the ship *Prince Albert* arrived at St. John's with a message to Lincoln that England had decided not to declare war but observe strict neutrality. This would cause Lincoln to say a prayer of thanks, and appeased Northerners demanding war against England, and could change the outcome of the Civil War.

Shortly after the packet arrived, the telegraph line across Newfoundland went dead. A young operator at St. John's, Thomas D. Scanlan, said the line probably was good beyond LaManche, over 100 miles away,

and he would get there somehow.

Scanlan left at midnight. At dawn a fisherman at Lance Cove took him in a skiff to Brigus. On foot, he went to Spaniard's Bay and New Harbour, crossed Trinity Bay by boat, and dashed on through swamps, underbrush, and heavy woods to Rantem, where he fired shots to·call a ferryman.

Finally reaching the telegraph office at LaManche, he found the telegraph key out of order. While two boys brushed mosquitoes and flies from him with boughs, he repaired the key and telegraphed the message that may have prevented a war.

Dangerous Work of the Telegraphers

There were heroic exploits in the Civil War by telegraphers on both sides. Sometimes the fate of an army depended upon a message from the commander-in-chief, a quick flash from an operator with the advance guard, or information gained by an operator tapping enemy lines.

Operators, unarmed and unmounted, often were in greater danger than the soldiers. They strung wires between units on the field of battle, sent urgent dispatches from dictation, and received messages while crouched behind rocks and trees in the midst of fighting. During the siege of Charleston, West Virginia, they were so close to Confederate skirmishers that their wires were frequently cut by rifle fire.

William B. Wilson said in his pamphlet *Civil War Personal Narratives*:

> Ofttimes they were sent where the sky was the only protecting roof over their heads, a tree stump their only office, and the ground their downy couch. Provisioned with a handfull of hard bread, a canteen of water, pipe,

tobacco pouch, and matches, they would work an office at the picket line, in order to keep the commanding general in instantaneous communication with his more advanced forces, or to herald the approach of the enemy. When retreat became necessary, it was their place to remain behind and to announce that the rear guard had passed the danger line between them and the pursuing foe.

Conte de Paris described the work of the Corps in his *History of the Civil War*:

As soon as a marching army had gone into bivouac, the telegraph wires established a connection between all the general headquarters; the tent where Morse's battery was hastily set up became the rendezvous of all who under any pretext whatever could obtain access to procure the latest news... A corps of employees was organized for this service, selected with care and sworn to secrecy, for upon their discretion depended the fate of the armies.

The field telegraph was composed of a few wagons loaded with wire and insulators, which were set up during the march, sometimes upon a pole picked up on the road, sometimes on the trees themselves which bordered it; and the general's tent was hardly raised when the operator was seen to make his appearance holding the extremity of that wire. An apparatus still more portable was used for following the troops on the day of battle. This was a drum, carried on two wheels, around which was wound a very slender copper wire enveloped with gutta-percha. A horse attached to the drum unwound the wire, which, owing to its wrapper could be fastened to the branches of a tree, trailed on the ground, or laid in the bottom of a stream.[8]

Many operators penetrated enemy territory, tapping lines for days and even weeks. One operator learned Confederate plans from the Charleston line, and enabled Union Generals Gilmore and Terry to foil an attack, but the operator was captured and died in prison.

Another operator stuck to his wire when the army left the field, although heavy fire killed his messengers, and obtained reinforcements for General Fitz-John Porter in the battle of Gaines' Mill. Operator Jesse Bunnell noticed armies lining up for battle in front of and behind him. Instead of running, he sent the news to McClellan's headquarters. Hiding behind a tree for hours in the midst of the battle, he received news of reinforcements, maintained contact with Porter by courier, and handled reports and orders.[9] Sounders were not used on battlefields; the messages were read by listening to the soft tick of the relay.

When the Confederate ironclad Merrimac destroyed the Cumberland and other warships off Newport News, Virginia, Operator George Cowlan watched from his office window and sent a detailed description to Washington until shells wrecked his office. To avoid Southern territory, his report had to travel a circuitous route, over a cable the Corps had laid to Ft. Monroe, across Chesapeake Bay, up the Del-Mar-Va Peninsula to Wilmington, and south to Washington.

Equally important were orders directing the movement of huge quantities of quartermaster's supplies. The Corp's maintenance of lines behind the Union armies was dangerous in areas inhabited by Southern sympathizers who were not deterred from breaking lines by threats of death. In fifteen states, where Southern feeling was strongest, some operators were captured at small stations, and repair forces met capture or death.[10]

One of the daring Southern telegraphers, George A. Ellsworth, served with General John H. Morgan, Confederate cavalry leader,

9.4 IN THE CIVIL WAR, A UNION TELEGRAPH OPERATOR PUT HIS MORSE KEY ON A BOX UNDER AN AVAILABLE COVER TO KEEP IT DRY, CONNECTED IT WITH A LINE HASTILY STRUNG ON TREES, AND PROVIDED THE COMMANDING GENERAL WITH CONTACT TO LINCOLN'S TELEGRAPH OFFICE, AND WITH SCATTERED UNITS OF HIS ARMY.

whose raids caused the Union Army no end of trouble. Ellsworth's fabulous exploits became the basis for novels and short stories. On numerous occasions he tapped Federal lines and obtained valuable information from orders he overheard.[11] Frequently he captured a telegraph office, cleverly impersonated the distinctive style of the sending of some Union operator, and got news of the movement of Union forces. Then Ellsworth would send misleading reports and orders, signing the names of Northern generals and causing great confusion.

An operator with General Robert E. Lee's army was behind Union lines several days, listening to messages between the War Department and General Burnside's headquarters. C. A. Gaston, another operator with Lee, was behind the Union lines for six weeks and heard all messages that passed over General Grant's wire. Although he could not read messages in cipher, he obtained much information from those in plain text. Emmett Howard, a mere boy, repeatedly crawled through Federal lines and told Confederate troops in Kentucky of the approach of Union troops by using a pocket telegraph set. Another operator, with Mosby, sent Secretary of War Stanton a message full of insults from a seized telegraph station.

Samuel H. Beckwith, General Grant's chief operator, tapped a wire and learned the hiding place of Wilkes Booth. A. Harper Caldwell served as chief cipher operator of five successive commanders of the Army of the Potomac: Generals McClellan, Burnside, Hooker, Meade, and Grant. He shared the responsibility with Beckwith after Grant took command.

When the Chambersburg, Pennsylvania

office was burned by Confederate raiders in 1864, the manager, Thomas F. Sloan, had hidden an extra set of telegraph instruments, and when they left, he was able to inform Federal troops of the route they took. Confederate Cavalry General Forrest captured several operators at towns along the Mobile and Ohio Railroad, but at Kenton, Tennessee, Operator Stephen Robinson, fifteen, was so small that the Confederates considered him a little boy and let him walk through their lines. Then he climbed a pole and wired the news to the Union Army.

The Confederacy took control of the telegraph lines in name only. Unlike the North, the South had no military telegraph corps; it relied on the telegraph companies.

Usually only two operators were attached to the staffs of Confederate generals. They were men who volunteered as soldiers, but were detailed to do telegraph work. When a Confederate army camped near a regular telegraph office, soldier operators helped the office handle the unusual load of messages.

R. O. Crowell, a young Atlanta telegrapher, used old telegraph equipment to "close the key," and blow up the U.S. Gunboat *Commodore Jones* in the James River near Richmond.[12]

Military telegraph operator J. Emmet O'Brien said General Palmer credited the telegraph with warning him of the approach of Pickett's force against New Bern, North Carolina, in February 1864, enabling him to concentrate his forces to meet the attack.[13]

In March 1865 O'Brien ran the line along with the troops in General Schofield's advance on Kinston and Goldsboro, lying in Gum Swamp—where the enemy struck us—two days and nights with the relay to his ear, transmitting dispatches. The Signal Corps cooperated handsomely, and ten picked cavalrymen rode right and left under fire with the dispatches. A whole regiment of ours was captured almost beside us. The morning after this affair, General J. D. Cox called at our post and courteously said that he wished "personally to thank the chief operator for the service rendered at the front." He seemed astonished at finding only a boy of fifteen, muddy and haggard, lying on the ground and too exhausted to care even if the president called.

In South Carolina, O'Brien said, two military telegraph men had climbed alternate poles, restoring a wire which had just been cut by a shell, when "another shot struck the pole between them and brought poles, wire and men in a tangle to the soft sand." Three linemen were killed by guerrillas on the Fort Smith line in March, 1864.

The turning point of the Civil War was brought about in part by the heroism of a telegraph operator. Lee had invaded Pennsylvania, was turning upon General Meade's army, and seemed about to concentrate his scattered forces in order to cut Meade's army to pieces and place the North at his mercy. In desperation, Secretary of War Stanton at Washington had Eckert wire a 17-year-old telegrapher, Ten Eyck Fonda, at Frederick, Md. to dash through the country at Meade's rear, strewn with Confederate raiders, with a message to attack Lee's divided forces before General Early's army, making a forced march to join Lee, could reach Gettysburg. Fonda and two other men, Rose and Hardesty, started for Meade's headquarters by different routes, each with a copy of the message. Fonda got through and delivered the message.

A band of Union spies stole a passenger train at Kennesaw, Georgia in 1862, and started north cutting telegraph lines and burning bridges. The conductor of the stolen train accompanied by soldiers went in hot pursuit on an engine. At Calhoun they picked up Edward Henderson, a fifteen-year-old telegrapher, who jumped from the engine at Dalton and wired ahead to Chattanooga. The spies were captured before reaching that city. The engine of the stolen train, named the "General," was exhibited in the railway station in Chattanooga for more than a century, and the pursuit engine, the "Texas," is in the Civil War Cyclorama in Atlanta.[14]

In Grant's final campaign, wires radiated from his headquarters to every salient point, enabling him to manipulate every movement of his troops in concert with others. A field telegraph system, developed by Eckert, used insulated cable strong enough to resist cannon wheels. The cable was carried on reels mounted on backs of mules and payed out on the field or on trees. Portable batteries were transported in ambulances.

Even with sixteen-shot automatic rifles and artillery the South never had, Sherman's Army was stalled for two months at Kennesaw Mountain, Georgia on its march to Atlanta by a small Southern force in trenches. When Sherman broke through and burned Atlanta, the North, weary of war and criticizing Lincoln, cheered and reelected him, though he expected defeat. With his large reinforced army, Grant practically exterminated General Robert E. Lee's men, who stood their ground.[15]

In the War Department telegraph room, the sounder on the line from Fort Monroe gave a staccato call. Operator William E. Kettles, fifteen, answered:

> Here is the first message for four years. Richmond, Virginia, April 3, 1865. Hon. E. M. Stanton, Secretary of War. We entered Richmond at eight O'clock this morning. G.

Weitzel, Brigadier General, commanding.

Kettles took the message quickly into the office of Assistant Secretary of War Eckert, where President Lincoln was talking with Charles A. Tinker, one of the cipher men. (Tinker had first demonstrated the telegraph to Lincoln in Illinois.) Kettles handed the message to Tinker. Lincoln rushed to Stanton's office. The great news was announced, and Stanton lifted little Kettles high, so the cheering crowd in the street could see him. The Union had won the war!

When Lee surrendered at Appomattox Court House April 9, 1865, it was only necessary to lay a line in two hours to Appomattox Station to the North. Grant then gave Operators Laverty and Shermerhorn the message, which they transmitted to Petersburg for retransmission to Washington.

Telegraph Problems in the South

Many Southern lines were useless during the war because of the lack of wire, supplies, and no men out of the army to repair them. Much to the discomfiture of the Southwestern officials, Dr. Morris, president of Southern Telegraph, was placed in charge of all military and commercial lines. When the Confederate Congress declared the Morse patent void on May 21, 1861, and had the lines south of Louisville destroyed, they suspected that Dr. Morris meant to ruin them. Southwestern and American officials feared their property would never be returned. When the Confederate government seized Southwestern's lines in September 1862 to sequester the stock of Morse and other "enemies," General Superintendent John Van Horne went to Richmond. He protested that the company's officers were

Southerners, and the lines were returned. It was an unequal war between the industrial North with factories providing arms and ammunition and the agricultural South with farms growing food and cotton. That made the troubles of the Southern telegraph lines more severe.

9.5 NORVIN GREEN, PRESIDENT OF NEW ORLEANS AND OHIO LEESEES, JOINED HIS COMPANY WITH WESTERN UNION AS ALL OTHER IMPORTANT LINES DID IN 1866, LEAVING W.U. THE ONLY NATIONAL TELEGRAPH COMPANY. GREEN LATER BECAME PRESIDENT OF WESTERN UNION.

Possession of the southwestern lines alternated between Confederate and Union forces. After the war, the government re-stored the broken lines in Texas and built new ones from Vicksburg to Shreveport, Hempstead to San Antonio, and Brazos to Brownsville, thanks to the influence in Washington of Dr. Norvin Green.

On December 1, 1865, the government returned all commercial lines in the South to the companies. In 1866 the military lines south of the Ohio River were turned over to the companies as compensation for their losses under government control. The few remaining military lines north of the Ohio River were sold to the private companies.

The Union forces had telegraph service almost everywhere they went. Even Sherman's Army, looting and burning homes occupied by women and children on its "March to the Sea," could wire its orders for supplies.

A movement for the government to take over the northern lines was halted by a report by Stager to Quartermaster General Montgomery C. Meigs in 1864. The companies had faithfully provided all of their service and facilities, the report said, and the Government's telegraph use was not large enough to justify assuming the expense of operating the lines.

In a report to Stanton November 3, 1864, Meigs praised the "zeal, intrepidity, and fidelity" of the 1,200 men in the Military Telegraph Corps:

> I have seen a telegraph operator in charge of a station in a tent, pitched from necessity in a malarious locality, shivering with ague, lying upon his camp cot, with his ear near the instrument listening for messages which might direct or arrest the movements of mighty armies. Night and day they are at their posts. Their duties constantly place them in exposed positions, and they are favorite objects of rebel surprise.

9.6 The Telegraph Construction Train (shown here at Richmond), under the direction of A. Harper Caldwell, chief operator of the Army of the Potomac, was used in the construction of field-telegraph lines during the Wilderness campaign and in operations before Petersburg. Following the capture of Richmond, Superintendent Dennis Doren used the train to restore the important telegraph routes of which Richmond was the center. In Virginia during 1864–1865, Major Eckert made great efforts to successfully provide Meade's army with ample facilities. A well-equipped train of thirty or more battery wagon and construction carts was brought together under the skillful and energetic Doren.

More than 200 Northern operators were killed, wounded or captured.[16] There were only fifty-one Confederate operators, but their casualties were high, partly because of the scarcity of clothing, food, and arms. They had only 4,061 miles of line and their annual cost was about $91,000.

In contrast, the Military Telegraph Corps built and operated 15,389 miles of line, and handled more than 6,000,000 military telegrams at a cost of $3,219,000. When the war ended, the corps operators were dismissed and without the many honors, payments, and pensions given to soldiers.

When President Lincoln was fatally wounded by John Wilkes Booth in Ford's Theatre in Washington on April 14, 1865, all wires out of Washington were cut within fifteen minutes, with the exception of a secret government wire to Old Point, through which the news finally reached the fortifications around Washington and was passed on to the outside world. Little of the sad news reached the North that night. It brought sorrow especially to the South because of Lincoln's desire for kindness in healing the wounds of war. He was regarded by thoughtful Southerners as their only hope of escape from the horrors of "reconstruction," with its harsh rule over a crushed people with no vote, by greedy carpetbaggers and ignorant former slaves.[17]

The Use of Cipher Systems

Ciphers were used in military telegraph messages. Only operators with the generals and at the War Department had copies of the

9.7 TELEGRAPH CONSTRUCTION CORPS STRINGING WIRE IN THE FIELD. THE CORPS WAS COMPOSED OF ABOUT 150 MEN, WITH AN OUTFIT OF WAGONS, TENTS, PACKMULES, AND PARAPHERNALIA.

ciphers. When an operator was captured, another cipher was adopted, and no less than twelve were used.

Early in the war, the Confederate cipher was letters of the alphabet horizontally and vertically from the letter "a." Under each horizontal letter the alphabet was written, vertically, beginning with that letter. The operator would take the first letter of a received message and find it in the horizontal column and then obtain the first letter of the decoded telegram by finding the point in the table at which the two columns intersected. After all letters of the key words were used, the same process was repeated. Solution of this simple cipher gave Union forces important advantages.

The Federal ciphers, devised chiefly by Stager and Eckert, became more and more complicated. "Commencement" words of different lengths indicated how many lines the message contained. Meaningless check words were inserted after each sixth word, and arbitrary words were used for various places and names. To decode, the message was written six words to the line in columns, using as many lines as the code word indi-

cated. The message was then read in accordance with the meaning of the key word in the message; for example, up column five, down column two, up column six, down column one. Later Union codes used more arbitraries to vary the number of column and other elements; girls' names indicated the time. When important messages were being decoded, Lincoln often read them word-by-word over the operator's shoulder.

The Signal Corps in the Civil War

Shortly before the Civil War, a young army assistant surgeon, Albert J. Myer, patented a system of wigwag signals which he demonstrated to the War Department. He used flags, torches and telescopes to signal between Fort Hamilton in New York Harbor, Fort Wadsworth, Staten Island, and Fort Hancock, Sandy Hook, New Jersey. As the first signal officer of the Army, and a major, he opened signal schools at Fortress Monroe and Georgetown, and in October 1862, there were 198 signal officers, with two soldiers detailed to assist each of them. The Signal

Corps cost about $2,000,000 during the war.

Experimental signal "trains" in 1862 consisted of a light wagon carrying ten miles of corded insulated wire developed by Henry L. Rogers, a hand-operated magneto generator, invented by George W. Beardslee, and a friction electricity generator. The Beardslee sending instrument was operated by moving the dial to a letter of the alphabet, and hand cranking the magneto with the number of pulses needed to move a dial on the distant receiving instrument to the same letter. This frictional-electricity system that Myer championed, worked so poorly that Frederick E. Beardslee, son of the inventor, was inducted to get someone who could repair it, but damaged instruments still had to be returned to New York for repair. One "train" was used at the battle of Antietam, but during the Fredericksburg campaign the wires were often broken.

The Military Telegraph men so far excelled that some generals, disgusted with Signal Corps failures, ordered them to take possession of the Signal Corps' wires. The shirt-sleeved, bareheaded, and often ragged-pants boys of the Military Telegraph refused to take orders from the "wigwag" officers of the Signal Corps, and infuriated Myer by not leaping to attention and saluting when he appeared in his spotless uniform.

Myer tried to get rid of them by organizing a Military Telegraph Department within the Signal Corps in 1863, but Stager pointed out to Secretary of War Stanton the danger of having two organizations providing the service. On November 19, 1863, Stanton relieved Myer of command, assigned him to duty at Memphis, and ordered the Signal Corps trains turned over to the Military Telegraph. The Beardslee machines were used no more.[18]

Major William J. L. Nicodemus, next in charge of the Signal Corps, recommended in 1864 the return of the field lines to the Signal Corps, but the report was circulated before approval by Stanton, who had all copies destroyed.

With Colonel B. F. Fisher as its next commander, the 168 officers and 1,350 enlisted men of the Signal Corps served with signal flags, rockets, fires and signal pistols to indicate the effect of artillery fire on hostile positions, signal the approach of the enemy, and send messages on the field of battle. A 42-mile series of visual signals was established between General Pope's headquarters at Culpepper and General Banks at Fairfax Court House, Va. Observations by Signal Corps men with telescopes in captive balloons were reported to the ground over insultated wires.[19]

Myer was again appointed Signal Corps commander in 1867 and remained until his death on August 24, 1880.[20]

After the Civil War the Signal Corps began building the communications system of the U.S. Army. It also distributed meteorological reports in storm areas, operated lines to lighthouses and life-saving stations, and constructed lines to frontier posts and settlements.

Censorship in the Civil War

General Winfield Scott, commander of the Union Armies, would not permit the transmission of war correspondents' stories, reporting the rout of his army at Bull Run.[21] False stories of victory were sent. Victory headlines, ringing bells, booming cannon, cheering, and celebrations turned into cries of indignation when the truth became known.

Public indignation, however, did not halt strict censorship, not only of news, but all private telegrams. Although the government announced it was taking over all telegraph lines, it only placed censors in telegraph offices in some cities. President Sanford of the American Telegraph Company, in charge of censorship, had orders to stop any message containing information of use to Southern agents or sympathizers. The press was threatened with severe penalties for any violation of secrecy, such as military intelligence "leaked" by Army or Government personnel.

The Government confiscated five years of telegraph company files of old telegrams, went on fishing expeditions to learn who had expressed sympathy for the Confederate cause in the past, and made charges against many people.

Changes in stories by censors producing false news from the front were "to boost public morale." Correspondents tried to escape such "editing" by using the mail, but it also was censored.

Conte de Paris said the messages of war correspondents often angered Union generals, who stopped providing news. "One day Sherman drove off all correspondents . . . but at the end of one month they were all back." The *New York Herald* had sixty-three war correspondents, he said. One was killed on the field of battle, one wounded six times, five wounded, two died of exhaustion, and seven or eight were captured. The Confederates treated captured *Herald* correspondents very well, but the sufferings of Richardson of the *Tribune,* an abolition paper, "in Southern prisons is one of the most affecting narratives that one can read," the *Tribune* said.

After the War

After the war, telegraph people in the South faced wholesale rebuilding in the midst of chaos, with no representation in government to aid them.

Sanford, relieved of his duties as censor, continued as president of American. On his board were Morse, Kendall, Field, Lefferts, and other industry leaders. American merged with Southwestern and the Washington and New Orleans Company, doubling its capital to $4,000,000.

Although Stager had kept his wartime duties as head of the Military Telegraph Corps and his Western Union job separate, the influence of his war post was fully recognized, and Western Union used that advantage to bring other lines into its system.

Western Union earned and paid its share owners fabulous returns. Cash dividends paid in 1858 were as follows: nine and three-quarters percent (February 1), five percent (April 1), eight percent (July 6), sixteen percent (July 24); and four percent (August 16). Stock dividends paid in the same year were thirty-three percent on August 19, and 414.40 percent on September 22.

Dividends of five percent each were paid in 1860 and 1861; in 1862, nine percent and a stock dividend of 27.2588 percent. In 1863 five dividends totaled nine percent, plus a 100 percent stock dividend March 15, and another of thirty-three and one-third percent December 23. In 1864 five dividends totaled eleven percent, and a 100 percent stock dividend increased capital stock to $22,000,000.[22]

The 1864 stock issues represented actual value of company holdings, the Western Union directors stated, pointing out on October 1, 1865, that since May 1864, the company had acquired companies with 3,800 miles of wire, and increased its own facilities by 8,593 miles. Some people, however, thought the latest 100 percent was based on nothing but hopes for future growth, and the company had become dangerously overcapitalized. "It was clear and unmixed water," said Reid.[23]

The dividends strengthened popular belief that Western Union was an Aladdin's lamp for a quick fortune. The stock soared as high as $225 a share, and owners of many competing lines were glad to receive its gold-tinged certificates in exchange for their own. "Stocks went up with every smile on Sibley's face," Reid said. Sibley paid tax of $8,390 on an income of $103,000 in 1865, the largest in Rochester.[24]

When failing health made it necessary for Sibley to resign the presidency of Western Union,[25] Jeptha H. Wade succeeded him on July 16, 1865.

Western Union's success also stimulated the formulation of new companies. This was made easier by the expiration of the Morse patents. Among the new companies were the United States, United States Extension, Inland, Inland Extension and Independent.[26] They merged on August 3, 1864 as the United States Telegraph Company, with James McKaye as president. It made low bids for railroad contracts, and soon had 16,000 miles of line. Backed by New York capitalists, it organized the United States Pacific Telegraph Company to construct a line from San Francisco to the Missouri River. With James Gamble in charge, it reached Salt Lake City in January 1866.

However, the United States company paid too high a price to outbid the old companies for contracts, and issued too much stock for the lines it merged. One stockholder pointed out in a *New York Tribune,* June 30, 1865, that the company was earning practically nothing, but paying liberal dividends. He asked whether a "ring" of insiders, composed of directors, had a contract to build the Pacific line at a huge profit.

President McKaye replied, calling attention to some errors in the stockholder's figures, but it was apparent that something was wrong. In seven months the stock dropped from $98 to $20 a share, business grew worse, and McKaye resigned. He was succeeded on November 1, 1865, by William Orton, who had been Lincoln's Commissioner of Internal Revenue.

Orton found that the company was losing $10,000 a month, although it was doing a large business. "He began to suspect that he had become the captain of a foundering ship," wrote Reid, secretary of the company.[27]

Orton had been president only three months when Wade offered a merger that followed on March 1, 1866. Western Union paid $7,218,500 in stock, a fourth of its capital, to get rid of a competitor that was failing. Evidently Wade decided the cure for overcapitalization was more of it.

Labor Forms Union

In the war years operator salaries averaged around $80 a month, but the cost of living soared, and the deluge of business required long hours of work by the limited number of qualified men. At the manual relay office in Cincinnati, Operator George Kennan said:

> Many a night I went on duty at 6 p.m. and never left my chair until I was relieved by a man coming from breakfast at 8 a.m. Once I had 10 hours sleep out of 72.[28]

Because of these conditions, the operators organized the National Telegraphic Union in November 1863. Its constitution stated their purpose to unite for mutual protection in adversity; elevate the standards of their profession; promote and maintain just, equitable, and harmonious relations with employers, "recognizing the principle that the interests of the employer and employee are identical, and firmly believing that a union among ourselves, established upon just principles, will result in manifold benefits, not only to us, but also to those who employ us."

The organ of the union was *The Telegrapher,* the "first telegraph newspaper in this country."[29] Its first monthly issue, September 26, 1864, reported the proceedings of the first convention at Philadelphia.[30]

The Telegrapher stated:

> We do not think strikes are a very intelligent way of effecting the object sought. It is a forcible way employed by those who are less intelligent, and is not becoming in a class of men such as the Union is composed of.

Notes

[1]Daniel Van Pelt, *Leslie's History of the Greater New York* (New York: Arkell Publishing Co., 1898) 1:395. Also, James Grant Wilson, *The Memorial History of the City of New York* (New York: New York History Co., 1893) 3:479. Before the war an abolitionist had no chance to win public office. Even Lincoln said the solution was to send the slaves back to Africa. The Southern states wanted to separate from the union peacefully and go their own way. Lincoln decided to fight the war to preserve the Union—not to free the slaves.

[2]Lincoln had many telegraph friends. He was pacing the floor of the news room of the *Illinois State Journal in* Springfield, awaiting news from the Republican convention in Chicago, when he received this telegram from J. J. S. Wilson, superintendent of the Caton telegraph lines: "You are nominated on the third ballot."

[3]See above, chaps. 4 and 5. Stager started work in O'Reilly's printing office in Rochester. When O'Reilly built the first line from Philadelphia to Harrisburg in 1846, Stager was the manager at Lancaster and then at Pittsburgh. He became general superintendent of the New York and Mississippi Valley Printing Telegraph Company and then of Western Union in 1856. When Southern Bell Telephone began operating in 1880, Stager was its first president.

[4]Stager to Wade, April 16, 1862. Wade Papers.

[5]After the war, when Clowry was mustered out of the service as a Brevet Lieutenant Colonel. He was placed in charge of Western Union's interests in the central and western states, and later served as president of the company from 1902 to 1910.

[6]Born April 23, 1826 at St. Clairsville, Ohio, Eckert was an operator before he was 21. In 1852 he helped Wade and Speed build the line along the Pittsburgh, Ft. Wayne and Chicago Railroad, and became the superintendent. When the line was taken over by Western Union, he remained with that company.

[7]After the war Eckert returned to Western Union as general manager of the Eastern Division. He became president in 1892, and after a decade, was chairman of the board until 1909. He died at his summer home in Long Branch, New Jersey on October 20, 1910 at 85 years of age.

[8]Conte Louis Philippe Albert d'Orleans de Paris, *History of the Civil War in America,* 4 vols. (Philadelphia: Joseph H. Coates & Co., 1876) vol 1, chap. 3, pp. 280-87.

[9]Bunnell, W. R. Plum and other operators linked the Union forces surrounding Atlanta when General Sherman destroyed that city by looting and fire. After the war Bunnell founded the J. H.

Bunnell Company, a New York electrical and telegraph supply house.

[10]Two large volumes, *The Military Telegraph During the Civil War*. by William R. Plum, an operator in the corps (Chicago: Jansen, McClurg, and Co., 1882), provided some of the following stories.

[11]His brother, Dick Ellsworth, served as a Union telegrapher. Dick had been the first operator at Fort Kearney when the transcontinental line reached that point.

[12]Fred L. Hester. *Atlanta's Telegraph History,* a mimeographed booklet.

[13]*Century Magazine*, September 1889.

[14]Fred L. Hester, "Atlanta's Telegraph History," manuscript.

[15]Desperate to save the Union, Lincoln chose Grant to destroy the smaller, poorly equipped opposition, instead of McClellan, who wanted to exhaust the South with less bloodshed and make peace by allowing it to keep its slaves.

[16]An article by General A. W. Greeley, "The Photographic History of the Civil War," *Review of Reviews* (1911), said:

> Scores of these unfortunate victims left families dependent on charity, for the Government of the United States neither extended aid to their destitute families, nor admitted needy survivors to a pensionable status." Plum, in his history of the Corps, said: "If one was killed, another took his place, and being a mere civilian, no notice was taken of his fate by the Government in whose service he died. No provision was ever made for his wife and little ones; no slab ever erected at Government expense; no military salute was fired over his grave.

William B. Wilson, pamphlet, *Civil War Personal Narratives*:

> Almost every field, almost every march numbered one of our telegraph boys among the fallen. A hundred nameless graves through the battle fields of the Union attest the devotion unto death . . . and yet the government they loved and labored for never

as much as thanked them for their services. They were disbanded without a word of thanks or a scrap of paper showing that they had honorably discharged their trying duties.

[17]William B. Wilson, head of War Department telegraph office, pamphlet, "A Few Acts and Actors."

[18]*The Telegrapher*, October 31, 1864.

[19]Conte de Paris described the Signal Corps men sitting in trees and on roof tops, sending wigwag signals, and the two balloons of the Army of the Potomac. He said the principal danger was from sentinels in Federal camps who "would be sure to fire at the indiscrete individual who thus hovered over their heads. A gas generator, a heavy machine composed of ovens, retorts, and pipes, which required 20 trucks to carry, followed the army at a distance, and the already inflated balloons, which a whole company controlled by means of strong ropes and strove to direct along the side roads of Virginia. At the least puff of wind each of these monsters would give a sudden jerk."

[20]"History of the U.S. Army Signal Center and School, Fort Monmouth, N.J.," a manuscript by Helen C. Phillips, museum director and Signal Corps historian. Fort Whipple, Virginia was renamed Fort Myer, and many Signal Corps men were trained there. Another Signal Corps station was Fort Wood, on the remnant of which the Statue of Liberty was erected in New York Harbor. Miss Phillips said Myer, who became a brigadier general, was a man of "considerable superiority," who not only founded the Signal Corps but "established 66 weather stations and gave the country its first storm signals."

In "Signal Telegraph," *Civil War Times Magazine* (May 1976), Raymond W. Smith said the battle for supremacy between Myer and Stager and their two corps was caused by Myer's ignorance of the telegraph and its equipment.

[21]In his booklet, *A Few Acts and Actors,* William B. Wilson described the scene at the War Department telegraph office after the battle of Bull run:

> General Scott could not understand that

a "hero of one hundred battles" could be beaten, and he only believed when the advancing hurricane of the flying, panic-stricken army sounded its approach to the Capitol. When the veteran at last believed, he gave me an autograph order to suppress all news of the disaster which might be offered for telegraphing to the country. Armed with this document I drove down Pennsylvania Avenue to the American Telegraph office and notified its manager of the commands of the General-in-Chief.

Piled upon the telegraph tables were "specials" from the field, describing in thrilling language—as only the "War Correspondent" could—the scenes and events of the day. All intimations of disaster were ruthlessly cut from the specials, and only the rose coloring permitted to be telegraphed. Thus it was that while the gloom of the darkest hour in the Republic's history hung like a pall over Washington City, throughout the North bells were ringing out rejoicing over the glad tidings of victory.

[22]From Western Union's official records.

[23]James D. Reid, *The Telegraph in America* (New York: John Polhemus, 1886) 484-85.

[24]Blake McKelvey, *Rochester. The Flower City, 1855–1890* (Cambridge MA: Harvard University Press, 1949) 115.

[25]Worry over a controversy with the Russian Government undermined his health. See below, chap. 10.

[26]Speed's obituary in the Portland (Maine) *Advertiser,* June 23, 1867, said that when Speed sold the Independent Telegraph Company, which he organized, he retired from the telegraph business.

[27]Reid, *The Telegraph in America,* 532.

[28]George S. Bryan, *Edison. The Man and His Work* (Garden City NY: Garden City Pub. Co., 1926).

[29]*The American Telegraph Magazine,* started by O'Reilly's family in 1852, evidently was not considered a "newspaper."

[30]*The Telegrapher*, Western Union Library.

The Russian-American Expedition

Peter the Great, Czar of Russia, wondered if Asia and America were joined by land. In 1724 he ordered an expedition to find out, and Danish navigator Vitus Jonassen Bering[2] rounded the corner of Siberia and told the Czar the two continents were separate. Bering returned in 1741 with 583 men, saw the American continent on July 16, and landed at several places. His ship was wrecked on an island, where he died December 8, 1741, but his men built a boat and returned to Russia.

In August 1784, Grigori Ivanovich Shelikof (or Shelekhov) founded the first permanent colony in Russian-America, now Alaska, taking possession of Kodiak Island and forcing the natives to hunt for him. Shelekhov put Alexander Andreyevich Baranof (or Baranov) in charge at Kodiak in 1791, and Baranof sent back rich cargoes of furs.

The next Czar, Paul I, merged small warring fur companies in the area into the Russian American Company, which Baranof and his agents ruled. One observer declared,

I have seen the Russian fur hunters dispose of the lives of the natives solely according to their own arbitrary will, and put these defenseless creatures to death in the most terrible manner.[3]

On May 25, 1799, Baranof established a trading fort at what is now Sitka. (The island on which Sitka is located is now named Baranof.) In 1810 John Jacob Astor, who immigrated to the United States in steerage with $25 and made millions in Manhattan real estate, sent the *Enterprise,* loaded with supplies, and Baranof bought a large part of the cargo. After that, American ships came to trade for sea otter and other furs. Baranof died in 1819. More coastal colonies were established, but little was learned of the vast interior.

On To Russia and Europe

Texas joined the Union in 1845, and the United States obtained what is now California, New Mexico, Nevada, Utah, Arizona, and Colorado from Mexico in 1848 and 1850. It was a period of expansion and great dreams, productive of colossal enterprises.

On the day the transcontinental telegraph linked the Atlantic and Pacific in 1861, Brigham Young's telegram pointed out a new goal:

> Join your wires with the Russian Empire, and we will converse with Europe.

Henry O'Reilly had suggested a line to Europe through British Columbia, Alaska, and Siberia, but the man who initiated the enterprise was Perry McDonough Collins.[4] He had made the long trip across Siberia in 1857 in an effort to start American commerce with Asia.

When an attempt to lay a transatlantic cable failed in 1858, Collins suggested to Western Union that a line westward to Europe be established. Encouraged by Hiram Sibley, Collins asked Congress for aid, and on February 18, 1861, the House Committee on Commerce proposed a $50,000 appropriation to find the route across Bering Strait. It might be many years, the committee said, before a permanent Atlantic cable could be developed. The Senate Military Affairs Committee also reported favorably on February 17, 1862, but Congress was busy with the Civil War.

The Senate report quoted a letter from Sibley to Collins October 16, 1861:

> Our men are pressing me hard to let them go on to Bering's Strait next summer, and (as you say to me) "if I had the money" I would go on and complete the line, and talk about it afterwards. If the Russian government will meet us at Bering's Strait, and give the right of way through their territory on the Pacific, we will complete the line in two years, and

probably in one.

Collins went to St. Petersburg. Aided by U.S. Minister Cassius M. Clay, he persuaded Russia to grant a charter for the line on May 23, 1863, and agree to extend its line from St. Petersburg to Khabarovsk and to Nikolayevsk at the mouth of the Amur, to meet an American line. It also granted a concession for thirty-three years of a 2,000-mile right-of-way from the Amur to Bering Strait, and forty percent of "the net produce" on messages.[5]

Western Union had provided $5,000 to Collins to negotiate with Russia and England, and he went on to London and obtained a right-of-way through Canada. He then made a formal report to Western Union on September 28, 1863, and renewed it by letter on March 9, 1864. On March 24, the company announced it had purchased Collins's rights, grants, and privileges for $100,000 in cash and 10,000 shares of Western Union Extension stock with a par value of $1,000,000.[6]

That corporation, headed by Sibley, issued a glowing circular, saying:

> On a line of two wires, one thousand messages a day can very readily be dispatched, i.e., twenty messages an hour day and night. The charge per message may reasonably be put at $25. This would give nine millions of dollars a year!

People clamored for the stock, and $8,434,000 was subscribed. Western Union appointed a committee on March 16, 1864, "to appear before Congress and secure necessary legislation and aid." Secretary of State Seward told the Senate Commerce Committee the project was rightfully entitled to government aid, and a bill was passed on July 1, 1864, granting right of way across

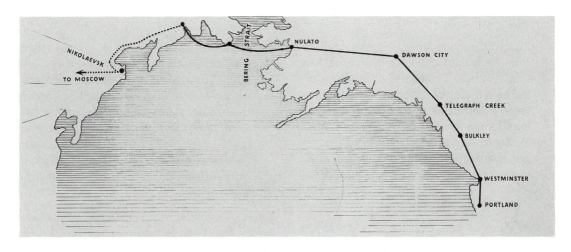

10.1 Route of the Russian-American Expedition to build line westward to Europe.

public lands, use of a naval vessel to make surveys and aid in the work, and compensation for government use of the line. President Lincoln signed it.

Expedition Organized

Organization of the expedition, popularly called the Collins Overland Telegraph, began at once. Collins was elected a Western Union director and managing director of the line. Charles S. Bulkley, who had constructed the New Orleans line in 1847, and headed the Gulf Department military telegraph in the War, was appointed engineer-in-chief, and arrived at San Francisco January 14, 1865. An order had been sent to Henly & Co. of England for 5,000 miles of number 9 galvanized wire and fifty miles of annealed tie wire to be shipped to Victoria, British Columbia.

The recruiting and outfitting office was on Montgomery Street, San Francisco. Several hundred adventurous and daring men, most of them just returned from the Civil War, flocked to the office in the spring of 1865, where Bulkley accepted only "live," "young" men. They were organized, uniformed, and disciplined in a manner typical

of the Army and Navy, and divided into Working Divisions, Quartermaster's Corps, and Engineering Corps. Working Divisions were divided into construction parties. Officers of the expedition were as follows.

Land Service. Frank N. Wicker, chief, who had participated in major Civil War battles as a member of the Signal Corps; Dr. Henry P. Fisher, surgeon-in-chief, who had been a surgeon in the war; Scott S. Chappel, chief quartermaster, a veteran; George M. Wright, adjutant and secretary, a lieutenant of artillery; John F. Lewis, chief draftsman; Frederick Whymper, artist; Eugene K. Laborou, chief interpreter; Lawrence Conlin, chief carpenter, a Mexican who was a Civil War veteran.

American Division. Edmund Conway, chief, who was in the Military Telegraph Corps during the Civil War; J. W. Pitfield, agent at New Westminster, British Columbia, likewise in the Military Telegraph Corps; F. A. A. Billings, assistant quartermaster; Henry Elliott, clerk; Franklin L. Pope, chief of explorations, British America; J. Trimble Rothrock, first assistant, a captain of cavalry; James J. Butler, second assistant and quartermaster; Ralph W. Pope, clerk and operator;

Robert Kennicott (many sources spell the name Kennicutt), chief of explorations, Russian America, a naturalist who previously had been in Alaska for the Smithsonian Institution and planned to collect specimens for the Chicago Academy of Sciences; William H. Ennis, first assistant; Thomas C. Dennison, quartermaster; Lewis F. Green, engineer in charge of the steamer *Lizzie Horner.*

10.2 When the great Russian-American Expedition was abandoned after the first successful transatlantic cable was laid, the Indians used the discarded wire to build bridges across rivers. One crossed the Hagwilget Canyon 500 feet above the roaring torrent of the Bulkley River.

Siberian Division. Serge Abasa, Chief, a Russian of noble blood; George Kennan, quartermaster and secretary; J. A. Mahood, chief of explorations, Lower Siberia; Richard J. Bush, secretary and quartermaster; Collins L. MacRae, chief of explorations, Upper Siberia; A. S. Arnold, quartermaster; Alexander Harden, interpreter.

Marine Service. Captain Charles M. Scammon,[7] chief of marine, commanding the flagship *Nightingale;*[8] Captain W. H. Marston (later J. W. Patterson), commanding the steamer *George S. Wright*; A. M. Covert,

engineer; Captain John R. Sands (later Charles Sutton), commanding the bark *Clara Bell*; Captain Matthew Anderson, commanding the bark *H. G. Rutgers*; Captain Thomas C. Harding, commanding the schooner *Milton Badger*; Lieutenant Davidson, commanding the bark *Golden Gate*; Captain Arthur, commanding the bark *Palmetto*; Captain Norton (later H. Tibbett), commanding the bark Onward.

There were also four material ships and three small river steamers. The *Evelyn Wood, Mohawk, Egmont,* and *Royal Tar* carried cargoes of wire, cable, and other supplies from the Atlantic to the Pacific.

Work in British Columbia

Since a line from San Francisco had reached Portland on March 1, 1864, work began there. The task of extending it about 300 miles to New Westminster, British Columbia, was assigned to R. R. Haines and James Gamble, who started work July 6 with a large party. A mast 180 feet high and a 200-foot fir tree were used to suspend the line over the Willamette River, and the mile-wide Columbia River was crossed by cable at Fort Vancouver, north of Portland. The party followed the Columbia to the Cowlitz River, and continued north, entering Olympia on September 4, after building 120 miles of line in sixty days. When the line reached Seattle October 25, John M. Lyon opened an office.

North of Seattle the virgin forest was interrupted by ten rivers which the line crossed on masts. The men slept each night on the small British steamer *Union*. The Snohomish River was so wide it required seven masts.

Gamble wired from New Westminster, British Columbia:

I have wired the left wing of the American Eagle to the tail of the British Lion. They work peaceably in harness. For 150 miles north of Seattle to New Westminster it was the most difficult and roughest extension of the telegraph ever made. . . . I went up with Colonel Bulkley and staff, to introduce him to Governor Seymour, and remained as guest of the governor while Colonel Bulkley explored the coast as far as Sitka.

Bulkley was on the revenue cutter *Shubrick*. Congress had voted to provide Naval assistance to the expedition, but numerous letters and visits failed to move the Navy until Congress passed a joint resolution citing its Act of July 1, 1864, promising naval aid.

The Secretary of the Navy is hereby authorized and required [it said], to detail one steam vessel from the squadron of the Pacific station or elsewhere.

It said Russia had (in addition to other ships) assigned the steam corvette *Variag,* a 2,156-ton ship with seventeen guns and 306 men.

The *Shubrick* was finally delivered, and work on the line began. The Fraser River was crossed by cable at New Westminster, where Governor Frederick Seymour of British Columbia piloted his own steamer to lay the cable.[9]

The Hudson Bay and Russian American Companies furnished maps, charts, boats, guides and interpreters, and the expedition hired natives in many areas.

When Bulkley returned to Victoria, B.C., on July 25, 1865, he reported that work in British Columbia was going ahead rapidly under Edmund Conway, with 460 miles of line north from New Westminster to the Cascade Mountains nearing completion. Conway was sending supplies ahead to Babine Lake, in preparation for winter work.

Exploration had gone ahead, under the leadership of Franklin L. Pope,[10] with thirty men and forty pack mules. Pope sent most of his party back in the fall of 1865, and pushed on to Fort Fraser, then on to Fort St. James, and through the lake region to Lake Takla. At the head of the lake the party built a long house, thirty-five by fifty feet, with a huge chimney. The windows were covered with caribou skins, scraped and oiled. They made tables, benches, and double-deck berths. The roof was of long grass, laid shingle-fashion and covered with clay. Except for explorations and surveys, the party spent the winter in that building, which they named Bulkley House. Supplies ran low, but hunting was good for rabbits, grouse, beaver, and other small game.

James J. Butler followed the Babine River to where it joins the Skeena, built a boat, and followed the Skeena through the wilderness to the Pacific Ocean.

Accompanied by George Benkinsop and two Indians, Pope tramped 500 miles northward on snowshoes to Lake Takla, up the Skeena headwaters and on to the Stikine River, which breaks through the coastal range of mountains to the Pacific, and was used later as an avenue for supplies. For seventy days the four men escaped starvation by trapping small animals in the snow.

The construction party headed by Conway built north along the Fraser River, reached Quesnel, B.C. on September 14, and continued northwest 400 miles up the river, then named for Bulkley to a point near the Indian village of Hazelton. The *Prince George Citizen* (British Columbia), August 25, 1866, said the line neared Hazelton on July 27, and the party built a fort they named for Anson Stager, stations and bridges, and a rough road that became known as

"The Telegraph Trail."

Conway did a tremendous job. His report to Bulkley on February 19, 1867, said:

By the 1st of June I had 150 men; 86 in construction camp, 26 packers with 160 animals, 38 white men and Indians transporting supplies in bateaux between Quesnel and Fort Fraser. . . . We constructed the telegraph road and line 378 miles, fifteen stations with a long house, chimney, door and windows, 25 miles apart. We built bridges over all small streams that were not fordable, corduroyed swamps. All hillsides too steep for animals to travel over were graded, from three to five feet wide. The average width of clearing the wood for the wire is, in standing timber, 20 feet. . . . All underbrush and small timber is cleared to the ground, thus leaving the road fit for horses. . . . Number of poles put up 9,246.[11]

A veteran of the Military Telegraph Corps, Conway maintained military discipline. His first group cut right of way, the second dug holes, the third cut poles, and the fourth, headed by Frank H. Lamb, constructed the line.[12]

Kennicott's Explorations

The steamer *George S. Wright* sailed on July 12, 1865, without the U.S. Navy officer assigned to command it. Bulkley wired New York that the officer left the vessel, objecting that "She would keep his patent leathers too dirty." On board was Major Robert Kennicott, a serious, dreamy-eyed young man of twenty-nine with a drooping moustache. He wore the uniform of the expedition's Scientific Corps, with Union Army style cap, epaulets, and two rows of brass buttons. His party included an English painter, Frederick Whymper, and two twenty-year-old San Franciscans, George R. Adams

and Fred M. Smith, who kept diaries of their experiences.[13]

Major Abasa's Siberian party left on July 3 for Petropavlovsk on the Russian Brig *Olga,* followed by the *Golden Gate* with men and supplies for Norton Sound, and the *Anadyr,* and the *Milton Badger* with supplies for Anadyr Bay.

10.3 MAJOR ROBERT KENNICOTT.

The *Wright* and *Golden Gate* arrived at St. Michael on Norton Sound late in September. Their cargoes were taken to shore in small boats, one of which capsized, ruining guns and ammunition. Curious Indians swarmed on the dock with sled dogs. The visitors feasted at the home of Post Commander Sergei Stepanoff in the Russian stockade, and pitched their tents beside it.

Bulkley sailed on to Bering Strait where he decided on a 178-mile cable route to a point south of the Strait in Siberia, and 209 miles across the Gulf of Anadyr, into the mouth of the Anadyr River. The water along that route was so deep, he said, the sheet ice in winter would not injure a cable, and the northward flow of the current through the Strait would prevent icebergs from moving south across it.

There was so much snow and ice in the Anadyr River in September, Bulkley feared his ship would be frozen in for the winter. MacRae's party had been landed by the schooner *Milton Badger,* and was preparing camp for the winter in order to explore the route with reindeer, west along the river and south across the base of Kamchatka Peninsula.

Leaving the Anadyr on October 13, 1865, Bulkley sailed to Petropavlovsk (Russian for "Village of St. Peter and St. Paul") on the Kamchatka Peninsula. He learned Abasa had divided his Southern Siberian forces into two parties, and was preparing to build north from the Amur River. Bulkley reached San Francisco on November 20.

Bulkley had left Kennicott[14] at St. Michael on Norton Sound to explore the "Kwichpak" [*sic*] River 500 miles into Alaska, as the proposed telegraph route, but Kennicott found to his consternation that his small river steamer, the *Lizzie Horner,* would not run. Adams wrote in his diary that in spite of days of effort, Lizzie would not go, "and our intended expedition up the Kivpak [*sic*] is broken up." Without the steamer, the river would freeze before the party could explore it and establish supply bases.

Kennicott moved his party east a few days travel to the village of Unalaklett, also on Norton Sound, and built log cabins for the winter. On October 22 the party started up a small stream, portaged to the "Kwichpak" River, and then went up that frozen stream to Nulato, a Russian trading post.

The party made several exploration trips by dog sled, seeking a route through the mountains and mapping the area. When Kennicott returned to Nulato in the spring, Adams's diary said Kennicott was "a good deal troubled" by failure to map the river, and these frustrations had "broken him down."

On May 13, 1866 Kennicott suddenly died. That page was torn from Smith's diary, but Adams wrote that the night before Kennicott came in and stood beside him for a minute. The next morning he was found dead on the beach.

> The last seen of the major was when he got up [Adams wrote]. He made a remark that he could not sleep. He got up and commenced writing. This was at 2 or 3 o'clock in the morning. After writing for a few minutes, he got up, lit a pipe, and went outside into the fort yard. He walked there for a few minutes, and then the watchman opened the gates, and he went out. . . . By his side was an open compass.[15]

Work in Alaska

The explorers were followed by a large construction party, and, on New Year's Day 1866, both joined near Nulato in erecting the first telegraph pole. Nulato was the most interior and northern Russian American Fur Company fort. A thirty-two-gun salute was fired—one for each state—and the pole was ornamented with flags of the United States, the expedition, the Masonic Fraternity, and the Scientific Corps. The salute so frightened Indian spectators that they fled.

A station was erected at Nulato to house the builders and supplies. A mile away the scientific party built Fort Kennicott, with buildings forming a 100 by sixty-five foot enclosure.[16]

Progress was slow because of deep snow and cold, sometimes 45° to 68° below zero. The only beasts of burden were dogs; supplies and poles were hauled on sleds. Axes and other tools were dulled cutting frozen poles. The ground was frozen to a depth of five feet, and only a few holes could be dug in a day.

Another party built a forty-five-mile section of line south from Port Clarence along the coast, to extend around Norton Sound to St. Michael, and then up the river. At Port Clarence they printed a little monthly news sheet, *The Esquimaux,* the first publication in what is now Alaska. One story described progress "in fierce and blinding" storms at 55° below zero. Another described a visit to an isolated Indian village, where a frenzied, elemental dance was held in their honor in the one large hut.

The Westdahl Diary

Experiences of the construction party that moved from St. Michael to Unalakleet were recorded in the diary of Ferdinand Westdahl,[17] a navigator and master mariner from Wisby, Sweden. He had been first officer of the bark *Golden Gate* when it and six other barks, a ship, and a steamer sailed from San Francisco in June and July 1866 loaded with supplies. All met at Plover Bay, Siberia, when an English ship arrived with the Bering Sea cable. The *Golden Gate* was wrecked in the Anadyr River, and Westdahl was appointed astronomer and surveyor for a party that went on to Alaska.

At Unalakleet, Westdahl's party lived in three houses built in a square. The fourth side was a "palisade" fourteen feet high, of pointed logs set upright in the ground. About 500 Indians lived at Unalakleet and used a large house for conferences and dances. The men danced stripped to the waist, with a wolf tail at the belt, feathers on the head, and white fox-skin gloves. Seven or eight men would form a circle and tap the floor with their right feet, twisting their bodies. One or two Indian women, fully dressed, would slide into the circle and stand in one spot swinging their arms and bowing rhythmically to the music.

Exploring the area and laying out the line, Westdahl sometimes slept in huts with fifteen or twenty Indians in a space about fifteen feet square, dug into the ground. The entrance was a tunnel. The Indians slept on benches with their faces turned toward a fire on a flat stone in the center. The smoke escaped through a hole in the roof. The Indians lived on fish, seals, reindeer, and birds.

That party erected its first telegraph pole on January 22, 1867. Their leader, William Ennis, made a speech, and everyone cheered. The camp moved as the line progressed, with all sleeping in the snow in temperatures 40° and 56° below zero. Pancakes fried in seal oil were eaten by men wearing large reindeer gloves, because forks and spoons would freeze to bare hands. Making camp for the night, they spread bear skins on the snow, wrapped themselves in heavy fur coats, drank a final cup of hot tea, and, with their feet toward a camp fire, covered themselves with blankets lined with rabbit, and slid into reindeer-hide sacks.

The party slept in the open for three months, building forty miles of line. Then they had to patrol the line to stop the Indians from cutting it down, because their wizards blamed it for a shortage of reindeer. Four Indians were shot before the vandalism stopped.

Westdahl said work could be done only in winter when supplies could be hauled on sleds on ice and snow. In summer, mosquitoes were so fierce they could drive a man insane.

The Yukon River was believed to empty into the Arctic Ocean. English traders and trappers had descended the Porcupine River and founded the trading post of Fort Yukon where the Porcupine joins the Yukon River. Since the lower Yukon had been explored only by the Russians, the maps showed two rivers. Traveling the "Kwichpak" River, Frank E. Ketchem and Michael Lebarge of the expedition reached Fort Yukon and learned it is one river. The Yukon flows north from Canada, but swings southwest at Fort Yukon, hundreds of miles to the Bering Sea. When they arrived at Fort Yukon, they were greeted with cries of, "Where under the sun did you come from?" They again made the trip to Fort Yukon by dog sled in March and April 1867, and continued by canoe to Fort Selkirk in Canada.

William H. Dall, who arrived on the clipper ship *Nightingale,*[18] and the artist Frederick Whymper also ascended the river, for 1,200 miles they estimated, to Fort Yukon. Whymper declared the Yukon navigable for 1,800 miles.[19]

Hudson Bay Company officials in London laughed at Hiram Sibley when he said the Yukon emptied into the Pacific. They produced maps and a man who had lived on the Yukon. It was years before the map makers were convinced.

Work in Siberia

A graphic story of the Siberian party's exploration from the Amur River to the Gulf of Anadyr was written by George Kennan,[20] who went to Petropavlovsk with Abasa's party in August 1865. He said James A. Mahood and R. J. Bush crossed the Okhotsk

Sea to Nikolayevsk, at the mouth of the Amur, to prepare for building northward.[21]

Abasa and Kennan, with an American fur trader named Dodd as a guide, travelled 700 miles up the Kamchatkan Peninsula to Gizhiga. The simple natives of the Kamchatkan Peninsula provided horses, river boats, and guides from village to village. The head of one village thought the word "operator," following Kennan's name in a message sent ahead, meant "imperator" and held a reception in the belief that he was the Russian Emperor.

10.4 Two unidentified members of the Siberian expedition. (Loaned by Frederic Bronner, Victoria, British Columbia.)

The three men passed through a great treeless expanse of tundra that melts to a depth of two feet in summer. The water, standing in twenty-four hours of daily sunshine on solidly frozen ground, produces a soft, dense moss about eighteen inches deep, like a great sponge, and breeds hordes of mosquitoes.

From Tigil, a seaport on the Okhotsk Sea, they pressed on to Lesnoi, crossed the mountains with dog sleds, and arrived in the land of the Wandering Koraks. They travelled from tribe to tribe on sleds drawn by reindeer. The Koraks, who resembled American Indians, lived in conical tents, surrounded by herds of reindeer, pawing up the snow to eat the moss beneath. When the moss was exhausted, the tribe would move. The reindeer provided clothing, tents, food implements, and transportation.

The Koraks rarely saw anyone from the outside world, and writing was inconceivable to them. They worshipped evil spirits, appearing to them as diseases, storms, famines, and brilliant Auroras, and publicly burned their old alive when infirmities made them unfit for the hardships of nomadic life.

After traveling three months dressed in heavy furs, sleeping on the ground or snow, never seeing furniture, never undressing, grimy from climbing down chimney entrances to Korak huts, the party reached Gizhiga.

Abasa then went southwest to meet Bush and Mahood at the village of Okhotsk, while Kennan and Dodd started northeast on December 13, 1865. Mirages mocked their eyes: luxuriant foliage overhanging blue, tropical lakes, or the walls, domes, and minarets of an oriental city. A long line of dog sleds moved through the air, upside down. Traveling on, because they had to find wood before camping, or die, their beards became tangled masses of frozen iron wire; eyelids grew heavy with long white rims of frost and froze together when they winked.

On stormy days progress was impossible: no tent would stand; no fire would burn. The party would dig in, and wait. At Anadyr, Kennan heard from wandering Chookchees that a party of Americans had landed months before near the mouth of the river, more than 200 miles away.

Natives declared the trip impossible, but Kennan obtained eleven men by threats. On the night of the tenth day, they were barely able to find enough wood to boil tea, and went on twenty-four hours without a stop. Intense cold, exhaustion, and lack of warm food weakened the party until their lives depended upon quickly finding a hut buried under the snow that a wandering tribe heard of months before. Dodd had been overcome by stupor when the hut was found. After three days of rest, all returned to Anadyr.

Kennan returned to Gizhiga and rejoined his party. The two groups had traveled about 10,000 miles in seven months, and explored the route from the Amur to the Anadyr. The *Nightingale* arrived loaded with sixty-five men, wire, poles, and supplies. The first pole was erected in Siberia on August 24, 1866.

The *Variag* of the Russian Navy and the *Clara Bell* brought men and supplies for the southern section. Col. Frank Knox, a *New York Herald* correspondent, was on the *Clara Bell*. Then the *Onward* and *Palmetto* unloaded cargoes that were distributed along the line.

Returning to the Anadyr, Kennan found the party there had been increased to forty-seven when the *Golden Gate* was wrecked by ice four miles off shore on October 4, 1866.[22] The crew had hauled some of the wrecked ship's supplies across the ice but when they were used up the forty-seven men had lived for months on such fish as they could catch through the ice until a ship took them away.

The Wanless Diary

10.5 LIEUTENANT J. W. WANLESS.

The hardships in Siberia are confirmed by the diary of Lieut. J. W. Wanless, in charge of surveying and preliminary work from Gizhiga to Hamsk.[23] Wanless, Lieut. J. Leet, and twenty-two other members of the Siberian party arrived on the *Onward* on September 26, 1866, and constructed a building for supplies and men. When snow fell on October 22, several groups started for their posts. Wanless and his men moved through reindeer country, where life often depended on the health of dogs, undependable native guides and workers, and fish and deer for food.

Collins wrote to President Andrew Johnson on November 27, 1865, that only 1,700 miles of the Russian line was not completed; the American line had reached Fort St. James, B.C., about 1,699 miles from Bering Strait; the 2,000-mile route in Siberia had been surveyed and work begun; and negotiations were underway to extend the line to China.

On January 18, 1866, Abasa reported:

Notwithstanding the scarcity of laborers in the country, I have commenced preparatory works in Anadyrak, Jijiginsk, Yamsk, Tao-usk, and Okhotsk.

In the summer of 1866 the expedition seemed to have won its race against the Atlantic cable, but when a successful cable was laid on July 27, 1866, the line westward to Europe was not needed. On February 27, 1867, the company ordered work stopped.[24]

The Siberian party received the order on July 15, and, from Yamsk, on August 2, 1867, the Wanless group sailed for home on the *Onward*. The *Clara Bell* took the men at St. Michaels, picked up Abasa's party, and arrived at San Francisco on October 7. Kennan, Mahood, and two others returned across Russia and Europe.[25]

One party in Siberia obtained $150 in rubles for $20,000 worth of supplies by persuading natives that glass insulators could be used as drinking cups, and wire for fish nets. Most materials were left as they were, but 1,000 miles of wire was returned to San Francisco. Western Union sent the Bering cables back to Henley and Company, and in return received 4,000 miles of landline wire.

In a farewell address to his men Bulkley said:

Over nearly one quarter of the circumference of our globe in frozen wilds, among savage tribes, and in unknown regions, you have steadily pursued your way. Although the telegraph is unfinished, the world will recognize and applaud the knowledge you have added to its stores, and the daring spirits who have accomplished so much.

Western Union paid the loss of $3,170,292, although under no legal obligation to do so, because the line had been incorporated separately. It issued bonds in exchange for the worthless stock. The "moral obligation" the directors felt, no doubt, was strengthened by the fact that they owned over $3,000,000 of the extension stock themselves.

Although Leonard Bright, eighty-seven, of Brooklyn, a survivor of the expedition, and others in 1932, believed that the Bering and Anadyr cables were laid, they were mistaken. The company agent in London reported on March 28, 1866, that the *Egmont* sailed with 500 miles of cable it was to lay. However, newspapers reported the *Egmont* had to put into Mauritius for repairs. The British brig *Ann* took 100 miles of cable from the *Egmont,* and the British ship *Evelyn Wood* took the remaining 400 miles. A clipping based on a letter by W. W. Smith, on the *Nightingale* in September 1866, said the *Evelyn Wood* arrived at Plover Bay, Siberia, with the cable, but because of forming ice, it was too late to lay it.

Russia completed the line across Siberia later and extended it to provide the first rapid communications with the Orient. For his diligence in furthering the telegraph, the Emperor made Tolstoy a Count.

Purchase of Alaska

Although American histories practically ignore the Russian-American Expedition, the purchase of Alaska, now the largest of the United States, was a direct result.

When Western Union decided to build the line in 1864, Sibley had Seward and Russia's minister to Washington, Baron Edward de Stoeckl, write to Prince Alexander Gortchakoff,[26] then minister of foreign affairs, and later chancellor of the Russian Empire, to pave the way. Sibley then traveled to Russia to replace Collins's thirty-three-year grant with a perpetual lease. He thought a lease would avoid trouble with the Russian American Company which held exclusive rights in that area.[27]

Sibley took a letter of credit for $750,000 with him to St. Petersburg, where he and Collins were presented to the Emperor on November 1, 1864.[28] The Emperor "promised his cordial cooperation in this great enterprise," and Sibley was entertained as a distinguished guest of the Empire.

During the discussions, Sibley told Prince Gortchakoff he would pay $750,000 for the rights to the strip of land desired for the line's right of way. "Why pay $750,000 for the rights of the Russian-American Company," Prince Gortchakoff asked, "when for that sum you can get the fee simple to the tract you want?"[29] That led to a discussion of the purchase of the entire territory, but Sibley did not want Western Union to own and rule that huge wild area, so he urgently proposed the purchase to Washington.[30]

The prince asked what would be the probable cost, in answer to which Mr. Sibley mentioned a considerable sum. This drew the remark that it was not worth any such sum, and Russia would sell the whole Alaskan area for a sum not much more. At the end of the interview, Mr. Sibley asked the prince whether he intended his words to be taken seriously and whether he might bring it to the attention of the United States government. The Prince replied that he was quite serious and had no objection to the suggestion being made to the U.S. Government. Sibley lost no time in communicating this to General Cassius M. Clay, then minister of the United States to Russia, who at once sent the information to Secretary Seward at Washington.

During the negotiations, Russia offered Western Union forty percent of the *net* tolls "after all expenses" on the part of the line in Russia, while Sibley argued for forty percent of all tolls. After months of frustrating delays, Sibley declared further discussion was seemingly useless, and he announced his intention to leave for Washington. Count Tolstoy urged him to delay his departure and hinted that a compromise might be reached, but Sibley did not intend to compromise.

In the morning Sibley left for Berlin in a conveyance provided by the Russian Government. He had not been in Berlin twenty-four hours when Count Tolstoy arrived and announced that he had been directed by his government to concede to Western Union's demand.[31] The concession may have resulted from a vigorous letter of protest by Sibley and Collins that Clay sent to Gortchakoff on March 21, 1865. The letter said the United States held the imperial government responsible for all losses resulting from Russia's failure to live up to its pledge.

Clay wrote to Seward on April 2, 1865[32] that the agreement had been signed, and confirmed by the emperor, and orders given for admission of men and material.

The long controversy—in which the entire investment in the expedition was in danger of being lost—had undermined Sibley's health and so weighed on his mind, that he followed the advice of physicians at Berlin to put business cares aside and rest. He sent his resignation as president to Western Union on July 26, 1865, but the compa-ny retained him "on the board as first vice president."

Western Union gave 100 shares of extension stock ($10,000 par value) to Russian Minister de Stoeckl as a token of appreciation for his cooperation. De Stoeckl had advised his government that since Russian America was of little value to his country, it should be sold to the United States. He said he believed Canada would then join the United States, and Great Britain would withdraw from North America.

Russian officials believed the remote Russian America area could not be defended in a war. Russia, as a good friend, had stationed its navy at New York and San Francisco in the Civil War to warn England and France not to aid the Confederacy, and the United States had aided Russia in the Crimean War in 1854–1856. Grand Duke Constantine, brother and chief advisor of Czar Alexander II, believed the United States, as an ally owning Alaska, would guard Russia's Siberian back door.[33]

Emperor Alexander II's advisors met in the palace on December 16, 1866, and told Grand Duke Constantine that Russia needed the money, and the sale would strengthen the United States as an ally and be a blow to England and Canada's future. The emperor instructed de Stoeckl to negotiate the deal, and the grand duke handed him a map with the Alaskan area marked.

When de Stoeckl returned to Washington in March 1867, Seward was delighted. He was determined to succeed Johnson as president, and believed the deal would help. He and de Stoeckl worked until 4 a.m. drafting the treaty, and hours later Seward persuaded President Johnson to sign it and send it to the Senate.[34]

The price was $7,200,000, less than two cents an acre for 375,000,000 acres—almost

10.6 CERTIFICATE FOR STOCK IN WESTERN UNION EXPEDITION TO EXTEND THE TELEGRAPH WESTWARD TO EUROPE. WHEN THE FIRST SUCCESSFUL TRANSATLANTIC CABLE WAS LAID, THE EXPEDITION WAS CANCELLED, BUT WESTERN UNION PAID THE FULL VALUE AT WHICH SHAREHOLDERS HAD PURCHASED THE STOCK, SAYING THE COMPANY HAD A MORAL OBLIGATION TO THE SHAREHOLDERS.

as large as all states east of the Mississippi River. Horace Greeley, editor of the *New York Tribune,* attacked the proposal. "Our splinter of the North Pole," he said, seemed valueless. Other papers called it "Seward's Folly," "Seward's Icebox," "Walrussia," and "Iceburgia."

In a 380-page address, Senator Charles Sumner, chairman of the Senate Foreign Relations Committee, declared the nation was indebted to the Western Union expedition "for authentic evidence with regard to the character of the country and the great

rivers which traverse it." The Senate then ratified the treaty on April 9, 1867 by a vote of 39 to 12. Alaska[35] was transferred to the United States with a ceremony at Sitka on October 16, 1867. The House finally appropriated money to pay for it July 14, 1868.

The voluminous Senate and House documents relied heavily on discoveries and reports of the telegraph expedition. "If it had not been for the interest excited by the expedition, and the information which its members were able to furnish," Dall asserted, "the proposition would have failed to win approval."[36]

U.S. Minister Clay wrote:

Hiram Sibley was the first to talk of buying a part of Alaska for the placing and management of the telegraph line and plant. Under his instruction, I was sounding the Russian Government. The Western Union Extension was first in suggesting to the Russian Government the sale of the Province of Alaska and the possession of the land for telegraph purposes in perpetuity.

After the expedition, Western Union continued operating the line 378 miles to Quesnel until February 1871, when it was leased to British Columbia. When British Columbia became a Canadian province that July, the Canadian Government took over the lease, and bought the line from Western Union in September 1880 for $24,000.

North of Quesnel the 400 miles of line were abandoned, but prospectors who refused to pay high steamer prices plodded along "The Telegraph Trail" in 1896, en route to the Klondike gold fields. The Yukon Telegraph Company extended the line north from Hazelton via Telegraph Creek, Atlin, and Whitehorse, reaching Dawson in 1901.[37]

Now all that remains of Bulkley House on Lake Takla is the ruins of the huge chimney. The slender bridges, built by the Indians with wires from the abandoned telegraph line, stood for many years. One crossed Hagwilget Canyon, on the Bulkley River, 200 feet above the roaring torrent, near where it joins the Skeena. The span, 150 feet long and six feet wide, had suspension cables made of the old wire anchored on each bank. The timbers were telegraph poles, dovetailed and held together by wire.

When the Indians built that bridge, seventeen of the fattest squaws reportedly were sent to its center to do a potlatch (cere-

monial feast) dance. When the bridge stood that test, the "braves" decided it was safe for men and pack horses. The British Columbia Bureau of Mines report of 1905 mentioned four loaded horses crossing it.[38]

Some names on Alaska maps date from the expedition: Telegraph Mountain Range, the towns of Telegraph Creek and Bulkley, the Bulkley River, and Bulkley Lake. A city, a lake, a glacier, and a mountain are named for Kennicott. Fort Stager is now the town of Kispiox. More than a dozen stations along the wilderness route of the line were used as post offices and roadhouses for bed and booze for many years.

The heroism of the line builders was commemorated by a memorial cairn and tablet at Quesnel, unveiled in 1932 by Western Union Superintendent F. C. Coles of Seattle.

Perry McDonough Collins lived in New York City until he died at age seventy-seven on January 18, 1900. His niece and heir, Kate Collins Brown, bequeathed $550,000 to New York University in 1917 as a fund in his memory to provide scholarships.[39]

Tribute of Hiram Sibley

Sibley was one of the builders who made America great. He was rough, tough, and scheming, but had the vision to see great needs, the courage to try what others believed impossible, and perseverance to succeed. When getting the job done required walking roughshod over others, he did not hesitate.

After he retired as Western Union's president, Sibley continued to serve as a director and vice president, but other interests gradually absorbed his energies. He bought two railroads and huge tracts of timberland on which coal and oil were found. His Sibley and Company seed house trademark was painted on countless barns and stores. He had a huge acreage of cultivated land, mostly in New York, Illinois, and Canada.

10.7 Hiram Sibley, whose strategy and determination led to the formation of Western Union as the national telegraph company, and his presidency of it. Sibley arranged the building of the first transcontinental telegraph line and negotiated the purchase of Alaska from the Czar of Russia.

Sibley was complex. He made many enemies, but to those he chose, no one could have been a better friend. To intimates his conversation was "fascinating." He was unostentatious in contributing a million dollars to education and art, but gave nothing to church or charity. Remembering his own needs as a boy, he preferred to give young men the opportunity to help themselves.

When Sibley died on July 12, 1888, he left an estate of ten to twelve million dollars, of which the *Chicago Tribune,* of July 18, 1888, said between six to eight million dollars went to his son Hiram Watson Sibley,[40] who was successful in railroad and other enterprises. Hiram's grandson Harper was a large owner of farm and ranch lands, and president of the U.S. Chamber of Commerce. Hiram's great grandson, Harper Jr., a real estate and insurance man, was Public Safety Commissioner of Rochester, N.Y., and succeeded his father on Western Union's Board of Directors.

The Alaska purchase agreement included about seventy percent of the Bering Sea. In the 1980s Russia wanted to move the boundary line east to add to its area 13,200 nautical miles of sea, including islands and large fishing and oil/gas reserves. In 1991, the dispute had not been settled.

Notes

[1]This chapter is the first substantial account of the expedition that resulted in the purchase of Alaska. Gathering these facts required years of research because the sources generally were journals, diaries and letters of participants in the hands of descendants, bits of information in rare, old books and magazines, Bulkley's papers in the Portland, Oregon Library Association, sketches and maps at Cornell University, House and Senate documents in the Library of Congress, State Department correspondence on microfilm in the National Archives, and minutes of Western Union board meetings during the 1860s.

[2]Behring was the original spelling, but modern spellings are generally used in this chapter.

[3]Hubert Howe Bancroft, *History of Alaska, 1730–1885* (San Francisco: A. L. Bancroft & Co., 1886).

[4]Sources agree it was spelled "McDonough." He was named for two naval heroes of the War of 1812: Commodores Oliver Hazard Perry and Thomas Macdonough. Born at Hyde Park, New York, in 1813, he worked at New Orleans for a Mississippi steamship company, joined the gold rush to California, became a partner of General Grant's father-in-law as a banker and dealer in gold dust, and, on March 24, 1856, was appointed U.S. Commercial Agent for the Amur River.

[5]From Clay's June 17, 1863, report to Secretary of State Seward. National Archives microfilm roll 20, vol. 20, U.S. ministers' dispatches from Russia to Department of State. Secretary of State microfilm provided the basis for much information in this chapter.

[6]From company records, including a 165-page booklet collated from official documents of Western Union's "Russian Bureau."

[7]A copy of Charles M. Scammon's private journal, 129 typewritten pages, was given to the author by Joel W. Hedgepeth of Rockport, Texas, in 1946. It provided many facts in this chapter. Selected because of his long experience as commander of a cutter in the U.S. Revenue Service, Scammon was granted repeated leaves of absence to serve with the expedition, and was permitted to fly the flag of that service on his flagship, which was the *Golden Gate* until the *Nightingale* was purchased in February 1866.

[8]The commanding flagship *Nightingale* was one of the great clipper ships which have a supreme place in the tradition of the sea. Named for Jenny (Johanna Maria) Lind, the "Swedish Nightingale" (operatic soprano), this beautiful ship carried as a figurehead a carved portrait bust of the great singer. In her staterooms and about the quarter deck were rare woods and fine carvings. Built at Portsmouth, N.H. in 1851 by Samuel Hanscom, Jr. for a Swedish count, and 160 feet long, she was a sight "so beautiful and dainty as to make a sailor weep for joy." She was so fast in the China tea trade that her owners offered to race her against any ship to China and back for $50,000 stakes.

Sold by a firm in Canton, China in 1859, she engaged in the African slave trade with a New Yorker, Francis Bowen—"The Prince of Slavers"—in command. On one trip, the boat played hide-and-seek with the slower U.S.S. *Saratoga*. The warship finally slipped a crew aboard the *Nightingale* at 1:00 a.m. in a bay on the coast of Africa and nearly 1,000 slaves were found in the hold, intended for sale in the United States. Bowen escaped. The ship was confiscated, armed with six guns, and served as a man-of-war in the Navy.

Then *Nightingale* was sold to Western Union. After the Russian-American expedition, she was engaged in the West Coast and China trade, and finally was a transatlantic lumber ship.

[9]*Harper's Weekly,* August 12, 1865, the first page of which was devoted to two large illustrations and the first part of an article on the arrival of the line at New Westminster. The illustrations showed the harbor and town and the three-story telegraph building.

[10]After the expedition, Pope returned to New York and became a partner of Thomas A. Edison. Pope became president of the American Institute of Electrical Engineers, made valuable improvements in telegraphy, and was on the Western Union staff handling patent matters. About 1880 he resigned to practice as a patent agent and expert witness. His death in 1895 resulted from accidently coming in contact with electrical wires used in telegraph experiments at his home in Pittsfield, Massachusetts.

[11]Article by Corday Mackay in *The British Columbia Historical Quarterly* (July 1946). Mr. Mackay had access to Conway's diaries in the Provincial Archives.

[12]Lamb, a branch office manager for American Telegraph at Williamsburg, New York at 17, enlisted, was captured twice, exchanged for Confederate prisoners, and served in the Military Telegraph Corps. He was with Western Union in Cincinnati when he joined the expedition.

[13]Their diaries were found in a barrel of papers in a San Francisco junk shop, bought by Charles S. Hubbell of Seattle, and acquired in 1954 by the University of Washington Libraries.

[14]Kennicott, born in New Orleans November 13, 1838, passed his boyhood at "The Grove," the Kennicott homestead at Desplaines, near Chicago. In 1859 he collected zoology specimens north of Lake Superior for the Northwestern University museum. In 1862 he explored the valley of the McKenzie River from its mouth on the Arctic Ocean.

[15]Some reports said Kennicott died trying to save a member of the party from drowning. Another account, in the *San Francisco Call,* September 19, 1897, under the headline, "Bearing the Body of a Suicide Six Hundred Miles in an Open Sailboat," said: "There were evidences of strychnine poisoning and, as Major Kennicott was known to have a quantity of that drug in his possession, the act of self-destruction was easy of performance."

[16]William H. Dall, *Alaska and Its Resources* (Boston: Lee and Shephard, 1870).

[17]Westdahl's grandson Dr. Philip R. West-

dahl of San Francisco, made the diary available to me.

[18]William H. Dall, "Alaska as It Was and Is, 1865–1895," *Bulletin of the Philosophical Society of Washington* 13 (Washington D.C.): 123-61.

[19]Frederick Whymper, *Travel and Adventure in the Territory of Alaska* (New York: Harper, 1869).

[20]Born at Norwalk, Ohio, in 1845, Kennan had been an operator at Cincinnati during the Civil War and on the transcontinental line at Sacramento.

[21]George Kennan, *Tent Life in Siberia* (New York: G. P. Putnam's Sons, 1870; rev. 1910).

[22]From Captain Scammon's private journal (see above).

[23]Frederic Bronner of Victoria, B.C. saw the importance of the diary of his father-in-law, J. W. Wanless, and presented it to the University of Washington Libraries, Seattle in 1964. I corresponded with Mr. Bronner for three years. Miss Isabella Sims, Gifts and Exchanges Librarian, University of Washington, generously provided copies of the entire diary and accompanying maps.

[24]*The Telegrapher,* December 12, 1868, quoted an unnamed director of Western Union as saying:

> We had no doubt they would lay the Atlantic cable; we thought it might be operated, but that its liability to accident would prevent permanent success. But when they went into mid-ocean, recovered the lost cable, spliced and completed it, we concluded it was time for us to stop.

[25]After the expedition, Kennan lectured on his travels and became the Associated Press reporter at the U.S. Supreme Court. He also was a war correspondent in the Spanish-American and Russo-Japanese Wars. In 1885 he visited Russia, and wrote articles for *Century Magazine* in 1888 and 1889 on the horrors of the prisons, convict transfer stations, and mines of Siberia. These articles became a book in 1891, *Siberia and the Exile System.*

[26]Alexander Michailovitch Prince Gortcha-

koff, chancellor of the Empire, was a clean-shaven, "evening-clothes" diplomat who had served in embassies 33 years before becoming Minister of Foreign Affairs in 1856, vice chancellor in 1862, and chancellor the next year. His desire for friendship with the United States was evident from the report of Gustavus Vasa Fox, assistant secretary of the U.S. Navy. From the *Journal and Notes of J. F. Loubat* (New York: Appleton & Co. 1883).

Loubat headed a good will mission to present a Resolution of Congress congratulating Emperor Alexander II upon being saved from the attack of an assassin on April 16, 1866.

[27]Letters between de Stoeckl and Seward, September 20 and 29, 1864. Notes from Russian Legation, microfilm vol. 29, roll 3; and from State Department to Foreign Legations, vol. 99, roll 82.

[28]*St. Petersburg Court Journal,* November 2, 1864.

[29]Jane Marsh Parker, "A Defeated Success," *Overland Monthly Magazine* (July 1886).

[30]Another version of this conversation is that Gortchakoff asked how Western Union proposed to acquire right-of-way through British Columbia. Sibley replied that he thought there would be little difficulty except in the case of the Hudson Bay Company. He said that while in London he submitted the matter to that company's directors, who did not welcome the proposition with enthusiasm and it might be necessary to acquire a considerable interest in the Hudson Bay Company. From Hiram W. Sibley (Hiram Sibley's son), "Memories of Hiram Sibley," in *Publication Fund Series* 2 (Rochester Historical Society): 127-34.

[31]Based on a copy of a document I found in Sibley's Rochester mansion on September 30, 1965, evidently written by Hiram W. Sibley when his father died July 12, 1888.

[32]Clay's reports to the secretary of state from November 14, 1864 to April 2, 1865 provide a running account of the negotiations. National Archives Microfilms.

[33]Canada was formed in 1867 of several independence-minded regions but not regarded as a nation.

[34]Hector Chevigny, *Russian America* (New York: Viking, 1965); Frederic Bancroft, *The Life of William H. Seward* (New York and London: Harper & Bros., 1900); and Frederick W. Seward, *Seward at Washington as Senator and Secretary of State, 1861–1872,* vol.3 (New York: Derby & Miller, 1891).

[35]Alaska got its name from the Aleut Indian word "alyeska," meaning "great land." The transfer took place on Castle Hill in Sitka on October 18, 1867 when the double eagle standard of imperial Russia came down and the Stars and Stripes went up the flagpole.

[36]Dall, "Alaska as It Was and Is, 1865–1895," 123-61.

[37]Letter from W. E. Ireland, provincial librarian and archivist of British Columbia, September 25, 1964.

[38]W. E. Ireland, who provided some of the information in this chapter, pointed out that this bridge, replaced by a highway bridge in 1910, collapsed in 1913.

[39]Corday Mackay, *British Columbia Historical Quarterly* (July 1946); Philip B. McDonald, *Sketch of Collins's Life* (New York: New York University, 1928).

[40]The above biographical sketch of Sibley is based on a college thesis (Princeton, 1949) by Sibley's great-grandson, Harper Sibley Jr., who loaned it to me.

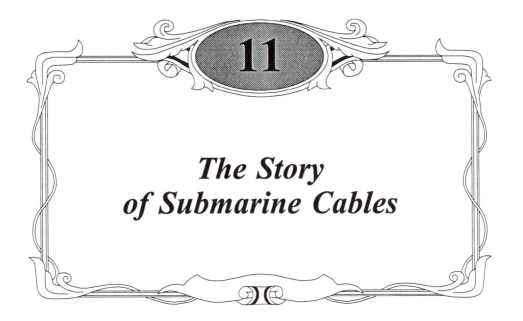

The Story
of Submarine Cables

In Shakespeare's play *A Midsummer Night's Dream,* Puck said, "I'll put a girdle round about the earth in forty minutes." It really required about 300 years. Girdling King Neptune's domain with cables was one of man's most dramatic communications conquests.

On October 18, 1842, loafers on the lower tip of Manhattan Island were startled to see two men in a rowboat, one rowing and the other, wearing a top hat, paying out what looked like a rope on a reel. It was Morse laying what apparently was the first cable in America. Enclosed in rubber and hemp and coated with tar and pitch, it was laid from the Battery to Governor's Island. After only a few signals were exchanged, however, a ship caught it in raising its anchor; the sailors cut off 200 feet as a souvenir, and sailed away.[1]

Morse repeated the experiment at Washington, D.C. two months later and suggested that the Government lay a cable across the Atlantic, but Congress had not even acted on his proposal for an experimental telegraph line then, and his idea was soon forgotton.[2]

Later a number of short cables were laid across rivers, but none served long until the sap of the sapodilla gutta-percha (latex tree) was discovered. The tree grows in the Maylayan Peninsula, the Indian Archipelago, Borneo, and Ceylon. Dr. W. Montgomerie, an East India Company surgeon at Singapore, noticed that the handle of a tool used by a native woodchopper became plastic in hot water, but congealed quickly when cooled. He took samples to England in 1843, and its importation began.

Samuel T. Armstrong of New York had five tons shipped to the United States in 1848, and established a factory in Brooklyn to insulate cables. Dr. John J. Craven used it in a cable at Bound Brook, near Elizabeth, New Jersey.

Armstrong laid a cable insulated with gutta-percha across the Hudson River at New York late in 1847[3] to provide direct service to New York City. In the *New York Journal of Commerce* he then proposed to lay a cable across the Atlantic for $3,500,000.

Dr. Ernst Werner von Siemens developed a heat machine in 1847 that eliminated leaky seams in insulation as the gutta-percha was applied. The next year he and his brother, Sir William, laid a gutta-percha-covered wire in the Port of Kiel to explode submarine mines.

The biographer of Samuel Colt, the pistol manufacturer, said Colt laid a cable in New York Harbor in 1842. Since Colt and Morse had become friends when both were lobbying in Washington, he must have been the man rowing the boat from the Battery.[4] Colt laid a cable across the East River at Fulton Ferry in 1845 for the New York and Offing Magnetic Telegraph Company which provided ship reporting and telegraph service between New York, Brooklyn, and Coney Island. That cable was relaid in 1846 above Blackwell's Island, and the wire was extended to Williamsburg and Brooklyn, where offices were opened.[5]

Colt wrote to O'Reilly on March 27, 1846, that, in crossing Hell Gate, he used one pipe of 2,000 feet without making a joint. He filled the space around the wires in the pipes with a composition of asphaltum, and offered to supply a mile a day to O'Reilly. He laid another across the East River in 1868 when Western Union constructed a sixty-mile line along the South Side Railroad on Long Island, and opened the first offices at College Point, Whitestone, Flushing, Jamaica, Hempstead, and Rockville Center. Offices were opened at Astoria and Fort Hamilton in 1871.

Jacob Brett and his elder brother, John Watkins Brett, a wealthy dealer in curios, laid the first cable across the English Chan-

nel from Dover to Cap Gris-Nez in 1850. It failed after only a few messages were exchanged, but a permanent channel cable was laid by T. R. Crampton, an English engineer, on September 25, 1851.

By 1852 a cable containing six lines connected England, Holland, Germany, Denmark, and Sweden, and another linked Italy with Corsica, Sardinia, and Africa. In the Crimean War a 300-mile cable was laid in the Black Sea from Varna, placing England and France in touch with their armies before Savastopol.

When storms and floods wrecked mast crossings of the Mississippi and Ohio Rivers in 1850 and 1851, Shaffner replaced them with cables coated with gutta-percha, but sand wore it off. Jeptha Wade remedied that by lashing lateral wires over the gutta-percha at St. Louis. After that numerous river cables were laid. They were made by S. C. Bishop at New York, using machines developed by Charles T. Chester and J. N. Chester to cover the gutta-percha with hemp and iron wires.

The Telegraph in Canada

Two years after Morse built the first line in 1844, T. D. Harris and associates formed the Toronto, Hamilton, and Niagara Electro Magnetic Telegraph Company. Suspending a wire across the Niagara River from Ontario to Lewiston, New York, Samuel Porter connected that line with Faxton's New York, Albany, and Buffalo line on January 14, 1847.

That first Canadian line was bought by the Montreal Telegraph Company, organized in 1847. Its superintendent for the next eighteen years was Orrin S. Wood, who had built lines with his brother-in-law, Ezra Cornell. The line reached Montreal on August 3 and Quebec in October 1847. Branches served points in Maine, New Hampshire, Vermont, New York, and Michigan.

The policy of Sir Hugh Allan, who became Montreal Telegraph's president in 1851, was to absorb any competitors that appeared, by stock control, lease, and purchase, as Western Union did.

The British North American Electric Telegraph Association, organized by Frederick N. Gisborne in 1847, started building from Quebec to the Atlantic, but advanced only 112 miles to River du Loup. When Lawson R. Darrow completed a line from Calais, Maine, to Saint John, New Brunswick on January 1, 1849, John A. Torney of New Brunswick extended the British North American line east to connect with it. When it reached the Nova Scotia border, the province had Gisborne extend it to Halifax, where transatlantic ships touched to drop news. British North American fell into the hands of the Montreal company, which leased the New Brunswick line to American Telegraph in February 1856.

The Montreal company bought the northern part of a line, built by Ezra Cornell and his son, Alonzo, from Montreal to Troy, New York, and the Canadian part of the Vermont and Boston line from Rouses Point, at the border, to Montreal and Ogdensburg, New York. It then bought a 125-mile line from Montreal to Ottawa, then named Bytown, and secured control of a Grand Trunk Telegraph Company line from Buffalo to Quebec. Another line was built between Buffalo and Quebec by the United States Telegraph Company, but, when that company was absorbed by Western Union in 1866, the line was sold to the Montreal company.

Montreal Telegraph lines covered a large part of Canada, connecting here and there with Western Union, of which its president,

Sir Hugh Allan, was a director. It had 20,000 miles of wire in 1875, and in 1876 reported 1,507 offices, 2,330 employees, and 12,044 miles of pole line.

The Wisconsin State Telegraph Company absorbed a Minnesota company in 1865, and named the combination Northwestern Telegraph Company. Its president, Zalmon G. Simmons, who later founded a bedding company, extended lines from Chicago to five states and the Province of Manitoba, before it was leased to Western Union in 1871.

A company named the Great Western Telegraph Company, formed by Josiah Snow and David C. Gage, gave a schemer named Selah Reeves a contract to build a line for $300 a mile in the same midwestern area. Subscribers to the stock became suspicious of the contract price, and the Illinois Supreme Court declared it fraudulent. Finally the line was saved from bankruptcy by lease to Western Union.

Reeves then went to Canada and organized the Dominion Telegraph Company in 1868. It gave him a contract to build a 2,000-mile line at $250 a mile, but it was cancelled when his past was exposed. Others reorganized the company, built from Detroit and Buffalo to Quebec and soon had 3,660 miles of line and 366 offices. Dominion was leased in 1879 by the American Union Telegraph Company, which was acquired by Western Union two years later.

Western Union encouraged the formation in 1880 of the Great Northwestern Telegraph Company, incorporated by an act of the Canadian parliament with its principal office at Winnepeg. A year later Great Northwestern leased the Montreal and Dominion lines, providing Canada's first big system.[6]

Great Northwestern, headed by Erastus Wiman,[7] took over 22,000 miles of pole line. It agreed to pay the Montreal company eight percent and the Dominion seven percent on their capital stock, but Western Union had to guarantee payment. The ninety-nine-year contract, signed July 1, 1881, provided for exclusive interchange of business with Western Union, which owned fifty-one percent of its stock.

Headquartered at Toronto, Great Northwestern made huge profits and paid dividends plus stock extras, but when the Canadian Pacific Railway built its transcontinental line in 1885 and 1886, it cut heavily into Great Northwestern's profits.

Two additional railway lines were built: the Grand Trunk Pacific from Quebec, in conjunction with the National Transcontinental, to the Pacific Coast; and the Canadian Northern from Montreal and Toronto to the Pacific. Both operated their own telegraph services[8] competing with Great Northwestern and Canadian Pacific.

In 1915 Western Union turned Great Northwestern over to Canadian Northern, but the railway defaulted on its bonds, and the Canadian Government assumed ownership in 1917, also taking over the telegraph in May 1919. In that year the Grand Trunk Railway and its telegraph were unable to meet their obligations, and the government likewise became their owner.

The government then consolidated Canadian Northern and Grand Trunk, adding Inter-Colonial Railway from Montreal to Halifax, Nova Scotia, and Saint John, New Brunswick, to form Canadian National Railways and Canadian National Telegraphs. Western Union sold its remaining Canadian landlines to Canadian National in 1924, 1925, 1927, and 1929. With the exception of 1,185 miles connecting with Western Union's transatlantic cable stations in Canada, the international boundary then became the dividing line in telegraph operations.

The Battle for European News

Disputes between editors and telegraph men were frequent. Stories that passed over three or four lines often never reached the papers, and editors could not prove who was at fault. Stories that did arrive often were delayed and contained numerous errors.

Under the "first come, first served" policy, a reporter could dash to the telegraph office one jump ahead of competitors and tie up the wires for hours. Sometimes he sent chapters of the Bible while he prepared his story, and men from opposition papers stood helplessly, shouting insults. Allowing each newsman fifteen minutes in rotation did not work either; short "takes" of different stories became intermingled at relay points.

Editors tried to cut costs by keeping dispatches brief and using codes, but coined words, meaningless to operators, multiplied errors. Speculators wanted the news first. Immediate publication protected the public.

Before "Fog" Smith built the first line to Boston in 1846, enterprising men got European news from steamers arriving there and rushed it to New York. One of them was Daniel H. Craig, a crafty, shrewd, imperious New Hampshire native with clean-shaven upper lip, close-cropped sideburns, and pointed beard. Craig's homing pigeons often beat competitors by hours, and forced speculators and editors to pay high prices for the news.

When the telegraph reached Boston, Smith incensed the press with poor service and full rates for news stories, plus fifty percent for each additional paper receiving copies. The press retaliated by encouraging Downing in 1847 and O'Reilly in 1848 to build House and Bain lines to Boston.

The high cost of receiving news separately, and the success of an association of papers between Buffalo and Albany, led six fiercely competitive editors in May 1848, to form the New York Associated Press. On January 11, 1849, the same papers formed the New York Harbor News Association to obtain foreign news from ships arriving.[9] (Before that, separate news boats of the papers met ships at New York, and the public watched strong-arm battles between their crews.) The *New York Courier and Enquirer, Express, Herald, Journal of Commerce, Sun,* and *Tribune* were the Harbor News Association members. Papers in other cities soon began subscribing to their service.

Facing stiff competition from other lines, "Fog" Smith agreed on May 18, 1848, to give the newly born New York Associated Press[10] priority on foreign news from Boston to New York. European ships already were dropping news at Halifax, from which Craig and his homing pigeons raced the A.P. to Boston. The A.P. chartered a steamer, the *Buena Vista,* to rush the news to Boston, but its speed was no match for Craig's pigeons.

Determined to control the news route, Smith built a 110-mile line from Boston to Portland, and in 1848 licensed the Maine Telegraph Company to build to Calais, Maine, at the Canadian border. Lawson R. Darrow organized the New Brunswick Electric Telegraph Company and completed the line from Calais to Saint John, New Brunswick on January 1, 1849, when the A.P. guaranteed at least 3,000 words from each steamer.

Smith refused to handle the dispatches to New York over his lines, however, unless the Boston papers could use the news, and the New York A.P. had to agree. Smith's friendly gesture to the Boston press was puzzling until it became known that he had created a foreign news business with John T.

Smith of Boston as his "front." Craig was charged with threatening to cut Smith's lines, and the A.P. then gave up the battle and hired Craig as its European news agent.

Craig organized a Pony Express over the 150 miles between Halifax and Digby, Nova Scotia, and an express steamer for the fifty miles across the Bay of Fundy to the telegraph at Saint John. The pony race took place every fortnight when an English mail steamer arrived. About 3,000 words were sealed in a bag carried by the riders. The first Pony Express left Halifax on February 21, 1849, and reached Digby in about eleven hours.

One night a rider, thundering through drowsy Nova Scotian villages at breakneck speed, felt his horse give a tremendous leap. The next day he learned the animal had jumped a twenty-foot hole in a highway bridge. Each trip of the Pony Express and Bay of Funday steamer cost about $1,000, not counting telegraph expense from Saint John to New York, but pigeons were sometimes used.[11]

"Fog" Smith wrote to the A.P. demanding Craig's dismissal:

> Until he totally abandons the use of carrier pigeons, I shall refuse transmitting any dispatches from him. . . . If the A.P. will employ an agent of his . . . they must expect counter measures of defense will be adopted.

Smith also demanded that the Nova Scotia line discharge their chief operator. The A.P. refused, declaring that

> All we could ascertain was that a man named Anderson, once in his [Craig's] employ, was detected in Saint John in the act of cutting the wires.

The A.P. offered to aid anyone who

would build a line to Halifax, and the Nova Scotia Government hired Gisborne as Government Telegraph Director. The chain of lines between Halifax and New York was then completed. The first dispatch from Halifax was in New York papers on November 15, 1849.

When the line was ready, John T. Smith arrived at Halifax and announced he was in charge of communications to New York. Realizing it was a plot by "Fog" Smith to get rid of him, Craig issued a twenty-nine-page pamphlet saying "Fog" Smith

> assumed to dictate to the A.P. in a very offensive manner who they should employ at Halifax as their correspondent—his wish being to place a convenient tool of his own here in the person of one John Smith.

Craig appealed to Gisborne, who replied that the rule was "first come, first served."

Craig then let a "secret" leak out that he had arranged for his papers to be thrown overboard as the ship approached and rushed to the telegraph office. He made a "show" of galloping a fast horse up and down the street, and made up a bundle of old London papers, with the heading but not the date visible. When the steamer entered the harbor, Craig's agent dipped the parcel in water, rushed to the telegraph office, and placed it on a table. John Smith raced to the telegraph office as soon as the steamer docked, but the clerk pointed to the wet package of papers. Smith left swearing, and caught the steamer to Boston. Then Craig leisurely prepared the news for New York.

"Fog" Smith was infuriated and refused to handle Craig's messages.[12] He controlled the only line from Portland to Boston, but Craig induced the Maine company to hold all messages at Portland until the A.P. stories arrived. Then he sent them on express

trains to Boston, and on to New York over the House or Bain lines.[13] The *Daily Globe* charged Craig with delaying the news and permitting swindlers to use the wires. For once, the papers were not unanimous in denouncing Smith.

The "Boston A.P.," upset by the New York Association, charged that it

> cut off the evening press of this city entirely from the receipt of the foreign news, on the ground that the express [train] was run solely for the benefit of the regular editions of the six morning papers of New York! . . . Protection of the public from speculators seems to have been forgotten all at once.[14]

Quick to aid any fight against Smith, O'Reilly rushed completion of his Bain line from Boston to Portland. Smith had the U.S. Marshal attach its instruments and supplies, but the Marshal accepted proof of ownership, and the line continued in business. On various occasions someone cut the line when steamers arrived, but it broke Smith's strangle hold on European news.

After Craig's unsuccessful battle for the European news, he was made general manager of the New York A.P. in 1851. He persuaded papers in other cities to buy the reports and pay two-thirds of the cost of the A.P.'s operation, and they organized the New England Way Paper Association, the Western A.P., New York State A.P., and Southern A.P.

When the Government sold the Nova Scotia line to the Nova Scotia Electric Telegraph Company, the A.P. refused to honor its guarantee to pay for 3,000-word dispatches, and in July 1858 began sending a news boat to Cape Race to meet the steamers. In retaliation, the line began delaying A.P. news.[15]

As noted earlier, American Telegraph

gained control of all Atlantic seaboard lines in 1859.[16] Deciding to end the A.P. foreign news monopoly, Robert W. Russell, secretary of American, announced that A.P. transmission priority was ended, and press rates would be substantially increased. A great cry of outrage appeared in the press, and Craig issued a pamphlet (*The Telegraph and the Press*), calling the American's officials monopolists, swindlers, cheats, drunken sots, and "superannuated political knaves." Amos Kendall declared Craig a "vulgar blackguard, wholly unapproachable by any gentleman."

American Telegraph wrote to its stockholders on June 22, 1860, denouncing the "despotic regulations" of the New York A.P. "calculated to create and perpetuate the most odious monopoly that the wit of man could devise. It also refused to elect five editors the press proposed to its board. To the editors' statement that the press paid $100,000 annually for telegraphing and were considering building a competing line, American replied that amount would not even maintain a line.

Russell had decided to make American the newsgathering company instead of the A.P., but strong public reaction forced him to give up that idea. He then promoted the organization of a rival press association, managed by George W. L. Johnson, Maturin Zabriskie, and James N. Ashley. Cyrus W. Field and Abram S. Hewitt, of his board, opposed that, and Russell bought their interests in American, and had them replaced by Edwards S. Sanford, manager of Adams Express, and Cambridge Livingston. But they also wanted no battle with the press, and after ousting Russell and his friends from the executive committee, made peace with the A.P. Sanford was elected American's president; Livingston, secretary; Francis Morris, treasurer; and Marshall Lefferts,

general manager.[17]

Newfoundland the Next Objective

After reaching Halifax, the telegraph was still far from St. John's, Newfoundland, on the southeastern point of Canada and closest to Europe. Unpopulated except for St. John's and a few coastal fishing villages, Newfoundland had frowning cliffs on its rockbound coast which was constantly enveloped in fog and mist. Ice floes and giant icebergs added their dangers to shipping.

The man who did most to extend the telegraph to St. John's was Gisborne, in charge of the Nova Scotia line which he extended to Sydney, Cape Breton, in 1852, crossing the strait of Canso with tall masts. Joseph Howe, Secretary of State for Canada, said Gisborne presented a plan to the government in 1850 to connect Newfoundland with England, and obtained leave of absence to go to New York and promote it.

The Newfoundland legislature granted 500 pounds sterling to survey a 400-mile route from Cape Ray to St. John's, and Gisborne completed it with four Indians in three months, narrowly escaping starvation. He then went to New York, obtained the backing of Horace B. Tebbets and Darius B. Holbrook, and formed the Newfoundland Electric Telegraph Company. On November 22, 1852, he laid a nine-mile cable from the mainland at Cape Tormentine, New Brunswick, to Cape Traverse, Prince Edward Island.

Gisborne then started building a subterranean telegraph across Newfoundland. He had laid about forty miles when his financial sponsors at New York disagreed and failed him. He gave all of his own property to his workers and returned to New York to seek new capital.

Cyrus W. Field's Decision

11.1 CYRUS W. FIELD, WHOSE PERSISTENCE RESULTED IN THE LAYING OF THE FIRST SUCCESSFUL TRANSATLANTIC CABLE.

In New York City, in January 1854, Gisborne met Matthew D. Field, railroad and bridge engineer, who introduced his brother, Cyrus West Field, a wealthy New York paper merchant and warehouseman, who had retired at thirty-five years of age. Gisborne spent an evening at Cyrus Field's home at 1 Lexington Avenue, in Gramercy Park.

The two were opposites in many ways, but each had a combination of dogged determination and practical idealism. Field, born in Stockbridge, Massachusetts, in 1819, was a small, nervous man with chin whiskers. He was cautious, conservative, not disposed to risk his fortune. Gisborne was older, large, powerful, daring, and enthusiastic. He told

Field a final link was needed to reach St. John's, where European news could be first received.

Histories portray Field looking at a globe of the world and exclaiming, "Why stop where you do? Why link America only with Newfoundland? Why not lay a cable on across the Atlantic Ocean and link us with Europe?"

Field wrote to Morse and Lieutenant Matthew F. Maury, U.S. Navy, head of the National Observatory at Washington. Maury's reply was to cite a survey just completed by the U.S. brig Dolphin. He wrote:

> From Newfoundland to Ireland, the distance between the nearest points is about 1,600 miles; and the bottom of the sea between the two places is a plateau, which seems to have been placed there especially for the purpose of holding the wires of a submarine telegraph.

The depth of the plateau, Maury said, gradually increases to fifteen hundred to two thousand fathoms as Europe is approached. Hearing that, Morse went to Field and urged him to go ahead.

Field asked four wealthy men to his home: his neighbor Peter Cooper, who later founded an educational institution in New York City; Moses Taylor, capitalist; Marshall O. Roberts, an Erie Railroad founder; and Chandler White, a retired merchant. They met in Field's dining room several evenings, studying maps, and sent Field to Newfoundland to negotiate for a charter.

Field left on March 14, accompanied by White, his own brother David Dudley Field as legal counsel, and Gisborne as engineer. At St. John's, Governor Kerr B. Hamilton of Newfoundland called a special session of the Council, which offered the following: exclusive cable landing privileges for fifty years;

11.2 LEFT TO RIGHT, CYRUS FIELD JUDSON PRESENTING THE TABLE AROUND WHICH HIS GRANDFATHER CYRUS W. FIELD AND ASSOCIATES ORGANIZED THE FIRST TRANSATLANTIC CABLE COMPANY, CHAIRMAN NEWCOMB CARLTON, SENIOR VICE PRESIDENT J. C. WILLEVER, AND PRESIDENT A. N. WILLIAMS OF WESTERN UNION.

5,000 pounds toward building a road along the telegraph line; guarantee of interest for twenty years on bonds (50,000 pounds); and fifty square miles of land on completion of the line.[18] The Newfoundland legislature incorporated the New York, Newfoundland, and London Telegraph Company. Chandler White drew $50,000 and paid all debts of Gisborne's old enterprise.

Meeting at Field's home at 6:00 a.m. on May 6, 1854, because one member had to leave town that morning, the five men orga-

nized the company with Cooper as president, White as vice president, and Taylor as treasurer. They then subscribed $1,500,000 capital and went to breakfast.

White and Matthew Field went to Newfoundland to construct the landline. They had 600 men building across Newfoundland, while Cyrus Field went to England to buy eighty-five miles of cable to be laid from Newfoundland to Cape Breton Island. The single-conductor cable was made by Glass, Elliot, and Co., and shipped to America.

11.3 THE *AGAMEMNON* AND THE *NIAGARA*, THE TWO SHIPS THAT ATTEMPTED TO LAY CABLE ACROSS THE ATLANTIC OCEAN, BUT FAILED.

The steamer *James Adger* left New York City on August 7, 1855, with Cooper, Morse, Field, Bayard Taylor of the *New York Tribune*, Fitz-James O'Brien of the *Times*, John Mullaly of the *Herald*, and others to see the cable laid. The trip was described in a book by Mullaly.

The cable end was floated ashore near Port aux Basques, on Cape Ray, and the *James Adger* began towing the bark *Sarah L. Bryant*, on which the cable was loaded, south toward Cape Breton Island. A violent storm forced them to cut the cable and abandon the forty miles that had been laid. Field ordered another cable which was laid by the ship *Propontis* and connected with the Nova Scotia line.[19]

Forming the Atlantic Cable Company

Field returned to England to seek financial support. Charles T. Bright, the twenty-year-old chief engineer of the Magnetic Telegraph Company, with lines throughout the United Kingdom, John Brett, and Field signed an agreement September 29, 1856 "by which we mutually, and on equal terms, engage to exert ourselves" to lay an Atlantic cable. They were joined later by Dr. Edward O. W. Whitehouse, a physician experimenting with cable transmission.

The British Government agreed to furnish ships for soundings and laying the cable, and to pay 14,000 pounds a year for government messages. Field then organized the Atlantic Telegraph Company, and 350,000 pounds of capital stock was over-subscribed at meetings in Liverpool, Manchester, Glasgow, and London. Field confidently reserved one-fourth of the stock for American investors, but could sell only twenty-seven shares in the United States and was forced to use most of his own fortune.

Field also had difficulty in getting Congress to pass a bill to aid. It finally passed the Senate by a majority of one, and was barely saved in the House in a storm of criticism because the route was between two British territories. President Pierce signed it on March 3, 1857, his last day in office.

The company had 2,500 miles of cable made with seven copper wires twisted spirally to form the conductor. It was covered with three layers of gutta-percha, tarred hemp, and a spiral sheathing of eighteen strands of iron wire.

Paying-out machinery was installed in the British battleship *Agamemnon* commanded by Captain Noddal, and the U.S. frigate *Niagara,* commanded by Captain William L. Hudson. Half of the cable was stowed in each of their holds. As engineer-in-chief, Charles Bright assembled a staff of engineers and electricians, and on July 30, 1857, the two ships and their escorts assembled at Valentia Bay.

The heavily armored shore end was landed from the *Niagara,* which was to lay cable to mid-ocean. At that point her cable was to be spliced to that on the *Agamemnon,* which was to lay it on to America. After thirty miles of cable had been laid, it snapped when the strain on it was too great and was lost.

Notes

[1]House document 24, second session, 28th Congress; p. 6 quotes Morse.

[2]House document 24.

[3]*The Telegrapher,* December 26, 1864, said this cable, laid across the Hudson by Thomas M. Clark and John W. Norton, "was the first gutta-percha-insulated wire manufactured and used in the United States."

[4]Henry Barnard, *Armsmear, A Memorial* (commissioned by Colt's widow and privately printed in 1866); and Jack Rohan, *Yankee Arms Maker: The Incredible Career of Samuel Colt* (New York: Harper Brothers, 1935). *The Electrical Review,* June 20, 1885, quoted Captain Moses Tower of Hull, Massachusetts, as saying he built a line for Colt from Hull to Boston and a message was sent, he thought, on December 10,

1843, but a storm wrecked the line that night.

[5]*The Brooklyn Eagle,* October 26, 1916.

[6]Reid, *The Telegraph in America,* 339.

[7]Wiman had been director of the Ontario branch of R. G. Dun & Co. of New York, named Dun, Wiman & Co. The New York company is now Dun and Bradstreet.

[8]The Canadian Northern Telegraph Company was incorporated on May 15, 1902, and the Grand Trunk Pacific on July 13, 1906.

[9]Frederic Hudson, *Journalism in the United States from 1690 to 1872* (New York: Harper & Bros., 1873), said representatives of six papers met at The Sun "and formed 1st, the Harbor News Association and 2nd, The New York Associated Press. . . . This was in 1848–1849."

Richard A. Schwarzlose, "Harbor News

Association; The Formal Origin of the AP," *Journalism Quarterly* (University of Minnesota, Summer 1968). The agreement, dated January 11, 1849, was found in the Henry S. Raymond's papers, Manuscript Division, New York Public Library. Raymond, managing editor of the *Courier and Enquirer,* on September 18, 1851, founded the *New York Times,* which became the seventh AP member.

[10]Oliver Gramling, *A.P. The Story of News,* credited David Hale, coeditor of the *Journal of Commerce,* with convincing enterprising, volcanic James Gordon Bennett of the *Herald* that if the publishers would not cooperate, the telegraph companies would gather the news and "make it virtually impossible for any news but their own to move on limited wire facilities. . . . So in the *Sun* office in May 1848, the first real cooperative newsgathering organization was formed."

[11]*New York Tribune,* January 25, 1850, said the A.P. also cut costs when the line reached Halifax by withdrawing the *Newsboy,* a boat it had cruising off Sandy Hook, New Jersey, at a cost of $20,000 a year to get news from inbound European steamers.

[12]Kendall and Morse wrote to Smith on April 1, 1850 that his position was "greatly and eminently wrong" and "nothing but evil" could flow from it. Smith's reply was an offer to buy Kendall's interest in the Maine line. Kendall refused, and copies of the letters were in the *Daily Mail* on April 15.

[13]*New York Journal of Commerce,* February 9, 1850.

[14]*The Boston Journal,* September 28, 1849.

[15]*New York Herald,* June 4, 1859.

[16]See above, chap. 6.

[17]James D. Reid, *The Telegraph in America,* 424-26.

[18]Printed charter, 14 pages. New York Public Library.

[19]A tablet commemorating the laying of this first ocean cable in North America was unveiled at the Western Union office at North Sydney, Nova Scotia on September 24, 1930.

The First
Transatlantic Cable

It was decided to try again. The *Niagara* and *Agamemnon* would meet in mid-ocean, and lay to opposite sides of the Atlantic. With their escort ships they sailed from England on June 10, 1858.

A great storm raged for ten days. Water poured in through gaping seams, and the heavily loaded *Agamemnon* almost capsized in gigantic waves.[1] At last the battered cable fleet met, spliced the cable between the two ships on June 26, and started. The cable broke when they were only three miles apart, but both put about and repeated the splicing. Again they started, and all went well until the next morning when the cable broke again.

The ships returned to the rendezvous, a third splice was made, and laying was resumed on June 28. This time the *Agamemnon* had traveled 112 miles when the cable snapped. The storm had broken the flooring in the tank and damaged the cable.

Field returned to London and planned another try, but was backed only by his

more daring directors. The chairman and vice chairman resigned, but Thomson and Curtis Lampson supported Field, and he finally prevailed.

A solemn procession of ships sailed from Cork on July 17, 1858 in heavy rain. Twelve days later, on July 29, they met at mid-ocean, spliced the cable, and started for their respective sides of the Atlantic. That afternoon, a great whale passed astern of the *Agamemnon,* grazing the cable where it entered the water. Everyone breathed more easily when the cable did not break. That night a damaged portion of the cable was observed, the ship was slowed, and workmen raced to repair it before it would pass overboard.

When the *Niagara* entered Trinity Bay, Newfoundland on August 5, Field rushed ashore at 4:00 a.m. to send the news to the Associated Press at New York. At 6:00 a.m. the *Agamemnon* signaled that she had reached Valentia. Queen Victoria received the news the next day as she banqueted with the emperor of France on a ship in Cherbourg Harbor, and immediately knighted Bright, then only twenty-six years old.

This central copper wire carries the electric current. If it breaks, these flexible copper tapes carry the current around the gap.

This is the permalloy tape whose wonderful magnetic qualities keep the signals from jumbling. A thick covering of gutta-percha holds the currents to their path.

A wrapping of jute cushions the pressure of three miles of sea-water.

Eighteen steel armor-wires protect the cable from injury.

Last of all a wrapping of tarred hemp cords; then the soft ooze of the ocean's floor.

12.1 THIS ILLUSTRATION SHOWS HOW CABLES LAID ON THE OCEAN FLOOR ARE BUILT TO WITHSTAND THE PRESSURE OF AS MUCH AS SEVEN MILES OF OCEAN ABOVE.

The *London Times* asserted that

Mr. Bright, having landed the end of the Atlantic cable at Valentia has brought to a successful termination his anxious and diffi-

cult task of linking the Old World with the New, thereby annihilating space. Since the discovery of Columbus, nothing has been done in any degree comparable to the vast enlargement which has thus been given to the sphere of human activity.

In America, buildings were ablaze with lights, banners and flags; there were fireworks displays, guns were fired, and bells rang by the hour. In Cincinnati, 100 barrels of tar were set on fire in public places. President Buchanan, governors, and others wired congratulations to Field.

After eleven days of testing, the first transatlantic message was received from the company directors in England to those in the United States:

Europe and America are united by telegraph. Glory to God in the highest, on earth peace, goodwill toward men![2]

Queen Victoria then sent to President Buchanan the following message:

The Queen desires to congratulate the President upon the successful completion of this great international work, in which the Queen has taken the deepest interest.

The Queen is convinced that the President will join with her in fervently hoping that the electric cable which now connects Great Britain with the United States will prove an additional link between the nations, whose friendship is founded upon their common interest and reciprocal esteem.

Only the first sentence of the message arrived on the sixteenth because the cable was in trouble.[3] Buchanan replied:

The President cordially reciprocates the congratulations of her Majesty the Queen, on the success of the great international enterprise accomplished by the science, skill and indomitable energy of the two countries.

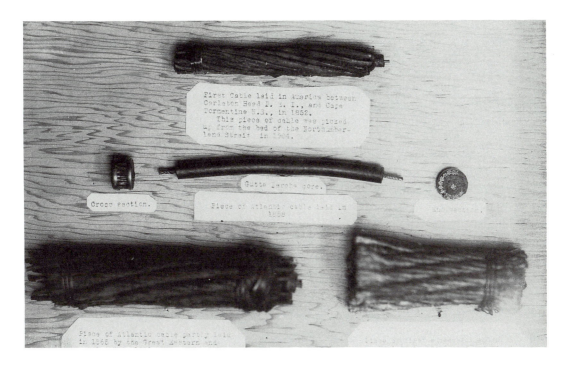

12.2 Section of first submarine cable (top) laid in America between Carleton Head, now Borden, Prince Edward Island, and Cape Tormentine, New Brunswick, on November 22, 1852, together with sections (center) of the Atlantic cable of 1858 and 1865–1866. This first cable consisted of a single copper conductor encased in gutta-percha (center cable shows only this core) with an armored covering (top). It has given place to a more modern cable, which now forms part of the circuit for the Canadian National Railways connecting Prince Edward Island with the mainland. Bottom left is a piece of the Atlantic cable partly laid in 1865 by the *Great Eastern* and completed in 1866. Bottom right is a piece of the first successful Atlantic cable laid by the *Great Eastern* in 1866. (Photo courtesy Canadian National Railways.)

Upon the arrival of Queen Victoria's message, the celebration started again. New Yorkers were awakened August 17 by "the thunder of artillery," wrote Reverend Henry M. Field, brother of Cyrus:

A hundred guns were fired in the Park at daybreak, and the salute was repeated at noon. At this hour, flags were flying from all public buildings, and the bells of the principal churches began to ring. . . .

That night the city was illuminated. . . . It seemed as if it were intended to light up the very heavens. Such was the blaze of light around the City Hall, that the cupola caught fire and was consumed, and the Hall itself narrowly escaped destruction. Similar de-

monstrations took place in other parts of the United States. . . . Nothing seemed too extravagant to give expression to the popular rejoicing.

In a formal celebration on September 1, Field and the ships' officers were met by dignitaries at Castle Garden,[4] an amusement place at the Battery, and paraded up Broadway bowing from their carriages. More than 15,000 members of the Army, clubs, societies, and other organizations were in the procession to the Crystal Palace, a big glass and iron building at Sixth Avenue and Forty Second Street, now Bryant Park.[5] A truckload of leftover cable[6] and a model of the *Niagara* were in the parade, escorted by the crew. Stores, banks, offices, and courts were closed.

At Crystal Palace addresses with grandiloquent rhetoric were delivered. David Dudley Field, jurist and brother of the honor guest, was "orator of the day." The next evening a banquet was held at the Metropolitan Hotel, with twenty toasts and responses, ninety delicacies, and elaborate confectionery ornaments.

Headlined "The First News Dispatch by Ocean Telegraph," the *New York Daily Tribune* on August 27 carried the story of the signing of the peace treaty between China, England, and France.

The Cable Fails

More news arrived from Europe, but delays became increasingly frequent. The *New York Leader* of September 26 asked, "Have we a pack of asses among us and are they specially engaged in electrical experiments over the Atlantic cable?"

Because the signals were weak, Dr. Whitehouse had increased the power "till the induction coils about five feet long yielded electricity that was estimated by experts . . . to have an intensity of about 2,000 volts!" That ruined the insulation and the cable went to sleep forever on October 20, 1858. Only 732 messages had been sent during its three-month life.

Many people believed a great hoax had been perpetrated; that the cable had never worked at all. Field, the national hero, was now denounced as a swindler.

Colonel Taliaferro P. Shaffner advocated a shorter route. He surveyed a route to northern Scotland and published a book

about it in London in 1861: *The North Atlantic Telegraph via the Faroe Islands, Iceland, and Greenland. The Results of the Surveying Expedition of 1859.* He proposed to lay four cables 1,755 miles.[7] Impressed, the British and Danish governments had soundings and surveys made. Prospects for Shaffner's project seemed bright, but investors, remembering the 1858 cable failure, were not interested.

Field's wholesale paper business in New York faced bankruptcy. He called his creditors together on December 27, 1860, made a frank and lengthy statement of his firm's misfortunes, and asked to be freed from debt so that he could go on with his work.[8]

The public seemed willing to forget the transatlantic cable, but Cyrus W. Field was not. Although he was seasick on each voyage, he crossed the Atlantic sixty-four times.

Secretary of State Seward again aided Field in 1862. With the approval of President Lincoln and his Cabinet, he wrote to England that the United States would join in another cable expedition. The British Government increased its subsidy from 14,000 to 20,000 English pounds, guaranteeing eight percent on 600,000 pounds of new capital for twenty-one years.[9] Having recently laid a Mediterranean–Red Sea cable for others, Glass, Elliot, and Company, the cable manufacturers, now were so sure of success that they offered to make the cable for only the cost of labor and material, and receive twenty percent of that in stock.

To obtain subscriptions, Field addressed meetings in the United States and England, but in 1864 the Atlantic Cable Company was still far short of the capital required. Finally, Thomas Brassey, a London capitalist, declared a great public enterprise was at stake, and he would find the money. He was joined

12.3 THE CABLE SHIP *GREAT EASTERN* LANDING THE FIRST TRANSATLANTIC CABLE.

by John Pender, a member of Parliament; John Chatterton, inventor of an insulating material; Willoughby Smith, a famous electrician; and R. A. Glass, who became managing director of the company. These five subscribed the remaining 315,000 pounds of stock and also 100,000 pounds of bonds. This gave the enterprise so British a complexion that until 1910 the Encyclopedia Britannica did not mention Field.

Gutta Percha Company and Glass, Elliot, and Company united in 1864 as the Telegraph Construction and Maintenance Company (T.C.&M.). It manufactured the cable with four layers of gutta-percha and spiral armor of ten iron wires, making it three times the size of the 1858 cable.

Only one ship in the world was large enough to carry the 2,300 nautical miles—9,000 tons—of that cable. Originally named the *Leviathan,* the *Great Eastern,* a British cargo ship of 22,500 tons, 700 feet long, with six masts, side paddle wheels, and a screw propeller, was repeatedly jinxed by fatal accidents and regarded as a white elephant.[10] A company headed by Daniel Gooch bought the ship and fitted it with cable machinery. Captain James Anderson was obtained from the Cunard Company.

Preparation of the cable required eight months, and coiling it in three great tanks on the ship, from January to June 1865. Also loaded on it were 8,000 tons of coal, twelve oxen, twenty pigs, and a cow. Nearly 500 men were needed to operate the ship and lay the cable, two hundred to raise the anchor.

The cable laying was in the hands of Chief Engineer Samuel Canning, Chief Electrician C. V. De Santy, and the T.C.&M. staff. Field and his advisors went as observers, as did W. H. Russell of the *London Times*,[11] and Robert Dudley, an artist and writer of the *London Illustrated News*.

Two British warships, *Sphinx* and *Terrible,* escorted the *Great Eastern* to Valentia,[12] and on July 23 she headed into the sunset for Newfoundland. At 3:15 the next morning a fault developed. The cable was attached to a "picking up" machine, which drew it in at the rate of a mile an hour for ten hours, until the fault came on board. A piece of wire had pierced the cable to the core. A splice was made, and laying was resumed.

On July 29, another fault developed after 716 miles had been paid out. Another piece of iron wire was found through the core, as though it had been done intentionally. Then on August 2, a third fault was discovered, after 1,186 miles of cable had been laid. The cable had been drawn in for a mile when it broke and sank to the bottom 2,000 fathoms below. After fifteen hours the cable was hooked, but the wire rope broke and the grapnel was lost. The cable was hooked again on August 7 and 11, but the second and third grapnels went to the bottom. Sadly the expedition turned homeward.

A Successful Atlantic Cable Is Laid

The promoters formed the Anglo-American Telegraph Company,[13] raised more capital, and prepared for another try in 1866.

The T.C.&M. made 1,600 miles of new cable with nearly three times the tensile strength of the old one and improved the cable laying and picking up machinery. The *Great Eastern* then started west, flanked by Telegraph Ships *Medway* and *Albany*, with the *Terrible* leading the way.

People in England knew of the ships' progress because signals were sent back through the cable as it was laid, but the fate of the expedition was unknown in America. After laying 1,896 miles of cable in fourteen days, the *Great Eastern* arrived off Heart's Content, Trinity Bay, Newfoundland on July 27. Gooch cabled the news to Foreign Secretary Lord Stanley, and wrote in his diary that the inhabitants seemed mad with joy, fired guns, rang bells, waved flags, cheered themselves hoarse, and even jumped into the water.[14]

While the *Medway* laid the shore end, Field rushed ashore to wire the news to New York, but the cable across the Gulf of St. Lawrence to Breton Island was not operating. As soon as repairs could be made, on July 29, he wired New York:

> Heart's Content, July 27. We arrived here at nine o'clock this morning. All well. Thank God, the cable is laid and is in perfect working order.

The first official message from England was from Queen Victoria to President Andrew Johnson:

> The Queen congratulates the President on the successful completion of an undertaking which, she hopes, may serve as an additional bond of union between the United States and England.

Johnson's reply cordially reciprocated this wish.

12.4 LAYING AT NEWFOUNDLAND THE SHORE END OF ONE OF WESTERN UNION'S TRANSATLANTIC CABLES.

The cable fleet remained at Heart's Content until August 9, and the cable men celebrated. Sixty-five years later, in 1931, I located four survivors of the *Great Eastern.* Each remembered the celebration at Heart's Content where the crew was not paid so they would remain sober, but the survivors agreed, "brandy and beer flowed like water—free to the entire crew."

Chief carpenter of the *Great Eastern,* William N. Napper, at his hilltop home at Lake Geneva, Wisconsin, enjoyed telling about the 1865 and 1866 trips. His daughter Annie brought out pieces of the cable, the grappling rope, and his "statement of wages," dated September 20, 1866.

Napper also recalled the excitement in 1865 when faults developed, and said that when the cable broke it "was a bitter disappointment to everybody on board. Four full weeks we grappled for the end that slipped into the water, but without success, and finally it was given up for the time being, and we went back to Sheerness."

Of the 1866 trip, Napper said, in one place the water was so deep it required fourteen miles of cable to lay four surface miles. He said, "a paper was published on board the ship, and bulletins were tacked on the bridge every twenty minutes, when important European news arrived over the cable, while it was being laid."

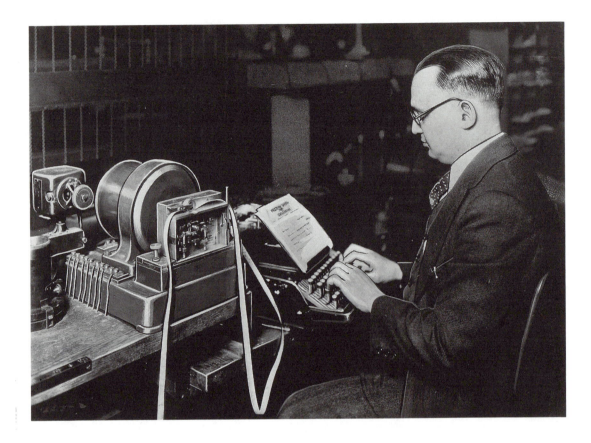

12.5 A CABLE OPERATOR PREPARING A CABLEGRAM FOR TRANSMISSION.

William Thomson of Emporia, Kansas, a seaman, said, "One of the prettiest sights I ever saw was the slow settling of the cable one-half mile behind the ship. It looked like a fine thread as it disappeared into the water." Thompson said the ship stayed about a mile out of Heart's Content while the shore end was laid.

Waldemar Peterson, of Long Beach, California, was a fifteen-year-old cabin boy, but remembered the leading men on the trip.

Martin Cahoon of Onset, Massachusetts, in the Navy during the Civil War, joined the *Great Eastern* as an able seaman on June 28, 1866. He described Field as a middle-aged man, always busy in the little testing room on deck, where he could keep in touch with England as the cable was laid. Once Cahoon came in contact with a live wire, while helping in the testing room, and danced around until Field shut the current off. He said the ship preceding the *Great Eastern* made soundings of the depth and nature of the ocean bottom, and two gun-boats kept astern to prevent any vessel from crossing the cable before it sank.

The celebration at Heart's Content was cut short on August 9, Cahoon said, when they put to sea to find the 1865 cable about 600 miles east of Newfoundland. After three weeks, the ship hooked the cable on the thirtieth attempt, and laid it to Newfoundland. The *Medway* laid the shore end on September 8, and an additional cable to

Breton Island. With two cables across the Atlantic, service became dependable.

At a banquet years later, Field[15] said, "I am a man of few words. Maury furnished the brains, England gave the money, and I did the work." This gave Maury and Field more credit than was due; many others made important contributions. Bright was a prime mover in the project.[16] Although Morse posed as a leader of the enterprise and craved more fame, only his name was used. The cable was made and laid by the British T.C.& M., and Lord Kelvin had the scientific knowledge.[17]

As to Maury, it was Lieut. O. H. Berryman, U.S.N. who made the soundings of the cable plateau in 1856, and reported them. Berryman's observations were referred to Maury as chief of the U.S. National Observatory, and Maury just sent the report to the Secretary of the Navy.

The American Telegraph Company, which owned the landlines handling the cable traffic, was bought by Western Union on June 12, 1866, a month and a half before the successful cable was laid.

The Battle between Press Associations[18]

At first the cables were operated at eight words a minute. The price set by the British was $100 in gold or $150 in greenbacks for twenty words or less, including address, date, and signature; additional words at five dollars gold. Press rates were the same, and the high cost resulted in only a brief, daily news report.

Some publishers blamed Associated Press General Manager Craig, and Western Union offered him a job to set up a commercial news service, receiving news from Western Union managers.[19]

Peter R. Knights wrote:

> Discovering Craig's plan, which seemed to them the rankest perfidy, the NYAP [New York Associated Press] board of directors fired him summarily November 5, despite the fact that it was election eve. Dropping his plan to cooperate with Western Union, Craig . . . organized a press association of his own, the United States and European News Association. Some newspapers, dissatisfied with the NYAP's domination of the news-gathering system, rallied to Craig, and a press association conflict began.[20]

Papers had been discontented with brief reports during the Civil War and were more so with the short, expensive cable reports, primarily of interest in New York. The Western A.P., led by Joseph Medill, *Chicago Tribune* publisher, seized upon Craig's expulsion to demand an equal voice. The New York A.P. refused, and the Western A.P. switched to Craig's service. The Southern and Eastern papers were divided; Washington, D.C. papers and *New York World* joined Craig. The New England and New York State associations remained with New York A.P.

After two months, the publishers realized service cost more with fewer members to share it. Knights wrote:

> The Western Union Telegraph Company, it would seem, forced peace upon the press associations. If so, it did this from operational necessity. In some areas of the country, the telegraph system could not bear the strain of delivering the reports of the two press associations. A telegraph company official stated in May, 1866, that the NYAP's reports to the New York State Press Association occupied the wires in that state ten hours a day.[21]

With Anson Stager of Western Union as mediator, an agreement was reached January 11, 1867, for mutual exchange of news, more news for the West, and fairer division of cost. New York A.P. and Western A.P. contracted to use Western Union, and it gave Craig a job with its Gold and Stock Telegraph Company (stock ticker service). Telegraph employees were not to gather news where the A.P. had coverage, but the relationship remained so close that congressional hearings were held to determine whether a "monopoly" protected the A.P. from competition and the telegraph company from press criticism.

The French Duxbury Cable

Not to be outdone by England, France had an Atlantic cable laid in 1868 by Baron Emile d'Erlanger and Julius Reuter, who formed the French Atlantic Cable Company. The *Great Eastern* was used again to lay 2,685 nautical miles of cable from Brest, France to the tiny island of St. Pierre, south of Newfoundland. Together with Miquelon, St. Pierre was the last rocky outpost of France's western empire. From St. Pierre, the Cable Ship *Chiltern* then extended the cable 750 miles to Duxbury, Massachusetts.

When the cable fleet reached Duxbury on July 23, 1869, the town's residents cheered and helped pull the cable up the beach. Messages were sent to Emperor Napoleon III. There were flags and streamers, triumphal arches, and wreaths of flowers, artillery volleys, and a procession led by a brass band to the crest of Abraham's Hill where a banquet for 600 was held under tents.[22]

Since Anglo-American had exclusive American landing rights, the French had no authority to land, but the Massachusetts legislature granted a charter to the Ocean Telegraph Co., a new company created to own the few miles of cable from shore to a point in the ocean outside the jurisdiction of the United States. The French company reduced rates but competed only two months before joining the cable pool. In 1873 Anglo-American absorbed the French company.[23]

When the 1866 first cable failed in 1872, only two of its transatlantic cables remained, so Anglo-American laid two from Ireland to Newfoundland in 1873 and 1874. In 1868–1871 Great Northern Telegraph Company of Denmark had built a landline across Siberia to China and Japan connecting it with the new Anglo cables from England to the United States.

Back in England, Field raised more money and had 700 more miles of cable made. He borrowed William E. Everett, chief engineer of the *Niagara,* from the Navy and took him to England, where he developed self-releasing brakes for the cable-laying machinery while Professor William Thomson, later Lord Kelvin, developed a mirror galvanometer which improved testing and signalling between ship and shore.

The Florida–Cuba Cable

In his *Personal Reminiscences,* James A. Scrymser, father of the cable business between the Americas, said he and Alfred Pell decided in May 1865 to lay a West Indies cable and persuaded Robert B. Minturn and Moses Taylor to back them.[24]

Governor Concha of Cuba sent their application for exclusive cable landing rights to the Spanish Government. After months of waiting, the Royal Decree arrived, but gave the permit to six influential Spaniards. Scrymser persuaded the *New York Herald* to use his statement that a Spanish cable monopoly would control the sugar markets of the world, and the United States should prohibit the cable unless Spain granted reciprocal rights for an American cable landing in Cuba.

On the advice of Secretary of State Seward, Scrymser organized the International Ocean Telegraph Company, with Major General William F. "Baldy" Smith, to whom Scrymser had been aide-de-camp during the Civil War, as its president. On May 5, 1866, Congress gave the company the exclusive right for fourteen years to establish cables between Florida and the West Indies.

General Smith went to Havana, exposed the fraud, got the royal decree annulled, and obtained an exclusive cable landing permit on June 17, 1866. Florida granted an exclusive cable permit for twenty years, and the right to construct landlines through the state.

The 235-mile cable was laid from Punta Rassa, Florida to Havana, via Key West, with a 400-mile landline from Punta Rassa to Lake City, in northern Florida. Operation began September 10, 1867, with rates $1 per word, and it was an immediate success.

American Telegraph demanded that International Ocean not extend its lines north of Gainesville, in North Georgia, but exchange traffic with it there. Scrymser agreed to that gladly because he foresaw that the Georgia and Florida business would become very valuable.

Later International Ocean backed the West India and Panama Telegraph Company, which laid cables from Cienfuegos to Santiago, Cuba, to Colón, Panama. It could not go on to Brazil because the British-controlled Brazilian Submarine Telegraph Company, later Western Telegraph Company, had exclusive landing rights from Brazil.

Jay Gould obtained control of International Ocean in 1878, increased its capital to $1,500,000, and leased it for ninety-nine years to Western Union, which absorbed it.

The Mexican Telegraph Co.

A distinguished-looking gentleman with pointed beard, Scrymser obtained a contract with Western Union to develop its business with Mexico. At that time, messages were exchanged with Mexican government landlines at the border, at Brownsville, Texas. In 1879, Scrymser persuaded General Diaz, president of Mexico, to permit a new firm, the Mexican Cable Company,[25] with Scrymser as president, to establish cables and lines connecting with Mexico City, Vera Cruz, and Tampico.

Western Union agreed to give the company all Mexican business for fifty years, and the first cable was laid 738 miles from Galveston to Tampico to Vera Cruz in 1881. Landlines were constructed from Galveston to Brownsville, and Vera Cruz to Coatzacoalcos, now Puerto Mexico. That provided a direct transmission route into Mexico "via Galveston," because of the increasing commerce with Mexico. Western Union acquired control of Mexican Telegraph in December 1926, and laid another cable from Galveston to Vera Cruz in 1929.

Central and South America

Success of Mexican Telegraph encouraged Scrymser to extend cables to Central and South America. He went to J. Pierpont Morgan, a director of Mexican Telegraph, and remarked casually that he thought he would go to London to raise funds for his new venture. Morgan invited him to dinner,

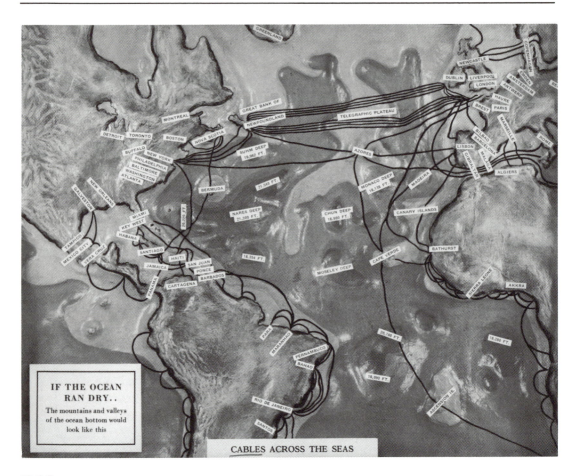

IF THE OCEAN
RAN DRY..
The mountains and valleys
of the ocean bottom would
look like this

CABLES ACROSS THE SEAS

12.6 CABLES ACROSS THE SEAS.

attended by Edward D. Adams, Charles Lanier, John W. Ellis, and others. Scrymser explained his plan, named $5,000,000 as the sum he needed, and it was subscribed in ten minutes.

The Central and South American Telegraph Company (C.S.A.T.) was organized on May 29, 1879, with Scrymser as president. It built a line from Vera Cruz to Coatzacoalcos, and across the Isthmus of Tehauntepec to the Pacific Ocean, and laid cables down the Pacific coast to Salvador, Nicaragua, Panama, Colombia, Ecuador, and Peru. When Lima was reached on October 2, 1882, service started. The company's cables and landlines then totaled 4,637 miles.

After waiting eight years for an exclusive contract between Chile and the West Coast of America Telegraph Co. (British) to expire, C.S.A.T. laid a cable from Chorrillos, Peru to Valparaiso, Chile in 1890, and purchased the Transandine Telegraph Company, with 1,200 miles of lines over the Andes to Argentina.

Eliminating the former Brazil–London–New York route and providing direct service, Scrymser radically speeded service and reduced rates. As commerce with South America increased, the company's business grew from $89,311 in 1882 to $614,489 in 1888, and a duplicate series of cables was added in 1893.

12.7 THE WESTERN UNION CABLE SHIP *LORD KELVIN.*

After the Galveston tidal wave in 1900 halted the company's service for twenty-six days, an alternate cable route was built in 1907 and 1915 from New York to Cuba to Colón. The 1915 cable was needed also because of World War I and growing communications with Panama after the canal opened in August 1914.

Blocked from the east coast of South America by Brazil's exclusive permit to Western Telegraph, Scrymser renewed his application, with the encouragement of every U.S. Secretary of State from Seward on. Now he applied again for a permit to lay cables from Buenos Aires to Santos and Rio de Janeiro, and the supreme court of Brazil

approved on January 27, 1917. However, the United States refused to permit the landing of a Western Union cable at Florida to complete the route because it was to be connected at Barbados with the British-owned Western Telegraph cable to Brazil.[26] Scrymser's cable to Brazil was not laid until 1920.

All America Cables, Inc.

When Scrymser died on April 21, 1918, he had joined the three Americas in rapid communications. He had started All America Cables, Inc. in two small back rooms at 55 Broadway, and shifted it to 37-39 Wall Street, which boasted two open fireplaces, oil lamps, and a staff of eight, with the addition of John L. Merrill in 1884 as office boy. Headquarters was moved to 66 Broad Street in 1903, and the company's first operating room was opened at 64 Broad Street in 1907 to serve the new New York–Cuba–Colón cable. The company had three cable ships, the *Relay, Guardian* and *All America.* In 1919, All America Cables opened its own headquarters building at 89 Broad Street. On February 16, 1920, Central and South American Telegraph was merged with All America Cables, Inc., with John L. Merrill as president.

The Mexican Telegraph operated as one system with Central and South American until Western Union bought control of it in 1926. Then All America Cables became a subsidiary of International Telephone and Telegraph Corporation in 1938, and its name was changed to All America Cables and Radio, Inc.

England laid cables in 1870 to India, via Lisbon, Alexandria, the Suez Canal, the Red Sea, and Aden, a route extended later to China, New Zealand, and Australia. At the same time, the Indo European Telegraph Company completed a landline to India, across Germany, Poland, Southern Russia, and Persia. That route also connected London with Karachi, India—more than 5,000 miles, about the same length as the Great Northern route via Leningrad to Peking.

The Atlantic Cable Pool

The next Anglo-American competitor was the Direct United States Cable Company, called D.U.S., which had the Siemens Company lay a cable from Ballinskelligs Bay, Ireland, to Newfoundland, but could not land there because of the exclusive Anglo-American charter. The landing was switched to Halifax, and a short cable was laid from Halifax to Rye Beach, New Hampshire. When the 1865 cable failed in 1877, Anglo-American added the D.U.S. cable to its pool.

Since its first cable had passed into British hands, the French, in 1879, formed a new company, La Compagnie Française du Telegraphe de Paris a New York, nicknamed "PQ" in honor of its founder, M. Pouyer-Quertier. Its cable, from Brest to St. Pierre to Orleans on Cape Cod, started out bravely to do battle with Anglo-American, but in 1880 it too was drawn into the pool.

Jay Gould attempted to obtain Atlantic cable connections from Anglo-American in 1880, but was rebuffed. Irritated, he had Siemens Brothers lay two Atlantic cables between Penzance and Canso in 1881 and 1882 for $7,000,000, and formed the American Telegraph and Cable Company.

Upon completing the first cable in 1881, Gould bought Anglo-American stock which had been depressed by news of his competition. He then sent it soaring by announcing higher tariff rates and inspiring rumors that his cable had broken and could not be repaired. Gould gave credence to the rumor by allowing his cable to lie idle. He then sold his Anglo-American stock at an excellent profit, and had his second cable laid. Gould then arbitrarily doubled the A.T.&C. stock and leased the cables to Western Union on May 12, 1882 for fifty years at five percent of $14,000,000. Western Union finally bought A.T.& C. for $2,030,000.[27] In 1911, all of the Anglo-American cables had been leased by Western Union for ninety-nine years.

12.8 DEEP-SEA CABLE AMPLIFIER BEING LAID ON OCEAN BOTTOM BY WESTERN UNION'S CABLE FLAGSHIP *LORD KELVIN*. DEVELOPED BY W.U. ENGINEERS, THE NEW AMPLIFIER WAS DESIGNED TO INCREASE A CABLE'S CAPACITY FROM 500 TO 1,500 LETTERS PER MINUTE. CONTAINING DELICATE ELECTRONIC EQUIPMENT, THE AMPLIFIER WEIGHS MORE THAN HALF A TON AND IS ABOUT FOUR FEET HIGH. THESE AMPLIFIERS WERE INSTALLED IN W.U.'S TRANSATLANTIC CABLE SYSTEM TO PROVIDE ADDITIONAL ALL-WEATHER FACILITIES FOR MAXIMUM-SECURITY, HIGH-SPEED MILITARY AND DIPLOMATIC COMMUNICATIONS AND FOR CABLE USERS GENERALLY.

Operating the Early Cables

The early cables suffered from weakening and distortion of the electrical impulses during their long trip. Instead of sending dots and dashes by closing a key to shut off the current for short and long periods, as in landline telegraphy, the early cable operator would strike a key at his left to connect the cable with the positive pole of the battery and send a dot, or the key at his right to connect the negative pole and send a dash.

The receiving instrument was a mirror galvanometer, a coil of fine wire wound in a circle, suspended between the poles of a permanent magnet, attached to a tiny mirror. A positive or negative signal caused the mirror to swing slightly to left or right. A beam of light from an oil lamp was directly focused through the open center of the coil so that the light moved left or right of a line in the center of a small stand. A receiving operator called out the dots and dashes, and another operator wrote and translated them.

Since that produced no written record,

Lord Kelvin developed a siphon recorder in 1870. He attached to a galvanometer coil a tiny hollow glass tube not much thicker than a human hair. One end of the tube rested lightly on a moving paper tape, and the other end in a small tub of ink. A fine ink mark was made down the center of the moving tape. When positive or negative impulses arrived over the cable, the coil swung the siphon above or below the "zero" line, making small mountains and valleys. After the advent of typewriters, operators "read" the siphon tape as it moved across the tops of their typewriters and typed the messages.

To keep pace with the siphon recorder, an automatic transmitter was devised. As the operator sent dots, dashes, and spaces, three plungers perforated a paper tape, and metal pins under each side of the tape made electrical contact through the holes.

When the "duplex" system was introduced in 1871, simultaneous operation in both directions required a "balance" with the real cable, which was obtained by adding resistance coils to provide a "phantom cable." The fastest cable before 1923 handled only 300 letters a minute. A typewriter keyboard receiver was then developed so each combination of pulses would cause the distant recorder to print the desired letter.

A drum relay to repeat an arriving message into another cable was introduced in 1899, and E. S. Heurtley's magnifier in 1912 increased the power of weak incoming signals. After 1921, rotary regenerators, also called regenerative repeaters, automatically strengthened the weak arriving signals and sent them on. That eliminated manual relaying of messages at such cable stations as North Sydney and Bay Roberts, and in two years all cables were operated direct to destination.

Cable Enemies

Far down beneath the ocean waves, about 400,000 miles of ocean cables carry world communications. They are the real heavy traffic workhorses, rarely mentioned by the satellite-minded media. Their enemy is a tiny marine borer, the teredo or eunicid worm, that burrows its way to the gutta-percha of older cables, which it considers the dessert supreme. One little worm hole can cost $100,000 in repairs, but brass stops the teredo, and cables now include a layer of brass.

Another enemy is fishermen with steam trawlers that draw big fish nets in shallow areas along the ocean bottom, with steel runners, known as otterboards, to hold the nets open. Since they break cables, the companies produced a trawl that would turn over and release the cable, but many fishermen believed they would lose the fish and did not use them.

Western Union engineers invented a plow to bury cables in the ocean bottom. The ten-ton plow, 25 feet long, was first used by Western Union Cable Ship *Lord Kelvin* to bury cables 100 miles southwest of the Irish coast where trawlers were costing the cable companies about $500,000 a year. The plow, pulled by a 4,200-foot nickel steel chain attached to the ship, cut a trench 24 inches deep in the ocean bottom. As the cable passed into the trench, the rear part of the plow filled the trench.[28]

Cable enemy number three is the seaquake which causes great bodies of underwater rock and land to slide and snap steel-armored cables like the threads of a spider web. The solution to that is to avoid epicenters around the earth's "faults" when laying cables.

Notes

[1]Nicholas Woods of the *London Times,* on the *Agamemnon,* wrote a realistic description of the ship's groaning and laboring as it rolled in the storm. Charles Bright's diary provided many details of the same trip. John Mullaly of the *New York Herald* was on the *Niagara.*

[2]Phineas T. Barnum, the great showman, offered $5,000 for the privilege of sending the first message, but was refused. Harvey W. Root in *The Unknown Barnum* (New York: Harper, 1927) quoted Barnum as saying: "If I had secured the notoriety of sending the first words through the cable, instead of five thousand dollars the message might have been worth a million to me."

[3]*New York Tribune,* August 18, 1858.

[4]Castle Clinton, built as a fort in the bay during the War of 1812, was converted into Castle Garden, a civic hall and amusement place in 1824. Connected with Battery Park by a drawbridge, it became the Immigrant Station in 1855, and in 1896 the Aquarium, now with the original, circular fort walls.

[5]The park was named for the poet William Cullen Bryant. Beside it on Fifth Avenue was the 44.5-foot high Croton Reservoir where the Public Library now stands.

[6]Tiffany & Co., then at 550 Broadway, bought the leftover cable, cut it in four inch pieces, with silver bands around them, suitably inscribed, and sold thousands as souvenirs.

[7]*Harper's Weekly,* August 4, 1860: article on Shaffner's plans, map of route, and sketches illustrating the expedition arriving at Greenland and camping in Labrador.

[8]Field Papers, Manuscript Division, New York Public Library.

[9]Charles Bright, *The Life Story of Sir Charles Tilston Bright* (London: Constable & Co., 1899) 160-61.

[10]Francis Rowsome, "The Strange Story of the *Great Eastern,*" *Harper's Magazine* (April 1939), lists misfortunes of the jinxed ship from the day it refused to be launched in 1857, until the wreckers fifty years later found the skeleton of one of the riveters who built it in a compartment in the hull. The *Great Eastern* was almost six times the bulk of any previous vessel so it could carry enough coal for a round trip to the Far East for indigo, silks, spices, and tea.

Afraid because of her fatal mishaps, only 36 sailed on the maiden voyage of the 4,000-passenger ship which arrived at New York on June 27, 1860. About 150,000 New Yorkers paid to visit the ship, and 2,000 who went on a cruise to Cape May, New Jersey, rioted because there was no food, and looted it. Beginning in 1865, the *Great Eastern* was a cable ship for about nine years. In 1888 she was a showboat attraction for the Liverpool Exposition. She was sold to a shipbroker in that same year. The *Great Eastern* lost millions, but without her the Atlantic cable would have been delayed for years, and the first telegraph route to Europe would have been through Alaska.

[11]William Howard Russell, a great journalist, had distinguished himself covering the Crimean War (1853–1856). He witnessed the famous charge of the Light Brigade at Baclava. His book, *The Atlantic Telegraph* is one of our sources.

[12]Acting Secretary of the Navy, Fox wrote to Field on June 7, 1865 that it was impossible to send a U.S. naval escort because the English government had not modified its "twenty-four-hour rule" concerning "war vessels of this country."

[13]Anglo-American laid and operated the cables as agent of the Atlantic Telegraph Company, which paid Anglo-American 125,000 pounds a year. It also operated the lines and cables from Heart's Content, Newfoundland, to Nova Scotia for the New York, Newfoundland, and London Telegraph Company, which paid Anglo-American 25,000 pounds a year.

[14]David Lindsey, "Lightning through Deep Waters," *American History Magazine* (February 1973).

[15]Honors were showered upon Field until he died in 1891. At his golden wedding anniversary in 1890, his five children and sixteen grandchildren were present. It was Field's grandson, Cyrus F. Judson of Ardsley-on-Hudson, New York, from whom I obtained Field's dining room furniture, including the table at which the group planned the cable. The table was buried in hay in the loft of his barn, but was moved to the Western Union Museum in New York.

[16]Lord Kelvin (then Sir William Thomson) said in his presidential address to the Institute of Electrical Engineers in 1889: "To Sir C. Bright's vigour, earnestness, and enthusiasm was due the successful laying of the cable." When Sir Henry Mance delivered his I.E.E. presidential address in 1897, he said "Bright was the one individual who preeminently distinguished himself in the development of oceanic telegraphy."

[17]A monument marks the place in St. Johns where the first cable was landed. The 1866 permanent cable was laid in a more sheltered location at Heart's Content. That station, with its old equipment, is a historical museum now. Photos of it and the monument were taken by Col. James Trigg Adams in 1986, and given to me.

[18]Sources: Documents in Western Union's files; Oliver Gramling, *AP, The Story of News* (New York: Farrar and Rinehart, 1940); Victor Rosewater, *History of Cooperative News-Gathering in the United States* (New York: D. Appleton and Co., 1930); Peter R. Knights, *The Press Association War of 1866–1867,* Journalism Monographs no. 6 (Minneapolis: The Association for Education in Journalism, December 1967).

[19]Craig's testimony before the Senate Education and Labor Committee in October 1883.

[20]Knights, *The Press Association War of 1866–1867,* 18.

[21]Ibid., 50.

[22]*The Old Colony Memorial* (a Plymouth, Massachusetts newspaper), July 1, 1937.

[23]*Journal of the Telegraph* (June 16, 1873).

[24]Scrymser's privately printed booklet in the Western Union Library.

[25]The Mexican Cable Company was later renamed Mexican Telegraph Company.

[26]See "The Miami Incident," below, chap. 16.

[27]Western Union records.

[28]About 30 years later, in 1967, the Bell System developed a fourteen-ton undersea plow, and began burying the shore ends of its new transistorized transatlantic cables.

The Industry Expands (and Gould Robs It)

In 1866 Western Union's empire extended to the Pacific, and covered all of the territory north of Fort Smith, Arkansas, and Santa Fe, New Mexico. It was ready to achieve Sibley's goal—a unified, national system.

One major obstacle remained—the American Telegraph Company lines from Newfoundland to New Orleans. When American's southern lines were returned after the war, it began acquiring other lines with money raised by selling stock. It boasted that it had twice as much property in proportion to its capital stock as Western Union. At its annual meeting January 31, 1866, American's President E. S. Sanford reported the purchase of Southwestern, the purchase and rebuilding of the Washington and New Orleans line, and the increase of its capital to four million.[1] With a board including Morse, Kendall, Lefferts, Field, Morris, Swain, Wilson G. Hunt, Hiram O. Alden, Cambridge Livingston, J. G. Bennett, Jr., and John McKesson, who could doubt its success?

One month later, however, Western Union tossed a bomb into American's ranks by taking over the other large company—the United States Telegraph Company.[2] That frightened Judge Caton's Illinois and Mississippi[3] and several other lines into merging with Western Union, and convinced American that further struggle would be futile. They knew that if Western Union refused to accept their messages, they would be ruined, and that Wade would not hesitate to do so.

Although Wade held the trump cards, Sanford bargained like a Yankee horse trader. Because of Western Union's high capitalization, it needed the economies of a merger, and Wade finally offered a liberal deal June 12, 1866. Western Union issued three shares for each American share, with forty million dollars capital, 27,380 miles of line, 75,686 miles of wire, and 2,500 offices.

Wall Street financiers bought control as Western Union's Rochester founders retired with their millions, and the company's headquarters was moved on July 1, 1866, to the

American Telegraph building at 145 Broadway, and two adjoining buildings on Liberty Street. When it moved from its birthplace in Reynolds Arcade, Western Union's officials evidenced a sentimental attachment for old room 22 by preserving it in every detail, even to the iron cuspidor in the form of a tall silk hat.[4]

In poor health, Wade refused reelection, and Vice President William Orton succeeded him July 10, 1867. A former school teacher whom Lincoln had appointed as Internal Revenue collector at New York and later as head of the Internal Revenue Department, Orton had great executive ability, keen understanding of public affairs, and fluency as a public speaker.

Orton bought an interest in the *New York Tribune,* but problems with the newspaper, in addition to the company, were too much. When its famous editor, Horace Greeley, died in 1872, Orton sold his interest, and Whitelaw Reid became the editor.[5]

Orton consolidated duplicating lines and offices, and increased Western Union's business to justify its large capitalization. He had Cromwell F. Varley, an English electrician, examine the company's lines and equipment, and learned that half required replacement. He promulgated standard operating practices through *The Journal of the Telegraph,* which he established soon after taking office.

Many lines had only one wire, and when it was down, the operator would leave a sign on the office: "Closed. Gone to find the break." Building usually was by contract until 1881, when a general superintendent of construction was appointed, with superintendents in several areas.

A few areas still had no service. In 1873 Congress appropriated $50,000 for a military line from San Diego to Tucson because of

Apache Indian trouble in Arizona Territory. R. B. Haines supervised the construction by soldiers. The 540-mile line connected Tucson, Prescott, Wickenburg, Maricopa Wells, Yuma and Phoenix with Western Union at San Diego.

On July 5, 1876, J. M. Carnahan, the lone operator at Bismarck, North Dakota, sent the startling news flash: "All the Custers are killed."

An expedition had left Fort Lincoln, across the Missouri River from Bismarck, to round up hostile Indians under Sitting Bull. Its forces were divided to search for the Indians, and those commanded by Custer were killed in the Battle of Little Big Horn on June 25. An Indian scout with Custer escaped and told the other troops, who reached Bismarck July 4.

Carnahan "raised" Fargo when the office there opened the next morning, was "cut through" to St. Paul, and sent his "flash." Although newspaper correspondents pled with him to send their stories, an official Army report to Washington kept Carnahan sending for forty-eight hours. After a short rest, he sent the stories for twenty-four hours. His seventy-two hours of almost uninterrupted transmitting was one of the great feats of pioneer telegraphing.

Farewell to Morse

With his long white beard and chest covered with medals of many nations, Morse acted his role as a venerated national figure. In 1848 he had married Sarah Elizabeth Griswold of Utica, New York, his second cousin, thirty-one years his junior and two years younger than his daughter Susan.

Morse lived at Locust Grove, his 100-acre estate on the Hudson River, two miles south of Poughkeepsie. His bride had been deaf and dumb, he said, but "has been gradually recovering both hearing and speech." There were four children from that second marriage: Arthur Breese, Cornelia Livingston, William Goodrich, and Edward Lind Morse. In 1854 Morse figured he had received nearly $200,000 for his invention. He was a member of the first Board of Trustees of Vassar College, which opened in 1865.

13.1 SAMUEL F. B. MORSE

Telegraph people from all parts of the nation gathered on June 10, 1871, for the unveiling of a bronze statue of Morse and the 1844 telegraph instrument in Central Park, New York by Governor Claflin of Massachusetts and President Orton. William

Cullen Bryant delivered the principal address, and Cyrus W. Field read messages that were received.

A reception that evening was linked by telegraph with cities throughout the nation, Canada, and Europe. Operator Sadie E. Cornwell transmitted a farewell message from Morse "to the telegraph fraternity throughout the world." Morse took the key, signed his name in dots and dashes, and President Orton said: "Thus the father of telegraph bids farewell to his children." Morse died the following year at eighty-one.

Gould Plunders Western Union

Consolidation of telegraph lines, spectacularly profitable when combined, set a pattern for later trusts in steel, oil, railroad, beef, sugar, tobacco, whiskey, wire nails, electrical supplies, gas, electricity, telephone, and other industries. Some eliminated wasteful competition, and improved research, efficiency, and production, but others brought favorable laws, ruined small competitors, and issued stock far beyond the value of the properties. Then they charged high prices to justify their "watered" stock.

During the Civil War fortunes were made by men whose families for generations were society leaders of New York and Newport. Some paid others to substitute for them in the Union Army, then sold the army guns that would explode when fired, food unfit for human consumption, and rotten troop ships that would sink in a stiff breeze.

One of the worst pirates of that era was Jay Gould, a furtive, beady-eyed little man with a curly black beard who decided to chisel his way into Western Union. At twenty-one he had persuaded Zadoc Pratt, a retired tanner, to finance a new tannery. Pratt sold out after discovering Gould was putting the firm's funds in his own bank account. Gould then got Charles M. Luepp, a fine old New York leather merchant, to become his partner, but Luepp learned Gould had bought another tannery and was using Luepp's name and money in an effort to "corner" the market in hides. After a stormy interview with Gould, Luepp, facing bankruptcy, shot himself.

132 Jay Gould, unscrupulous manipulator of Western Union and railroads, obtained control of the telegraph company with the aid of Russell Sage. He sold cable and telegraph companies to Western Union at high prices, making a large fortune.

At twenty-six Gould married the daughter of Philip Miller, a wealthy grocer. With Miller's money, he gained control of two small railroads, and sold them to larger ones. Accumulating $100,000 in a year, he opened a Wall Street office.

Daniel Drew, a cattle drover turned Wall Street operator, speculated his way into control of the Erie Railroad, and took into its management Gould and Jim Fisk, former circus laborer and peddler, and now a loud-mouthed, flashy playboy and speculator. Erie stock earned the name "The Scarlet Woman of Wall Street" because this unholy trio forced it up and down so rapidly.

Stockholders finally rebelled, and Gould was arrested for appropriating $12,800,000 on Erie stock sales. He agreed to restore $6,000,000 in securities if the charges were withdrawn and he was given an option of 200,000 shares at 30. Gould announced the restitution, the price of the stock rose rapidly, and he sold at a large profit.[6]

In 1869, Gould attempted to corner the gold market. Entertaining President Grant on Fisk's yacht and the Erie's private car, he told Grant a higher price for gold would benefit the nation. He gave Grant's brother-in-law, Abel R. Corbin, a free contract to buy $1,500,000 of gold at 133, and also gave favors to others close to Grant. When the corner squeezed the national money market, Grant delayed ordering the sale of government gold until the market reached 144¼ and his brother-in-law and friends could sell. Even then General Daniel Butterfield, Federal Subtreasurer at New York who also had a $1,500,000 contract, notified Gould an hour before announcing it. That enabled Gould to sell as high as 165, while Fisk, in the adjoining room, continued to squeeze the trapped speculators by buying.

When the government announced it was selling, the price plunged. Thousands were ruined, and September 24, 1869, became known to history as Black Friday.[7] Gould gained $11,000,000, and Fisk avoided ruin by repudiating his oral purchases and bankrupting his brokers. Fisk was killed on January 6, 1872 on the grand staircase of the

Grand Central Hotel by Edward Stokes, a blackmailer who had stolen the affections of Fisk's mistress, Josie Mansfield.

With public affairs in the hands of Boss Tweed's Tammany judges, legislators and officials, the financial jungle was perfect for Gould. He bought control in 1870 of the Atlantic and Pacific Telegraph Company, with lines along the Union Pacific Railroad, of which he and Cyrus W. Field were directors. He expanded its lines to 5,097 miles, and tried to sell it to Western Union, but Cornelius Vanderbilt and his son William Henry wanted no dealings with him.

Edison— Telegraph Operator, Inventor[8]

Gould tried to use Thomas Alva Edison, who applied in his lifetime for 1,328 patents, thirty-nine of them telephone and 146 telegraph, but who is best known for the electric light.

Edison's father, Sam Jr., had joined an insurrection to establish a Canadian republic, planned by William Lyon Mackenzie, later Henry O'Reilly's pal. He fled across the border with soldiers and hounds at his heels, and settled at Milan, Ohio, where "Alva" was born February 11, 1847.[9]

The Edisons moved to Port Huron, Michigan in 1854, where Alva made his own telegraph equipment and practiced operating at eleven years of age. Working as a newsboy and candy butcher on the railroad to Detroit, he snatched the three-year-old son of James Urquardt Mackenzie, stationmaster at Mt. Clemens, from the path of a boxcar. In 1862, Mackenzie gratefully taught Edison to be an operator. Edison was hard of hearing, but could hear the clicking of the telegraph.[10]

From 1864 to 1869 Edison worked in Western Union offices at Toledo, Fort Wayne, Indianapolis, Cincinnati, Nashville, New Orleans, Memphis, Louisville, back to Cincinnati, and Boston. He once told this writer many roving operators were heavy drinkers, broke, happy-go-lucky, and undependable, but he was just broke. At most places he was fired for letting messages pile up unsent, while he read technical books, made drawings, or carried out unauthorized experiments that damaged operations. After receiving a half-month's pay, he would buy a bundle of electrical gadgets, be broke again, and borrow money for food.

In the Cincinnati office, Edison said he rigged up metal plates connected with batteries as a "rat paralyzer," and at Boston used strips of tinfoil to electrocute cockroaches. At Memphis he rigged up a repeater to connect New York with New Orleans and eliminate manual retransmission, but a jealous manager fired him. In Louisville, he developed his distinctive copperplate style of fast handprinting.

Always broke, Edison arrived in Boston without having been in a bed for four days. The Western Union manager decided to have some fun with the ragged hick. He assigned him the number 1 New York wire to the *Boston Herald,* and had one of the fastest operators in New York send. Edison amazed the jokers by keeping up. Finally he broke in and sent, "Say, young man, change off, and send with the other foot."[11]

Birth of the Stock Ticker

Speculation in gold was so active that a Gold Room was established at the New York Stock Exchange, and hundreds of

13.3 Edison experimented at Newark between 1870 and 1876 to increase telegraph transmission speed. He ran a paper disk perforated with dots and dashes to transmit them at 200 words a minute to his automatic repeater. Sound vibrations as the current made contact through the perforations led Edison to invent the phonograph to reproduce sounds. (Courtesy of the U.S. Dept. of the Interior, National Park Service, Edison National Historic Site.)

"runners" carried quotations to brokers' offices. Fastest of the runners was William Heath, nicknamed the "American Deer."

Dr. S. S. Laws, vice president of the Gold Exchange, replaced the blackboard and chalk with a gold indicator dial on the wall in the Exchange, and another facing the street. Deciding to telegraph gold prices simultaneously to brokers' offices, he placed in them a disc indicator operated over two wires by the step-by-step principle. It showed three figures, through a long slot in a box containing three overlapping dials.

Fifty brokers subscribed in 1866; 300 in 1867. Dr. Franklin L. Pope,[12] engineer of the Gold Indicator Company, substituted drums with figures on them. Edward A. Calahan, a telegraph operator, made the first ticker. It used three wires and two type wheels—one for figures and the other for letters—to print the prices on a paper tape. The ticker was so named because of the noise it made. A glass dome was placed over each ticker to deaden the sound.

The Gold & Stock Telegraph Company was organized August 16, 1867, on the strength of the Calahan ticker, with $200,000 capital, and the Gold Indicator Company was consolidated with it.[13] In 1869 its capital was increased to $1,250,000, and Marshall Lefferts, Western Union chief engineer, was elected president.

Edison went to New York in 1869. Penniless and hungry, he was sleeping in the ticker company's battery room when a break in the master transmitter threw the financial district into shouting confusion. Edison saw what was wrong, fixed it, and Dr. Laws hired him for $300 a month.

While holding that job, Edison joined Pope and J. N. Ashley, publisher of *The Telegrapher*,[14] in forming the first electrical engineering firm in America, and the Pope, Edison & Co. "card" appeared in the October 1, 1869, issue, just six days after Black Friday. *The Telegrapher* constantly criticized Western Union, which tried to offset that by establishing *The Journal of the Telegraph* in 1867 with James D. Reid as editor. Edison boarded at Pope's home in Elizabeth, New Jersey, worked in New York all day, and then devoted most of the night to ticker experiments in a small shop in Jersey City.

Edison improved the Calahan instrument, and produced the two-wire Universal ticker, used later on all major ticker systems. Others made improvements, and eight different models were used, but Edison's ticker was unchanged in basic principle.

An Edison invention in 1869 permitted tickers to be brought to unison from the central office, instead of by sending men frequently to reset each ticker. Calahan praised Edison's Unison, which stopped the type wheels of all tickers on the circuit at the same point to synchronize them.

General Lefferts called Edison to his office late in 1869 and asked how much he would take for his ticker inventions. Edison started to answer $3,000, but didn't have the nerve to name such a large sum. He asked Lefferts to make an offer, and when the General said, "How would $40,000 strike you?" Edison nearly fainted. He signed a contract and received a check for $40,000—the first one he ever had. Edison went to a bank, and was dismayed when the teller handed the check back to him. Because of his deafness, he did not hear the request to endorse the check and identify himself. He returned to Lefferts, who sent a clerk to the bank with Edison. As a joke, the teller gave Edison small bills that he gravely stuffed into his pockets. He sat up all night guarding his wealth, and returned to Lefferts, who arranged for him to open a bank account.

Edison used the $40,000 in January 1870 to rent and outfit a shop at 15 Railroad Avenue, Newark, New Jersey, hired fifty men and began building 1,200 tickers. He used the top floor as an experimental laboratory. That is how the round-faced young operator, with gray-blue eyes and a quizzical mouth always seeming about to smile, got the first solid start in his career as an inventor.

Ashley, Edison, and Pope signed an agreement with Gold and Stock on April 18, 1870, to invent and improve printing telegraph instruments, and not to engage for ten years in any activity rivaling the company's business. The company was to pay $5,000 to $10,000 when patents were obtained.

William Unger, an associate of Lefferts, was superintendent of the Edison shop, which operated like the manufacturing arm of Gold and Stock. Pope left December 1, 1870, and the firm was named Edison & Unger, but the shop, known as the Newark Telegraph Works, was moved to 4-6 Ward Street in 1871, and 10-12 Ward Street in 1872, when Edison bought out Unger.

With 800 subscribers and annual gross receipts of $236,215 in 1871, Gold & Stock took over the Commercial News Department (CND) of Western Union.[15] It doubled its capital to $2,500,000, and was leased to Western Union for ninety-nine years. That lease was suggested by Tracy R. Edson, president of the American Bank Note Engraving Co., and on the Gold & Stock executive committee, because Western Union could extend the CND ticker system to other cities. James H. Banker of Western Union's executive committee, and vice president of the Bank of New York, worked out the plan. Revenues rose to $623,900 in 1873.

In the meantime, the ubiquitous D. H. Craig got Edison to sign an agreement in 1870 to develop and manufacture a telegraph

13.4 THOMAS A. EDISON EXPLAINING HIS PERFORATOR (THE FOURTH ONE MADE) TO HENRY FORD, IN THE EXHIBIT OF THE WESTERN UNION COMPANY AT THE PANAMA-PACIFIC INTERNATIONAL EXPOSITION IN SAN FRANCISCO, ON MONDAY MORNING, OCTOBER 20, 1915. LEFT TO RIGHT: FORD, EDISON, AND J. G. DECATUR OF WESTERN UNION.

perforator-repeater. He was to pay Edison $1,300 in cash, $3,700 in stock, and receive half of the proceeds from selling the apparatus and patents to the "National Telegraph Company."

Edison also signed a contract two months later with George Harrington, a former Assistant Secretary of the Treasury and Minister to Switzerland, who had organized an Automatic Telegraph Company. Edison did not realize he was signing away his rights to future automatic telegraph inventions, or becoming the unwitting tool of

a scheme by Jay Gould. First to approach Edison was Edward H. Johnson of the Denver & Rio Grande Railway. Directors of Automatic included Colonel Josiah C. Reiff, Washington lobbyist for Gould's Kansas & Pacific Railroad, and General W. J. Palmer.

Edison's agreement was to perfect an automatic system Palmer and others had bought from George Little, an English inventor. Little's method was a modification of Bain's chemical telegraph. Electrical contacts through the perforations in tape sent the signals, and a machine at the other end of the line recorded the dots and dashes. With $40,000 Harrington provided, a shop, known as the American Telegraph Works, was set up on October 1, 1870, at 103-109 Railroad Avenue, Newark, and Edison gave Harrington a power of attorney on April 4, 1871.

Craig believed his perforator agreement gave him control of Edison, and he was playing ball on both the Automatic and Gold & Stock teams, hoping to share whatever Edison received from either. He constantly harassed the inventor. Two of his letters said Harrington believed the perforator was a failure, and money was being spent on "useless" experiments. Writing to Edison October 3, 1871, Craig said all he, Lefferts, and Little had wanted for a year was a rapid perforator and a good copying printer, and they wanted it without a day or an hour of delay. Three days later, Craig wrote: "The W. U. Co. are *very anxious* to know our secrets, and I suggest the extremest care and prudence to keep every fact of possible value from their legions of *spies.* They begin to see that we are not the d---d fools they said we all were."

Still another Edison shop was established in 1872 in partnership with Joseph T. Murray of Newark at 115 Railroad Avenue, and later moved to 39 Oliver Street. J. T. Murray & Co.[16] produced the keyboard perforator, so arranged that the depression of a key would perforate in tape the combination to send any letter of the alphabet. It stepped up the perforating speed to twenty-five words a minute, but the keys had to be forced down about two inches, and the operator had an exhausting job.

Patrick B. Delaney, working with Edison, reported fairly good results to points as distant as Charleston, South Carolina. Harrington had it demonstrated for Postmaster General A. J. Cresswell, senators and congressmen on January 27, 1874, by sending the 12,000-word president's message from Washington to New York.

Quadruplex Telegraphy

In 1873 Edison obtained President Orton's approval to use Western Union wires in experiments to improve that company's duplex system, invented by Joseph B. Stearns. Edison agreed to sell these improvements to the company, and Orton instructed Eckert and George B. Prescott, the company electrician, to provide all possible assistance. Edison then produced the quadruplex, his greatest telegraph invention, which made one wire do the work of four.

Using a wire to Albany and a large room in the Western Union Building, Edison constantly worked all night, testing with the assistance of four Western Union operators at each city. His method then, as it was later when he invented the electric light, was to try many ways until he found the right solution.

He succeeded by combining a modified single current system with a double current duplex. To permit two messages to travel in each direction, he arranged two relays at each terminal to be unresponsive to outgoing signals. One responded only to reversals of current from a double current transmitter. The other responded only to increases and decreases in current strength, controlled by a single current key at the distant station.

Edison demonstrated the quadruplex to Orton, W. H. Vanderbilt and other Western Union directors July 9, 1874, and was given $5,000 as part payment for the invention and $25,000 later.[17] Royalty of $233 a year was to be paid for each quadruplex circuit used. Edison applied for ten patents to be assigned to the company, and Orton left on a trip.

Production was started at once, and Western Union's annual report for 1874 said the quadruplex was in operation between New York and Boston.

Hearing of the sale to Western Union, Harrington sent a note to Edison:

> Midnight. I returned this afternoon. Having learned what was going on, have been all evening investigating. . . . Beg of you to see me before you sign any papers. Come to 80 Broadway. I am in hopes I can relieve you. At the moment, adverse action will cause a loss of $100,000."[18]

Josiah Reiff, president of Automatic, also urged Edison to make no deal with Western Union, and his company loaned the inventor $10,000 to pay creditors pressing him.

Gould's Atlantic and Pacific was fighting Western Union. General Eckert, Western Union's Eastern Division general superintendent, called Edison into his office, Edison said, "and made inquiries about money matters. I told him Mr. Orton had gone off and left me without means, and I was in straits. He told me I would never get another cent, but that he knew a man who would buy it."

"If I got enough for it," Edison replied, "I would sell all my interest in any share I might have."[19]

Eckert secretly brought Gould to see the quadruplex in Edison's shop at 10-12 Ward Street, Newark. The following evening Eckert took Edison to Gould's home, using the servants' entrance for fear of being watched.[20] Gould paid Edison $30,000 for the quadruplex, and made him chief electrician of Atlantic and Pacific, which then bought the Automatic Telegraph Company. Edison also gave Gould a power of attorney.

Edison then wrote to Orton that his agreement with Western Union was a mistake, and revoked his request to the Com-

missioner of Patents to issue the patent to him and Harrington. After his double-crossing deal, Eckert left Western Union January 14, 1875, to be president of Gould's A.&P.[21]

Edison had sold the invention "twice over, once to President Orton and again to his rival Jay Gould," *The Electrician* of January 1886 said. "The sudden defection of Edison to the lines of the enemy in February 1875 occasioned a feeling little short of consternation in the executive offices of Western Union." Western Union brought suit, tying up use of the quadruplex by the A.&P.

Fearing they would lose out, Harrington, Reiff, and associates formed the Electric Automatic Telegraph Company August 31, 1875, listing Edison's duplex and quadruplex patents as assets and claiming that Harrington had a two-thirds interest in the patents and power to sell the remaining one-third.

Craig wrote to Reiff December 24, 1875, that he had put $50,000 into Automatic Telegraph on his recommendation, and it was Reiff's personal obligation! Actually, Edison's personal list of Automatic Telegraph owners did not mention Craig, but credited Reiff with investing $176,000; Harrington, $106,350; and Edison himself, $75,000 (probably credited for services).[22] Craig wrote to Edison that Harrington and associates had tried to rob Craig after he "had put him [Harrington] in control of Automatic Telegraph and brought him and you together. . . . In trusting to Harrington and his associates, I have trusted to a crowd of incompetent rogues." A more accurate statement would have been that Craig and the others were fighting for possession of the fortune Edison's wizardry was producing.

"In hurriedly casting about for means to protect the interest of his company," *The Electrician* of January 1886 said,

President Orton was informed of an invention by one Dr. [Henry C.] Nicholson . . . of a quadruplex, and acquired his patent, then pending in the Patent Office. . . . Interference proceedings were instituted by the Patent Office between the application of Edison and the two applications of Nicholson, in order to determine the question of priority of invention. In June 1878 . . . Judge Sanford confirmed the title to Western Union, which thus became the owner of all the inventions involved in the interference.[23]

In 1876 Edison learned, as others had, that Gould would not hesitate to rob his associates. He had been paid in Automatic Telegraph stock, but, when that company was merged with A. & P., its stockholders got nothing. They brought suit, but it was not until thirty years later, in 1906, that Judge Hazel in the U. S. Circuit Court, ordered an accounting for the patents and wires Gould had taken over. When the damages were announced, they were for only one dollar! Edison said he never received a cent for those three years of hard labor.

Edison told me that he always "considered Western Union his company," and returned to work for it in 1876. Of Gould, he said, "I never liked his face. It was dark, and covered all over with whiskers so you could hardly see him."

Western Union maintained a direct wire from its New York offices to Edison's laboratory for a number of years; he had only to tap his Morse key to "talk" with company officials. His delight in telegraphy never ceased. I spent two long Saturday afternoons alone with Edison in telegraph offices where he "talked" with boyish delight over a "round robin" circuit with retired Morse Club men nationwide and stayed after the last old-timer signed off, reminiscing about his early years.

Gould's Atlantic and Pacific was a thorn in Western Union's flesh. A.&P. constructed lines from New York to Buffalo on October 15, 1867; to Cleveland by January 11, 1868; and to Chicago on November 2, 1868. A.&P. slashed rates by twenty-five percent, and obtained contracts with the Baltimore and Ohio, Pennsylvania, Northern Central, Erie, and other railroads.

On January 1, 1877, Atlantic and Pacific had 17,759 miles of pole line, 36,044 miles of wire, and 1,757 offices. Gould organized a stock pool in which James R. Keene, a rich mining operator, and Major A. A. Selover, a muscular California stock gambler, joined in selling Western Union stock "short" to hammer its price down. Learning that Gould was double-crossing his associates by buying while they sold, Keene found Gould in Sage's office and brandished a pistol in his face. On August 2, 1877, Selover caught Gould on the street, slapped his face, and threw him into an areaway seven or eight feet deep at Exchange Place.[24] After that Gould was accompanied by a bodyguard.

Western Union cut rates, putting the A.&P. under heavy pressure. Gould then agreed to sell 72,502 shares of Atlantic and Pacific stock at a large profit, and requested membership on Western Union's board. William H. Vanderbilt, who had succeeded his father, Commodore Cornelius Vanderbilt when he died, and the other directors quickly refused.

Western Union's board had plagued Orton with demands for increased rates and higher dividends. Instead, he devoted earnings mainly to reconstruction of lines and improvement of service. He worked day and night to carry out his ideal of public service. To halt his heavy spending, the board created a committee on expenditures which took over some of his powers. Orton considered resigning, but decided to take a three-month leave of absence in order to go abroad and try to improve his health. He was about to leave when he died suddenly on April 22, 1878.

Dr. Norvin Green

Dr. Norvin Green became president of Western Union November 12, 1878. It was an odd twist of fate that made a Kentucky physician, violinist, and state legislator, the head of the world's largest telegraph company.

A great grandson of Peter Cooper, Gould had, at sixteen, a flat boat on the Mississippi as a floating store. While practicing medicine, Dr. Green and others bought the combined Morse and O'Reilly lines between Louisville, Nashville, and New Orleans at a sheriff's sale, and he became president of the Southwestern Telegraph Company when it was consolidated with American Telegraph January 2, 1866. When Western Union absorbed American a few months later, Dr. Green became a vice president.

When Gould sold Atlantic and Pacific to Western Union, he expected that Eckert would at least be appointed general manager, but the board no doubt saw the danger of placing Gould's henchman in a high position. Eckert was ignored again when Dr. Green was elected president, and Gould and Eckert angrily resumed their fight. They organized the American Union Telegraph Company on May 15, 1879, with David Homer Bates as president; Giovanni P. Morisini, Gould's secretary, as treasurer; and Russell Sage, John B. Alley, F. L. Ames, and Gould as directors.

With capital of $10,000,000, the company leased several lines, elected the president of Baltimore and Ohio Railroad to its board, connected with the French Cable, and created a construction company to build lines. Western Union obtained a court injunction to block American Union lines from entering New York City, but that caused only a brief delay, after which it established its headquarters at 135 Broadway.

Eckert's enmity for Western Union made him the ideal president for American Union, and Gould elected him to that office January 1, 1880. Gould bought more Western Union stock and carried on a raid for proxies to gain control, but Vanderbilt and other directors began a "war of extermination" against American Union. How savagely American Union fought back is indicated by the fact that in one year it had 10,701 miles of pole line, 46,422 miles of wire and about 2,000 offices.

It slashed rates, and thugs cut down Western Union poles and carried off wires. Western Union bought rights-of-way ahead of American Union line builders to block construction. American Union threw Western Union lines off railroad rights-of-way, and lawyers and courts in many states were busy!

Gould loaned money to Whitelaw Reid, publisher of the *Tribune*,[25] on the controlling stock in that paper, and when Thomas A. Scott, president of the Pennsylvania Railroad, died, Gould bought control of the *New York World* from the Scott estate. Thereafter, both papers championed Gould as a "public-spirited benefactor" for his "altruism," and the *World* denounced Western Union as "blood-sucking, un-American."

In 1880 Western Union net earnings sank to $5,919,019.31. In fourteen years its profit had been $45,350,241.48; it paid $23,103,492.99 in dividends; and it spent

$8,012,676 for construction of lines. However, competition and rate cutting caused its revenues to fall $804,000 in six months.

An unhappy William H. Vanderbilt sent for Gould on January 9, 1881, to come to his mansion at 640 Fifth Avenue. "The Great Mogul" went at once. They agreed to merge, increase Western Union stock from $38,926,590 to $80,000,000, give Gould's stockholders 84,000 Western Union shares valued at $8,400,000, and Western Union shareowners 72,502 shares as a stock dividend. At the same time Gould doubled the price of his two Atlantic cables and leased them to Western Union.[26]

With Sage, Addison Cammack and others selling its stock and the *New York World* attacking it, Western Union stock dropped, and Gould secretly bought large blocks, leaving his associates with heavy losses. Then he surprised the Vanderbilt group by taking control of Western Union, and they resigned.

Challenging the "holdup" price paid to Gould, stockholders asked in court how much it would cost to reproduce the American Union. Gould said he did not believe it could be done.

> "At any price?" the court asked.
> "No, because you could not get control of the railroads. You would have to invest $40 million or $50 million to get control of them. It is the railroads that furnish the foundation of the plant."
> "And you were able to give these facilities to the American Union Company?"
> "I have been all my life in the railroad business," Gould replied, "and I own and control large amounts of roads. It was that power which enabled the American Union to make a success."

While hearings were held on a law to block consolidation, the company rushed it to completion, and on the day the law was to be offered to the legislature, the morning papers announced the merger already was a fact.

Gould's secret manipulation of this deal was described in the autobiography *The Americanization of Edward Bok*. Bok, who was a native of Holland and became a famous magazine editor, was later memorialized by a park with gardens and carillon. This was the "Singing Tower" at Lake Wales, Fla. Bok started work as a Western Union office boy and then became a stenographer. Gould "happened in" daily at "195" to dictate to Bok. However, Bok became disgusted when Gould invited others to lunch and omitted him. The final straw came after Bok had attended a Sunday meeting at the financier's Fifth Avenue home, and delivered forty-five pages of notes to Gould at 9:45 the next morning without receiving a word of thanks. Bok left the company in 1882, and began a publishing career.

Reviewing the "Telegraph War," *The Operator*,[27] with tongue in cheek, held up Gould, Bates, Tinker, and Morisini "to public admiration as natural-born philanthropists." "These trials," it said,

> were remarkable. . . . Brother Jay Gould unblushingly calling on his Maker to witness that he could not remember whether he had written his check for five million or ten million dollars; and his crony, Mr. Giovanni Purissimo Morisini, swearing in set phrases that he was treasurer of the American Union Telegraph Company; that he never drew a check, never saw the books of the company, and never received any salary.

Cyrus W. Field   Rufus Hatch   D. O. Mills   August Belmont   Jas. R. Keene
          Russell Sage   Jay Gould   Sidney Dillon   Wm. H. Vanderbilt   Geo. Wm. Ballou

13.5 WESTERN UNION'S DIRECTORS WERE FOR MANY YEARS AMONG THE WEALTHIEST FINANCIERS IN AMERICA. THE BOARD OF DIRECTORS WAS CALLED "THE BILLION DOLLAR BOARD."

The Western Union board elected to its board the directors Gould proposed: Dr. Green, Eckert, Edwin D. Morgan, John Van Horne, Augustus Schell, Harrison Durkee, Russell Sage, Alonzo B. Cornell (son of Ezra), Sidney Dillon, Cyrus W. Field, Edwards S. Sanford, James H. Banker, Moses Taylor, Robert Lenox Kennedy, Hugh J. Jewett, J. Pierpont Morgan, Frederick L. Ames, Edwin D. Worcester, William D. Bishop, C. P. Huntington, George B. Roberts, Zalmon G. Simmons, Samuel Sloan, Erastus Wiman, Amasa Stone, George J. Gould (son of Jay), Chauncey M. Depew, and James W. Clendenin.

Sage's biographer Paul Sarnoff said Gould was Russell Sage's "messenger boy" in his sales to Western Union. In the September 1873 panic, Sarnoff wrote, Sage loaned money for Gould's market operations. Gould was caught "short," and his creditors held a meeting to bankrupt him. Sage walked in, placed his check for two million dollars on the table, and rescued Gould, who was deeply in debt to Sage until he died.[28]

Cyrus W. Field joined Sage and Gould in raiding New York's elevated railway companies. Field leased his New York Elevated Company to the Manhattan Railway Company which Sage created. With ex-

Governor Samuel Jones Tilden[29] as counsel, they pushed the Manhattan stock from fourteen to fifty dollars, and then beat it down to twenty dollars by having Judge Theodore R. Westbrook throw it into receivership. Gould then installed Westbrook in an office next to his in the Western Union Building to sign court orders as needed.

Sage and Gould then doubled the fare to ten cents, an action the *New York Times* exposed under the headline, "Public Trusts Betrayed." Fearing his public image as a cable hero would be ruined, Field, as president of Manhattan, cut the fare back to five cents, and tried to freeze Sage and Gould out by cornering the Manhattan stock. That raised the price to $175, whereupon Sage and Gould dumped their heavy holdings and sent the stock down. Field was forced to sell 75,000 shares to Gould at $120. Field lived on until 1892, a comparatively poor man.[30]

A mob of 5,000 marched on the Western Union building in 1884 because Gould had aided Republican James G. Blaine's candidacy for the presidency. The vote was so close that it was days before the Associated Press reported Grover Cleveland's election. The *New York World,* now anti-Gould since he sold control in 1883 to Joseph Pulitzer, claimed Western Union delayed election reports until Gould could sell stocks. The crowd wanted to hang Gould, but he was on his 150-foot yacht in the Hudson River in front of his Tarrytown mansion.[31] Western Union was Gould's one solid business success, his constant source of dividends and capital gains, and it also was one of Sage's major interests.

Sage fostered the story that he would haggle over a penny on the price of an apple and eat it for lunch at his desk while making a million dollar loan, thinking a reputation of penny-pinching would give people confidence in him. Actually his lunches were free—at Western Union. There he would feel a director's coat, ask what was paid for it, pretend amazement over such extravagance, and brag that he paid only one dollar for his hat. Although he was one of the world's richest men, Sage wore frayed and baggy garments in contrast with the morning coat, striped trousers, and frilled shirts of the other directors.

Aristocratic directors observed Sage's satisfaction in receiving his five-dollar gold piece for attendance at board meetings and carrying away the cheap palmetto fan, writing paper, and pencils at each director's seat. Once Colonel Robert C. Clowry mischievously told Charles M. Holmes, in charge of the board room, he should watch Sage and protect the company's property. He whispered to other board members who waited to see the fun.

When Sage filled his pockets and started out with the fan, Holmes stopped him at the door and offered to take the fan. Annoyed, Sage refused, and Holmes declared, "That is Western Union property." Glaring at the man, Sage roared, "I'll have you know I am Western Union." With that the man worth a hundred million, clutching the five-cent fan, tried to dart past the attendant, but was collared. Holmes emerged from the struggle with the fan, and Sage left, snorting in disgust, while others tried to control their laughter.[32]

The 30,000 miles of wire of Northwestern Telegraph Company was added to the Central Division, headed by Colonel Clowry in 1881. Western Union extended the Northwestern lines to Washington and Oregon along the right-of-way of the Northern Pacific and other railroads.

Private Wires

It might surprise officials of thousands of companies with private wire systems to learn how long private wires have been leased. In 1867 Harrison Brothers, sugar refiners, leased a wire between New York and Philadelphia from the Franklin Telegraph Company. Edison's partners, Pope and Ashley formed the American Printing Telegraph Company in 1869 to provide leased wires. One of Western Union's earliest leases linked Charles Pratt's oil company on Long Island with his New York office.

The Associated Press leased its first wire from Western Union in 1875 between New York and Washington. The Manhattan Telegraph Company entered the private wire business in New York in 1871, and six years later it was leased to Gold and Stock, serving 300 companies in New York and other cities. The Universal Private Line automatic printer developed by Elisha Gray of Chicago and produced by Edison, was used on most early leases.

General Lefferts died July 3, 1876, and George Walker, a Western Union vice president, took over Gold and Stock operations. G&S retained its corporate identity as an underlying company until Western Union's Commercial News Department absorbed it around the end of the century. Then H. M. Heffner was appointed to run it, and for many years was one of the best-known telegraph men in the financial world.

Introduction of the Typewriter

Christopher L. Sholes produced a typewriter in 1867 in crude, experimental form. Realizing its first major use would be in telegraphy, Sholes offered a typewriter to E.

Payson Porter of Western Union if he would demonstrate it. Since Porter was a House Printing telegraph operator, he found typing easy, and used the typewriter in the company's Chicago office.

Porter exhibited the typewriter early in 1869 in a test of fast Morse receiving in the office of General Stager, then general manager of the Lake Division. Stager was so impressed that he introduced typewriters in other telegraph offices. *The Journal of the Telegraph,* July 1, 1869 said, "the recent invention of the typewriter enables an operator to produce more and better copy."

Since Western Union transmitted 369,503,630 words, received in longhand for newspapers alone in 1869 (for $883,509), it ordered so many typewriters, that in March 1873 Sholes and his backer, James Densmore, had to get E. Remington & Sons, a firearms, sewing machine, and farm tool maker, to manufacture them.[33]

Morse Operators

The typewriter so increased their speed and efficiency that operators began using abbreviations published in a "Phillips Code" manual. Using that code, first-class operators often handled 20,000 words of news in one night. In Phillips Code "wi efy dmz ay osn" meant "will effectually demoralize any opposition;" "dbf," "destroyed by fire;" "fapib," "filed a petition in bankruptcy," and "Scotus," "Supreme Court of the United States."

In the jargon of the Morse operators, 25 meant busy; 73, my compliments or best regards; 13, understand; 4, where was I?; 95, the following is urgent.

Sending to the wind meant futile transmission; *round robin,* a circuit between a number of places, but circling back to the

13.6 IN 1875, WESTERN UNION MOVED ITS HEADQUARTERS FROM 145 BROADWAY TO THE NEW, LARGEST TELEGRAPH BUILDING IN THE WORLD AT 195 BROADWAY. THE AIR OVER BROADWAY WAS BUSY WITH WIRES CONNECTING THE BUILDING WITH LINES TO ALL PARTS OF THE COUNTRY. THE WIRES WERE PLACED UNDER THE STREETS AFTER FIRE DESTROYED THE UPPER FLOORS IN 1890.

starting point; *mill,* a tyewriter; *ops,* brass pounders or operators; *ham,* an inept operator; *fist* or *mitt,* the characteristic "touch" of each operator; *copperplate,* good handwriting; *query,* a message from a correspondent offering a number of words on a story; *service,* a company message; *bust,* error in transmission; *GA,* go ahead; *GN,* good night.

"OK," first used in January 1846 by the Magnetic Telegraph Company to acknowledge receipt of messages, came into general public use. It was derived from the expression "Oll Korrect"* in the 1840 political campaign.

Operators used "30" to say "the end" as early as 1858. George E. Allen said in the Utica, New York *Press* in May 1895, that "30" meant "collect pay at the other end," and, since all stories were sent to newspapers collect, press dispatches ended with "30." Also, thirty sheets of stories were sent on many AP news wires every night; "30" at the end of the last story meant the end of the day's work. Still another version was that a New York editor named G. W. Thurtee always signed the final sheet with his name before going to press.

Morse operators competed in national tournaments from 1868 to 1915, achieving average speeds of 43.8 to 52 words a minute, using straight Morse. At a San Francisco meet, the winner averaged 53.7 words a minute, sending 1,500 words in Phillips Code. The greatest Phillips speed apparently was 326 words in five minutes by F. M. McClintic of the AP, Dallas office, in the 1902 Atlanta tournament.[34]

At news events the operator sat beside the newsman whose story he sent. Many operators spent a lifetime working at major news events. Since the reporter's work was all in vain unless it "made" the next edition,

he cherished friendships with operators. For example, when the Lindbergh baby kidnapping trial jury rendered its death verdict, one major press service flashed an erroneous verdict worldwide, while another sent the correct story, signalled by Emil ("The Count") Visconti, a veteran news telegrapher. He slid an ace of clubs playing card under the locked courtroom door.[35]

Francis A. Gribbon was Western Union manager—supervisor at United Press New York headquarters from the day UP was formed in 1907. Listening to a German radio station before the United States entered World War I, "Grib" heard the dots and dashes say a British vessel had fired on a German ship. He called Roy Howard, president of UP, and began copying the story. That is how UP scored a "beat" on the battle of Jutland. When "Grib" retired in 1939, after fifty-one years of telegraph service, UP officials held such a ceremony as they might have for one of their founders.[36]

The Washington press corps loved Carroll S. Linkins, Western Union's press man in the White House for thirty years, who travelled a million miles with presidents. The side of the press car on presidential trains bore his name in large gold letters.[37]

Many a news story was sent by an operator when his reporter teammate was absent, and countless errors by young reporters were corrected by telegraphers with long experience. When Operator Edson S. Brewster retired, he was such an authority on baseball that he toured the country for years lecturing as a baseball league employee.

Wherever major news stories broke, telegraph men rushed to the scene, strung wires, set up emergency press rooms and sent the news at top speed. The sound of dots and dashes, boosted by the resonance of

13.7 Frank Leslie's *Illustrated Newspaper,* February 12, 1881, showed the main operating room of Western Union at 195 Broadway, with women operators wearing voluminous skirts.

an old Prince Albert tobacco can, was heard wherever there was news, and in newspaper offices where it was received.

Around 1934 the Associated Press discontinued the last Morse operation and used printers.[38] Those colorful, dedicated, and wise Morse men are gone now, but their deeds will live as long as old newsmen reminisce.

The number of words Western Union telegraphed from a few major events were as follows: 1924 Democratic National Convention, 9,705,603; 1925 Scopes Trial, 1,716,000; 1927 Hall-Mills Trial, 4,698,311; 1929 Snyder-Gray Trial, 1,318,385; 1933 California earthquake, 15,000,000; 1936 Hauptmann Trial, 4,620,263; 1945 United Nations Conference at San Francisco, more than 6,000,000.

Old 195 Broadway

The five-story Western Union building, used at 145 Broadway when it moved from Rochester in 1866, soon was too small. In 1872 the block on the west side of Broadway between Fulton and Dey Streets was selected for a new headquarters. The company paid $900,000 for the site and $1,300,000 to erect a massive ten-story building, with huge first floor columns and an iron steeple 230 feet above the street. Because of its close relations with Western Union, the Associated Press also moved to 195 Broadway.

Operators moved into the main operating room on the seventh floor on February 1, 1875. Since it was considered proper to keep men and women separated, the lady operators worked behind an eight-foot canvas fence dividing the room. Strict orders were posted, but adventuresome operators risked a reprimand from the stern Ladies' Chief, Miss L. H. Snow, by standing on chairs and peeping over the fence. The girls' sly smiles seemed encouraging, but in that Victorian period, there wasn't much else to see. Officials finally realized the fence was ridiculous and had it taken down.

Visitors daily marveled at the scene of activity, with the four long rows of operators at tables and a switchboard along the north wall. *Frank Leslie's Popular Monthly* featured "Our Monster Telegraph System," and called the structure "magnificent."

In the operating room in 1885 were 459 operating tables, 410 Morse sounders, six Phelps motor printers, thirty sets of duplex, forty-nine sets of quadruplex, five automatic repeaters, eight "typewriting machines," and ten Wheatstone perforators. The number of telegrams handled in the room daily had grown to 93,127, and the day and night operating forces totaled nearly 1,000.

Another famous sight was a huge time ball mounted on an iron framework atop the building. Exactly at noon a signal from the United States Naval Observatory at Washington, D.C. caused the iron ball to slide to the bottom of its tower. People in the streets, on ships in the harbor, in New Jersey, and on Long Island set their timepieces daily by watching the ball. Its use continued until 1913, when other skyscrapers obstructed the view.[39]

The "fireproof" building caught fire on July 18, 1890, on the sixth floor where numerous electric batteries provided current. The fire ate its way to the roof, leaving only walls and wreckage. Employees carried their equipment to lower floors, where carpenters

set up rough tables, and electricians connected outside lines. Wire Chief Fred W. Baldwin, struggling through water and wreckage to keep communications going, contracted a severe cold and died.

13.8 CLOSE-UP OF THE "TIME BALL" FORMERLY ATOP THE OLD WESTERN UNION BUILDING AT 195 BROADWAY, NEW YORK, WHICH WAS DROPPED AT NOON UPON RECEIPT OF TIME SIGNALS FROM THE U.S. NAVAL OBSERVATORY.

It required a year and a half, to February 22, 1892, to restore the building, increase its height, and extend it fifty feet to the west.

13.9 Western Union headquarters building, 195 Broadway, New York. The gold-winged figure at the peak of the tower is *Genius of Electricity*, erected October 24, 1916, the 55th anniversary of the completion of the first transcontinental telegraph. When W.U. sold its half ownership of the building to AT&T, AT&T renamed it the *Spirit of Communications*.

When Dr. Green and General Eckert visited the new operating room, telegraph keys were thrown open, and men and women jumped to their feet and cheered. The batteries on the sixth floor were replaced by dynamos in the cellar, and pneumatic tubes were installed through which messages sped between floors.

New York traffic in 1892 averaged 105,000 telegrams and cablegrams a day, exclusive of the press. Telegraph offices often handled a sudden large volume of business, such as on July 2, 1881, when the Washington office sent 275,000 words of press on President Garfield's assassination, and Cleveland had 292,000 words on the funeral.

Early Labor Troubles

In the early 1850s, $400 a year was top salary for operators, and $500 to $600 for managers in cities. In the late 1850s first class operators received fifty and sixty dollars a month, and chief operators seventy-five dollars. The Civil War demand for operators, and an increase of fifty percent in telegraph rates resulted in a raise of twenty percent, but operators worked nine hours a day and a part of every other night.

A few operators at New York City organized the Telegraphers' Protective League in 1866, and its membership spread to other cities. On New Year's Day 1870, the League wired other cities that because of a "general reduction" in salaries on the Pacific Coast they had stopped work. Workers in a number of cities joined in the work stoppage, and demands were presented that no salaries at San Francisco be reduced, and "dismissed" operators be reinstated.

The fact was that no "general reduction" in salaries had even been contemplated; two men had been dropped because there was no

work for them. Faithful employees handled the business, and in ten days all vacancies were filled.

In 1881 the Brotherhood of Telegraphers of the United States and Canada was organized at Pittsburgh, and by July 15, 1882, claimed a membership of 15,000 linemen, operators, battery men, and clerks. Business had increased 160 percent between 1867 and 1882, partly due to the addition of 300,000 miles of wire and 9,503 offices, and many operators were needed, but the union demanded that all telegraph schools be closed. Telegraph rates had been reduced in the 1882–1883 depression, and the company sought to cut operating costs by employing eight percent of inexperienced but competent operators.

The Brotherhood presented a list of demands on July 16, 1883, for a fifteen percent raise, five hours Sunday to count as a day, two days pay for linemen on Sunday, an eight-hour day, seven-hour night, and equal pay for both sexes.

General Manager Eckert sent a message to Clowry at Chicago which operators construed as preparations to refuse their demands. At 12:12 p.m. on July 19 whistles sounded in operating rooms of principal cities, and numerous operators walked out. In New York 300 left, and 161 remained. The company at once closed branch offices in New York and other cities, and the managers went to work at the main offices. By 4:00 p.m. more than 200 operators and officials were at work "pounding brass" at New York, and the business was all cleared before midnight. Employees were hired to fill vacancies, but in many cities not one operator deserted his post. The strikers appealed without success for government intervention, claiming that government messages were being delayed.

Some strikers made a nightly practice of cutting telegraph wires, and the *New York Times* and the *Herald* supported them, but the consequences for those who walked out were so disastrous that unions were unable to invade the industry again for years.

The 1883 strike was largely the result of dissatisfaction over a sliding scale of wage reductions by Vice President H. McK. Twombly, son-in-law of W. H. Vanderbilt, over the opposition of other telegraph officials. Control of the company was then gained by Gould, who suffered the consequences. The company later granted some of the reforms the employees wanted.

Notes

[1]*The Telegrapher,* March 1, 1866; also see above, chap. 9.

[2]See above, chap. 9.

[3]In a circular, June 28, 1867, Caton told employees that on July 1 all lines and property would be transferred to Western Union, and they would report to General Superintendent Stager of that company.

[4]When the old Arcade was torn down in 1932, Western Union's offices were shifted to the first section of the new building to preserve its unbroken tenancy. Room 22 was taken down in sections and stored in packing cases. When Rochester celebrated its centennial in 1934, the old room, including furniture and ledger, became a permanent exhibit in the Rochester museum.

[5]Royal Cortissoz, *The Life of Whitelaw Reid Charles* (New York: Scribner's Sons, 1921) 245.

[6]Matthew Josephson, *The Robber Barons* (New York: Harcourt, Brace, 1934); Richard O'Connor, *Gould's Millions* (Garden City NY: Doubleday, 1962).

[7]However, the article "Western Union," in *Fortune Magazine* (November 1933) said "Jay Gould, by selling 5,000 [Western Union] shares at the opening on Sept. 18, 1869, broke it seven points and started the general collapse that is still recalled as Black Friday" (p. 93).

[8]Major sources for this section include personal conversations with Edison; his papers at Edison National Historic Site, West Orange, New Jersey, in an underground vault with 3,400 notebooks, 100,000 letters and contracts, patents, drawings, and ledgers, which had not been sorted, made available by Norman R. Speiden, supervisory museum curator, Harold S. Anderson, museum curator, and Mrs. Kathleen McGuirk, archivist.

[9]Frank Lewis Dyer and Thomas Commerford Martin in collaboration with William Henry Meadowcroft, *Edison, His Life and Inventions* (New York: Harper & Bros., 1910 and 1929).

[10]Above, chap. 4, and Matthew Josephson, *Edison* (New York: McGraw-Hill, 1959).

[11]Ibid.; and Dyer and Martin, *Edison, His Life and Inventions.*

[12]Pope had been chief of explorations, British America, of the Russian-American Expedition. See above, chap. 10.

[13]H. A. Calahan of Mamaroneck, New York said his father received $100,000.

[14]Improvements were made by George M. Phelps, who had supervised the production of House printers at Troy, New York. In 1855 American Telegraph bought his shop, and in two years as its superintendent, he made the Hughes printer practical and successful. The Phelps Combination Printer, introduced in 1859, was used for the next twenty years, sending and printing 2,800 words an hour. He became superintendent of Western Union's machine shops at New York in 1866, and developed widely used tickers and private wire equipment. His automatic printer, driven by an electromagnetic motor,

received sixty words a minute.

[15]CND connected with the New York Associated Press to provide reports from foreign and domestic markets to thousands of firms nationwide.

[16]The Edison and Murray firms were merged in 1873; Edison & Murray of 10-12 Ward Street was dissolved on July 13, 1875, and Edison went on alone.

[17]Edison wrote to Orton on December 16, 1874:

> This company has over 25,000 miles of wire which can now be profitably "quadruplexed." Considering these 25,000 miles to be already duplexed, the quadruplex will create 50,000 miles additional. For all our patents and efforts in protecting the company in the monopoly of the same during this life, we will take one-twentieth of the average cost of maintenance of 50,000 miles of wire for 17 years.

Before receiving Edison's letter, Orton wrote to Edison on December 17, 1874:

> This will authorize you to make 20 sets of quadruplex complete for 20 circuits, all to be finished and delivered within 55 days from date. The finish and workmanship to be equal to those made by G. M. Phelps, and the cost of each set not to exceed the cost of the sets already made.

[18]Matthew Josephson. *Edison*, 115-16.

[19]*Telegraphers of Today*, a biography of Eckert by John B. Taltavall (New York: Club Press, 1894).

[20]*New York Tribune*, July 2, 1877.

[21]George S. Bryan, *Edison—The Man and His Work* (Garden City NY: Garden City Publishing Co., 1926).

[22]List found among Edison's Papers, Edison National Historic Site, West Orange, New Jersey; researched with Edison's personal permission.

[23]Proceedings in suit by A.&P. vs. Western Union, Edison, and others, vol. 4.

[24]H. D. Northrop, *Life and Achievements of Jay Gould* (Philadelphia: National Publishing Co., 1892) 172-83.

[25]When Horace Greeley, who founded the *Tribune* in 1841, returned after his disastrous campaign for the presidency, he found his managing editor Whitelaw Reid in control. The shock was too much, and he died on November 29, 1872. In 1875 Reid built one of the most impressive skyscrapers in the city for the paper on Park Row.

[26]See above, chap. 11.

[27]*The Operator,* a telegraph journal, January 1, 1882.

[28]Paul Sarnoff, *Russell Sage, the Money King* (New York: Obolensky, 1965) 190.

[29]James Melvin Lee in his *History of American Journalism* (Boston/New York: Houghton Mifflin Co., 1917 and 1923) said Gould's company was charged with helping prevent Tilden's election as president and Democratic Congressmen had Western Union's record copies of political telegrams seized. To their consternation, the messages related to purchases of election boards and votes by their own party. They then claimed company officials had hidden similar messages by Republicans.

[30]Richard O'Connor, *Gould's Millions* (Garden City NY: Doubleday, 1962).

In *The Life of Whitelaw Reid* (New York: Charles Scribner's Sons, 1921) 402, Royal Cortissoz said "cipher dispatches" were found indicating an effort to purchase Oregon's electoral vote for Tilden. He said interested parties saw that the telegrams got into the hands of the *New York Tribune,* which solved the cipher and exposed the whole scheme.

[31]The home was received by the National Trust for Historic Preservation in 1965 under the will of Gould's daughter Anna, the Duchess of Talleyrand-Perigord.

[32]This incident was described to me by Lewis McKisick, a Western Union vice president, who heard it from those present. The Russell Sage Foundation, established by his widow, carried out her dream of a garden city for white-collar workers. Started in 1908, the 175-acre Forest Hills Gardens in New York City is the result. After her husband's death in 1906, Mrs. Sage

gave sixty million dollars to set up Russell Sage College in Troy, New York, the Sage Foundation for Social Betterment, and other good works.

[33]Sholes sold his interest for $12,000, but Densmore received $1,500,000 in royalties, *News Front* said (May 1963). Mark Twain bought a typewriter in 1874 to use in writing *Life on the Mississippi.* In 1886, Remington sold its typewriter business to Henry Benedict for $186,000.

*O.K. began as an abbreviation for "Old Kinderhook," a nickname for Martin van Buren (U.S. president, 1837–1841, who was born in Kinderhook, New York). The initials were taken up by the Democrat's "O.K. Clubs," "secret" political clubs, in 1840. Apparently the Whigs were unable to penetrate the meaning, and gave the initials a new and humorous twist when they attributed them to Andrew Jackson's creative spelling. Jackson was said to have told Amos Kendall that he marked certain papers with the initials "O.K." to signify they were "oll korrect." Subsequently, Whig cider barrels were marked with the same "O.K." seal of approval, and O.K. came into everyday speech.

[34]This information was provided personally by McClintic, who won many speed medals. *The Journal of the Telegraph, Telegraph Age,* and *Telegraph and Telephone Age* reported the tournaments.

[35]I was present at the trial.

[36]I attended the ceremony.

[37]Linkins was my longtime friend. General Grant's Civil War telegrapher Thomas Morrison was the first telegrapher of the Department of State, 1867 to 1884.

[38]*The A.P. World* (Autumn 1967).

[39]C. H. Murphy, in charge of Time Service for many years, said the time-ball service was discontinued in 1913 when the operating rooms were moved to 24 Walker Street and the 47-story Singer Building, then the world's tallest, was built south of it, at 149 Broadway, in 1908. See "Time Service" in chap. 18, below.

The Telephone

It is difficult to imagine how people lived before the invention of the telephone. Its service is so much part of our daily lives that we take it for granted, like electricity and water.

Like the telegraph, the telephone did not burst full-blown from the mind of one person. Lincoln once said, "I have not made events; events have made me." The needs of the times produce persons who can synthesize earlier discoveries. Efforts to improve the telegraph led to "sound telegraphs," and many claimed the honor for inventing the "telephone." Alexander Graham Bell's right to the telephone patent was contested in 600 law suits, and ultimately barely survived in the Supreme Court. Bell later admitted that Elisha Gray's telephone preceded his, and that he used the confidential information in Gray's caveat to produce his first telephone.

Sounds made by a magnet when its magnetism is suddenly changed were noticed in 1837 by a Dr. Page at Salem, Massachusetts. He called the tones produced by rotat-ing a horseshoe magnet in a strong magnetic field "galvanic music."

Philip Reis, a physics teacher in Germany, produced a "telephone" with a "make-and-break" transmitter in 1861, fifteen years before Bell. Reis's transmitter was a thin sausage skin stretched over the open side of a wooden box, with a small platinum wire suspended over the skin. Each vibration of the voice would throw the platinum away from the skin, momentarily breaking the current.[1] His receiver was a knitting needle inside a coil of wire mounted on a violin for a sounding board. Each pulsation of current over the line magnetized and demagnetized the needle, causing the sounding board to reproduce tones. But the small piece of platinum wire never became stuck accidentally to the skin and vibrated with it.

Reis's telephones were brought to America and tested by physics professors of Cornell, Columbia, Northwestern, and Yale. They said they exchanged words and sentences. Joseph Henry bought a Reis "telephone" for the Smithsonian Institution, and Bell said one was shown to him in November 1874 by a professor at the Institute of Technology.

However, in October 1887, U.S. Supreme Court Chief Justice Waite stated:

> It was not contended that Reis had ever succeeded in actually transmitting speech, but only that his instrument was capable of it if he had known how.

Reis died in 1874 before he was forty, an unrecognized, disappointed man, but his countrymen in Germany built a monument honoring him as the inventor of the telephone.

The telephone transmitter is a mechanical ear. When someone speaks to you, the sound sets in motion the tiny molecules that make up the air and these pulsating molecules push and pull on your eardrum, causing it to vibrate thousands of times a second. Inside our eardrums are chambers of liquid and tiny bones that move with the vibrations of the eardrum, signaling the message to the brain. Likewise, air waves, set in motion by your voice, start the mechanical eardrum, or diaphragm, of a telephone transmitter vibrating, pressing inward, springing outward, and exerting varying pressure on a chamber of tiny grains of coal, through which electric current is flowing. Thus your voice becomes variations in electrical current, flowing over the wire.

The telephone receiver is an electrical mouth. When your voice causes variations of electrical current to arrive over the line, an electromagnet in the receiver exerts a greater or lesser magnetic force on a thin iron disc, which bends back and forth many times a second. The disc's vibrations set the molecules of air in motion against your eardrum and enables you to hear what is being said.

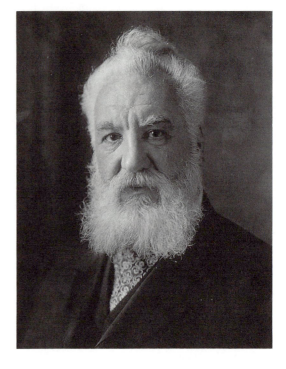

14.1 ALEXANDER GRAHAM BELL. (COURTESY AT&T.)

The public attitude toward the telephone in 1874 was indicated by a Boston newspaper:

> A man about forty-six years of age, giving the name of Joshua Coppersmith, has been arrested in New York for attempting to extort funds from ignorant and superstitious people by exhibiting a device which he says will convey the human voice any distance over metallic wires so that it will be heard by the listener at the other end. He calls the instrument a "telephone" which is obviously intended to imitate the word "telegraph" and win the confidence of those who know of the success of the latter . . . were it possible to do so, the thing would be of no practical value. The authorities who apprehended this criminal are to be congratulated, and it is to be hoped that his punishment will be prompt and fitting.

Alexander Graham Bell

If ever a child was prepared by birth and environment for his role in life, it was Alexander Graham Bell, born in Edinburgh, Scotland, on March 3, 1847. His mother was deaf, and his grandfather and father were teachers of elocution. Bell's father originated a method of teaching the deaf to read movements of the lips, called "visual speech."

Threatened by tuberculosis, from which two brothers died, Bell moved to Brantford, Canada, west of Niagara Falls, with his father in 1870 to seek a better climate. Bell became a teacher in a Boston school for deaf mutes, and started his own school for teachers of the deaf at Boston in 1872. He soon became Professor of Vocal Physiology at Boston University.

Bell then went to Salem, Massachusetts, as the private teacher of the five-year-old deaf-mute son of Thomas Sanders, a leather merchant. There Bell fell in love with another deaf pupil, fifteen-year-old Mabel Hubbard, whom he married four years later. Her father, Gardiner G. Hubbard, a Boston lawyer, was impressed by a prediction by Bell that a "musical telegraph will enable us to send as many messages simultaneously over one wire as there are notes on the piano." Hubbard and Sanders became Bell's financial backers.[2]

In a cellar workshop at the Sanders' home, Bell experimented with a harmonic telegraph. Vibrations sent over a wire produced sounds on the reeds of a harp. He stretched a membrane over the end of a speaking trumpet. Words caused the membrane to vibrate and a stylus to record the wave pattern on a smoked glass.

Bell experimented in 1847 with an ear taken from a dead man. Trying to imitate the human ear, he used iron discs to vibrate an iron rod, and connected them by wire so that each would cause the other to vibrate. He told Sanders and Hubbard a telephone could be produced in that way, but Sanders and Hubbard told him to work only on telegraph improvement because that might be of practical use and profitable. To please them Bell continued the telegraph experiments and teaching during regular hours, but worked at night on the telephone.

Early in 1875 Bell went to New York to test his telegraph development on Western Union wires, and learned Gray already had patented a harmonic telegraph. Gray told the American Electrical Society at Chicago on March 17, 1875, in a paper entitled "Transmission of Musical Tones," that he noticed his nephew playing with a small induction coil, taking shocks for the amusement of younger children. One end of the secondary coil was attached to the zinc lining of the bath tub, and when the boy ran his hand along the tub there was a sound with the same pitch as electrical vibrations. It led Gray to find that composite tones could be sent over a wire, and his 1875 patent followed on sending a variety of musical tones simultaneously over a wire.

While installing his harmonic telegraph on the line from Milwaukee to Chicago in 1874, Gray wrote, he entertained friends at the Newhall House in Milwaukee by stretching a wire across the street from the telegraph office, over which his assistant, W. M. Goodridge, transmitted tunes to the group. The music somehow got onto the regular telegraph lines, and inquiries came in from other cities where astonished operators had heard "Yankee Doodle" and other numbers on their relays.

When Gray noticed two boys about 100 feet apart on the street talking with a "lovers' telegraph"—a thread connecting the center of the bottoms of two fruit cans—he examined one of the cans, and "suddenly, the problem of electrical speech transmission was solved in my mind."[3] At lunch he sketched a telephone and told Goodridge he would make one over which people could talk a long distance.

14.2 THOMAS A. WATSON, SKILLED MECHANIC AND ELECTRI-CIAN, IN 1874 WHEN HE MET ALEXANDER GRAHAM BELL. WATSON MADE BELL'S TELEPHONE APPARATUS, AS VAIL DID MORSE'S TELEGRAPH KEY. WATSON INVENTED THE SWITCH HOOK AND OTHER TELEPHONE DEVICES. HE HEARD THE FIRST WORDS TRANSMITTED ELECTRICALLY: "MR. WATSON, COME HERE, I WANT YOU." (COURTESY OF AT&T ARCHIVES.)

At that time Bell, a tall, slender man with black side whiskers and drooping mustache, big nose, and bushy black hair, was experimenting with the assistance of Thomas A. Watson, an apprentice in the workshop of Charles Williams, Jr. at 109 Court Street, Boston. Williams manufactured telegraph instruments. Bell worked in the garret above Williams' four-story workshop. Seven years younger than Bell, Watson's role as maker and developer of instruments, was much like that of Alfred Vail with Morse.

Using two "harps," each with a row of steel reeds having different pitches, Bell tried to tune pairs of reeds so that when one was vibrated the other would have the same speed of vibrations and tone. He wished to establish a number of tones and send telegrams simultaneously over each tone.

14.3 THE DEVICE WITH WHICH BELL HEARD THE SOUND OF WATSON SNAPPING A METAL SPRING, ON JUNE 2, 1875.

On June 2, 1875, Watson repeatedly snapped a transmitter reed in the hot garret workshop. At the other end of the wire in a separate room, Bell listened intently, tuning one of the metal strips to vibrate in response. There was a low sound, and Bell dashed into the room.

"What did you do then?" Bell demanded. "Don't change anything. Let me see!"[4]

The reed had accidentally been placed in permanent contact with a pole of the electromagnet, and the current had been flowing over the line continuously while Watson varied the intensity of the flow by plucking the reed. This had caused the receiving reed Bell held to his ear to act as a diaphragm and produce a sound. They realized they were on the right track to transmit the human voice. Bell at once drew plans for an instrument that was both transmitter and receiver, with goldbeater's skin (the outside membrane of a cow's large intestine) stretched over it as a drum, and a piece of magnetized iron fastened to the skin but free to vibrate when current flowed through an adjacent electromagnet. When Watson had the instrument ready, he reported hearing sounds over it, but not words. Reis had accomplished that much.

Bell demonstrated his apparatus to Western Union President Orton in 1875 at Washington and wrote to his parents that Orton said he would "be glad to give me every facility in perfecting my instruments." He also invited Bell to New York where he would have the assistance of the best electricians. At Western Union, Chief Electrician Prescott aided Bell in a series of tests over 200 miles of wire. Then Orton learned Hubbard was involved and said, "Western Union would never take up a scheme which will benefit Mr. Hubbard, who had done so much to injure the company."

On February 14, 1876, Bell filed a patent application for an "improvement in Telegraphy." It described his unsuccessful method of transmitting sounds telegraphically. By coincidence, on that same day Gray personally filed his caveat for an invention to transmit the human voice. Gray's caveat said he had invented "the art of transmitting vocal sounds or conversations telegraphically through an electric current," and described the variable resistance method and the use of acidulated water.

Trying to make his instrument work, Bell tried the method Gray's caveat described, attaching a small platinum wire to the goldbeater's skin, and the other end of the wire in contact with acidulated water. This permitted sound waves to produce vibrations of the drum, varying the depth and area of contact of the wire with the water. The transmission of speech was then possible because the vibrations energized an electromagnetic coil and produced an "undulatory" current, varying in intensity as the air varies in pressure during the transmission of sound.

14.4 ALEXANDER GRAHAM BELL IN 1876, THE YEAR THE TELEPHONE WAS PATENTED. (PHOTOGRAPH BY HOLTON, COURTESY OF THE BELL FAMILY AND NATIONAL GEOGRAPHIC SOCIETY.)

It was charged that Bell's lawyers "were enabled to have unlawful and guilty knowledge of Gray's papers as soon as they were filed in the Patent Office," and that Bell's lawyer, Anthony Pollock, "boldly interpolated" the contents of Gray's caveat into Bell's application on February 14 and 19. Zenas F. Wilbur, the Patent Office examiner, testified years later that he had illegally notified Bell's lawyers of the conflict.[5] Bell admitted under oath years later that Wilbur pointed out the "particular clause" to him and "I

14.5 Thomas A. Watson with the first telephone he constructed for Bell in 1875. (Courtesy AT&T.)

therefore knew it had something to do with the vibration of a wire in liquid."

Hubbard had written to Bell on July 2, "I am very much afraid that Mr. Gray has anticipated you in your membrane attachment."[6] "The rough draft of Bell's applica-tion shows the single paragraph on the varia-ble resistance principle written into the margin of a page, apparently as an after-thought," said John Brooks in his book *Tele-phone* (New York: Harper & Row, 1975).

With a new instrument Bell and Watson

still struggled without success. Testing the instrument March 10, 1876, with a tuned-reed receiver at his end of the wire, Watson heard Bell's voice say:

"Mr. Watson, come here; I want you."[7] Watson dashed down the hall shouting: "I heard you. I heard you."

Bell had upset acid on his clothes, and it was his cry for help that Watson heard. Amazingly, Bell already had been granted three days earlier on March 7, 1876, probably the most valuable patent ever issued,[8] for what he now found with different instruments. His telephone patent was granted in three weeks, while Edison's waited for fifteen years and Berliner's fourteen years. It was four days after Bell's twenty-ninth birthday.[9]

Edison credited Gray with producing the first telephone. "I went to work long before Professor Bell," Edison said. "Elisha Gray turned in at it and got out the first machine. Bell's and mine came out about the same time. The machines were different. Were it not for my deafness I would have discovered it eight months before." The caveat for his analyzer was filed January 14, 1876, a month before Bell and Gray, but he did not realize it would carry a voice.

Bell's telephone combined the transmitter and receiver, and his transmitter was poor. Edison separated the two, and had an efficient transmitter. After Bell obtained his patent, "the fun began." Supporters of Reis, Dolbear, Blake, Gray, Drawbaugh, and others fought in the courts for eleven years in 600 law suits to prove they were the real telephone inventors.

It was frustrating to the professional electricians and inventors to hear that Bell, an outsider, had succeeded. Moses Farmer, a telegraph inventor, could not sleep for a week. The professional journals backed Gray, the man most qualified by experience and knowledge to have invented the telephone.

Did Drawbaugh Invent the Telephone?

Who today has heard of Daniel Drawbaugh? In 1888, the U.S. Supreme Court came within one vote of declaring he invented the telephone, and one justice who favored Bell died a few days later.

Whether Drawbaugh was one of the world's greatest inventors or a clever impostor was once debated throughout the nation. Many disinterested witnesses identified "wordable" phones in court as the ones they had helped test as early as 1867, nearly a decade before Bell.

After hearing 7,000 pages of evidence by 200 witnesses, the U.S. Circuit Court, Southern District of New York, decided against Drawbaugh. When this decision was appealed, four Supreme Court justices affirmed the decision for Bell, while three favored Drawbaugh.[10]

The majority opinion of Chief Justice Waite said:

> We do not doubt that Drawbaugh may have conceived the idea that speech could be transmitted to a distance by means of electricity, and that he was experimenting upon that subject, but . . . [his] conduct was so inconsistent with his pretensions [that the court rejected the testimony as incredible].

The People's Telephone Company, formed in 1880 to use Drawbaugh's telephone, saw in him a rustic genious in Cumberland County, Pennsylvania, who invented many excellent devices, but so poor he had been unable to patent the telephone.

A neighbor testified that Mrs. Drawbaugh once smashed some of her husband's equipment "to stop him from fooling with them." When Drawbaugh's son Charles was born in 1870, his wife, who was confined to bed, said she had listened to an instrument and could hear what her husband said in a distant part of the house.

Mr. Justice Field said to Bell's attorneys:

I want you to explain the possible occurrence of two or three hundred witnesses in regard to a fact which they could not be mistaken about, that he heard speech, because there are some facts that are so striking themselves as that once seen they are never forgotten.

Mr. Justice Bradley, speaking for the dissenting justices, said the evidence was

so overwhelming, with regard both to the number and the character of the witnesses, that it cannot be overcome. . . . We are satisfied from a very great preponderance of evidence that Drawbaugh produced and exhibited in his shop, as early as 1869, an electrical instrument by which he transmitted speech.

He added, "There can hardly be a reasonable doubt" that Drawbaugh produced it, but that Bell realized its importance, while Drawbaugh "looked upon what he had made more as a curiosity than as a matter of financial, scientific, or public importance."

In the *Journal of the Franklin Institute* (1885), Professor Edwin J. Houston surveyed the numerous steps Drawbaugh testified he took in building and improving his instruments from 1861 to 1875. He concluded they apparently were overwhelming evidence of his success.

Before the court decided against him, Drawbaugh was reported to have refused to accept $1,100,000 from the Bell company to settle the case.

Bell's Exhibit

Bell demonstated his telephone before the American Academy of Arts and Sciences in Boston in May, 1876, and then at the Centennial Exposition in Philadelphia. Few were willing to "make fools of themselves" by talking into the funny little toy. At last the exposition judges, including Elisha Gray, reached Bell, on their official inspection of exhibits. They were tired and ready to leave when Don Pedro do Alcontara, Emperor of Brazil and his Empress entered the room. The Emperor exclaimed, "Professor Bell! I am delighted to see you."

The Emperor had met Bell at the School of the Deaf in Boston. Newspapermen crowded around, as did the judges and exposition officials. Don Pedro held the receiver to his ear as Bell spoke at the distant end. "My God! It talks!" he cried.

Sir William Thomson (later Lord Kelvin), scientist and submarine cable expert, then tried the telephone, and declared it "the most wonderful thing I have seen in America." Joseph Henry, the leading American scientist, also tried it, and congratulated Bell. The "toy" of a few hours before was moved to the judge's stand and was the sensation of the exposition.

The First Telephone Business

Like the early telegraph, few men with money could see the possibility of financial gain in the telephone business. The New York Tribune asked "Of what use is such an invention? Well, there may be occasions of state when it is necessary for officials who are far apart to talk with each other without the interferences of an operator."

The telephone enterprise now occupied

part of the third floor of Williams' electrical workshop at 5 Exeter Place. Watson was busy experimenting and improving the instruments. Hubbard took over promotion. Hoping to raise money by putting on shows, he had Bell give a series of lectures, with conversations over telegraph wires between Boston and Cambridge, October 9, 1876; Boston and Salem November 26;[11] Skiff's Opera House, New Haven and Robert's Opera House, Hartford, April 27, 1877; and other New England cities. At each place Bell lectured and talked with Watson in the workshop at Boston.

Before 500 persons at Lyceum Hall in Salem February 12, 1877, Bell asked Watson to sing, and, after considerable protest, he sang "Yankee Doodle" and "Auld Lang Syne."[12] News of this transmission of music was sent by telephone after the program by Henry M. Batchelder to the *Boston Globe*—the first news story dictated over the telephone.

Bell also gave demonstrations at the Hotel St. Denis at Broadway and Eleventh Street in New York on March 11, 1877, and at Chickering Hall on Fifth Avenue on May 18 and 19. Watson's music at the Atlantic and Pacific Telegraph office in New Brunswick, New Jersey was heard by the audiences in New York.

The first regular use of an outdoor line began April 4, 1877, between Williams' home in Somerville to his workshop in Boston, three miles away. The first customer, a Mr. Emery, leased telephones to connect his home in Charlestown with his brother's across the street, and other early telephones were leased in pairs; there were no exchanges.

Edwin T. Holmes, whose father was in the electrical burglar alarm business, installed telephones in five banks and stores in May 1877 to connect them with the E. J. Holmes & Co. headquarters at 342 Washington Street, Boston. They were used for alarms at night, and calls for messenger and express wagon deliveries by day. Holmes later provided regular telephone service and printed a telephone directory listing sixty subscribers.

On August 4, 1877, the associates promoting Bell's "new and useful methods and apparatus for telegraphing,"[13] agreed to form the Bell Telephone Company with Hubbard as treasurer. The only employee, Watson, was given a one-tenth interest in the patents, and the rest was divided between Hubbard, Bell, and Sanders.[14] Bell then went abroad for more than a year. Sanders' small fortune of about $110,000 had gone into the enterprise, and the bank would extend no more credit.

One widely quoted source said Hubbard, Sanders, and Watson offered to sell the telephone to Western Union for $100,000, and that Orton scornfully replied, "What use could this company make of an electrical toy?" However, that quotation was obviously not true. Two years before that in July 1875, Orton had given Edison reports of the Reis experiments and put him to work on a telephone transmitter at a salary of $500 a month.[15]

In August 1877, Orton decided to have his Gold and Stock subsidiary go into the telephone business. Despite a report by Pope that Bell's patent was valid and should be purchased,[16] Orton decided to use receivers based on the patents of Elisha Gray, Amos E. Dolbear,[17] and George M. Phelps, and a "carbon telephone" transmitter developed by Edison.

One reason Orton repeatedly refused the Bell telephone was his dislike of Hubbard, who had attacked Western Union and lobbied in 1868 to create a government-financed telegraph postal system, incorporating Hubbard and friends. Orton was busy fighting Gould's Atlantic and Pacific raid. He advised Chauncey M. Depew, a Western Union director, against it when Hubbard offered Depew one-sixth interest in his company for $10,000.

14.6 EDISON'S CARBON TRANSMITTER AND CHALK RECEIVER. BELL'S EARLY TELEPHONE HAD A WEAK RECEIVER AND USERS HAD TO SHOUT TO BE HEARD. IN 1877, EDISON COMBINED A BATTERY CURRENT WITH A CARBON MICROPHONE, AND HIS TRANSMITTER TEAMED WITH BELL'S RECEIVER AS THE BASIS OF TODAY'S TELEPHONES. (COURTESY OF U.S. DEPT. OF THE INTERIOR, NATIONAL PARK SERVICE, EDISON NATIONAL HISTORIC SITE.)

Edison developed the first good telephone transmitter. Finding that a sensitive diaphragm was not necessary, he removed the spring and installed a thin metal disc fastened to a carbon button. The carbon was obtained by scraping it from smoking lamp chimneys and pressing it together tightly. It

was battery powered, and for the first time it was not necessary to shout into the mouthpiece.

The American Speaking Telephone Company was organized by Western Union on December 6, 1877, with capital of $300,000. George Walker was president, Norvin Green vice president, and Orton a director.

Tests of the Phelps and Edison telephone were made early in 1878 between New York, Philadelphia, and Washington, with Edison, Orton, and W. K. Vanderbilt participating. A concert was held with the artists in New York, the audience in Philadelphia, and Edison listening at Menlo Park.

Edison wanted $25,000 for his telephone invention, but asked Orton to make an offer. Orton promptly named $100,000, but Edison agreed to accept $6,000 a year during the seventeen-year life of the patent.

American Speaking Telephone Company contracted with Gold and Stock to manufacture telephones for use on Western Union's lines, and the telegraph company went into the telephone business in 1878, establishing exchanges in its offices in many cities.

"The Telephone Department (of Western Union) was put in the hands of Hamilton McK. Twombly, Vanderbilt's ablest son-in-law, who made a success of it," Edison said. "The Bell Company of Boston also started an exchange, and the fight was on, with Western Union pirating the Bell receiver, and the Boston company pirating the Western Union Transmitter."

At the same time, George W. Coy, the telegraph manager at New Haven, Walter Lewis, and H. P. Frost organized the New Haven District Telegraph Company, with $600 capital. Coy mailed circulars to 1,000 New Haven residents, but obtained only one subscriber—the Reverend John E. Todd, pastor of the Church of the Redeemer. After months of effort, twenty subscribers were enlisted at $18 a year. The first commercial exchange exclusively for telephone use was then opened on January 28, 1878.

Adapting the telegraph messenger call box system at New Haven, Coy devised a switchboard to connect the subscribers and printed a directory. In three weeks the list had increased to 125 subscribers, and in another month to 227. Within a year the New Haven company's name was changed to Connecticut District Telephone Company; in 1880, to Connecticut Telephone Company; and in 1882, to Southern New England Telephone Company, with 3,634 telephones. To finance expansion, it sold control to Jay Gould, who used it as a pawn in his fight to control Western Union.[18]

Additional Bell exchanges were opened in Meriden, Connecticut on January 31, 1878, and in San Francisco on February 17. The Albany, New York exchange, fourth in the nation, was opened in a room over the Van Heusen and Charles store at 468 Broadway on March 18, 1878, by Charles H. Sewall, manager of the District Telegraph Company at Albany. In 1879 the 100 telephones at Albany were linked through the switchboard, and programs, featuring such songs as "Seeing Nellie Home," were broadcast from 9:00 to 11:00 p.m. each Tuesday and Sunday.

Exchanges were opened in Western Union offices in April 1879 at Wilmington, Delaware and Lowell, Massachusetts; on April 1 at St. Louis; June 26 at Chicago;[19] July 15 at Hamilton, Ontario; August 2 at Portland, Oregon; August 5 at Detroit; in September at Manchester, New Hampshire, Keokuk, Iowa, Cincinnati, and Boston. On September 1 the first woman telephone operator, Miss Emma M. Nutt, was hired at

Boston. Also in 1879 Marjorie Gray was hired at Bridgeport. The first exchange in Pennsylvania was opened at Philadelphia on November 1, 1879.

On December 1, 1878 George C. Maynard, in the Signal Corps during the Civil War, opened the first exchange in Washington, D.C. with five telephones; at the White House, Capital, Associated Press, Treasury Department, and Institute for the Deaf and Dumb.

If ever a company needed a good general to take command, the Bell Telephone Company did. This need was met by Theodore N. Vail.

Theodore N. Vail

Theodore N. Vail's great-uncle was Judge Stephen Vail, father of Alfred Vail, the partner of Morse. Vail's father was employed at the Speedwell Iron Works at Morristown, New Jersey where Alfred Vail built the first successful Morse instruments. Working in the drug store in which the telegraph office was located, he learned the Morse code. His uncle, Isaac Quinby, with Western Union at Rochester, obtained a job in 1864 for Vail, who worked at several New York telegraph offices.

Vail was a telegraph operator at Pinebluff and other points in Wyoming when the Union Pacific Railroad was under construction. Uncle Quinby then got him a job as a mail clerk on western trains. Vail systematized the routing of mail so well that he was called to Washington in 1873 to improve the railway mail. In 1876 he was appointed General Superintendent of Railway Mails.[20]

Vail became enthusiastic over the possibilities of the telephone, borrowed money to buy telephone stock, and quit his government job to join the Bell company in 1878. The company was on the verge of bankruptcy. Hubbard, always an optimist, refused to take seriously the fact that Sanders had put $100,000 in with no return, and had no more money to pay for telephones. Hubbard's appointees in various cities had made little progress in startup service. In New York they had not started at all. Then on August 1, 1878, Western Union's American Speaking company opened a central exchange at 198 Broadway. Page one of the January 10, 1880 *Scientific American,* showed a general view, and said the telephone's

> adjustable arm carries an Edison carbon button transmitter. . . . A receiving phone, which is connected with the line wire, hangs upon a switchboard at the opposite end of the desk. Above the instrument is an ordinary single-stroke electric bell.

Other illustrations showed boys with "French" handset telephones plugging into annunciator-type switchboards in the telegraph company's American Telephone office.

When boys called "switchmen" saw annunciator discs flop down indicating incoming calls, they plugged in their portable telephones and asked, "Well, sir?" Upon hearing the name of the desired party, they connected the caller by using a jackknife switch with a rod in the switchboard to make the connection. The boys made the scene chaotic by running and shouting.

The Bell Telephone Company gave General Manager Vail the territory within a radius of thirty-three miles of New York, and on June 18, 1878 he incorporated the Bell Telephone Company of New York with headquarters at 518 Broadway. Edwin Holmes, the burglar and fire alarm man, was president. It agreed to pay the Bell company

an annual rental of ten dollars for business and four dollars for home telephones, and gave it forty percent of its capital stock for the franchise.

14.7 THEODORE N. VAIL, THE ORGANIZER AND LEADER RESPONSIBLE FOR TRANSFORMING THE STRUGGLING SMALL TELEPHONE COMPANY OF BELL AND ASSOCIATES INTO THE GREAT AT&T TELEPHONE SYSTEM. (COURTESY AT&T.)

An experimental office was used at the Holmes Burglar Alarm Company at 194 Broadway, and in March 1879 its first New York Bell Telephone Company exchange was opened at 82 Nassau Street, followed a few months later by one at 97 Spring Street.[21] After much sales effort, a directory of 243 subscribers was issued. This was quite an accomplishment because the opposition exchange had the advantage of Western Union's more convenient and efficient instruments.

A third New York Exchange was opened at 140 Fulton Street, in the spring of 1879, by the Law Telegraph Company, connecting law firms. It was a forerunner of modern private wire systems. Its printers, somewhat like typewriters, were in 400 offices, and were used to communicate by typing messages.

14.8 A DRAWING OF THE NEW YORK CITY TELEPHONE CENTRAL OFFICE OF WESTERN UNION AS IT APPEARED IN 1880. THE BOY OPERATORS ARE USING PORTABLE HAND TELEPHONES WHICH WERE THE FORERUNNERS OF TODAY'S EFFICIENT HAND SETS. INSET: A HAND TELEPHONE USED BY BOY OPERATORS IN NEW YORK CITY IN 1879.

At that time the Bell instrument was in one box with a large mouthpiece, to which the subscriber would place his lips and shout. Then he would quickly turn the same mouthpiece to his ear to hear the reply, which often was indistinguishable. Often people became confused and tried to listen with the mouth.

The Western Union exchange, however, had both Gray's receiver and Edison's transmitter conveniently mounted on a curved metal bar. The Western Union chief operator at New York, Robert G. Brown, had developed what is now called the French telephone handset in 1878. The central office boys had found it difficult to hold the re-

ceiver and separate transmitter in their hands and wire cords around their necks, while rushing to switchboards to plug in connections. To remedy this, Brown's instrument could be held in one hand, while the boys plugged the cords into the Western Electric "Universal" switchboard.

The fame of Brown's handset and the efficiency of his exchange resulted in his being called to France the next year. There he established a similar system in Paris and other cities for the Societe General des Telephones. That is how an American handset became known as the French telephone. Brown employed a girl operator, a Mlle. London, in Paris in 1880, and girls were used on the Paris exchanges thereafter.

Western Union opened exchanges in many cities in 1878 and 1879. Some managers reported sales resistance by people who believed a telephone would attract lightning, transmission of a voice was unnatural and unholy, or it was a novelty of no real worth.

Many companies rented private lines before there were exchanges. J. L. Haigh connected his office at 81 John Street, New York by private phone on August 5, 1877, with his factory in South Brooklyn, where he was manufacturing steel suspension cable wire for the Brooklyn Bridge, under construction from 1870 to 1883. The line ran across the partly finished bridge. One of the first women subscribers was Miss S. A. Duryea, whose name is listed in the first Brooklyn directory, issued in 1879. Exchanges were opened in Astoria, Flushing and Babylon, Long Island in 1883; Glen Cove, 1884; Islip, 1886; and Hempstead in 1889.[22]

Charles F. West, chief train dispatcher of the Northern New Hampshire Railroad, designed a switchboard for the first private exchange, connecting his office and the offices and homes of the railroad's officials at Concord, N.H. in 1879.[23]

The Edison Telephone Company of Great Britain, Ltd. was formed in 1878. George E. Gouraud represented Edison in introducing his carbon transmitter telephone in England, as he had done with Edison's telegraph inventions. The Bell people threatened to sue if their receiver was pirated. Facing the threat of a Bell suit in England, Edison quickly produced a new receiver in 1879, with a cylinder of chalk connected with a diaphragm by a thin spring. It was used in the first London exchange.

Gouraud's secretary, Samuel Insull, helped open the London exchange and later became a power utilities magnate in the United States.[24] Another employee of Edison Telephone was George Bernard Shaw, who assisted in demonstrating the telephone, and later became a novelist and playwright. For the English telephone rights, Edison accepted a cable offer of $30,000" and was surprised when he received 30,000 pounds, and not dollars.

The Phonograph

Many of Edison's inventions flowed from his developments in automatic telegraphy. His attempt to improve the speed of a repeater to relay telegrams and avoid manual retransmission resulted in the phonograph invention at his laboratory in Menlo Park, New Jersey, in 1877. "I was experimenting," he said,

on an automatic method of recording messages on a disk of paper laid on a revolving platen, exactly the same as the disk talking-machine of today. The platen had a spiral groove on its surface, like the disk. Over this

was placed a circular disk of paper; an electromagnet with the embossing point connected to an arm travelled over the disk; and any signals given through the magnets were embossed on the disk of paper. If this disk was removed from the machine and put on a similar machine provided with a contact point, the embossed record would cause the signals to be repeated into another wire. The ordinary speed of telegraphic signals is thirty-five to forty words a minute; but with this machine several hundreds words were possible.[25]

14.9 THOMAS ALVAH EDISON (1847–1931).

When Edison tried to run this telegraph machine faster, the point bounced on the indentations and made "a light musical, rhythmical sound, resembling human talk heard indistinctly." He reasoned that if he could record the vibrations of the voice through a diaphragm in the form of indentations on the foil, he could run them through another diaphragm and reproduce the voice. He tried this with a strip of paraffined paper, shouting "Halloo!" and then pulling the paper so that the indentations actuated another diaphragm. He heard "a distinct sound, like 'Halloo'."

The phonograph made Edison world famous in 1878.[26] He then worked to invent the electric light. He and his lawyer, Grosvenor Lowrey, realized that would be a lengthy job, requiring assistants and money, and he turned to Western Union as he had so often in the past. President Norvin Green, W. H. Vanderbilt, Vice President Hamilton M. Twombly, Tracy Edson, Vice President James Banker and J. P. Morgan, through his partner, Eggisto Fabbri, at once put up the initial $50,000. S. L. Griffin of Western Union was sent to serve as Edison's private secretary. Fourteen months later Edison produced the first incandescent lamp at Menlo Park.[27]

The Struggle for Control

The telegraph company's competition was a blessing in disguise. It dignified the telephone for what was called the nation's largest corporation to consider it worth controlling. Orders for telephones came in so fast that Sanders could not pay for the sets produced at the Williams shop. Williams accepted stock, but no longer could pay his workmen. The company frequently did not have a dollar to pay creditors.

The New York company's finances were no better. One day Vail sent to the L. G. Tillotson & Company store at 15 Dey Street for supplies costing less than $10, and received the reply that they could be provided only after cash was paid.

When the Edison transmitter appeared, the gloom at Bell headquarters was described by Herbert N. Casson in his *History of the Telephone*:

Lessees of Bell telephones clamored with one voice for a transmitter as good as Edison's. . . . The five months that followed were the darkest days in the childhood of the telephone. How to compete with the Western Union, which had this superior transmitter, a host of agents, a network of wires, and a first claim upon all newspapers, hotels, railroads, and rights of way—that was the immediate problem that confronted the new general manager.

The spirits of the Bell group were too low in 1877 to be lifted by the first telephone music: Freeman's "Telephone Polka," Moses' "Telephone Gallop," Westendorf's "The Wondrous Telephone," and telephone marches by both Mack and Turner. In 1891, the "Hello Central, Give Me Heaven" song cover showed a little girl with long hair, white ruffled dress and hair ribbon, standing on a stool at a wall telephone, to call Heaven "because my Daddy is there."

The company was reorganized July 20, 1878 with Hubbard as president; Sanders, treasurer; Watson, general superintendent; and Vail, general manager. Capitalization was $450,000. Sanders sold 500 shares to friends at $50 to pull the little company through its perilous infancy. Even then it probably would have failed if Emile Berliner,[28] a German emigrant, then a Washington, D.C. drygoods clerk, had not developed a sensitive, variable pressure contact transmitter in 1877, and Francis Blake, Jr., of the U.S. Geodetic Survey had not improved it.[29]

It was similar to Edison's, and at last Bell could offer a telephone over which a conversation could be held without shouting. Since Blake's transmitter would operate only in a vertical position, vertical wall and desk telephones were used. At this critical point, Western Union President Orton died of a stroke in April 1878 and was succeeded by Dr. Norvin Green, who was kept busy fighting Jay Gould.

The Bell company brought suit in the U.S. Circuit Court at Boston September 12, 1878, against Peter A. Dowd, agent of American Speaking, charging that Gray's patent infringed Bell's. Western Union's defense was that others invented telephones before Bell, and the apparatus described in Bell's patent could not transmit speech. Court hearings during the following year convinced George Gifford, Western Union's eminent patent counsel, that further litigation would be a waste of time and money. He did not know that Bell would admit later his use of confidential information of Gray's invention enabled him to produce his first telephone.[30] By concealing that fact, Bell won fame and changed the future of communications. Western Union offered to compromise out of court, and suggested that it would take the long distance, and leave the local telephone business to Bell. This offer was debated repeatedly, and finally an all-night conference between officials ended in an historic contract on November 10, 1879.[31] The Bell company pledged to stay out of the telegraph business, and Western Union agreed to give up the telephone. The Bell company agreed to pay Western Union a royalty of 20 percent on all telephone rentals or royalties from licenses or leases. Western Union gave Bell the use of its telephone inventions, and use of its rights of way for pole lines for proper compensation. Western Union sold to Bell its telephone exchanges in 55 cities and 56,000 subscriber telephones[32] but retained the right to use telephones in its own business and to sublet their use on private wires leased to firms or individuals.

Bell agreed to turn over to Western Union all telegraph messages it received, and not permit any person to use its facilities to transmit "messages for hire," news for sale, general business messages, market quotations, or telegraph messages in competition with Western Union or its subsidiaries. AT&T did not live up to the agreement.[33]

AT&T Is Born

The November 10, 1879 agreement was with the National Bell Telephone Company—created on February 17 that year by consolidating Bell with the New England Telephone Company. William H. Forbes, Civil War veteran and son-in-law of Ralph Waldo Emerson, was president, Bradley treasurer and Vail general manager. Headquarters at 95 Milk Street, Boston, was moved to 66-68 Reade Street, New York City.

When the contract with Western Union became known, Bell stock, which had gone begging at $50, climbed rapidly to $995 a share. This suddenly made Hubbard, Sanders, Watson and Williams rich, and they retired. National Bell then bought enough stock to control the New York company—750 shares of it from Vail, greatly to his profit—and reorganized the New York company as Metropolitan Telephone & Telegraph Co. March 20, 1880, with capitalization increased to $10 million.[34]

Bell then stopped using the mouthpiece for both transmitter and receiver, and began providing a receiver attached to a flexible cord. Because electrical roaring and sputtering made telephoning difficult over a distance, John J. Carty, who had climbed from switchboard boy to chief engineer, tried using telephone wires in pairs in 1880, in-

14.10 <u>Top</u>: A telephone construction crew, engaged in building the first transcontinental telephone line in 1914, moving camp in some of the desolate western territory through which the line runs. <u>Circle Inset</u>: The line was laid out in a straight line across countless miles of barren wilderness. <u>Middle</u>: Alexander Graham Bell, inventor of the telephone, occupied the post of honor in New York at the formal opening of service over the transcontinental telephone line on January 25, 1915, when his voice was the first to reach the Pacific Coast, at San Francisco, by wire. <u>Bottom Inset</u>: Pole setting in Humboldt Lake, Nevada.

stead of the single, grounded wire previously used.[35] The roaring ceased instantly; the metallic circuit worked so well that his method was followed worldwide. He also transposed wires to avoid electrical induction from nearby power lines, and made many other improvements to eliminate the babble of voices that often made conversation impossible. He recruited and trained the first college-bred staff of Bell engineers that laid the foundation of the telephone plant, from switchboards to cables.[36]

In the year following the agreement with Western Union, the number of telephones jumped from 62,000 to 134,000, and the number of exchanges to 408. Boston could call seventy-five cities and towns.[37] On January 1, 1881, the company paid its first dividend—$178,500; in 1882, two dividends totaling $416,500; and a million dollars in 1885. Exchanges were opened in cities all the way to Seattle between 1880 and 1883.

Control of the Western Electric Manufacturing Company was purchased from Western Union. It was renamed Western Electric Company, incorporated November 26, 1881, and became the manufacturing arm of the Bell System.[38]

In the 1880s instructions to patrons were as follows:

> To call: Press the button and turn the crank briskly; unhook the listening telephone (receiver) and put it close to your ear. When Central Office will inquire, "What Number?" give your number and the number of person wanted . . . hang up the receiver and wait until your bell rings, then . . . address person called. When through, do not fail to hang up the receiver, and call off, pressing in button and turning crank briskly.

Long distance telephoning made a giant advance when Thomas B. Doolittle, telegraph manager at Bridgeport, Conn., discovered the conductive superiority and strength of "cold drawn" copper wire over iron wire. He experimented, drawing cold copper rods through a series of dies, at the Ansonia Brass and Copper Company mills in 1877 and 1878. Vail ordered 500 miles of the wire and completed the first line between Boston, Providence, New Haven and New York on March 27, 1884.

Based upon Vail's vision of creating a nationwide toll system, four men incorporated the American Telephone and Telegraph Company on February 28, 1885. The four were Edward J. Hall, Jr. of Elizabeth, New Jersey, who became general manager; Doo-little; Joseph P. Davis, of the New York Company; and A. S. Dodd of New York. Vail was the first president and Angus S. Hibbard, who had established long distance service in Wisconsin, general superintendent. AT&T, called the long distance company, was a subsidiary of American Bell for fifteen years, and then absorbed American Bell in 1900, giving two of its shares for one.

Hibbard gave people exclusive local territories, and rented the instruments to them. The local companies had little money and often paid the Bell company in stock. Vail foresaw the advantage of having AT&T serve as a holding company, provided money, accepting stock in return, and took over the franchises of the operating companies as they expired. The local companies evolved into units serving larger areas, avoiding much of the warfare of the telegraph's infancy, when rival promoters fought over territory.

The Bell System that resulted was likened to a tree. The branches were 21 operating companies, each serving telephone subscribers in one or more states. The trunk of the tree was American Telephone and Telegraph Company, which provided the financing, research, development, patents, and operation of the long lines interconnecting the operating companies. The roots of the tree were Bell Telephone Laboratories, Western Electric Company, Long Lines Department, and the Advisory Staff, which handled financing and advised the operating company managements.

General Manager Hall extended the Boston-New York wire to Philadelphia in 1885, and then to Washington, Albany, Buffalo and other cities. As long distance service grew, Angus Hibbard designed the familiar blue bell symbol, and Hall okayed it on January 5, 1889.

Dissatisfied because President Forbes and the directors, representing Boston capitalists, voted annual $18 a share dividends instead of expanding the plant and improving service, Vail resigned the AT&T presidency on September 19, 1887. He was succeeded by John E. Hudson, the reserved conservative general manager and solicitor of American Bell. The antagonism of the Forbes group was revealed by the absence of any mention of Vail's resignation in the 1887 annual report.

Americans welcomed the long distance telephone as lines spread to distant parts of the nation. By 1892 there were 240,000 telephones and 10,000 employees. In Europe the business was crippled by government rules and regulations, and telephones were few. Use of the telephone somehow conflicted with the temperament of people in Latin countries. The British considered it no substitute for personal conversation. In Germany and most other countries, the telephone was a government monopoly which lacked the profit incentive provided by private enterprise to expand and improve service.

Notes

[1]The International Telecommunication Union centenary book said Reis transmitted music in 1860. "He stretched an animal membrane, to which a platinum wire was fastened by means of sealing wax, over a small cone in the shape of a human ear, and inserted this into the bunghole of a beer barrel. The platinum wire formed part of the battery circuit, and as the sound vibrated the animal diaphragm, the wire would make and break a contact in the circuit with a corresponding frequency."

Silvanus P. Thomson, *Philip Reis, Inventor of the Telephone* (London and New York: E. & F. N. Spon, 1883), said Reis wrote in 1860 that he had invented an instrument with "the function of the organs of hearing, which could reproduce tones of all kinds at any distance by galvanic current, and named it the 'Telephone'." Reis exhibited his telephone before the Physical Society of Frankfort on October 26, 1860, and before other groups.

[2]Hubbard had attempted to make a quick fortune in communications in 1868 through an act introduced in Congress. *The Telegrapher* (November 21, 1868) said the act, to create a United States Postal Telegraph Company, had proposed "to incorporate Mr. Gardiner G. Hubbard of Bos-

ton, Massachusetts, and certain other gentlemen, who are to build an entirely new system of telegraph lines, and contract with the Postmaster General to carry on the telegraph in connection with the Postal business of the country and proposed a foreign, government-owned system, but the union replied that holding telegrams for postal mail delivery would render many of them valueless. At House Post Office Committee hearings, Orton testified that Hubbard's scheme would make him a millionaire at the expense of the taxpayers. *The Telegrapher* said, "We regard the scheme as wild and impracticable." Evidently Congress agreed.

Later, Hubbard was a founder of the National Geographic Society. The Society's first president was Bell, whose son-in-law Gilbert Hovey Grosvenor ran it from 1899 to 1954.

[3]Quotation from an article by Gray, in a story by Charles D. Stewart in the *Milwaukee Journal,* November 23, 1939.

[4]Watson's address, "The Birth and Babyhood of the Telephone" before the third annual convention of the Telephone Pioneers of America, October 17, 1913.

[5]The Dayton, Ohio *Journal Herald,* May 2, 1964, said Wilbur permitted Bell to look at

Gray's caveat, which included drawings and descriptions and "Within two weeks after seeing the caveat, Bell returned to Washington and filed an amendment to his application, substituting a liquid transmitter for his original one."

Horace Coon, *American Tel and Tel* (New York/Toronto: Longmans, 1939) said: "But there is one claim that will not die. Again and again it has been raised. In the minds of many who have examined the evidence there is considerable doubt, and among several of the experts there still exists a conviction that the claim to the invention rightly belongs to Elisha Gray."

[6]Alvin F. Harlow, *Old Wires and New Waves* (New York: Appleton, 1936).

[7]Frederick Leland Rhodes, *Beginnings of Telephony* (New York: Harper & Brothers, 1929).

[8]Patent number 174,465, "Improvement in Telegraphy."

[9]Thoroughly confused by so many applications for the telephone patent, and Hubbard's demands that only Bell's application be considered, the patent examiners issued the patent to Bell.

[10]Only seven justices were sitting at that time.

[11]*Boston Globe,* November 27, 1876.

[12]*Boston Globe*, February 13, 1877; and *Frank Leslie's Weekly*, March 31, 1877.

[13]At that time "telegraph" was a synonym for all electrical communication.

[14]William C. Langdon, "The Early Development of the Telephone," *Bell Telephone Quarterly* (July 1923); also a capsule *History of the Bell System*, an AT&T booklet, p. 12.

[15]Edison's papers include his progress reports to Orton and assignments of his caveats on "accoustic instruments" to Western Union. Edison also had "passes" from Orton to test his telephone instruments in the operating rooms on various nights in 1876.

[16]Herbert N. Casson, *The History of the Telephone* (Chicago: A. C. McClurg & Co., 1910).

[17]Amos Dolbear, a Tufts College student, developed a telephone in 1876, and visited Bell and Hubbard and asked that his contribution be recognized. Bell and Hubbard gave him a cold recep-

tion. Then he became professor of Physics of Tufts College, applied for a patent, and assigned his rights to the telegraph company. Dolbear used a condenser-receiver, with two thin metal discs which would attract one another and vibrate as varying currents flowed into and out of the condenser.

Bell wrote angrily to Eli Whitney Blake and John Pierce, Brown University professors, when he heard of their improvements. One was a small metal diaphragm with a bell-shaped sunken mouthpiece, to replace the large opening Bell used, and another was a tuning fork to call the subscriber, where Bell had nothing. They sent two of their telephones to him and wrote that he was welcome to use the improvement; they wanted no money. Bell promptly claimed he had just made the same improvements.

Watson said people "couldn't be expected to keep the telephone at their ears all the time waiting for a call, especially as it weighed ten pounds and was as big as a small packing case," so he devised a buzzer to signal the person being called. Users forgot to hang up the receivers and disconnect, so he produced a hook on which the receiver was hung.

[18]*The First Century of the Telephone in Connecticut, Southern New England Telephone,* booklet by Reuel A. Benson, Jr., January 1978.

[19]The Chicago exchange, in the American District Telegraph Company office at 118 LaSalle Street, was known as the Edison Exchange because his transmitter was used. The first subscribers already had ADT call boxes and, by turning a handle, sent signals over the wire for messengers, police, etc. When telephones were installed, they sent a signal over the separate call box line to notify the central office that they wished to make a telephone call. The name of the subscriber was written and taken to the telephone operator who plugged a metal pin into a socket connected with the caller's line. Upon learning the name of the person being called, the operator could plug a wire, with a pin at each end, into two sockets to make the connection. Another signal by call box notified the central office that

the conversation had ended.

As subscribers multipled, Charles Ezra Scribner of Western Electric invented the "jack-knife" switch so that the insertion of a plug would open the contact. The switchboard was arranged in the following year so that the arrival of current from a caller would cause one of 25 annunciators to drop its shutter, notifying the operator to plug in. Scribner followed with 440 other inventions, probably more than anyone in the electrical industry except Edison.

[20]Albert Bigelow Paine, *In One Man's Life (Theodore N. Vail)* (New York: Harper & Bros., 1921).

[21]In the fall of 1888 an exchange was established at 18 Courtlandt Street to serve telephones in "downtown" New York. By that time there were about 10,000 telephones in New York City.

[22]*The Brooklyn Daily Eagle,* May 30, 1936.

[23]Charles F. West's mimeographed "Personal Reminiscences" of telephone and telegraph experiences from 1876 to 1926, in my possession.

[24]Insull came to the United States on March 1, 1881 and was Edison's private secretary and financial manager for about a decade.

[25]Dyer, Martin, and Meadowcroft, *Edison—His Life and Inventions,* 206-207.

[26]*The Telegraphic Journal and Electrical Review of London,* January 1, 1878, said Edison's phonograph was "not strictly a telegraphic instrument; but if used in conjunction with the articulating telephone, it will become so, since it will then be possible to record speech at a distance by permanent marks." The model of his "embossing telegraph" is in his laboratory at Ft. Myers, Florida, now a museum.

[27]At the site where the Menlo Park Laboratory was operated from 1876 to 1886, the Edison Tower was erected in 1937. The 131-foot tower, on a hill, is surmounted by a 13-foot 8-inch replica of the original incandescent lamp, weighing more than three tons.

The laboratory was moved to Greenfield Village, Dearborn, Michigan by Henry Ford, who spent $3,000,000 to gather early models of telegraph, phonograph, electric light and other Edison inventions and establish an Edison section in his huge village of Nineteenth Century memorabilia. The old laboratory was reopened as a museum in 1929, with President Herbert Hoover, Edison, Ford and other notables participating. The General Electric Company, which had taken over the Edison lamp business, made elaborate plans for the celebration of the "Golden Jubilee of Light" without even consulting Edison, so he went instead to Dearborn with Ford.

[28]An article by Charles Askowith in the *Boston Jewish Advocate,* January 30, 1964, credited Berliner with filing a caveat for a telephone on April 14, 1877. Vail bought Berliner's patent and placed him in charge of production at the Williams shop. He and his brother, Jacob, organized the Berliner Telephone Company of Germany. They also formed the Victor Gramophone Company to manufacture a talking machine at a plant in Camden, N.J.

[29]Rhodes, *Beginnings of Telephony,* 79.

[31][30]In the *New York Times,* January 14, 1940, Lloyd W. Taylor of Oberlin, Ohio said the transmitter through which Bell first spoke on March 10, 1876 "was of an utterly different type from that described and illustrated in his patent, and the transmitter he used had previously been described by Elisha Gray in a confidential document filed in the Patent Office." Mr. Taylor said: "Bell subsequently repeatedly acknowledged in court having received information about the contents of this document two weeks before he made his own transmitter." Assuming Bell's first successful sentence was with a transmitter quite different from his own, Gray conceded priority to Bell, and Western Union's agreement to settle out of court was under this misapprehension.

[32]Gould's raids in an effort to gain control of Western Union may have caused its officials to welcome a peaceful settlement at that time.

[33]In New York Western Union turned over to Bell 1,892 miles of wire on poles on the streets, and 970 miles of wire attached to racks on the roofs of buildings.

[34]This is an important point overlooked by AT&T historians.

[35]Control of the Canadian patent was assigned to Alexander Melville Bell, father of the inventor, in 1876. Bell of Canada was incorporated in 1880.

[36]Two-wire metallic circuits had been used on telegraph lines, but Bell applied for a patent in which two wires would be twisted around one another for telephone use, and was awarded the patent July 19, 1881.

[37]*John J. Carty*, a biography by Frederick L. Rhodes, privately printed at New York in 1932.

[38]American Bell annual report for 1880.

[39]Western Electric and Graybar are well-known names. Western Union consolidated its manufacturing shops in 1869, retaining those at Ottawa, Illinois and New York City. It sold its Cleveland shop to the foreman, George Shawk and Enos M. Barton, Western Union's chief operator at Rochester, joined Shawk in the electrical equipment business. Shawk's interest was bought by Elisha Gray, and the company, named Gray and Barton, was a large producer of Western Union telegraph and telephone equipment, and burglar and fire alarms.

Leaving Western Union's vice presidency, General Stager joined them as a partner, and became president. They formed the Western Electric Manufacturing Company in 1872 to take over Gray and Barton and buy Western Union's Ottawa shop. Headquarters was moved to Chicago, which was being rebuilt after the great fire of 1871, and demand for electrical equipment was tremendous. Western Electric manufactured the Edison and Gray telephone and printing telegraph equipment. The company also helped Western Union to organize local telephone companies.

When Western Union left the telephone business in 1879, President Stager of Western Electric merged the Chicago Telephone Company and several midwestern companies. In June 1880 he worked out a deal with Gould and Vail to sell Western Union's Western Electric stock to American Bell. Western Electric of Illinois, formed Nov. 26, 1881, took over the old company, the Williams shop in Boston, and Gilliland Electric Manufacturing Company in Indianapolis. In 1882 it became the manufacturing, warehousing, and supply arm of AT&T. Barton was president from 1885 to 1908. The electrical supply end of the business became a separate company in 1925, named Graybar Electric Company (after Gray and Barton), and Stager became president of Southwestern Bell Telephone Co.

Prelude
to Modern Communications

Communications systems and services important to us today took shape in the 1881–1910 period to meet the developing needs of the nation.

In 1881 our way of life had changed little since the Civil War. Kerosene lamps were in homes, horse cars in city streets, and carriages for the well-to-do. G. F. Swift's refrigerator car in 1875 forecast a revolution in daily diets. Edison's phonograph in 1878, and electric light in 1879 had reached few homes. Fewer homes had telephones, and none had radios or television.

Ivory Soap, Royal Baking Powder, Mellins Food, Scott's Emulsion, Doctor Pierce's remedies, Walter Baker Cocoa, Hire's Root Beer and W. L. Douglas shoes were popular. In 1882 Edison's power plant on Pearl Street in lower Manhattan began providing electrical illumination in homes. The Brooklyn Bridge, "the engineering wonder of the age," opened in 1883. In 1888 a former Edison worker, Nicola Tesla, devised the alternating current used in motors,

a forerunner of machine and assembly line manufacturing, that sped industrial growth, and doubled the capital invested in industry. The automobile was born. Aided by waves of immigration, population jumped from fifty to sixty-three million in the 1880s.

The bicycle was first made in the United States in 1877, and the major users were messengers delivering telegrams, which previously had awaited the convenience of the manager or the public. By 1896 there were some 100 manufacturers of bicycles, retailing from $100 to $150, and 200,000 were in use.

Under Gould's control, Western Union was powerful and ruthless. Its operations were in three geographical divisions in 1884. In the Central Division were 9,609 employees, 8,203 offices, and 87,117 miles of pole line. It handled annually 15,000,000 messages and 150,000,000 words of press. The Eastern Division had 8,219 employees in the 4,545 offices, 34,000 miles of pole line, 18,000,000 messages and 80,000,000 words of press. Southern Division had 2,227 employees, 1,321 offices, 57,000 miles of wire, 3,000,000 messages, and 8,000,000 words of press. It had 80.19 percent of the wire, 72.56 percent of the offices, and 92.15 percent of the telegraph business.

15.1 A WELL-KNOWN SIGHT IN AMERICA FOR A CENTURY WAS THE YOUNG TELEGRAPH MESSENGER. ABOUT 25,000 USUALLY WERE DELIVERING TELEGRAMS ON FOOT OR BICYCLE. THE TURNOVER WAS RAPID BECAUSE SO MANY WERE HIRED BY BUSINESSMEN TO WHOM THEY DELIVERED TELEGRAMS. ABOUT ONE MILLION BUSINESSMEN AT ANY TIME WERE FORMER MESSENGERS. ONE BECAME PRESIDENT OF U.S. STEEL, AND ANOTHER WAS THE MOVIE COWBOY SINGER GENE AUTRY.

Four competitors appeared in the early 1880s. First was the Baltimore and Ohio Railroad. The Western Telegraph Company, which had an 1847, thirty-year charter from Maryland, sold the line it had erected along the B&O right-of-way to Wheeling to American, which later was absorbed by Western Union. When the charter expired in 1877, the railroad refused to renew it and forced Western Union to remove its line from its right-of-way. It had decided to go into the telegraph business, and spent a million dollars buying and constructing lines along its right-of-way to handle its own messages and dispatch trains.

The railroad became a partner of Gould's Atlantic and Pacific, but withdrew when that company was absorbed by Western Union. It then joined its lines with the American Union, but withdrew when that company also was consolidated with Western Union in 1881. It organized the Baltimore and Ohio Telegraph Company in 1882, and hired David Homer Bates, assistant general manager of Western Union, to head it. Bates took with him a number of Western Union men, made deals with independent lines, and competed for agreements with railroads. Although Western Union bought land in the path of B&O's new lines, he added 28,000 miles of wire in a year and built up a system of 1,143 offices, 7,535 miles of pole line, and 54,972 miles of wire, operated by 2,419 employees.

Bates organized a pool with National Telegraph (a successor to the Ohio and New Orleans and People's line), Bankers and Merchants, American Rapid, and Postal. He slashed tolls to a cent a word, but that ruined the finances of all companies in his pool.

Western Union brought suit against the B&O in 1885, charging infringement of Edison's quadruplex patents, and circulated rumors it would buy the B&O lines. Bates quickly denied that, but Dr. Green stated:

> The Baltimore and Ohio is doing business at a large monthly loss with the sole purpose and expectation of compelling Western Union to either buy it out or form a pooling arrangement. . . . As to Robert

Garrett's[1] denials that he has ever offered to sell his lines to the Western Union, we refused overtures made before Mr. Garrett's departure for Europe.

The B&O sold bonds to pay debts, and the press rumored:

Jay Gould will get his share of the telegraph system. In fact, it is alleged that $500,000 has already been paid on account for the telegraph system to the syndicate. The whole sum asked was $6,000,000.

The rumor was true. Western Union bought the B&O telegraph system in 1887. *The Electrical World* reported the price was $5,000,000 in Western Union stock and $50,000 a year for fifty years.

The second competitor was Bankers and Merchants' (B&M), organized by Wall Street stock brokers Jeremy G. Case and J. Heron Crosman. B&M built a line from New York to Washington in 1882. A. W. Dimock, B&M's stock broker and vice-president, bought control, reorganized it, and began constructing lines to the west, but the panic of 1884 wrecked it.

The third competitor, American Rapid, was formed by ubiquitous D. H. Craig in 1879, when Western Union refused to purchase an automatic telegraph developed by George A. Hamilton. Craig claimed that with sixty-six girls perforating messages, 480,000 words could be sent over one wire in eight hours. He rashly predicted a rate of nineteen cents for 1,000 words, reduced gradually to six and one-half cents.

The B&M exchanged bonds for all American Rapid stock, mortaged the property for a $10 million bond issue, and failed. Seven thousand miles of B&M lines were on American Rapid poles when Western Union leased it. When a court order was served on him on July 10, 1885, to give up the Ameri-

can Rapid property, John G. Farnsworth, the B&M receiver, pretended the wires of one company could not be distinguished from the other. He left the office to call his lawyer, and the Western Union men promptly cut the wires, putting both B&M and Rapid American out of business.

B&M bondholders, facing heavy loss, bought the property for $500,000, and organized the United Lines Telegraph Company which combined its service with Postal Telegraph, under the guidance of Albert B. Chandler, receiver for Postal.

The Postal Telegraph-Cable Company

Postal Telegraph, the only important telegraph company opposing Western Union after the 1880s, was organized June 21, 1881. It erected a line from New York to Chicago, and entered that city in August 1883, after much delay because Western Union bought land it needed and raised legal obstacles.[2] Postal exchanged business with the Southern Telegraph Company for southern points, and American Rapid to reach Boston.

Postal used steel wire, developed by Chester Snow, covered with a heavy deposit of copper, and Elisha Gray's harmonic telegraph, that handled messages simultaneously on sixteen different tones. The separate tones were received by variously tuned "analysers" which were "metallic membranes placed free before the faces of an electromagnet so as to respond to a corresponding current."[3]

Postal's promoters attempted to justify its capital of $21,000,000 in stock and $10,000,000 in six percent bonds by claiming that, with a limited number of wires and low rates, they could handle all of the nation's telegrams at a good profit.[4]

Because of agitation in Congress for government ownership, the name "Postal" was selected to meet a supposed demand for combined telegraph and mail service. The organizers hoped the name would help them obtain government telegraph business. Fostering this idea, Postal established a uniform rate of twenty words for twenty-five cents, and delivered telegrams by mail from the nearest point reached by its lines which were only to large cities.

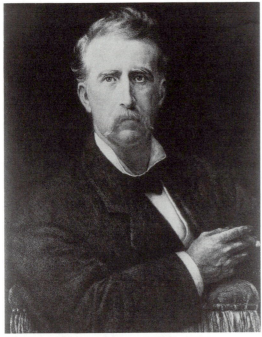

15.2 JOHN WILLIAM MACKAY (1831–1902).

Postal offered to handle government messages at twenty cents for twenty words between points not more than 1,000 miles apart. The Baltimore and Ohio at once made a similar proposal, but the Postmaster General refused both offers. He ruled that all telegraph companies could afford to serve for less and reduced the rate. That meant a loss, and Postal already was in financial trouble.

George D. Roberts of California, one of Postal's incorporators, was a friend of John William Mackay, a Dublin boy, who had immigrated to the United States, worked in New York shipyards, and joined the gold rush to the West.[5] Mackay had fabulous luck in staking out a claim that developed into the Bonanza mine. He and three partners—O'Brien, Flood, and Fair—took more than $175,000,000 from the mine, and Mackay became rich beyond his wildest dreams.

Roberts persuaded Mackay to head a syndicate that bought 12,000 of the 21,000 shares of Postal stock. Mackay and his associates reorganized Postal October 19, 1883, and named it Postal Telegraph-Cable Company because he was thinking of laying an Atlantic cable. Erastus Wimen, a Western Union director, told a Chicago Inter-Ocean reporter in 1884 that Postal had issued $24,000,000 in stock and bonds, "represented by an expenditure of less than $1,000,000."

Postal used the slow Leggo automatic system to record messages in dots and dashes on a revolving cylinder that held 1,200 words. Then an electrical stylus was placed on the cylinder to send the dots and dashes over harmonic tones. At the distant end, dots and dashes appeared on strips of chemical paper.

After its associate companies became bankrupt in 1884, Postal defaulted on its bonds February 1, 1885. Albert B. Chandler and Henry Rosener were appointed receivers, but Mackay went to the rescue with money to refinance it.[6]

Since Western Union had exclusive lines along the railroads, Postal had to condemn rights-of-way along highways, and obtain special permission from towns and cities to build along their streets. Since many states would give condemnation rights only to "local" companies, Postal became an association of about fifty corporations. In 1886 it obtained an agreement to exchange traffic with Canadian Pacific, spanning the continent in Canada.

Postal's receiver, Chandler, had been in charge of Western Union cable traffic from 1866 to 1875 when he went to Gould's Atlantic and Pacific Telegraph Company. He became Postal's president in 1886 and chairman in 1901 when Mackay succeeded him as president.

Commercial Cable Company

In the early 1880s James Gordon Bennett, the crusading editor of the *New York Herald,* frequently denounced the "cable trust," because his paper had to pay the pool's rate of fifty cents a word, and he persuaded Mackay to go into the cable business.

The result was the Commercial Cable Company, which had Siemens Brothers lay two cables from Ireland to Canso, Nova Scotia in 1884 for $7,000,000. Two feeder cables were laid from Canso to Rockport, Mass., and Rockaway Beach, Long Island, to connect with Postal. Service began December 24, 1884, with a small office at Wall and Broad Streets, where the tall New York Stock Exchange building now stands, and a Boston office at 4 Arch Street. Feeder cables were laid from Ireland to Havre, France, and Bristol, England, in 1885.[7]

Western Union also laid two feeder cables connecting the D.U.S. cable at Rye Beach with the French "PQ" cable at Cape Cod. This eliminated the need for connecting landlines through New England States and Canada.

Declaring war on the cable pool, Mackay lowered cable rates from fifty to forty cents a word, and Western Union retaliated by cutting them to twelve cents. Commercial had to go to twelve cents to compete, but all cable companies lost money, and finally agreed on a rate of twenty-five cents a word in July 1888.[8]

In 1894 the French Cable was amalgamated in La Campagnie Francaise des Cables Telegraphique, a combination of French companies, and withdrew from the pool. In 1898 it laid a cable direct from Brest to Cape Cod, and a connecting cable to Coney Island.

Commercial then laid a cable from Far Rockaway, New York, to Canso to Horta, in the Azores Islands, and another from Horta to England. Starting to lay another, from Canso to Ireland in 1905, the Cable Ship *Colonia* struck a rock off Nova Scotia, but was repaired in drydock at Halifax with the 2,300-mile cable on board, and finished the job on October 6, 1905.

In 1900–1904, D.A.T. (Deutsch Atlantische Telegrafengesellschaft) laid two Atlantic cables to Horta, where an international exchange of cable traffic was established.[9]

Two Decades of Rapid Growth

In the late 1880s and the 1890s, Postal grew by investing additional sums advanced by Mackay and his associates, while Western Union spent $65,181,357 in the 1866–1892 period to increase its system, and then issued a 10 percent stock dividend raising its capitalization to $100 million. It also invested more than eight million in telephone companies, the Gold and Stock Telegraph Company stock ticker and commercial news, and the American District Telegraph Company messenger, fire alarm, and burglar alarm business.

In the three and a half years after

Gould's seizure of control in 1881, Western Union had paid nearly $18,000,000 in dividends, and $4,693,495 to buy additional properties.

American Bell's business also grew rapidly. A home telephone in New York cost $180 a year in 1893. By 1899 Bell circuits totaled more miles than Western Union's, and its stock soared to $386 a share.

When Dr. Green was president of Western Union, crooked politicians harassed utilities with threatened legislation, and persons with secret motives demanded underground wires. Labor appealed for public support, and a public outcry was incited for government ownership. Dr. Green skillfully took the public into his confidence and helped block unfavorable legislation.

Telegraph use by the press grew rapidly in the 1890s. Led by such editors as Ochs, Dana, Pulitzer, Hearst, Medill and Reid, newspapers began covering more major stories. Many of the 1,600 small dailies in 1891 also began featuring national news.

When Jay Gould's man, Thomas T. Eckert, succeeded Dr. Green as Western Union's president,[10] the annual reports in his nine-year regime, which lasted until 1902, were a model of corporate evasion. Even the *Journal of the Telegraph* degenerated into a mere bulletin of tariffs and rules. Postal reports likewise gave little information.

The volume of Spanish American War news helped pull Western Union out of the depression with a $6,165,363 profit in 1900. The nation's population grew twenty-one percent during the 1890s to 76,304,799. The "horseless buggy" automobile industry and travel gave impetus to telegraph use.

As Eckert's regime drifted along, the stenographer in the executive office was J. C. Willever, later senior vice president. In the general business depression of 1894,

Coxey's army of the unemployed invaded Washington. Western Union profits sank from $7,496,037 to $4,793,484 in 1894, while Postal established impressive headquarters and operating rooms in a fourteen-story building Mackay constructed in 1894 at 253 Broadway,[11] facing City Hall. "253" was Postal and Commercial headquarters until 1928, when they moved to the International Telephone and Telegraph Company building at 67 Broad Street.[12]

Commercial's gross for 1895 was $2,009,738, and net income $1,215,397; Postal's gross was $4,321,296 and net income, $617,863. In 1886 Postal had 4,400 miles of poles and cables, 37,000 miles of wire, 400 offices, and handled three million telegrams. In 1895 it had grown to 19,500 miles of poles and cables, 117,000 miles of wire, 1,067 offices, and 12,500,000 telegrams. Postal collected and delivered nearly $600,000 in business for Commercial.

The Commercial officials then were as follows: John W. Mackay, president;[13] George G. Ward, vice president and general manager; C. R. Hosmer, Mr. Chandler and C. W. Mackay, vice presidents. It was consolidated with Postal in 1896, when Commercial exchanged $20 million in four percent gold bonds for $15 million of Postal's stock.[14] In 1898 Postal bought the Rocky Mountain Telegraph Company, which had 300 miles of line connected with the transcontinental Canadian Pacific line. That provided Postal with a route to send messages to the west.

Since Postal's lines usually were along highways, its general construction foreman, W. H. McCollum, constructed large boxes on wheels, which resembled railroad cars. In these cars, thirty-three feet long and eight and one-half feet wide, the workers ate, slept, and carried supplies.

15.3 Newsmen and Western Union operators at Flemington, New Jersey courthouse transmitted millions of words from this small room during the trial of Bruno Hauptmann, who was convicted of kidnapping Charles Lindbergh's baby.

Government Ownership Agitation

Agitation for government ownership blossomed at Senate and House committee hearings for fifty years, but most newspapers feared liberty of the press would be endangered and opposed it.

A bill had been introduced in 1865 to create a company to compete with Western Union, but Postmaster General Dennison helped defeat it. A substitute bill gave him the power to fix rates for government messages.

One Senate committee report charged that newspapers approving government ownership lost their telegraph service, or had their rates raised. In a San Francisco speech in 1875, Congressman Charles A. Sumner said:

> The Western Union has a twin connection with another incorporated thief and highway robber known as the Associated Press. They are banded together in the strong bond of mutual plunder and rapacity against the people.

Postal had a bill introduced in 1888 to establish a combined government postal and telegraph service. Opposing it, C. H. Jones of the Jacksonville, Florida, *Times* told the American Newspaper Publishers' Union that forty-seven million Western Union messages were sent by less than one million people in 1887, less than two percent of the population. Government operation, he argued, would produce another deficit of at least ten million dollars, to be paid by ninety-eight percent of the people, who did not use the telegraph.

President Green sent a memorial to Congress stating that the telegraph business was commerce, and the Constitution provided that the government might regulate, but not engage in it. The telegraph is not an "oppressive monopoly," he declared, but "competition was open to all." He cited the poor service and heavy deficits of the British Post Office Telegraph. With its most distant offices less than 600 miles apart and only 29,895 miles of line to maintain, he pointed out, the British Telegraphs handled 40,243,230 messages in 1887 at a deficit of $2,349,230.[15] He said U.S. government operation would produce a very large deficit. The bill was not passed.

In Frank Leslie's magazine in 1890, Erastus Wiman,[16] a Western Union director, argued that since railroads were more necessary to handle the mail than the telegraph, government ownership of the railroads should come first. Postmaster General John Wannamaker, President Harrison's appointee, was the last of a long line proposing government ownership, but in every case a majority in Congress refused.

Underground Wires and Tubes

Under the streets in every city now, an intricate network of pipes and cables provides communications, water, gas, electricity, and sewers. Paris put its telegraph wires in the gigantic system of sewers planned in 1865 by Baron Eugene Haussmann.

Here is how it started in America. By 1882 telephone and telegraph, news and stock ticker, district messenger, fire alarm, electric light, and other wires filled the air over some streets. A public desire "to brush away the gigantic cobweb of telegraph wires now spun from pole and housetop along the streets of large cities," the *New York World* said, grew in proportion to the number of lines.

Lurid stories appeared in the 1880s about the dangers of overhead lines. One story under large headlines told of the electrocution of a horse by a broken electric light wire. Women were afraid to drive in the business sections of New York, said another story, designed to arouse the merchants. Investigation revealed that no horse was injured. The man originating the stories was promoting legislation against overhead wires on behalf of an underground conduit company.

However, officials were convinced that they had to bury the wires. Unable to enter Washington, D.C., because of Western Union's railroad rights-of-way, Postal was refused permission by the city in 1883 to construct overhead lines in the streets. It then had Daniel N. Hurlburt of Chicago lay an underground cable 10,000 feet to its Washington office. The wire, in paper tubes, in asphalt in a brick trench, worked until lead-covered cables came into use.

A New York law enacted in June 1884 required utility wires be placed underground in large cities. Since only 1,000 of the 15,562 miles of line in New York City were underground, a Commission of Electrical Subways was created. The members went to see Edison at his New Jersey laboratory. Wearing an old linen duster and a battered

154 RIGHT: IN SEVERAL GREAT FLOOD DISASTERS, HEROIC LINEMEN RACED IN RAIL MOTORCARS ACROSS BRIDGES NEAR COLLAPSE TO SAVE LIVES AND RESTORE COMMUNICATIONS WITH THE OUTSIDE WORLD. LEFT: LINEMEN ALSO DID HEROIC WORK IN REPAIRING LINES DURING BLIZZARDS.

straw hat, Edison said: "All you have to do, gentlemen, is to insulate your wires, draw them through the cheapest thing on earth, lead pipes, run your pipes through channels or galleries under the street, and you've got the whole thing done." The Commission asked what his bill was. "Not a cent," Edison answered. "Do you suppose I'd stick you for so simple a thing as that?"

The carriers began moving the wires underground. Then came the Great Blizzard of 1888, the historical landmark from which telegraph and telephone men for the next fifty years dated their reminiscences. Communications workers struggled through mountainous snow drifts to reach their posts, and remained there for days. Most New York City lines were down in a tangled mess. Many ninety-foot poles had thirty crossarms and 300 wires. A newly built telephone line from Boston to a point near New York City was

saved, thanks to strenuous efforts by linemen. This inspired a famous drawing—"The Spirit of Service"—showing Angus Macdonald, a Boston lineman, on snowshoes in the storm, looking up at ice-laden wires.

New York was cut off from Boston, and telegrams were sent between the two cities over the cables via London. Linemen performed feats of endurance and bravery, pulling supplies through the deep snow and repairing breaks in the face of bitter winds.

Seeking to make political capital by a grandstand stunt the following year, Mayor Hugh J. Grant of New York on April 16, 1889, ordered men with axes to cut down the tall poles at Union Square, 14th Street and Broadway. As telegraph, telephone, and electric lines fell to the street, foolish citizens cheered the action that shut off their own electric lights and food supplies. Broadway returned to gaslight and candles, and the city was crippled until the damage could be repaired.[17]

Moving poles and hundreds of wires from Broadway to underground made a startling change in the appearance of New York City, as shown by before-and-after photographs of Broadway.[18]

15.5 AN OLD-TIME LINE GANG, TRAVELING TO SCENES OF EMERGENCY AND REGULAR LINE CONSTRUCTION JOBS.

In 1876 Western Union laid the first pairs of pneumatic underground tubes in America from its headquarters at 195 Broadway to the Stock and Cotton exchanges. One tube was for sending, and the other for receiving. Three years later it laid a single tube a few blocks long to speed news stories to six newspapers, The *Times, Tribune, Herald, World, Sun,* and *Staats Zeitung.* Tubes were laid in 1884 to the branch office at Fifth Avenue and 23rd Street, and 407, 599, and 844 Broadway.

Describing the tube system, designed by A. Brotherhood, Western Union mechanical engineer, the *New York Sun* said:

> To transmit a message the operator places it in a cylindrical leather box a trifle smaller in diameter than the tube, but having flanges at each end which make an air-tight joint. . . . When filled, it (the box) is popped into the tube, the air is exhausted before it and pumped in behind it by engines at each end,

and away it goes at a trifle over a mile a minute.

Scientific American reported that a box of vulcanized fibre was adopted, with a thick pad at one end to absorb the crash when it reached the end of its trip. Soon telegrams were shot beneath the streets of major cities, linking branch and main offices.[19]

Heroism in Major Disasters

The business world has not seen more faithful, self-sacrificing workers than the linemen, the shock troops in man's struggle to maintain service in the face of the worst nature can produce. Once line gangs lived in scores of camp outfits consisting of a railroad sleeping car, dining car, tool and supply car, pole car, and tank car to provide water.

When hurricane, flood, fire, or earthquake struck, communications men and women flashed the call for aid to the outside world. Linemen rushed in to restore lines[20] and operators stuck to their posts day and night, often in the face of danger, to send messages, directing the work of rescue and relief. When danger was past, they sent telegrams reassuring relatives of survivors, and orders for rebuilding supplies.

When the Conemaugh River dam four miles above Johnstown, Pennsylvania, broke on May 31, 1889, after three days of continuous rain, a wall of water twenty feet deep swept through the city, and 3,000 people drowned. Buildings, haystacks, wagons, and bodies of cattle and horses clogged the stone arches of the Pennsylvania Railroad bridge. Other bridges, railroad tracks, and telegraph lines were washed away.

To reach a line where he could send news of the disaster to the outside world,

15.6 THE "LINE GANGS"—THE SHOCK TROOPS IN THE BATTLE TO KEEP THE ARTERIES OF COMMUNICATION INTACT REGARDLESS OF WIND OR STORM—LIVED IN GROUPS OF SPECIAL RAILROAD CARS WHICH WERE MOVED AROUND THE COUNTRY AS EMERGENCIES AROSE. SHOWN HERE IS ONE OF THE CAMP CARS. WITH MICROWAVE NETWORKS AND SATELLITES HANDLING MUCH OF THE TRAFFIC NOW, FEW LINE GANGS REMAIN.

William J. Meloney, Western Union manager at Altoona, borrowed a railroad velocipede and carried thousands of words of news stories, 500 telegrams and a Morse key. Crews of trapped trains helped him cross the Juniatta and other rivers in rafts and boats. After two days and nights, he reached Lewistown Junction, found a wire working to Harrisburg, Pennsylvania, and flashed news of the disaster that aroused the nation to action.

More than 100 newspapermen dashed to the flood area. J. Hampton Moore, a *Public Ledger* reporter and later Mayor of Philadelphia, and Crute of the *Philadelphia Press,* reached Cambria City on the fourth day and walked on to Johnstown through the mud and debris.

"Our first move was to visit the telegraph office," Mr. Moore told the Johnstown Flood Correspondents' Association fifty years later.

A few hurriedly adjusted poles, bearing the weight of a number of wires, served as guide posts, and, following the line of these, we soon came to "headquarters." Headquarters, indeed. They harmonized fully with the scenes of misery we had just witnessed. The great Western Union Telegraph Company—like the mighty Pennsylvania Railroad—had been reduced to beggary.

The most available place it could find to run in its wires and put up its instruments was this "headquarters"—on the side of the mountain, just above the eastern end of the railroad bridge. The "headquarters" consisted of a frame shanty, the roof and sides of which were boarded in. Its dimensions were about ten by twelve feet. The floodwaters had not quite reached up to it, or it would have been swept away. It did not shelter its occupants against the rain, which dropped through the roof on the instruments, the writing paper, the oil lamps and candles, and the worn-out telegraphers and correspondents who sent out their news messages to the world day and night, for weeks. Evidently the shanty had been an oil-barrel repository, for there were barrels soaked and greasy inside and outside, but these were serviceable, for the weary correspondents who used them as desks and tables.

When Crute and I entered this forlorn

structure and asked for wires, the operator in charge, Mr. Munson, a polite but careworn fellow, informed us. . . . The early Associated Press and Pittsburgh boys were holding every wire, and we would have to wait until somebody dropped out, or more wires, instruments and operators were available.

15.7 AN OLDER TYPE OF POLE-CLIMBING EQUIPMENT.

Moore did get a wire, as did all of the newsmen arriving on succeeding days. As the stories of human suffering appeared in the press, carloads of food and clothing were shipped from all parts of the nation, and more than $3,000,000 was raised for relief.

The Galveston Flood

In September 1900, Galveston, Texas, built on a mile-wide sand bar, was cut off from the outside world, when waves of the Gulf of Mexico were swept in by hurricane winds blowing more than 100 miles an hour. These destroyed half of the city and drowned 6,000 people. Not a telegraph pole was left standing, and even the cables to the mainland were swept away. The nearest point at which messages could be sent was Houston, fifty miles away.

Mayor Walter C. Jones told Richard Spillane, telegraph operator at Galveston, "We're cut off as if on an island in the Pacific, and before night 30,000 will be starving. What under heaven can we do?"

"Do, man?" Spillane asked. "Get into communication with the outside world somehow, quick as heaven will let you. Give me a requisition to impress anything or anyone I want into my service, and I'll show you what to do."[21]

Spillane braved the house-high waves of Galveston Bay in a launch crossing the seven miles to the mainland, and ran the boat full speed into the wreckage littering the shore. Plodding through kneedeep water and mud, he found the right-of-way where the railroad had been, and set out for Houston. Almost mad with thirst, swimming streams, his bare legs swollen to twice their normal size, covered with mud and unrecognizable, his clothes in tatters, Spillane somehow reached Houston that night. Mumbling over and over, "Galveston is gone," he staggered to the telegraph office, seized the key of an amazed operator, and called St. Louis where President McKinley was visiting. His message was:

A hurricane and tidal wave destroyed Galveston. At least 10,000 are dead in Galveston and surrounding country. 20,000 to 30,000 are homeless. We need food, clothing, tents, doctors, drugs, and—above all—disinfectants.

A New York newspaper wired Spillane an offer of $5,000 for his story. He refused, replying that his first duty was to those in need. Then he pounded out the whole story on the key, without a word of copy written, to the Associated Press.

Western Union Construction Superintendent C. H. Bristol, Chief Electrician J. C. Barclay, General Foreman George Gudgeon, and District Superintendent P. T. Cook of St. Louis dashed to Houston. Assembling linemen and supplies, they started for Galveston, running a temporary line of four wires.

> The linemen of Western Union virtually laid wires over nine miles of dead bodies, beginning the work fourteen miles from the town [Barclay reported]. The wind and waves had carried bodies to this great distance from the island town. Thickly scattered about were the dead bodies of men, women and children, cattle, horses, dogs, hogs and sheep. . . .
>
> Of course, no telegraph poles remained standing and we had to string the wires as best we could. We had brought some supports with our supplies, but we also had to utilize the driftwood and broken timbers lying about. We got the wires up about three and a half or four feet from the ground on these temporary supports and made remarkable time over the country . . .
>
> It was by far the most sickening undertaking. Most of the time the men were stepping over dead bodies above which they had to stretch the wires.

The Baltimore Fire

When fire destroyed 2,600 buildings of Baltimore—the entire business section—on February 7-8, 1904, Western Union and Postal workers stuck to their posts until the upper floors of their main-office buildings were burning furiously. Taking their instruments and typewriters, they retreated, connecting their instruments with wires cut from overhead poles. Sitting on curbs, they worked on amid turmoil and confusion to send the greatest fire story since that of Chicago in 1871.[22]

About 100 operators from the main Western Union office in the Equitable Building, and Associated Press reporters from the Herald Building reached the branch office at Gay and Lombard Streets, where they wrote and transmitted feverishly. When flames almost surrounded that building at midnight, they ran. At dawn they moved to the third floor of the House of Welsh, a saloon at Guilford Avenue and Saratoga Street.[23] For three days and nights thirty-five operators there kept the world informed of Baltimore's fate. Postal obtained space at 110 Orleans Street as a makeshift operating room. Both companies connected every wire that could be made good to the outside world. Telegraph service also was used to organize relief and rebuild.

The San Francisco Earthquake and Fire

During the earthquake and fire that almost wiped out San Francisco on April 18 and subsequent days in 1906, telegraphers remained at their keys to send the news, despite imminent danger from falling walls and approaching fire. They stuck to their posts three, four, and five days without rest, until they fell asleep from exhaustion, not even going to the ruins of their homes or to find their families scattered upon the suburban hillsides.

15.8 THE GREAT SAN FRANCISCO EARTHQUAKE IN 1906 WAS FOLLOWED BY FIRE. WITH THE CITY WATER SUPPLY CUT OFF, MILITARY AUTHORITIES RESORTED TO DYNAMITE TO CHECK THE FLAMES. THE ABOVE VIEW WAS TAKEN FROM GRANT AVE. LOOKING DOWN PINE ST. THE LARGE BUILDING ON THE LEFT IS THE MERCHANTS EXCHANGE BLDG. ON THE RIGHT SIDE OF PINE ST. IS A BUILDING WHOSE TOP HAS FALLEN IN. THE MAIN OFFICE OF WESTERN UNION IS DIRECTLY ACROSS PINE ST. FROM THAT BUILDING. THE FIRE ADVANCED TO THE WESTERN UNION BUILDING BEFORE IT WAS HALTED.

The main offices of both companies and about fifty branches were destroyed. The first shock of the earthquake sent plaster and bricks pouring down in the Western Union building. The stunned workers picked themselves up and restored service.

In the face of advancing flames shooting high into the sky, they worked on in the belief that unless they summoned aid people in the city would die. When fire forced them from the building, they went to Oakland, across the bay. There Chief Operator H. J. Jeffs, climbed a thirty-foot telegraph pole, dropping wires to the street. The wires were run into a private residence where tables were surrounded by operators.

That emergency office was soon shifted to a poolroom across the street, where operators around pool tables kept in touch with the White House, War Department, and Western Union headquarters. Later a temporary office was established in the Oakland Ferry House at the foot of Market Street, San Francisco.

Company electrician Lewis McKisick was installing a dynamo at St. Paul when news of the earthquake arrived. He loaded the dynamo and a carload of other equipment on a special train and rushed to San Francisco. At Chicago, D. R. Davies, superintendent of construction, quickly gathered a force, loaded baggage cars with telegraph wire, motor generators, instruments, food, clothing, and other supplies, and sped 2,000 miles to Oakland. Other baggage cars took supplies from Topeka and Ogden, and the company's private Pullman "Electric" and other cars were sent to house operators and provide a company restaurant.

Working day and night, McKisick soon had a complete telegraph plant in operation at West Oakland, connected by cables under San Francisco Bay with ten emergency offices in the city. Those "offices" were election booths borrowed from the city and fragments of ruined buildings, hastily nailed together. In one week he constructed a main office fifty by one hundred feet, and had 170 operators working in it.[24]

15.9 Crowds of refugees besieging the temporary telegraph offices in Oakland after the San Francisco earthquake and fire, seeking messages from friends and relatives. The Western Union office was on the left and the Postal office on the right. Inside the offices, police with drawn clubs kept the crowd in check. Later, desks with boxes in alphabetical order were placed on tables along the sidewalk.

When the earthquake struck, W. C. Swain, Postal electrical engineer, had his company's officials at New York on the line until advancing flames forced evacuation of the building. He opened a temporary office in the Oakland Ferry House. Then Postal manager in Chicago, T. W. Carroll, later Western Union Eastern Division general manager, rushed to San Francisco.

Telegraph operators catching their "forty winks" in the parks, were summoned by buglers, sent by General Funston of the U.S. Survey. Martial law was clamped down, and messengers were not permitted to enter the city to deliver messages, but so many offices and homes had been destroyed and so many people had left the city, that delivery would have been like "hunting for a needle in a haystack" anyway. Tables were set up on the sidewalk; telegrams were placed in boxes in alphabetical order; and people found messages addressed to them.

When a grand jury charged later that the telegraph companies had not delivered many messages to survivors, President Clowry replied,

> Why, we did all we possibly could. We are charged with making a million dollars by sending messages. It should be that we lost

that much. They accuse us of not giving any contribution to San Francisco. Why, we gave everything we had— every message of relief was sent free. We served the newspapers of this country for two weeks with all the news we could get, free of cost. There were 23,000 points where news bulletins were sent free of cost. We forwarded all messages dealing with relief free absolutely, and all relief societies and private individuals sent messages free. . . . They complain about this lack of contribution. I wonder how much all this was worth to the people of the stricken city?

This charge by the twelve men of the grand jury is an outrage. It is worse—it is wicked. Our company did more to assist the people of San Francisco than any other company or organization in the United States.

The Dayton Flood

A flood spread ruin in Dayton, Ohio, March 25-26, 1913. Telegraph men plunged into the water to rescue the drowning, and sent out calls for rescue workers, boats, food, clothing, doctors, nurses, and medicine.

The main offices of Western Union and Postal were caught in the flood. Men on the roof of Western Union's main office fought a fire also spreading in downtown Dayton, and stopped it at their building. They found a live wire in the "Little Wolf Creek" telegraph station, and on March 26 it was working.

Both companies established temporary offices in the National Cash Register plant, the Beckel Hotel, and other high points. Postal workers shoveled aside the mud and set up an office on the sidewalk as the flood receded, moving back into their main office as soon as it could be cleared of mud and

debris. They set up emergency offices and worked long hours to handle the messages of survivors. Restoration of lines was difficult because the railroad tracks were washed out, and trains could not bring poles and equipment.

Col. John H. Patterson, president of National Cash Register, provided the entire facilities of his plant and organization for telegraph operation, hospital and relief work, and a lunch counter for communications workers and many others in a city where bread lines were long. Men who knew how to telegraph volunteered their services. Prominent men volunteered as messengers, plodding to relief stations with bundles of telegrams. Communications people worked all hours at Columbus, Hamilton, and other cities in the flood area.

Orville D. Parker, Western Union manager at Austin, Texas, for nearly half a century, saved thousands of lives when a dam broke in 1900, discharging Lake McDonald down the Colorado in a mighty flood. Cutting the Associated Press wire from the North, Parker sent warnings to every community in the path of the flood to flee to higher ground.[25]

The story of aid in disasters could go on and on. Equally heroic telephone men and women stuck to their posts and saved lives. The Vail Medal Awards, established in 1920, have been presented to numerous heroic Bell employees.

AT&T Violates Contract

In 1887 AT&T decided to ignore the contract it made eight years previously to leave the telegraph business to Western Union. President Green protested to President Hudson on January 15, 1891:

The most glaring violation of the contract is the rental of your long distance wires to brokers for the transmission of commercial news and market quotations for public exposure and use, and this violation is more aggravated by the fact that you actually furnish telegraph instruments to do the service by telegraph.

We have instituted legal proceedings to restrain this service, which, you know, we are loth [sic] to do, and was not done for about two years after it was ordered to be done, because of your promises to have something done about it.

Western Union's suit for violation of the contract, and all other efforts to maintain a separation between voice and record service, as agreed to under the contract, were of no avail. The telephone company increased its telegraph business.[26]

In a period of great business expansion, AT&T grew. A lawyer and Greek scholar, Hudson carried on hundreds of suits to protect Bell patents, which produced cries of "monopoly" in the press. When the Bell patents expired, independent exchanges sprang up in hundreds of towns, but their customers could not call Bell subscribers.

A private telephone cost about $240 a year, and a local call at a public booth fifteen cents. The booths often were mistaken for elevators or toilets. Most public places would not permit booths; it was not until 1899 that the Grand Central Railway Terminal in New York permitted the installation of one public telephone. The Metropolitan Company (later New York Telephone) had only 12,500 subscribers in 1896.

Almon B. Strowger, a Kansas City undertaker, invented the dial telephone in 1891. Angry because he thought operators were giving him busy signals and wrong numbers, and diverting calls to his business

rivals, he devised a "girless, cussless" central office switching system to eliminate the operators. He arranged electrical contacts in rows on the inner surface of a cylinder. A thin metal blade would move vertically step-by-step to a desired number and then horizontally to the next number, enabling the subscriber to select one of ninety-nine telephones.

The Automatic Electric Company of Chicago, later a division of General Telephone and Electronics Corporation, was formed to develop Strowger's patents. It installed the first push-button switching telephone exchange on November 3, 1892, at La Porte, Indiana. It also provided dial telephones in Milwaukee on August 27, 1896, and in other cities, but it was twenty-three years before the first large switching exchange was installed in the Bell System, at Norfolk, Virginia. (Dial installation at New York City was not completed until March 3, 1940.)

AT&T neglected the opportunity to absorb many telephone companies. Hudson's policy evidently was to bring law suits charging patent infringement, pressure bankers to refuse loans, and convince independents that fighting the Bell System was futile. Sometimes AT&T would buy an independent company, pile its apparatus in the street, and burn it as a public bonfire before the eyes of its unhappy local stockholders and subscribers. That policy made it almost compulsory for officials of independent companies to fight to the bitter end. They did not dare to deal with AT&T. One independent president arranged a midnight meeting at an obscure hotel with a Bell man, and arrived with his coat collar high, wearing a false beard.[27]

The result was public disfavor for the "telephone octopus." There were only

291,253 telephones in the United States in 1893 when the Bell patents expired.

New York and Chicago were linked by telephone on October 18, 1892, with Bell, the inventor, inaugurating the service in time for the first Chicago World's Fair. The first line from Boston to Chicago was opened February 7, 1893; New York to Cincinnati December 12, 1893; Nashville to Chicago November 12, 1895; Kansas City to Omaha May 9, and New York to St. Louis June 24, 1896; New York to Charleston, West Virginia June 1, to Minneapolis August 1, to Omaha September 15, to Norfolk October 4, 1897, and to Kansas City November 15, 1898. In 1898, long lines also reached Richmond and Atlanta.

A new policy—expansion—was adopted by Frederick P. Fish, another Harvard-bred lawyer and patent expert, who succeeded Hudson in 1900. To learn the problems of associated companies, Fish visited them, but spent Fridays at headquarters, where Friday was irreverently called "Fish Day."

Touring areas in which independent companies were strong, he established rival exchanges.[28] Fish rushed everywhere, catching a little sleep on trains between stops, and his health began to fail.

The George F. Baker–J. P. Morgan interests, with T. Jefferson Coolidge Jr. of Old Colony Trust Company, Boston, and John I. Waterbury of Manhattan Trust Company, New York, formed a trust designed to obtain control of AT&T and combine Postal Telegraph with it. Clarence Mackay of Postal backed out when the bankers asked him to provide $37.5 million in underwriting a $135 million bond issue. Fish and his directors also refused because the deal would have given the bankers control of the company. Mackay then bought 70,000 shares of AT&T and demanded a place on the board,

but was refused.[29]

On February 8, 1906 a $150 million AT&T bond issue was awarded to a syndicate headed by J. P. Morgan and Company. It remained unsold for two years. While expanding, the company had piled up bonded indebtedness of $202 million, and in 1907 increased its stock to $509 million. The bankers decided retrenchment was necessary to save AT&T. Fish resigned and resumed the practice of law in Boston.

In 1907 there were 6,118,578 telephones (2,986,515 by independent companies), 15,527 exchanges, and 140,000 employees. Many independents were installed by local electricians or handymen, and were operated by their wives and daughters. Usually several homes were connected on one line, and the crank turned five times causing all telephones to ring five times, and alerting all five households to listen in on the other subscribers' conversations.

The Pacific Cables

The British recognized the importance of having cables bind together its worldwide empire. Sanford Fleming, Canadian Pacific chief engineer, went to Australia in 1893 to seek support for a cable. A Pacific Cable Board was created in 1901, and Great Britain, Australia and New Zealand provided $10,000,000.

Since the first section of the cable, from Vancouver, B.C. to Fanning Island would be 3,457 nautical miles, the world's longest, the Telegraph Construction and Maintenance Company built a larger cable ship, the *Colonia,* 500 feet in length, with a capacity of 10,000 tons of cable. Loaded with 3,550 miles of cable, it laid the section to Fanning Island September 18 to October 6, 1901. Other ships laid the sections to Suva, Fiji

Islands, 2,043 miles; Norfolk Island, 981 miles; Southport, near Brisbane, Australia, 837 miles; Norfolk Island to Auckland, New Zealand, 718 miles, and to Sydney, Australia, 1,251 miles.

The Pacific cable opened November 1, 1902, with a message to King Edward from Sanford Fleming. The communications distance from London to Auckland or Melbourne was reduced to sixty minutes, and the rate from Vancouver was fifty cents a word.

The U.S. War and Navy Departments sponsored a dozen bills in Congress to lay cables to the Hawaiian and Philippine Islands and the Orient, but Congress refused because the estimated cost was twenty-five to thirty million dollars.

With Spain's approval, Eastern Cable Extension Company of England laid a cable from Hong Kong to Manila March 30, 1898, but thirty-one days later, after destroying the Spanish Fleet and Cavite Arsenal at Manila, U.S. Admiral Dewey cut the cable to Hong Kong to prevent the Philippines from reporting his raid to Spain. That also prevented the news from reaching the United States, and President McKinley had the cable repaired. The United States still paid about $400,000 a year for messages to and from Manila that had to travel around the world via London.

The United States needed a Pacific cable, and John W. Mackay organized the Commercial Pacific Cable Company on September 23, 1901. A few days before his death on July 20, 1902, he placed a $3,000,000 order for the first section with the Silvertown Cable Manufacturing Company of England. His son, Clarence, carried the project on, but the Eastern Extension and Great Northern companies owned three-fourths, and Commercial Cable only one-fourth.

The Cable Ship *Silvertown* laid 2,276 miles from San Francisco to Honolulu in 1902. Then the C. S. *Colonia* laid the 2,593-mile section from Guam to Midway Island; and C. S. *Anglia,* the sections between Manila and Guam, and Midway and Honolulu, 1,490 and 1,254 miles, a total of 7,614 miles.

On July 4, 1903, President Theodore Roosevelt sent the first message over the cable to Governor Taft at Manila. Roosevelt and Mackay, at Oyster Bay, Long Island, with the cooperation of other companies, then sent messages around the world to each other. Roosevelt's message required eleven minutes, and Mackay's nine.[30] That cable was extended to Shanghai in 1906 with a branch from Guam to Bonin Islands to connect with a Japanese Government cable to Tokyo. The Commercial Pacific system then extended 10,000 miles.

When Anglo-American's fifty-year monopoly in Newfoundland expired in 1904, Western Union, Commercial, and the D.U.S. lost no time in diverting the transatlantic cables from Newfoundland by laying connecting ones to the United States. In 1910 Western Union built a station at Bay Roberts, as Commercial did at St. Johns. Also in that year Western Union had the *Colonia* lay another transatlantic cable, from Hamel, Long Island to Bay Roberts and Penzance.

The Cable Pool Is Broken

"Trust busting" was a major policy of President Theodore Roosevelt, shown in cartoons by the press with his toothy grin and brandishing a "big stick." Criticism of the cable pool had grown since passage of the Sherman Antitrust Act in 1890, and the U.S. Attorney General, bowing to public sentiment, declared it was a "British Trust"

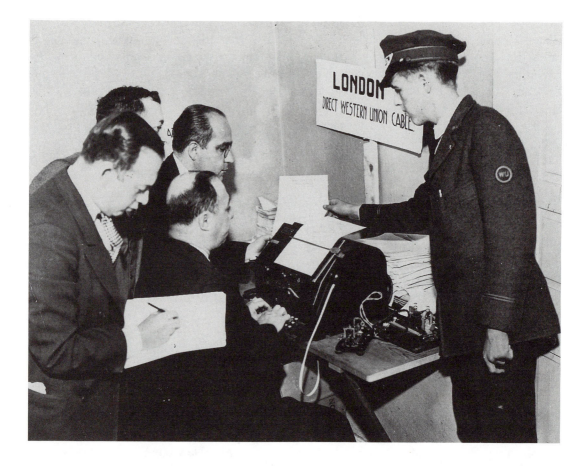

15.10 Messenger rushes cablegram reporting important news break to operator who transmits it to London. Newsmen stand around.

and notified Western Union in 1910 that it would have to withdraw.

When Western Union left the pool, Anglo-American and D.U.S. had no landlines to handle their business. Western Union at once laid a third transatlantic cable, sold it to Anglo-American in 1911, and leased Anglo-American's transatlantic system of five cables and the D.U.S. cable. With eight cables, Western Union was prepared to handle the heavy load of military, diplomatic, press and commercial messages during World War I.

Western Union's main cable office then occupied forty by 160 feet of the south end of the main floor and basement of the New York Stock Exchange. When it required more space in 1919, the company constructed its own cable building adjoining the south side of the Stock Exchange, at 38 and 40 Broad Street. It charged twenty-five cents a word for plain language messages between New York and London, but introduced two deferred classes of service at lower rates on December 6, 1911: night cable letters and weekend cable letters.

Notes

[1]Robert had succeeded his father, John W., as president of B&O in 1884.

[2]*The Operator* (May 1, 1883) 7.

[3]James D. Reid, *The Telegraph in America* (New York: John Polhemus, 1879, 1886) 772.

[4]Reid, *The Telegraph in America,* 773.

[5]George Dunlap Lyman, *Saga of the Comstock Lode: Boom Days in Virginia City* (New York: Scribner's, 1937).

[6]*The Operator* 17 (June 13, 1885) 95. Other issues of *The Operator, Electrical World,* and Postal's operating reports in 1883, 1884, and 1885 provided information on Postal's first years; *Telegraph Age* did on later years.

[7]*The Operator,* 15:107 and 152; 16:79.

[8]Testimony of Clarence H. Mackay, son of John W., before U.S. Senate Interstate Commerce Committee on January 10, 1921.

[9]Horta, a Portugese possession in the Azores, is 1,350 miles from Newfoundland, and 2,330 miles from New York.

[10]See chap. 9, above, for biographical notes on Eckert.

[11]*Telegraph Age* (March 16, 1894) 114-15; (June 1, 1894) 206-209, 212-13.

[12]Postal moved back to "253" in 1940 after being divorced by ITT.

[13]John W. Mackay died on July 15, 1902, and was succeeded by his son, Clarence W. Mackay.

[14]*Telegraph Age* (January 1, 1897) 6.

[15]Even with no competition or taxes, the British service lost an average of nearly three million dollars a year for many years.

[16]Wiman was referred to previously as a leader in developing the telegraph in Canada, and president of Great Northwestern.

[17]Harry M. Davis, "Under the City's Skin," *New York Times Magazine,* April 16, 1939.

[18]Most of the wires in New York City were moved into underground ducts of the Empire City Subway Company.

[19]By 1930 Western Union had 500 branch-office pneumatic tubes that totaled two million feet, and cost ten million dollars. In the 1930s teleprinters, connected by direct wire with main offices, began replacing branch office tubes, and reperforator switching made underground tubes obsolete in the late 1940s. Intrabuilding tubes, linking ground-level commercial offices with upstairs operating rooms, continued in use; a 500,000-foot interdepartment tube system was used in Western Union's building at 60 Hudson Street, New York City.

[20]The amount of open line wire they maintained was greatly reduced later by radio beams and cables.

[21]A. W. Rolker, "Heroes of the Telegraph Key," *Everybody's Magazine* (December 1909).

[22]Eliza Stone, heroine of the Chicago fire, died at 97 years of age on October 9, 1939, the anniversary of the fire. She stuck to her key in the Western Union office at La Salle and Washington Streets, risking her life to send out messages appealing for help.

[23]Arriving in Baltimore, Samuel R. Crowder, office layout engineer, broke into the saloon building, ran wires through a window, and set up operating equipment on "poker tables." He found the saloon owner, Martin J. Welsh, signed a lease, and paid the rent before divulging that he already had possession. Since the wires were not "good" to the south, Crowder set up a "jackass relay," with boys on mules carrying messages to the Camden station.

Crowder designed the layout and installed nearly every major office until he retired in 1940 after 56 years of service. At that time he told the above story to me. I visited the old saloon and restaurant, built in 1838. Many details were confirmed by Carroll Dulaney's column "Day by Day" in the February 8, 1937 *Baltimore News Post.*

[24]Later, as vice-president and secretary of Western Union, McKisick recognized the im-

portance of company history and told me several stories in the book, including this one.

[25]Austin, Texas *Tribune,* July 21, 1939.

[26]Private wire leasing had become an important Western Union service. In 1887 A.T.&T. leased a circuit between New York and Philadelphia to a broker that he could switch from voice to telegraph. It was the first alternate record-voice service.

[27]Albert Bigelow Paine, *In One Man's Life* (New York: Harper & Brothers, New York, 1921) 226.

[28]Many independent companies still refused to join. The United States Independent Telephone Association organized at Detroit on June 22, 1897, soon represented 2,300 companies.

[29]Horace Coon, "The Bankers Take Charge," in *American Tel & Tel* (New York/Toronto: Longmans, Green and Co., 1939).

[30]*Telegraph Age* (July 16, 1903) 360.

AT&T–Western Union
Merger-Divorce

The telegraph and telephone companies entered the twentieth century sadly lacking in leadership. Gone was the nation-building spirit and drive of the pioneers. The expense of plant expansion, hundreds of court battles to save the Bell patent, and competition with independent companies had forced AT&T to increase its bonded indebtedness and strain its finances dangerously.

At AT&T's bankers' suggestion, Vail was asked to return to the presidency. At first he refused, saying that at sixty-two he was too old, but he had just sold a South American transit development for $3,000,000, his wife and son had died, and he needed to keep busy. He was reelected president April 30, 1907, and his return after an absence of twenty years marked the real beginning of modern commercial communications. As an experienced organizer and business builder, Vail envisioned the industry's potential.

Vail extricated AT&T from immediate financial danger by selling 220,000 shares of stock at one hundred dollars a share, enabling it to survive the panic of 1907–1908. He reorganized the company and moved its headquarters to 15 Dey Street, across the street from Western Union. He made John J. Carty chief engineer, with the task of trimming sails to weather the panic. Carty cut and consolidated the scattered research, development and other engineering groups of Western Electric, American Bell, AT&T, and the operating companies. He brought these forces from various cities to Dey Street in 1907 as the Engineering Department of Western Electric.[1] Carty reduced construction costs from eighty million dollars in 1906 to about forty million dollars in 1907, and to $26,600,000 in 1908. In 1909, Vail established a uniform system of accounting for all Bell companies.

When an ice storm snapped all lines during President William Howard Taft's inauguration on March 4, 1909, Vail ordered underground cables between Washington and Philadelphia, and they were completed in

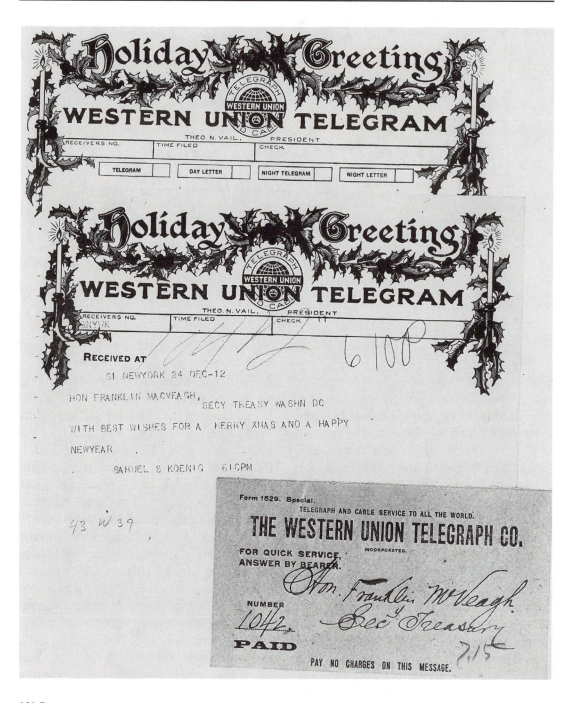

16.1 Colorful greeting telegrams, many painted by famous artists, were sent on holidays like Christmas and Easter, as well as on occasions such as marriages and anniversaries. Telegram blanks were originally white sheets with printed headings and later became bright items in family albums. The two top headings shown were when Western Union and AT&T were one company.

1913. Cables already linked Philadelphia with New York and New Haven, and Boston with Providence.

Vail's policies contrasted vividly with the jungle morality of Western Union in the Gould era and Eckert's incompetence. Building an outstanding public relations department, he cultivated the confidence of the public. A statement he issued in 1908 became the keynote of telephone policy. It was headed "One policy, one system, universal service."

Newspapermen liked Vail because of his availability, friendliness and frankness. He recognized the duty to keep the public informed and made telephone affairs an open book. "Take the public into your confidence and you win the confidence of the public," he said. His annual reports told what happened; if it was bad, they told what was being done to remedy it. His public relations policy also produced reform in the politeness of employees. In 1908 he hired the N. W. Ayer advertising agency to convince the public that the Bell monopoly was in the public interest.

Consolidation was offered to small telephone companies, and many joined "Ma Bell's" family. The 2,530,924 Bell telephones in 1905 grew rapidly to 5,882,719, including connecting telephones, in 1910.

Western Union had no Vail to modernize its operations and no public relations man to advise him. Eckert, Gould's henchman, had had one major purpose. That was to rob and ruin Western Union. He did little as president to repair the damage he and Gould had done, and left another Civil War telegrapher as his successor. Eckert became board chairman March 12, 1902; and Colonel Robert C. Clowry, president.[2]

Having directed the construction of many lines and offices in the West, Clowry

continued building them. He brought several of his Central Division men to New York and appointed C. H. Bristol, general superintendent of construction; George F. Swortfiger, widely known as "Mr. S.," Eastern Division construction superintendent; and J. C. Barclay, electrical engineer and assistant general manager. Clowry pushed the closing of contracts with railroads, and improved the quality of lines by adding copper instead of iron wire and increasing the number of poles per mile.

Western Union had a simple "one column" organization, with traffic, commercial, plant, and engineering reporting directly to its headquarters office, which J. C. Willever, then chief clerk, ran. Willever wrote his own letters and filed correspondence and reports by folding them, marking the subjects on their backs, and putting them in boxes stacked in an old wardrobe cabinet. J. B. van Every, the company auditor, placed reports in piles on the floor around his office, and no one else knew where to find anything.

In the 1900s little was done to sell the service; twenty to thirty million dollars a year was received only because the telegraph, symbolized by the yellow blank, was a business necessity. Clowry tried to destroy even that symbol by shifting to white blanks, but changed back when it aroused a storm of protest.

The Telegraph Messenger

Pickup and delivery was the weakest link in telegraph service. The last mile a telegram traveled, for delivery, took as long as its transmission from coast to coast. For example, in 1860, the Newark, New Jersey Western Union office, at the southwest corner of Broad and Market Streets, handled an average of 175 telegrams a day, with two operators, a day manager, a night manager, and four messengers riding ponies. By 1900, the Newark main office and twenty branch offices required forty adult employees, and about as many messengers, some using bicycles.

162 Turning the handle of a call box in most business offices for many years was the standard way to call a messenger to pick up a telegram or perform some errand. The desk fax replaced the call box and messengers for many companies.

The messengers were not telegraph company employees; deliveries were made by "district telegraph" companies, so called because the American District Telegraph Company (ADT) opened several offices in different districts of New York and Brooklyn in 1871, connected with call boxes in local offices and stores.[3]

By turning the handle of a "call box," the subscriber could call a messenger, a man with a fire extinguisher, or police. In ten years ADT had 5,000 subscribers and about 1,000 uniformed employees.

District telegraph companies were then organized in Chicago in 1872, followed by Philadelphia, Baltimore, and other cities. In 1877 there were twenty, and by 1900, messenger telegraph delivery had been established in about 500 cities. Using more elaborate call boxes, subscribers could turn the handle to different positions to signal for a messenger, firemen, police, doctor, laborer, expressman, coupe, or carriage. If it was fire, a messenger would run to the fire house. Each call box had a dial with a spring device that would send an identifying series of impulses which were recorded on tape in the central office. Impatient people, turning call box handles too vigorously, turned in so many false police and fire alarms, that it became general practice to send a messenger first.

After 1890, telephone use for non-telegraph messenger services caused the ADT companies gradually to set up wire systems for fire alarm, night watchmen, and burglar alarm services. By 1900 there were about fifty cities with protective signal services by Holmes Electric Protective Company.

Western Union officials organized the American District Telegraph Company of New Jersey in 1902 with Clowry as president. It acquired the capital stock of fifty-seven ADT companies, and contracted to collect and deliver Western Union telegrams for 25 years.[4]

Many former telegraph messengers have heard of James F. Smith, who delivered a telegram to President Paul Kruger of the Transvaal Republic, South Africa, in 1900. The telegram, signed by 29,000 Philadelphia schoolboys, expressed sympathy with the Republic's valiant defense of its independence from Great Britain. After delivering the message with the aid of Richard Harding Davis, a famous war correspondent, Jimmy was welcomed back to Philadelphia as a hero.

Western Union bought control of ADT

and took over the pick up and delivery of telegrams January 1, 1911,[5] with polite boys in pressed uniforms. An esprit de corps developed, with rivalry between boys to "find their man" on difficult deliveries.

16.3 WESTERN UNION MESSENGER.

Before unions became powerful, messengers were about 25,000 uniformed bright high school boys dividing their time between classes and work. In New York, Western Union operated a school for them. A majority were hired after a few months by the business people to whom they delivered telegrams. Telegraph officials encouraged that rapid turnover because those boys were future business users of telegrams. Some became famous actors, singers, and even presidents of corporations, including U.S. Steel.[6]

The uniformed messenger, rushing with messages on bicycle or on foot, was for generations a familiar American sight. Later, because of union pressure, he was not permitted to wear a uniform, and adults at union wages took his place, forcing the company to adopt electrical tieline instead of messenger pick up and delivery for most telegrams.

Western Union vs. Pennsylvania Railroad

Western Union refused to accept changes the Pennsylvania Railroad proposed when their twenty-year contract expired in 1902. Postal had offered $100,000 in free messages, and $70,000 a year for a contract. Also George J. Gould and associates, who controlled Western Union, had extended the Wabash Railroad into Pittsburgh, and planned to build east to the Atlantic through the heart of Pennsylvania's territory. Hot-tempered Alexander J. Cassatt, the railroad's president, ordered the telegraph company to remove its lines from the 4,745-mile right of way. Western Union refused.[7]

In May, 1903, the Circuit Court of Appeals at Pittsburgh ruled that the railroad had the right to remove the lines, and it set 1,000 men to work, chopping down telegraph poles. Newspapers printed graphic accounts of the destruction. Fortunately, Western Union had two old lines along highways between Philadelphia and Pittsburgh, and shifted traffic to them, while 150 line gangs improved them.

Conferences between Western Union general superintendent Belvidere Brooks and the Pennsylvania, resulted in the telegraph men removing remaining wires with the railroad providing locomotives and cars for the work. The railroad and Postal then rebuilt the lines. The unpleasantness between Western Union and the railroad vanished later when they agreed on another contract, and President Martin W. Clement of the Pennsylvania joined Western Union's board.

Clarence H. Mackay was elected president of Commercial Cable and Postal on October 14, 1902, when he was only twenty-eight, but he relied on the advice of Chairman Chandler of Postal, and George G. Ward, general manager of Commercial.[8]

Steady growth of the Mackay companies was shown by their annual reports. The 1903 report showed net income of $2,514,193 and a reserve fund of nearly $5,000,000. In 1906, $40,000,000 each of preferred and common stock was issued, and the North American Telegraph Company, which operated a telegraph system in Illinois, Wisconsin, Iowa, and Minnesota was acquired. In 1907 Mackay income had grown to $3,310,327.

In 1910 Western Union reported $33,889,202 in receipts, $7,274,900 net, and 214,360 miles of poles and cables, 1,429,049 miles of wire, 24,825 offices, and 75,135,405 messages, not including those on leased wires or railroad contracts.

The Strike of 1907

Western Union's employees were so loyal that its only labor troubles were the minor ones in 1870 and 1883. When Eugene V. Debs led the Pullman strike in Chicago in 1894, Western Union workers watched the rioting and destruction of railroad property from their ninth floor operating room at Jackson and Clark Streets. They saw numerous fires springing up in the railroad yards and knew the Governor was taking no action. J. C. Barclay got the White House operator on the wire and had him call in President Cleveland. The next morning Federal troops were on the scene.

The Commercial Telegrapher's Union, affiliated with the American Federation of Labor, was formed in 1902. After a boycott against the Associated Press in 1906, the union attempted to gain recognition and a closed shop at Western Union and Postal in 1907, but without success.

When a Western Union operator at Los Angeles was discharged on July 25, 1907, for "maliciously delaying traffic on important circuits," the superintendent refused a demand that he be reinstated. On August 8, Chicago operators refused to work circuits with nonunion men who replaced the strikers at Los Angeles. Many operators at fifteen other cities stopped work, but supervisory employees filled the vacant positions until operators could be hired. The union "suspended" the strike in November, but all vacant jobs had been filled.

AT&T and Western Union "Merge"

In the list of stocks Western Union owned in the 1890s were the names of many telephone companies. The company was larger and better known than AT&T. By continuing its purchases, it could have gained control of the telephone company in the small building across Dey Street, and become the Western Union Telegraph and Telephone Company. However, Eckert did not favor growth. He "milked" the company for profits, while its offices became shabby and uninviting, and his policies bred poor service and public disfavor.

Western Union disposed of most of its telephone holdings in the 1900s, and in September 1909 sold the remainder to AT&T. When the telephone company planned to finance acquisition of telephone properties by selling stock to its shareowners, Clowry refused to buy saying, "It was thought inadvisable to undertake to raise the large sum required for Western Union's proportion of the new issue."

Led by Vail, AT&T in 1909 was expanding with strong financial and popular support. Its founders had provided for telegraph as well as telephone operation in its name and charter. After the panic and depression of 1907–1908, Western Union's finances were in such poor condition that Vail decided to act.

AT&T used its net earnings of $30,190,000 in 1909 to acquire working control of Western Union, which had assets of nearly $500 million. George J. Gould sold his thirty percent holding of Western Union's stock to AT&T for $85 a share.

A contract was signed on December 15, 1909, providing for joint use of the plant and operating facilities of the two companies. At Western Union's annual meeting, Vail was elected president and Edward J. Hall, vice president of AT&T, chairman of the executive committee. Vail became the first man to head a company with capital of nearly a billion dollars.

Vail's 1909 annual report said telephone and telegraph services were complementary or supplemental, and not competitive. Telegraph service is not complete with collection and delivery of messages by telephone, he said, and the telephone was too expensive for correspondence that could be handled at the lower telegraph rates. He pointed out that great savings could be made in line construction and maintenance because the same right of way, poles, and wires could be used for telephone and telegraph.

Vail also pointed out that few of the 22,000 Western Union offices had enough telegraph business to warrant night service, but telephone subscribers could call night and day. He introduced the Night Letter on March 7, 1910, and the Day Letter a year later, inviting people to send fifty words or more at moderate cost. Business grew.

Vail revolutioned Western Union's public relations. One newspaperman wrote that

> When American Telephone took hold, telegraph people had the idea that the company was doing the public a favor in providing its facilities, but Western Union now sought to make every employee from the lowest to the highest feel, that in accepting a message, he wasn't doing somebody a favor, but was selling a commodity.[9]

Vail did not cover up the fact that he had had an appraisal made of Western Union properties which resulted in that company's surplus being slashed nearly $14,000,000.

Greater personal comfort was provided for employees, and cordial relations between the workers was encouraged by holding conferences in the six geographical divisions. A three-column organization—traffic, plant, and commercial—was established in 1911. The six divisions were headed by general superintendents, later called general managers, and finally division vice presidents. Each division had a number of districts headed by superintendents, and offices by managers.

Vail relieved Clowry of most of his duties, and appointed Belvidere Brooks general manager. A native of Wheelock, Texas, Brooks had fought his way up from messenger to Eastern Division general superintendent.

John Calvin Willever remained chief clerk at headquarters. He had learned Morse operation as a boy when his father was railroad station agent at Port Murray, New Jersey, and he entered Western Union's service at Asbury Park in 1880. He was secretary and executive clerk to Presidents Eckert and Clowry.

After the merger, Willever headed the company's cable business, which grew rapidly, and added the entire commercial organization in 1914. For the next four decades he was first vice president, a ruler whose instant snap judgment was law, and a relic of a long-gone era who shouted at subordinates instead of listening to reason.[10]

The years 1909 and 1910 marked the beginning of vigorous promotion in the telegraph business. As president of AT&T and Western Union, Vail established progressive policies and appointed experienced men to carry them out.

Telephone subscribers could dictate telegrams to the nearest office and charge them on their telephone bills. The public realized a revolution had taken place in telegraph service. Postal charged that when a caller wished to dictate a telegram, the telephone company would connect him with Western Union. It obtained injunctions in several states restraining the telephone company from favoritism, but it only served to promote calling Western Union.

In 1910 a Texas manager suggested the slogan "The Western Union Goes Everywhere," used later as "Western Union—Everywhere." Another well-known slogan, "Don't Write—Telegraph," was proposed in 1918 by Superintendent Charles A. Crane of Boston. A song with that title had Crane's photograph on the cover. "Don't Write" was dropped, however, after years of protest by paper companies.

Newcomb Carlton

Vail met Newcomb Carlton[11] on an ocean voyage in 1910, and selected him to reform Western Union. Carlton improved the appearance of telegraph offices with cleaning, painting, new floor coverings, plate glass windows, shiny new counters, desks, and chairs. Salaries were raised fifty percent in three years, and a pension and sickness benefit system was established for both companies—one of the earliest company-paid plans in industry. Better educated and paid employees were hired, and they developed greater company loyalty, meeting the public with a smile.

In 1913 the company spent $1,300,000 improving its headquarters building at 195 Broadway.[12] AT&T moved its headquarters into the Western Union building in 1914. Western Union made room by moving its main operating room to 32 Sixth Avenue, at Walker Street.

The new Western Union policy was to spend all net profits above a three percent dividend for improvements; in 1911 that was $7,105,000. Vail's 1913 report contrasted his three years of Western Union leadership with the preceding three. Gross cable and landline revenues of $125,190,000 were up forty-five percent; construction, not including real estate, up ten and one-half percent; and $22,624,000, set aside for renovation, improvement, and reconstruction, up fifty-seven percent. In three more years, Carlton said, Western Union would be in good condition.

Some Postal men joined Western Union, including F. E. d'Humy, later engineering vice president; T. W. Carroll, Eastern Division general manager; S. B. Haig, Traffic Department; H. C. Worthen, Southern Division general superintendent; and A. C. Cronkhite, Lake Division general manager who later became Commercial vice president.

Men obtained from the telephone company were as follows: G. M. Yorke, engineering vice president; R. E. Chetwood, engineer of construction; J. C. Hubbard, general supervisor of lines; M. C. Allen, superintendent Eastern Division plant; R. J. Meigs, valuation engineer; his brother J. H. Meigs, Plant Department; and J. C. Barclay, assistant general manager.

Carlton, an engineer, placed major emphasis on technical efficiency and speed of service. The time for telegrams and cablegrams to go from sender to addressee was no longer hours, but minutes. To the other improvements was added Vail's campaign to educate the public to use the telegraph to speed business and personal correspondence. Receipts from 1908 to 1913 were listed as follows: $28,582,000; $30,541,000; $33,889,000; $35,478,000; $41,661,000; and $45,783,000. Both companies were just attaining the ability to be a greater asset to the nation when political tinkering achieved a national loss.

The AT&T–Western Union Divorce

The cry of "communications monopoly" arose, and newspaper headlines forecast government action. Some who kept up the cry, like Mackay, had axes to grind or hoped for gain from government ownership. Under the Sherman Antitrust Act, the Standard Oil Company of New Jersey was forced in 1911 to give up its subsidiaries. Two years later the Department of Justice demanded that AT&T and Western Union be divorced. It also accused Northwestern Bell of violating the Antitrust Act by not connecting with independent companies.

AT&T had to yield to pressure and dispose of its $30 million in Western Union stock. N. C. Kingsbury, vice president of AT&T, wrote to Attorney General J. C. McReynolds on December 19, 1913:

> Wishing to put their affairs beyond fair criticism, and in compliance with your suggestions formulated as a result of a number of interviews between us during the last sixty days, the American Telephone and Telegraph Company, and the other companies in what is known as the Bell System, have determined upon the following course of action:
>
> First—The American Telephone and Telegraph Company will dispose promptly of its entire holdings of stock of the Western Union Telegraph Company in such way that the control and management of the latter will be entirely independent of the former and of any other company in the Bell System.
>
> Second—Neither the American Telephone and Telegraph Company nor any other company in the Bell System will hereafter acquire . . . control over any other telephone company owning, controlling or operating any exchange or line which is or may be operated in competition with any exchange or line included in the Bell System. . . .
>
> Arrangements will be made promptly under which all other telephone companies may secure for their subscribers toll service over the lines of the companies in the Bell System.

President Woodrow Wilson expressed gratification that the telephone company "should thus volunteer to adjust its business to the conditions of competition." Allowing independents to connect with Bell lines did open long distance service to millions.

Vail issued a statement that no real merger had ever taken place, that the companies had merely developed mutual relations of a complementary character, with each using the other's facilities, and now each would go its own way, with its organization which had never ceased to operate separately.

J. P. Morgan resigned from the Western Union board, and Kuhn, Loeb and Company underwrote the sale of the thirty million of

Western Union stock in the telephone company treasury. On April 5, 1914, Newcomb Carlton was elected president of Western Union and continued his program of modernization. Friendly relations between the two companies continued.

A fashion plate of sartorial elegance, with a derby cocked on his bald head, Carlton delighted in his collection of ship models in Western Union's executive offices, and answered questions from the press with short quotable sentences. For example:

> It's the breaks. Success depends on which side of the street you are walking at a certain minute of a certain day. There are a dozen men in our plant here in Kansas City who could fill my job as well as I can.

He never tired of visiting telegraph offices, shaking hands and asking about the comfort and welfare of employees. It mattered not to Mary Smith, operator, that the manager had briefed Carlton on names before entering the room; she would tell for years how President Carlton had held her hand and exclaimed: "Hello, Mary, it's so good to see you again!"

Carlton's replies to memos were a scrawled "No. N.C.," or "Yes" in colored crayon on a maroon sticker he pasted on correspondence. Some thought his quick decisions sometimes had bad results, but to most employees he was perfect. Did he not visit operating rooms on hot days to see that there were electric fans, and have the windows opened? For a generation, unions could not get a foot in Western Union's door. His Association of Western Union Employees "won" liberal employee benefits and a profit-sharing plan that distributed millions. Carlton "refused to have the operators unionized even when the demand was made by President Woodrow Wilson."[13]

16.4 NEWCOMB CARLTON, WESTERN UNION PRESIDENT WHO MODERNIZED AND MECHANIZED TELEGRAPH OPERATIONS AND DEFIED THE GOVERNMENT'S EFFORTS TO BLOCK THE COMPANY FROM LAYING THE FIRST AMERICAN CABLE CONNECTION WITH SOUTH AMERICA.

Under telephone control, Western Union's message business had increased seventy-six percent, while its leased private wire revenue, which AT&T wanted, decreased seventy-one percent. Western Union had seventy percent of the private wire business before merger, but only forty percent after it. Having gone into the private wire telegraph business in 1889 in violation of its contract, the Bell System now had sixty percent.

AT&T's annual report for 1917 reiterated its advocacy of a communications merger:

> A combination of like utilities under proper control and regulation, the service to the

public would be better, more progressive, efficient, and economical than competitive service given by competitive systems. . . . No properly regulated charge for utility service could be high enough to cover the cost of duplication and the increased cost incident to competition.

The Miami Incident

Carlton was not only friendly, likeable, and humorous, but he had the nerve to fight for his company's rights. Opening the Pan-American Financial Conference in 1916, President Woodrow Wilson deplored "the physical lack of means of communications" to South America which were "absolutely necessary, if we are to have true commercial relations." William G. McAdoo, Secretary of the Treasury, urged "direct cable communications," pointing to "the present roundabout and unsatisfactory facilities and the expensive cost of cable communications." The Secretary of Commerce echoed that, and an International High Commission meeting at Buenos Aires resolved that the existing companies be induced, if possible, "to extend their lines so as to render service to all of South America." McAdoo then urged Western Union to increase cable facilities to South America.

Western Union estimated a cost of $15,000,000 and a delay of years to provide a 7,500-mile cable to Buenos Aires and Rio de Janeiro. The best way to meet the government's request, it decided, was to lay a cable from Miami to Barbados, and connect there with a Western Telegraph cable to Brazil, a British company with a sixty-year monopoly from Brazil. The two companies signed a contract July 15, 1919, providing that All-America, with cables to the West Coast of South America, be invited to join in the

arrangement. All-America refused and protested to the United States Government.

To aid the direct service project, Brazil imposed a tax of three cents a word on cablegrams via the All-America Pacific Coast route. Western Union obtained a permit on May 12, 1920, from the United States Engineer's Office of the War Department for the cable to pass under the Biscayne Bay drawbridge in Florida, but on June 1 a permit for laying the shore end had not arrived. Carlton went to Secretary of State Bainbridge Colby, reminded him of the company's application, and was given the impression that it was being processed in a friendly spirit. No American company had ever been denied a landing permit. In July 1920 the cable was loaded aboard the cable ship *Colonia,* which left the short Miami shore end at Norfolk Virginia, for customs declaration. Then the ship went to a point in the Atlantic several miles east of Miami to start laying the deep sea section to Barbados. There it was halted on August 6 by United States warships.

At Colby's request President Wilson had secretly ordered the Navy, Army, and Department of Justice to prevent a cable route to South America that his administration had requested. A policy, originally stated by President Grant in 1875, was to refuse a landing permit to any foreign cable company that had a landing monopoly in any other country. However, this was refusing a landing to an American company for a cable to a British island with no landing monopoly. Also, All-America had monopolistic landing privileges in several Latin American countries.

The cable ship was halted by two destroyers and a submarine chaser, and boarded by Rear Admiral E. A. Anderson, who told Captain Campos he could not lay the cable. Acting on company orders, the *Colo-*

nia proceeded to lay the cable from a buoy outside the three-mile limit to Barbados. Colby demanded that Carlton order the *Colonia* back to Newport News, but was refused.

Admiral Anderson then halted the laying of the shore end under the Biscayne Bay drawbridge, and Western Union applied to the U.S. District Court, District of Columbia, for an injunction to restrain the Navy from interfering. The Government countered by applying for an injunction to restrain the company from laying the cable. Judge Augustus N. Hand (272 Federal 311) decided President Wilson had no constitutional power to prohibit the landing by Western Union, which had operated cables between Florida and Cuba for many years. The government appealed, but the U.S. Circuit Court of Appeals sustained Judge Hand's decision.[14]

Following the court decisions, Western Union cable ship *Robert C. Clowry*[15] went to hunt the cable outside the three-mile limit. As it began work on March 3, 1921, it was halted by shots from the submarine chaser on guard. That resulted in newspaper headlines.

The government then appealed to the Supreme Court, but feared it would sustain the lower courts. Since that would deprive the president of the power he already had exercised, Wilson appealed to Congress for help, and Senator Kellogg introduced an act to give the president that power.

The Submarine Cable Act was passed by Congress on May 27, 1921. It was another year before Western Union could persuade Brazil to grant All-America a license, and thus mollify Wilson and Colby. A license for the cable to Barbados was issued August 24, 1922, more than two years after it was laid.

The First Transcontinental Telephone

The first conversation between New York and Denver was on May 8, 1911, over a line of hard-drawn copper wire about as large as a pencil, with Pupin's loading coils every 7.78 miles. The coils, applied to practical service by George A. Campbell, enabled electric impulses to travel farther without fading.[16]

Then Lee De Forest invented the three-element vacuum tube, a triode, to repeat and amplify the voice at intervals on long lines. The vacuum tube amplifiers to which De Forest, Edison, Fleming, and Richardson contributed, were then placed at repeater stations along the line to restore the fading electrical waves, so that normal tones would reach their destination.

In building the line, Salt Lake City was reached in 1913. In the thirteen states crossed by the line, 130,000 poles were required to carry the wires, weighing 435 pounds a mile, a total of 5,920,000 pounds or 2,960 tons. The east and west lines were joined at Wendover, Utah, at the Nevada line.

World War I cannon had been booming in Europe for six months before New York said "Hello San Francisco." After the line was built, and the first trial conversation held July 29, 1914, it could not be used until that winter because of technical problems.

When the transcontinental telephone was ready January 25, 1915, notables gathered with Bell at 15 Dey Street, then New York Telephone headquarters. The first official conversation over the 3,400-mile line was with Watson in San Francisco.

"Hoy! Hoy! Mr. Watson! Are you there? Do you hear me well?"

"Yes, Dr. Bell, I hear you perfectly. Do you hear me well?"

"Yes, your voice is perfectly distinct. It is as clear as if you were here in New York."

Bell then repeated his first words by the telephone: "Mr. Watson, come here. I want you!"

He would be glad to come, Watson replied, but now it would take a week.

Vail, at Jekyll Island, Georgia, talked with Mayor Rolph of San Francisco. From the White House President Wilson spoke with the head of the Panama-Pacific Exposition and members of the National Geographic Society at San Francisco.

When service to the public began February 11, the Liberty Bell in Philadelphia was tapped several times, and the sound was heard in San Francisco. A transcontinental call cost $20.70. It required about twenty-three minutes to make the connection because operators had to call ahead to the switching stations at Chicago, Omaha, Denver, and Salt Lake City. It was like two persons shouting at each other across an open field. Before that, long distance operators had to relay conversations; repeating each sentence to the next station.

While private enterprise in the United States made tremendous progress in spite of government interference, government operation almost ruined the industry abroad. In England, the Telegraph Act of 1869 enabled the Post Office to take over the telegraph. On December 31, 1911, the British Post Office also acquired the telephone business. In France, Switzerland, Belgium, and other European countries, governments took over the telephone business in the 1880s and 1890s.

Vail resigned because of ill health in June 1919, and died April 16, 1920. Bell died August 2, 1922. Vail was succeeded as

president by his close associate Harry B. Thayer, president of Western Electric since 1908.[17] A second transcontinental line was completed in 1923, following a southerly route through El Paso, Texas to Los Angeles. A third, in 1927, was through Chicago, Minneapolis, Fargo, Bismarck, and Seattle.

16.5 THE *GENIUS OF ELECTRICITY* (LATER RENAMED THE *SPIRIT OF COMMUNICATIONS*) WAS A FAMILIAR FIGURE ON THE NEW YORK SKYLINE. (SEE ALSO FIGURE 13.9 [P. 207] AND FRONT COVER.) (COURTESY OF AT&T ARCHIVES.)

On January 7, 1929, Walter S. Gifford,[18] who became president of AT&T January 20, 1925, opened radiotelephone service between New York and London. The first direct call around the world, April 25, 1935, was from Gifford to Theodore G. Miller, Long Lines vice president, in another room of the Long Lines Building.

AT&T was Vail's monument; he was the leader most responsible for transforming the struggling little company in Boston into the great Bell Telephone System.

Western Union invited sculptor Evelyn Beatrice Longman to design a statue symbolizing the *Genius of Electricity* to top the Fulton Street tower on its 195 Broadway headquarters, built in 1875. The result was a twenty-four-foot-high bronze statue of a man with a twelve-foot wing span, standing on a large globe. One arm held cables; the other hand, held high, grasped darting lightning bolts, representing electricity that powers telecommunications. The statue was erected on October 24, 1916, the fifty-fifth anniversary of the first transcontenental telegraph line.

In 1930, Western Union sold its interest in the building to AT&T, which rechristened the statue the *Spirit of Communications,* gilded it with more than 12,500 pieces of gold leaf, and called it "Golden Boy." In 1980 AT&T moved the statue to its new headquarters at 550 Madison Avenue. For sixty-four years at 195 Broadway the statue was the second largest sculpture in New York, higher than the 151-foot Statue of Liberty, and a familiar sight to millions.

Notes

[1] Seventeen years later that group became the Bell Telephone Laboratories, with Dr. Frank B. Jewett as president. AT&T's Research and Development Department was consolidated with Bell Laboratories in 1934, with D. E. H. Colpitts as vice president. This became the world's largest engineering organization.

[2] Clowry started work as a messenger at Joliet, Illinois in 1852, became an operator, and was superintendent of the military telegraph in the southwest during the Civil War. In 1881 he became general manager at Chicago of Western Union's Central Division, which then extended to the Pacific Coast.

[3] In a privately published book, *A Wonderful Fifty Years* (1917), E. T. Holmes, president of the Holmes Electric Protective Association, said his father Edwin H. Holmes came to New York in 1858 with an electric burglar alarm, which was installed in a number of stores, restaurants, and homes. Charles F. Wood, Western Union Boston manager, told him ADT was being organized to place call boxes throughout the city, "which on the turning of a little crank will register your number in a central office where they propose to have many boys in uniform, one of whom will immediately proceed to your house, office, or store, ready to do any errand or carry any package."

From Western Union records:

The American District Telegraph Company—incorporated in New York, October 4, 1871. . . . On January 12, 1892, the American District Telegraph Company (of N.Y.) acquired the capital stock of the Mutual District Messenger Company, Limited, and the Mutual District Telegraph Company.

The principal business of the American District Telegraph Company (of N.Y.) is the collection and delivery of telegrams . . . on the island of Manhattan under agreement first entered into August 31, 1880.

[4] From ADT company records.

[5] ADT continued its protective services to stores and plants with 113 offices operated jointly with the telegraph company. (About 40 years later, Western Union sold its ADT stock at a profit of several million dollars, after collecting

millions in dividends.)

[6]One messenger named Steve Brodie became famous when he took a chance and jumped from the Brooklyn Bridge on July 23, 1866 to win a bet.

[7]Richard O'Connor, *Gould's Millions* (Garden City NY: Doubleday, 1962) 251.

[8]Ward had been superintendent of the Direct United States Cable Company from 1875 to 1883. He supervised the laying of the first Commercial cable in 1884 and built up its business.

[9]H. A. Bullock, *Boston Evening Transcript,* April 15, 1911.

[10]In 1929, President Walter S. Gifford of AT&T, said: "The pretentious conferences, the domineering executive, the stuffed shirt, are all going or gone. We really hardly give orders. We suggest certain courses of action, and the man who carries them out does his work better because he contributes something to the decision."

A 1936 magazine advertisement said Willever's name was unknown, though it had been printed more than six billion times on telegraph blanks, because it is not repetition that counts, but how a thing is said.

In 1935 Western Union President R. B. White told *Fortune Magazine* (November issue, p. 91) Western Union really was the "J. C. Willever Telegraph Company." A man of prodigious knowledge of the industry, Willever testified for hours without notes before the F.C.C. and Congressional committees. At 75, he described conferences between officials as far back as the early 1880s. He retired at 77, after 62½ years of service.

[11]Born at Elizabeth, New Jersey, February 19, 1869, Carlton was educated at Pingry School and Stevens Institute, and practiced mechanical engineering at Buffalo. He was vice president of the Bell Telephone Company of Buffalo from 1902 to 1904, following a period as director of works of the Pan-American Exposition. Joining Westinghouse in 1904 as a vice president, he was managing director of British Westinghouse in London from 1905 to 1910.

[12]After 38 years in the Western Union Building, the Associated Press needed more space and had moved to nearby 51 Chambers Street in 1913. It moved to 383 Madison Avenue ten years later, and to its headquarters at 50 Rockefeller Plaza in 1938.

[13]*Saturday Evening Post* (July 26, 1930) 117.

[14]Western Union's Annual Report for 1920, pp. 17-29.

[15]The *Robert C. Clowry* was sold in later years and led the U.S. prohibition agents a merry chase in its new role of rum runner.

[16]In another of the many simultaneous inventions, Campbell of Bell Telephone Laboratories arrived at the theory of the loading coil at about the same time Pupin did. The loading coil does not strengthen the impulses, as a repeater does, but reduces their tendency to grow weaker. Both applied for patents, but Pupin established a slightly earlier date of conception, and the Bell System bought his patent. Campbell did much to make the coils a success, formulating the rule for their design and spacing. In earlier years induction had raised havoc. To get rid of it, it was harnessed and put back on the line. Campbell also made major contributions in remedying "crosstalk," in telephone and telegraph carrier systems.

[17]H. B. Thayer was president of AT&T from 1919 to 1925, and chairman until 1928. He bagan work as a Western Electric shipping clerk in 1881.

[18]Born at Salem, Massachusetts on January 16, 1885, Gifford wrote to G.E. for a job, before graduating from Harvard. His letter was misaddressed and reached Western Electric, which hired him as a payroll clerk at Chicago. By 1908 Vail recognized his special abilities and had him examine competing companies and advise whether they should be purchased. He became controller in 1918, finance vice president in 1919, and executive vice president in 1923. To finance postwar expansion, he made every telephone office an agency to sell AT&T stock.

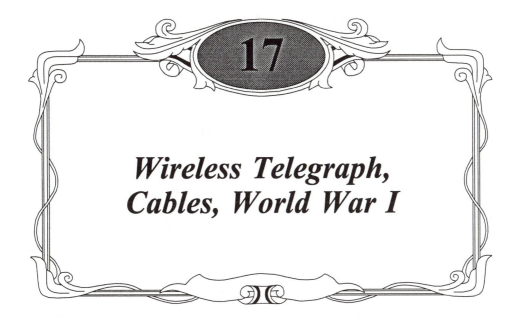

Wireless Telegraph, Cables, World War I

Someone wrote that nature has its own radio systems. A female moth, in a scent- and soundproof box, can call a male a mile away, and he will fly directly to her. Some such natural observation may have inspired Heinrich Rudolph Hertz, when he created radio waves in his laboratory at Bonn on the Rhine River in 1887. To receive the waves, he bent a wire almost into a circle, with a brass ball at each end of the wire. It looked somewhat like the antenna of a moth; in fact, he called it "the moth."

Hertz measured the electromagnetic waves which pierce mountains, walls, fog, and the darkness of night. James Clerk Maxwell, a Scotch genius, had demonstrated mathematically in 1865 that there are oscillating electromagnetic waves that travel through space at a speed equal to light waves.[1]

Joseph Henry had invented the essentials of telegraphy and radio long before that, and Mahon Loomis, a Washington, D.C. dentist, was on the verge of inventing a practical radiotelegraph when he staged a public exhibition in 1866, sending signals between two Blue Ridge Mountains, eighteen miles apart. His aerials (antennas) were kites with copper wires dangling from them.

Loomis obtained financial backers to develop his "aerial telegraphy," but Black Friday on September 25, 1869, ruined them. He went back to his dental office, and accumulated funds to go to Chicago and resume his radio work with new financial aid, but this time Mrs. O'Leary's cow kicked the lantern over, and his backers lost all in the Chicago fire in October 1871. However, he filed for a radio patent on July 30, 1872. Congress passed a bill creating the Loomis Telegraph Company, and President Grant signed it. Congress, however, forgot to appropriate $50,000 for the experiments. Ridiculed by the press and members of Congress, Loomis died brokenhearted in 1886. Writing to his brother, he summed up the tragedy of his life:

The time will come when this discovery will

be regarded as of more consequence to mankind than Columbus's discovery of a new world. I have not only discovered a new world, but the means of invading it. . . . My compensation is poverty, contempt, neglect, forgetfulness.

A telephone inventor, Professor Dolbear of Tufts College, obtained a patent in March 1882 for a radiotelegraph system, and Edison applied May 23, 1885, for one which was issued December 29, 1891. Edison sold his patent to Marconi in 1903. During his electric light experiments around 1880, Edison noticed a thin clear line in the soot deposited on the bulb, and later found that by inserting a straight wire into the lamp, he could create a current without a physical connection. That became known as the "Edison effect."

Guglielmo Marconi combined earlier inventions and produced the first practical wireless telegraph to achieve wide use.[2] He had a wealthy father and a mother who encouraged him to study physics.[3] At twenty-one he experimented with a Hertz spark transmitter and a "coherer" wave detector that Edouard Branly of France invented in 1891. With a spark induction coil and a telegraph key, Marconi sent dots and dashes across a room.

Alexander Stepanovitch Popoff reported that he also used Branly's "coherer" to record electrical disturbances in 1895. Repeating Hertz's experiments, he sent the Morse code 600 yards by wireless before a scientific group in 1896. The Soviet Union honored Popoff with an annual celebration on May 7 as the inventor of radio.

Oliver Joseph Lodge of England improved Marconi's system so that a transmitter and receiver could be tuned to the same wave length, and patented an adjustable inductance coil.

Italy was so skeptical of Marconi's

claims that he went to London in 1896. He so impressed Sir William Henry Preece, Chief Engineer of the British Post Office, that government aid was provided. In 1897 he sent signals several miles using an aerial on a kite. Asked how far a message might some day be sent through space, Marconi answered, "Twenty miles."

The Wireless Telegraph and Signal Company, Limited was organized on July 20, 1897, with 100,000 pounds capital. With the aid of leading British electricians, Marconi sent signals across the English Channel, and in October 1899, reported the International Yacht Races off Sandy Hook, New Jersey by wireless to the *New York Herald.* On November 22, 1899, Marconi Wireless and Telegraph Company of America, owned by London financiers, was incorporated.

The Marconi International Marine Communications Company was organized in 1900, and the U.S. Navy installed Marconi equipment on the battleship *Massachusetts,* the cruiser *New York,* the torpedo boat *Porter,* S.S. *Philadelphia,* and at the shore station at Babylon, Long Island.

Fifty years after Joseph Henry sent waves oscillating through space, Marconi, then twenty-seven, had signals transmitted from the southwestern tip of England on December 12, 1901. The man sending was John Ambrose Fleming, scientist and electrical engineering professor of University College in London, who had improved on the "Edison effect." At the same time George S. Kemp and others flew a kite to which an aerial wire was attached at Signal Hill, St. Johns, Newfoundland. The Morse code—three dots for the letter "s"—was received. Marconi verified the success by cable, and a story "Wireless Spans the Ocean" appeared in the *New York Times* three days later.

An American Marconi station was opened on Cape Cod on January 19, 1903, with an exchange of greetings between President Theodore Roosevelt and King Edward VII.

Marconi attempted to establish a British wireless monopoly. When American companies and the German firm Telefunken entered the business in 1903, he ordered Marconi International operators not to communicate with ships or stations equipped by other manufacturers. He was willing to lease his British wireless equipment to the U.S. Navy, but demanded a long-term contract, royalties, and relaying of messages only from Marconi-equipped ships. The Navy indignantly refused, bought a German system—the Slaby-Arco—and in 1904 equipped twenty-four vessels and twenty coastal stations with it. However, Marconi equipment was in most merchant ships, and by 1909, in United States coastal stations.

A nine-nation conference in Berlin in 1903 agreed that any shore station would handle telegrams to or from ships at sea, regardless of who made the equipment. England and Italy sided with Marconi and refused to sign. The conference adopted "SOS" as the standard distress call. At the next conference in 1906, the Bureau of the International Telegraph Union adopted worldwide frequency allocations and regulations.

Disturbed by the conglomeration of waves of wireless companies, President Theodore Roosevelt appointed a commission in 1904 to investigate. It recommended that the Department of Commerce and Labor control commercial wireless; that the Army oversee all interior government stations; and that the Navy control the link to all possessions with its own wireless system.

The naval battle of Port Arthur in the Russo-Japanese War was reported by wireless in 1905. Robert E. Peary radiotelegraphed, "I found the North Pole," in 1909. When the S.S. *Titanic* sank in 1912, the *Carpathia* heard its SOS and rescued 710 persons. Wireless ended the great silent communication void of the sea.

Radiotelegraphy is the simple making and breaking of current for short and long periods, but radiotelephony uses continuous waves of radio frequency modulated by voice.

Reginald Fessenden established a telephone transmitter at Brant Rock, Massachusetts for his National Electric Signalling Company, and his voice and music were heard by radio operators of ships at sea on Christmas Eve 1906. Later Fessenden produced a high-frequency continuous-wave generator in cooperation with General Electric.

Lee De Forest's vacuum tube, already used in telephone long lines, replaced Fessenden's equipment. De Forest organized companies and equipped ships and coastal stations, but his greatest contribution was to put a metal grid between the filament and plate in the radio tube, thus improving its ability to detect and generate electromagnetic waves. De Forest organized the Radio Telegraph Company of New York in 1907 and set up radiotelephone service between New York, Philadelphia, Albany, Cleveland, Chicago, Detroit, and Duluth. However, his company failed.

The Massie Wireless Telegraph Company, organized in 1905 by Walter Massie, an early radio experimenter, established New England coastal stations. In 1912 the Marconi Company bought the Massie company and the United States Wireless Telegraph Company. The latter had merged the American De Forest Company and the International Telegraph Construction Company.

The United Fruit Company bought an

interest in the Wireless Specialty Apparatus Company, and organized a subsidiary, the Tropical Radio Company. The Federal Telegraph Company established a radiotelegraph circuit between San Francisco and Honolulu in 1912.

Experimenting at Columbia University in 1933, Major Edwin Howard Armstrong noticed that radio signals were repeated in amplified form inside the De Forest audion vacuum tube. Connecting the input and output circuits in the tube, he produced a regenerative circuit to feed back part of the outgoing current. That stopped the howls and whistles, made the audion sensitive as a detector of radio signals, and improved radio's ability to select desired signals and shut others out. Closer coupling of elements in the audion, he found, produced a heterodyne receiver which would detect continuous-wave signals. The tube would transmit waves, as well as amplify weak signals. In another of those numerous cases where two men arrived at an invention almost simultaneously, Irving Langmuir of General Electric filed for a regeneration patent on October 29, 1913, the same day Armstrong did. Armstrong got there first, however, and received a patent for the regenerative receiver circuit on October 6, 1914.

World War I

When World I started in 1914, Atlantic Communication Company, a subsidiary of Telefunken Wireless, using Armstrong's regenerative circuit, operated between Sayville, Long Island and Hanover, Germany. The High Frequency Machine Company operated between Tuckerton, New Jersey and Nauen, Germany, while British Marconi operated between Wales and New Bruns-

wick, New Jersey and between Norway and Marion, Massachusetts.

Great Britain immediately cut the two German cables five miles from New York, to the dismay of Commercial Cables, which operated them under a forty-year contract. The British diverted the end of one cable to Penzance and the other to Brest, France. After the war the Brest cable was turned over to the French Cable Company. These cables had been laid in two sections, from Far Rockaway, Long Island to Horta in the Azores in 1900 and 1904, and from Horta to Emden in 1901 and 1903. Clarence Mackay protested that the seizure of the cables from the Azores, a Portugese possession, was illegal, but England and France called them "spoils of war," and France kept them.

The German cruiser *Nurnberg*, disguised by flying a French flag, raided Fanning Island on September 7, 1914, dynamited the British cable station, and smashed the instruments. When the raider left, the cable staff managed to raise the cable by using an ordinary pickaxe as a grapnel, to connect the broken ends long enough to report the raid and call for a cable repair ship.

While engaged in sinking troop transports from Australia to Europe, the German cruiser *Emden*, disguised with a dummy fourth funnel, wrecked the Eastern cable station on Cocos Island in the Indian Ocean on November 9, 1914. The staff managed to send a call for help before Count von Luckner, master of the *Emden*, could blow up the wireless mast and cut the cable. HMS *Sydney* arrived in time to engage the *Emden* in battle and sink it.

Use of radio by the armies in Europe early in the war gave the enemy so much information that its use was restricted to emergencies, but radio was used to report

17.1 STUDENTS IN ONE OF 112 WESTERN SCHOOLS LEARNING TO SEND TELEGRAMS.

sightings of German submarines, reroute convoys and give orders to other ships.

German submarines cut half of the Atlantic cables. Five were repaired off New York by cable ships, while submarine chasers circled them to ward off attack. Often cable ships made repairs in submarine-infested waters without escort, or diverted cables under the noses of hostile fleets. Cablemen also did minesweeping in the North Sea.

Major General G. O. Squier, chief signal officer of the U.S. Army, told Vail and Carlton in January 1917 that war was imminent, hundreds of communications technicians and operators were needed, and there was no time to train enlisted men. The men needed were selected from the two companies, commissioned as officers of twelve battalions, and went overseas with the first combat units of the American Expeditonary Forces. The Army cabled back a request for an additional 1,100 operators. Sixteen schools were established to train enlisted men for telegraph, telephone, and radio work.

Western Union also set up schools to train women operators to replace men entering the armed services. A good multiplex operator could be trained in four months and send as many as eighty-three messages an hour, while a Morse operator ordinarily had a year's training and averaged only twenty-four messages an hour.

When the United States declared war on April 6, 1917, the Navy took over all wireless stations. The Signal Corps then had fifty-five officers, 1,570 enlisted men, and 9,767 civilian employees.[4] Its old equipment was replaced with the latest AT&T and Western Union developments. The Signal Corps established a telephone network in France with 100 switchboards and 100,000 miles of wire. John F. O'Ryan, formerly of Western Union's Law Department, became commander of the 27th Division, American Expeditionary Forces.

The demand for special services by 110 government units at Washington, D.C., and the addition of hundreds of Western Union employees there forced T. W. Carroll, Eastern Division general manager, to seek more space, and President Wilson commandeered a four-story building for the company.

The explosion of a munitions ship in 1917 wrecked Halifax, Nova Scotia, killed several thousand people, and broke all landline and Atlantic cable connections. Western Union Traffic Vice President J. J. Welch and Thomas J. Foley, division traffic superintendent, rushed to the scene from Truro, fifty miles away, with a special train of doctors, nurses, and supplies. They worked for a week almost without sleep, handling all communications between Europe and the United States.

More than 25,000 Bell employees went into the U.S. Army. Bell Chief Engineer Carty joined the Signal Corps as a Major and organized the telephone volunteers into fourteen Signal Corps battalions. These men were already familiar with the Western Electric equipment used in France.[5] More than 200 French-speaking Bell women operators served there.

Taking advantage of the sudden shortage of telegraph operators, union organizers soon had the newly trained replacements in a turmoil, and the union called a strike on April 9, 1918. It was postponed, pending arbitration by the War Labor Board, but Western Union refused to accept a decision that included union membership for employees, and the strike was called for July 8. The strike was postponed at the request of the secretary of labor. On July 22, President Wilson announced that, on August 1, Postmaster General Burleson would take over the telegraph and telephone systems for the government.

A law was passed authorizing manufacturers to provide radio equipment to the government regardless of patent rights. With that protection, radio requirements for war use were met by using combinations of patented components.

Government, press, and commercial cable traffic grew so rapidly that it was necessary to cancel the night letter and weekend letter cable services, and increase operating staffs. European censorship prohibited private codes, and that lengthened the average cablegram from eight to twenty-two words. Military messages were sent by cable instead of radio for security reasons. To handle so much urgent traffic, operators stuck to their posts long hours, working at top speed.[6]

When Burleson, a government ownership advocate, took wartime possession, Vail told Carlton he feared government control would make it impossible to raise capital for needed construction and expansion. Carlton replied, "It's your salvation. The government will be able to raise your rates and get you new money."[7] That was what happened. Burleson appointed Vail, rated by Carlton as a genius, to manage the telephone, and Carlton to operate the telegraph.

Wages were increased, and government operation produced a deficit, although telephone rates were increased twelve percent, and telegraph rates twenty percent. Public protests mounted, and Congress threatened to restore private ownership. Former advocates of government ownership criticized government inefficiency. After nine months Burleson unhappily recommended that the companies be restored to their owners. "Since government ownership does not meet with the approbation of the incoming Congress," he said, "return is necessary." After one year of government possession, on August 1, 1919, the companies were returned and the government paid AT&T $65,148,641.

Before the companies were returned, the Commercial Telegraphers Union struck on June 11 for union recognition and another increase in wages. That strike against the U.S. Government failed to halt the business or win anything. As spoils of war, the Treaty of Versailles gave England and France the German transatlantic cables, and Germany formed a radio company called Transradio. The Dutch, Japanese, and United States divided the three German cables in the Pacific: from the Island of Yap to the Dutch East Indies; Yap to Shanghai; and Yap to Guam, an American possession.

The Radiotelephone

Carty had persuaded Vail in 1909 to spare no expense in developing wireless telephony. In 1915 AT&T succeeded in transmitting Carty's voice from Montauk Point, Long Island to Wilmington, Delaware and St. Simons Island, Georgia. It transmitted a voice from the Naval station at Arlington, Virginia with its 450- and 600-foot high antenna towers, to Honolulu on September 29, 1915. Finally, after weeks of failure, Bell engineers on the Eiffel Tower in Paris heard the words, "And now, good night, Shreeve." Those first words, heard across the Atlantic on October 21, 1915, were to H. R. Shreeve of AT&T in war-torn Europe.

The first military radiotelephone was tested May 6, 1916, when Secretary of the Navy Josephus Daniels at the Arlington station talked with the U.S.S. *New Hampshire* at Norfolk, Virginia. During the war, the Navy maintained contact with its ships through its fifty-eight coastal stations and AT&T's New Brunswick, New Jersey station, but the land forces relied on a network of cables and wires to coordinate their actions.

Using regenerative circuits to Honolulu and Europe in 1915, AT&T ignored Armstrong, its inventor. Instead, in 1917 AT&T bought De Forest's audion and other rights for $300,000. While Armstrong was giving important aid to the Signal Corps in France, De Forest was selling regenerative sets he invented to the Navy. The lone poor inventor, returning from the war, was faced by the huge corps of Bell engineers and lawyers.

In Paris Armstrong had developed superheterodyne circuitry, basic in modern radio and radar. Lieut. Lucien Levy of the French Army, who met Armstrong in France, was one of several who later claimed to be the inventor. AT&T bought Levy's claim and filed interferences with Armstrong's patent application, but the court awarded the patent to Armstrong.

To obtain money for his legal battle, Armstrong sold his feedback regenerator and superheterodyne patents to Westinghouse for $434,000, with $200,000 of it to be paid when the feedback patent was issued. Fortunately, he reserved the rights for amateur use. These later became valuable.

17.2 AT&T President W. S. Gifford opens New York–London telephone line, January 7, 1927. (Courtesy AT&T.)

The rosy dreams of the radio companies still seemed doomed because waves from the sputtering spark generator decreased in intensity and soon died out. A wave thousands of times faster was needed to handle commercial business with assurance. Then Ernest F. W. Alexanderson produced a high-frequency alternator or generator in the General Electric laboratories. Installed at AT&T's New Brunswick station in 1917, it was a major "breakthrough" for overseas radiotelephone service. G.E. installed a powerful radiotelephone transmitter tube on the S.S. *George Washington* in 1919, and conversations between the ship and New Brunswick and Belmar, New Jersey stations kept President Wilson in touch with Washington on his voyage to the peace conference.

In January 1923 telephone officials at New York spoke one-way to a group of engineers and scientists in London. On March 7, 1926, groups in the United States and England held a two-way conversation. Years of work in the Bell System Laboratories "paid off" when President Gifford inaugurated regular radiotelephone service between New York and London, on January 7, 1927, in cooperation with the British General Post Office. Critics had said people would not pay seventy-five dollars for a three-minute

17.3 PRESIDENT HOOVER INAUGURATING RADIOTELEPHONE SERVICE TO SOUTH AMERICA. LEFT TO RIGHT: AMBASSADOR CARLOS G. DAVILA OF CHILE; ACTING SECRETARY OF STATE JOSEPH P. COTTON; THE PRESIDENT; WALTER S. GIFFORD, PRESIDENT OF AT&T; SOSTHENES BEHN, PRESIDENT OF THE INTERNATIONAL TELEPHONE AND TELEGRAPH CORPORATION.

conversation, but the service was so popular that more circuits were needed.

Service to Hawaii was established on December 23, 1921, to Belgium in 1928, and then to Holland, Germany, Sweden, France, Denmark, Norway, Switzerland, Spain, Austria, Hungary, and Czechoslovakia. South America and Australia were added in 1930; South Africa, in 1932; and Java and Japan, in 1934. Finally the first around-the-world conversation took place on April 25, 1935. AT&T's 33,700,000 telephones, about ninety-three percent of those in the world, could be interconnected by wire and radio combinations in 1936, and the price of a three-minute call from New York to London was lowered to twenty-one dollars.

The lack of privacy that many people thought doomed the radiophone to failure gave birth to the new industry of radio broadcasting. Where the measure of man's progress had been the speed with which he could convey his written thoughts from one place to another, the ability to transmit the voice by radiophone gave promise that he could speak by radio to millions.

Using strong vacuum-tube transmitters at High Bridge, New Jersey, De Forest demonstrated the broadcasting of voice and music in 1916. The *Detroit Daily News* began broadcasting news bulletins for radio amateurs in August 1920.

Developments by the Westinghouse laboratories during its war work for the Signal

Corps led Dr. H. P. Davis to persuade the company to establish KDKA, the first public radio broadcasting station. Using the first condenser-type microphone, KDKA broadcast returns of the Harding-Cox presidential election in 1920, but only a few hundred people had receiving sets, sold by the Joseph Horne Department Store in Pittsburgh.

Westinghouse added Station WBZ, Springfield, Massachusetts, in September 1921; WJZ, Newark, October 12; and KYW, Chicago, November 11. They were on the air irregularly at first, but the public was delighted and other companies rushed to establish stations or manufacture sets. In 1922, WOR and Bamberger's Department Store in Newark, featured a morning exercise program by John B. Gambling. As a reporter, I was in the WOR studio with Sir Thomas Lipton when he talked during a program with the head of Selfridge's Department Store in London. What is now commonplace was news then. The Newark *Sunday Call* published the first weekly section devoted to radio.

How R.C.A. Was Born

The Government wanted American companies to own wireless telegraph circuits because cables cut during the war had caused undamaged cables to be overloaded. Learning that British Marconi was about to buy $5,000,000 worth of the new Alexanderson alternators from General Electric, Franklin D. Roosevelt, then acting secretary of the Navy, feared the result would be a worldwide British monopoly. Roosevelt asked G.E. to suspend the deal and meet with the Navy on April 4, 1919. From Paris, President Wilson then instructed Admiral W. H. G. Bullard, director of Naval Communications, and Commander S. C. Hooper to ask General Electric not to sell the alternators, but to establish an American-owned wireless company. Owen D. Young of G.E. agreed on the spot to do so.

G.E. then bought the American Marconi company from the British on November 21, 1919, and created the Radio Corporation of America, with Young as chairman and Edward J. Nally, president. It was agreed that RCA would operate the Marconi transatlantic wireless telegraph; G.E. would continue to do the radiotelegraph research and manufacturing; and AT&T, which bought $2,500,000 of RCA stock, would make the transmitters.

When World War I ended, only one house in four had electricity. The electric washing machine and vacuum cleaner were unknown. But a tuning device (variable capacitor) Armstrong developed during the war permitted a radio wave to be tuned in by turning only one dial, and that helped to speed radio growth. And radio advertising helped produce the modern consumer society.

Numerous companies manufactured sets, but none could produce up-to-date sets without infringing on the patents of others. Pressure by the Navy in 1919–1920, broke the patent deadlock, giving RCA cross-licensing agreements with G.E., AT&T, and Westinghouse. RCA planned to sell G.E. and Westinghouse radio equipment with a catalogue of "do-it-yourself" parts, prepared by its commercial manager, David Sarnoff.[8]

President Wilson, seemingly adopting RCA as his fair-haired child, appointed Admiral Bullard to represent the government at all RCA board meetings, and turned over to RCA the Marconi[9] stations federal authorities had seized during the war, with wireless telegraph circuits to England, Hawaii, and Japan. RCA established circuits to Norway,

Germany, and France. With Navy-arranged cross-licensing to use the Alexanderson alternator, RCA installed it at the old German wireless stations.

RCA began national radiotelegraph service on March 1, 1920, and built a "Radio Central" station in 1921 on a ten-square-mile tract at Rocky Point, Long Island. Lieutenant General James G. Harbord, Pershing's chief of staff in World War I, retired and became RCA's second president on December 29, 1922.

Radio Broadcasting

How communications advances changed peoples lives is illustrated by radio. The prophecy in 1904 by Nikola Tesla, an early alternator inventor, that radio broadcasting would entertain the masses, came true in the 1920s. Radio broadcasting ended the age-old solitude of people on farms and those in towns and hamlets with no available entertainment. It delighted city residents, hungry for up-to-the-minute news. Mass communication revived the home as the center for family life, but it reduced person-to-person conversation, and many lonely people substituted the soap opera dream world for friends.

Ironically, De Forest's vacuum tube, which provided the means for broadcasting, was called a hoax. He was snared by promoters, charged with stock fraud, and a judge actually ordered the great inventor to get a common garden-variety job and stick to it.

In the gay, roaring, prohibition twenties, a radio craze spread to homes and apartments throughout the nation, along with jazz music, the fox trot, the hip flask, "bathtub gin," homemade wine, raccoon coats, and Stutz Bearcat sports cars. Thousands of companies made radio parts, and millions of men and boys spread components over home

floors and pieced together radio sets using crystal detectors.[10]

People sat up all night with hands on dials and earphones on heads, tuning in distant stations and proudly listing them in their "logs." It thrilled them to leap a thousand miles by merely turning knobs. Weary at last from foregoing sleep to hear faint voices through the groans, squeals, and roars of nature's static orchestra, people began buying ready-made sets, at first with separate "horns," and later with speakers enclosed in cabinets. Among the 200 or 300 makers of sets were Atwater-Kent, Crosley, Emerson, Philco, and Zenith.

Those early radio sets were all powered by batteries, but in 1930 a battery eliminator was developed. Motorola was founded in 1928, and its first product was the battery eliminator. Around 1930 people began having expensive custom-made AC-powered radio sets installed in their autos. Motorola's President Paul V. Galvin had his staff develop a car radio that gave his plant more orders than it could handle. It made half of the seventy million radio crystals used in World War II, where two-way radio was an important asset in the South Pacific. By the 1980s Motorola was a $3 billion plus international corporation, and one of America's largest electronic product makers.

The 1920s also produced a hundred thousand "radio hams," who set up short-wave stations in their homes and talked nightly with dots and dashes or voice with other amateurs all over the world.[11] Many hams listened to the A.P. wireless report of the day's news, which was telegraphed to ships at sea by the 1903 Marconi station at South Wellfleet, Cape Cod. The American Radio Relay League (ARRL) provided communications in disasters when telegraph and telephone lines were down, and reported

SOS signals when a ship was in distress. ARRL's "official journal" was and still is *QST*. Call letters of "hams" appeared on automobiles and homes with tall antennas. Their short waves bounded up and down between the earth and the Kennelly-Heaviside layer or ionosphere, which hangs like a mirror across the sky.

In 1922 Westinghouse established a shortwave station at Pittsburgh, over which it transmitted the regular programs of KDKA. Shortwave was used also to beam programs to people behind the "Iron Curtain."

Until 1924 the transatlantic wireless stations were operated on wave lengths of 10,000 to 25,000 meters, but were crippled by static during the six warm months of the year. Then it was found that short waves beamed in the desired direction were comparatively free from static, though even they were prostrated by magnetic storms when there were spots on the sun's face.

A Commercial shortwave transmitter, tuned to 103 meters, was placed in service in 1924 by RCA. It would not perform well by day, but worked so well at night that in 1926 high-power vacuum tubes were put to work beaming waves as short as fifteen meters. Wave lengths under twenty meters proved best by day, and forty to a hundred meters best at night, but signals would fade for seconds or minutes on one receiver, and be strong on another one nearby.

That led to the adoption of "diversity reception," with three antennas as much as 1,000 feet apart, each with a receiver feeding its output into a common signal. It greatly improved the reliability of wireless operation, and short waves from vacuum tubes replaced long waves from the big alternators.

RCA established its first South American circuit on January 25, 1924, to Buenos Aires, Argentina. The British Government set up shortwave stations at London; Montreal; Cape Town, South Africa; Bombay, India; and Melbourne, Australia. By the end of the 1920s Great Britain and the United States dominated world communications. Of 245,000 nautical miles of cables, Great Britain had 144,000, the United States 85,000, France 20,000, and all other nations 16,000. In wireless telegraphy, both Great Britain and the United States served all parts of the world. In radiotelephony the United States led the world.

People with political, commercial, educational, or religious motives and amateur performers provided many early radio programs for years without charge in exchange for the publicity. The cost grew so large that a store or newspaper could not afford the cost for talent, writing, and other expenses to produce hours of programs.[12]

AT&T solved the cost problem by selling time on WEAF[13] to a number of advertisers. On August 28, 1922, WEAF broadcast the first commercially sponsored program—ten minutes for $100—for the Queensboro Corporation, a New York realty company. AT&T opened another radio station, WCAP in Washington, D.C. in 1923, and in two years owned twenty-six stations as far west as Kansas City.

David Sarnoff of RCA protested that the telephone company was prohibited by the pool agreement from being in the broadcasting business. He proposed that radio-set manufacturers and licensees contribute two percent of their gross sales to support broadcasting as a public service. He said broadcasting companies (AT&T) should sell no advertising. AT&T then disposed of its RCA stock, tried to block any other company from selling advertising, and said it would provide no telephone lines to radio stations without licenses to use Western Electric transmitters.[14] RCA also tried to stop G.E. from selling transmitters, as a violation of the pool agreement.[15]

AT&T wisely promoted the interconnection of stations over its wires so that production costs of unsponsored programs could be shared. The first multiple station broadcasts were in January 1923 through WEAF and WNAC Boston, and in June over WEAF, KDKA, KWY, and WGY (Schenectady). The first transatlantic radio program was in March, 1924, from Coventry, England, to Houlton, Maine, and then by wire to WJZ, which broadcast it from New York. The company saw the possibility of large revenues from leasing networks to broadcasters.

Troubled by public criticism, apparently engineered by RCA and G.E., that it was trying to establish a monopoly in broadcasting, AT&T adopted a liberal licensing policy in 1926 to encourage growth of the industry. G.E., RCA, and Westinghouse then made a new agreement with AT&T, and formed the National Broadcasting Company on September 9, 1926, with Merlin Hall Aylesworth, as president. NBC bought WEAF from AT&T for $1,000,000[16]; took over WJZ[17] and WRC of Washington, D.C. from RCA; and soon more than fifty stations were subscribing to its program service, which began on November 15, 1926.

Now RCA decided selling time on radio was not such a terrible thing after all, and began selling it. Sarnoff explained that it was necessary to have either public support or government subsidy and a sales tax on radios. Since there were two competing stations in some cities, NBC established two networks—the Red with WEAF, and the Blue with WJZ as the program-originating stations. Some popular stars were Jack Benny, Kate Smith, Fred Allen, Edgar Bergen with Charlie McCarthy, Burns and Allen, Fibber McGee and Molly, and Rudy Vallee. The NBC studios were at 711 Fifth Avenue until November 1, 1933, when they moved to the RCA Building in Rockefeller Center.

The number of radio stations increased rapidly to 733 in 1927. In that year the Columbia Broadcasting System was formed by William S. Paley and Philadelphia associates who owned the Columbia Phonograph Record Company. CBS operated nationwide from 485 Madison Avenue, New York City, and its net profits soon matched NBC's. A third network, the American Broadcasting System, bought the NBC Blue network July 30, 1943, and grew rapidly. The fourth network was a cooperative enterprise, the Mutual Broadcasting System.

The new agreement between AT&T and the radio group was cited by the government in 1930 as a violation of the Sherman Antitrust Act, and a new contract provided for telephone circuits to stations without licenses or equipment from the group.

As radio stations multiplied and the radio spectrum became jammed with manmade long and short waves, Congress, which had placed telephone regulation in the hands of the Interstate Commerce Commission in 1913, created the Federal Radio Commission in 1927 to reduce chaos and establish order. The commission's job of licensing and allocating wavelengths or frequencies to stations, became more complicated later when the police began "calling all cars," airplanes began reporting weather and altitudes, and some companies set up private, point-to-point radio systems.

In 1927 RCA's President (General) Harbord inaugurated a policy of licensing competitors to use RCA patents and manufacture radio sets, greatly increasing the number of sets sold and the listening audience. On January 3, 1929, RCA Communications, Inc. was organized to provide international radiotelegraph service. William A. Winterbottom was its head for the next fifteen years.

Radiotelephone and Radiotelegraph

After a radio signal for help saved 1,500 lives in the collision of the *Florida* and *Republic* in 1909, Congress required all vessels with more than fifty passengers to install wireless telegraph equipment. AT&T's radiophone service between New York and London was inaugurated January 7, 1927, and to ships December 8, 1929, with the S.S. *Leviathan* first. To make such a call, you asked the long distance operator to connect you with a passenger on the ship. The long distance overseas switchboard would connect you with a control terminal, say at Forked River, New Jersey. A technical operator there would connect the circuit with a nearby shortwave radio station and call the ship to get the passenger to the telephone. Propelled by 10,000 volts, your words to the ship would pass through a series of vacuum tubes, be amplified millions of times, and be hurled against the sky in the form of radio impulses. These impulses, weak from their long trip to the sky and back, would be converted into your voice by equipment on the ship. The voice coming back from the ship would travel over a parallel path to you.

Radiomarine Corporation of America, organized in 1927, agreed that all of its telegraph ship-to-shore communications would be handled to and from their land destinations by Western Union. By 1930 RCA had circuits to thirty countries, and on December 6 that year China agreed to permit service to Shanghai. Mackay Radio extended its telegraph service to Manila in the Philippines. When Mackay began operating to China in 1933, RCA charged it was a breach of its agreement with China, and for years the two battled for "exclusive" contracts with many nations.

The Federal Communications Commission was established on July 11, 1934, to succeed the Federal Radio Commission, and to develop an American-controlled world communications system with direct circuits to other countries. It gave RCA so many radiotelegraph circuits to foreign capitals in 1936 that Mackay charged it was virtually a monopoly. RCA claimed its expansion resulted from superior apparatus, but Ellery W. Stone, Mackay's operating vice president, testified at an FCC hearing on January 13, 1936:

> The modern generator of continuous waves, the vacuum tube oscillator, in use by all radio agencies in this country today, was the fruit of the early research of the Federal Telegraph Company, being invented by Dr. Lee De Forest while in our employ as an engineer in 1912.

General Harbord became chairman of RCA's board in 1930. Sarnoff, RCA's third president, at once began maneuvering to obtain RCA's independence from the companies that created and helped build it. The Department of Justice obligingly brought an antitrust suit and obtained a consent decree November 21, 1932. That freed RCA from General Electric and Westinghouse, which shared ownership in RCA Victor, RCA Radiotron, National Broadcasting, and RCA Photophone, which recorded sound on motion picture film. The radio set, talking machine, and record business became RCA Victor at Camden, New Jersey,[18] and E. T. Cunningham Company, a well-known tube maker, was acquired.

Broadcasting and manufacturing soon replaced wireless telegraphy as RCA's principal business. In 1931, RCA moved its headquarters from its first small office with Marconi Wireless in the Woolworth Building to entire floors in a forty-eight-story building at 570 Lexington Avenue. Two years later, it shifted to the RCA Building in Rockefeller Center. Sarnoff tried to make his predictions that radio would replace the telegraph come true. In April 1934, RCA established point-to-point radiotelegraph service between New York, San Francisco, Boston, Washington, Chicago, and New Orleans, and announced it would be extended to other cities. Western Union agreed to pick up and deliver the messages and receive four percent of gross revenues. In 1936 RCA's domestic radio network reached twelve cities, but costs of offices and personnel cooled Sarnoff's ambitions, and the service faded away.

FM Radio

Major Edwin Howard Armstrong, who invented the feedback regenerator (1912) and superheterodyne circuitry (1918), also discovered the "superregenerative effect" (1920). Superregeneration was produced by letting a regenerative circuit amplify beyond the point where its audion tube became a generator of radio waves, and a second tube cut in to suppress the oscillations at the rate of 20,000 times a second, allowing amplification to build up in intervals to 100,000 times the original signal strength.[19]

RCA paid Armstrong $200,000 and 60,000 shares of its stock for the use of his new invention. RCA then announced its "sensational improvement in home radio reception." Its superregenerator provided such sharp tuning and sensitivity, RCA said, that it "superseded the regenerative set as effectively as the 'regenerator' had sent the crystal detector into discard."

Armstrong also developed frequency modulation (FM), which had superior clarity because it is almost immune to static and fading. He received four patents on December 26, 1933 for FM. At that time he was reputed to have a fortune of $9,000,000. However, his FM was a thorn in the flesh of radio industry officials, and especially David Sarnoff, whose secretary, Marion MacInnis, Armstrong married. She championed Armstrong's cause and left RCA.

Before receiving FM patents, Armstrong demonstrated FM to Sarnoff and RCA engineers, and in 1934 installed it in the Empire State Building tower for RCA to test. However, Sarnoff kept delaying a decision for two years, and Armstrong finally decided Sarnoff was keeping him dangling while RCA struggled to develop a substitute. He then applied to the FCC for FM frequencies and found Sarnoff had hired the FCC chief engineer, Dr. Charles B. Jolliffe, to oppose him.

Armstrong believed the radio industry would have to adopt FM because it was superior. The manufacturers, radio networks, and hundreds of stations, however, had invested millions in amplitude modulation (AM) systems, and the public's home sets were all AM. Asking the industry to switch to FM was like waving a red flag at a bull.

A determined man, Armstrong persuaded some manufacturers to add FM on AM receiving sets, and convinced stations to broadcast FM. He spent $2,000,000 building and operating his own FM station, and was said to have helped finance the Yankee Network, using FM. That pressured RCA into making FM transmitters and receivers, but it refused to pay any royalties to Armstrong. Ignoring its Empire State tower and other tests of Armstrong's FM, RCA claimed it had made "early tests which afforded valuable information as to its advantages and potentialities, and its pioneering in this field has never stopped."

Armstrong was a stubborn genius. With bulldog tenacity, he carried on suits for infringements of his patents. After years of court battles, RCA finally offered in 1940 to pay $1,000,000 for Armstrong's patent rights, but would pay no royalties. He refused, saying it was his moral obligation to treat all licensees equally. Without paying anything, RCA again started operating an FM transmitter in the Empire State Building tower in January 1940. In October that year the FCC granted construction permits for fifteen commercial FM stations simultaneously. On May 29, 1941, WSM-FM, Nashville, was the first in operation. Western Union signed a license agreement in September 1945 to pay Armstrong for the right to use his patents in its carrier telegraph operations.[20]

Armstrong's legal battle with RCA and promotion of FM finally drained away his fortune. Believing he was persecuted with government help, he wrote a letter February 1, 1954, put on his overcoat, hat, and gloves, and stepped out of a thirteenth-floor window, a genius who despaired of ever winning the rewards he believed were rightfully his.

Armstrong's sudden end in New York at age sixty-four was a shocking tragedy. He was a tall, blue-eyed, partly bald, gray-haired man of erect military bearing, considerate and friendly. The world may never again see a lone inventor of such importance. In 1983 it was estimated that FM with its greater fidelity and use for stereo, served sixty-two percent of the listening audience in the United States.

Telecasts reported major events in 1939, but commercial television broadcasting began with WNBT, New York, July 1, 1941. As stations were added, the transmitters usually were miles away in open country to avoid electrical interference. The networks were permitted to own only a few stations; the others subscribed to network programs. All received them over circuits leased from AT&T.

For many years, the world's crossroads of speech, its Tower of Babel, was 32 Avenue of the Americas,[21] New York. The building not only contained Western Union's operating rooms, but was the chief control point of AT&T's broadcasting networks, radiotelephone circuits overseas, and the nation's long distance telephone and leased wire center. Vigilant men there made sure, for example, that telephone traffic flowed over an alternate route if needed, or a network program was fed to stations on time, in correct volume and quality.

As wireless circuits reached out to foreign lands in the 1930s, Sarnoff delivered funeral orations for the submarine cables. Cables once were useful, he said, but now would die in peace in their ocean beds. As though nature was answering, some of his pontifications were followed by aurora borealis or sunspots producing radio chaos. For days all radio carriers like RCA had to give their traffic to the cable companies. Another Sarnoff claim was that direct radio communication "cannot be broken by any other nation." Everyone in the business knew he was wrong. AT&T laid the world's finest cables, and RCA and all other international "radio" companies carried on business over leased circuits in those "dead" cables.

Ciphers and Codes[22]

Some people asked, "Why was there a continental or international Morse code?" The Morse code was modified to provide a thirty-two-letter alphabet (plus of course codes for numerals and punctuation) to accommodate other languages that require certain accented letters. The International Telecommunications Union in 1903 relegated the continental code to lines of moderate activity, but in World War I wireless telegraphers used it to send messages overseas or to ships. During World War II, automatic printers, like those on landlines, were installed on cables, and an improved siphon recorder could receive 100 words a minute.

There are other kinds of "codes." That man sending a cable, "KWJNZ RLXBP," is no enemy spy. He's a salesman reporting to his home office. The International Telegraph Conference at Brussels on September 10, 1928, reported that eighty-seven percent of all international messages sent by Americans were in code.[23]

The course of history has been changed by breaking secret ciphers. Failure in the German cipher system during September 1914 caused a French army to be rushed to a decisive victory in the Battle of the Marne. The United States entered World War I when the British solved a coded German message offering Mexico three American states to enter the war. American cryptoanalysts also broke the Japanese diplomatic, merchant marine, and naval codes during World War II.

A code may be simple transposition or substitution of letters, figures, or symbols for letters of the text. Extremely complicated codes, with pattern words, frequencies, and so forth, require months of work to "break."

In "Modern Cryptology" in the July 1966 *Scientific American*, David Kahn stated that "Lieutenant Joseph O. Mauborgne of the Army Signal Corps came to the conclusion" that the only unbreakable cipher is one "with a key that never repeats and contains neither meaning nor pattern." In 1918, Mr. Kahn wrote, Mauborgne "added the concept of a patternless, endless key" to a cipher machine invented by Gilbert S. Vernam of AT&T. The machine that enciphered and transmitted in a single operation, used teletypewriters that carried its key in the form of a perforated teletypewriter tape. The pulses, represented by the holes in the tape, were automatically added to the teletypewriter pulses of the plain text. At the receiving end the machine with an identical key tape subtracted the key pulses from those of the enciphered message.

During World War II Western Union manufactured and installed the "unbreakable" Telekrypton system for American and Allied governments. The president used it constantly, and about ten million words a month over it kept the State Department in touch with American representatives abroad. Before Telekrypton's automatic operation, the coding and decoding room was a "holy of holies," which only the most trusted experts could enter.

Congress crippled the ability of the FBI and CIA to protect Americans from attacks by terrorists and foreign enemies, but the little-known National Security Agency (NSA) largely escaped congressional and press inquisition. Its 20,000 employees maintained listening posts worldwide, pouring information into its headquarters at Fort Meade, Maryland to keep the White House and Pentagon informed.

NSA's encrypting devices mix a message, divide it in half, add the key to one half, mix the two halves together, run the message through protective "gates," and rapidly repeat the process automatically sixteen times. That "Data Encryption Standard" protects NSA messages and government data between distant computers.[24]

When the telephone is used by heads of state and diplomats, the voice frequencies are scrambled to make the words indistinguishable to a listener-in. One method inverts the tones of the voice, changing high notes to low, and low ones to shrill treble.

The purpose of business codes usually is not secrecy, but economy. The International Telegraph Union at first required the use of dictionary words in code. Later, when it permitted any pronounceable ten-letter word, new codes were constructed with two five-letter words joined to count as one word. That was reversed on January 1, 1934, when the ITU reduced rates by forty percent for five-letter code words.

A five-letter code builder, making each word different from any other by two letters to avoid errors, could construct 466,000 combinations. Some industry or company codes contained thousands of combinations. "PBXBD" might mean "private branch exchange board" in a telephone code. One general code book contained 100,000 phrases, and more than 50,000 copies were sold. One private code used 200,000 code words because the company sold many styles and models.

The British "All Red" Route

To provide a safe "All Red" route to Canada,[25] the British General Post Office (GPO) diverted the Irish terminus of an old German cable to Penzance, England. Chaotic conditions in Ireland resulted in so many wartime interruptions that Western Union

also shifted a cable from Valentia to Penzance in 1918. When Ireland declared independence in 1919, it sent censors to the Valentia station. The station was raided by guerillas and wrecked by soldiers. Western Union then laid a second cable to bypass Ireland. In 1923 Commercial Cables laid a "Jumbo" cable from Waterville, Ireland, to Horta, to Canso, to Far Rockaway, Long Island, and extended the European end to Weston and Havre.

The British laid a second chain of cables from Vancouver to Australia in 1926, increasing their Pacific system to 16,500 miles. The 509-foot Telegraph Construction and Maintenance Company (T.C.&M.) cable ship *Dominia* laid the Vancouver-Fanning Island section.

Establishment of Empiradio, a shortwave radio system between Montreal and Australia in the 1920s placed the GPO in open competition with the "All Red" route, and Eastern Extension Telegraph Company's cables to Australia via the Suez Canal, India, and China. An Imperial Wireless and Cable Conference proposed a merger of all British systems, and the result was Cable and Wireless, a holding company, and Imperial and International Communications, Ltd. (I.&I.C.), an operating company. Into this gigantic company was placed everything except radiotelephone. I.&I.C. formed the Canadian Communications Company to operate its Canadian cable terminals, but used Empiradio and Marconi for radio services.

Permalloy Cables

For sixty-six years only minor alterations had been made in the design of ocean cables. Then in 1924 Western Union laid a new type of cable 2,330 miles from New York to Horta. It carried 1,500 letters a minute, compared with the 150 to 300 letters of earlier cables. This was made possible by "loading" the cable, that is, by wrapping around the copper conductor core a continuous strip of permalloy, an alloy composed of nickel and iron, only six-thousandths of an inch thick and 10,000 miles long.

Bell Engineer G. W. Elmen discovered a cable "loaded" with permalloy tape covering the core would keep each signal travelling the electrical highway in its proper length and shape, and so sharply defined, that one signal could follow another in rapid succession without overlapping. After World War I, Bell suggested that Western Union use permalloy-loaded cables, and three years of experiments followed, in which hundreds of short sections of cable were laboriously tested at freezing temperatures under a pressure of 10,000 pounds per square inch, to simulate conditions on the ocean bottom.

Western Union Cable Ship *Lord Kelvin* then laid 120 miles of the cable off Bermuda. After four months of tests in 1923–1924, Western Union ordered T.C.&M. to make 2,400 miles of cable and lay it from New York to Horta. It was only an inch in diameter in its deep sea section, and three inches in diameter in the shore end, with brass tape under the sheathing to frustrate the teredo (shipworm). The cable cost about $2,000 a mile, and laying it cost $150,000. To utilize its unprecedented capacity, the cable was operated in four channels. Typewriter-like machines converted the messages into perforated tape to send, and distant printers received them on blanks ready for delivery.

When Mussolini became dictator of Italy in October 1922, he had Italcable, a private company, organized to establish an Italian international communication system. Since neither Italy nor Spain had been connected with America by cable, Italcable and Western Union agreed on July 24, 1923, to lay cables to Horta and exchange traffic with them in mid-Atlantic. Western Union then laid a cable from Hammels, Long Island, to Horta in 1924. In 1925, Italcable laid in two sections, from Anzio, near Rome, to Malaga, Spain, 997 miles, and from Spain to Horta, 1,347 miles. Service began over the 4,674 miles of cable on March 16, with President Coolidge and the king of Italy, and Secretary of State Kellogg and Mussolini exchanging messages.

Commercial Cables, which had operated the German cables from America to Horta before World War I, was anxious to restore communications, and agreed to meet a cable laid by DAT of Germany at Horta, but instead it leased channels in Western Union's new 1924 cable to Horta. Since the DAT cable was loaded, both Western Union and Commercial could again operate direct to Germany.

Because world commerce had more than doubled its cable traffic (from 37,972,000 words in 1913 to 92,375,000 in 1927), Western Union laid a second loaded cable in 1926[26] from Penzance, England to Bay Roberts, Newfoundland to New York. It operated at 300 letters per minute on each of eight channels, which were extended to connect New York with Amsterdam and London, and Montreal and Toronto with London. In 1928, the world's fastest cable, with a speed of 2,800 letters a minute, carrying four messages simultaneously in each direction, was laid by Western Union between Bay Roberts and Horta. There were then twenty-one transatlantic cables.

The Great Earthquake of 1929

On November 18, 1929, a giant earthquake recorded its path of destruction in twenty-eight rapid, successive breaks in thirteen of the twenty-one Atlantic cables, southeast of Newfoundland and east of Nova Scotia. The quake cast a wave fifty feet high on the coasts of Newfoundland and Nova Scotia. A continuation of the fifty-mile-wide Cabot Fault in the Atlantic had suddenly slumped, as I immediately notified the press.

Seven cable ships rushed to the scene, and began raising to the surface the steel-armored cable, shredded almost beyond belief. One mangled cable was 3,190 fathoms below, several were around 2,900 fathoms. Rough seas often forced the ships to ride helplessly, waiting for the weather to clear. Huge waves, giant icebergs, ice floes, fog, snow, rain, bitter cold, and furious winds added to the danger. In numerous breaks, long sections were buried, necessitating splicing in hundreds of miles of new cable. In the French cable to St. Pierre, 130 miles of cable were buried. It was a long, freezing winter on the stormy Atlantic, never to be forgotten by cable men.

ITT Enters the Battle Royal

Two adventurous brothers, Sosthenes and Hernand Behn, natives of St. Thomas in the Virgin Islands, were sugar brokers in Puerto Rico after the Spanish-American War when a friend took over the primitive Puerto Rican Telephone Company as payment for a bad debt. He sold it to the Behns "for a song," and they turned it into a modern telephone system. At Cuba's invitation, they installed a telephone system there also, and persuaded AT&T to lay a cable to Cuba and connect with them.

The Behns obtained a concession to install a modern telephone system in Spain, but had to make the equipment there, so they bought International Western Electric from Western Electric in 1925 for thirty-three million dollars, renamed it International Standard Electric Company, and suddenly became manufacturers. Next was a telephone system for France, then Mexico. In 1927 they bought into All America Cables, Incorporated, with 30,000 miles of cables to the West Indies, and Central and South America.

In 1928 the Behn brothers acquired control of Clarence Mackay's Postal Telegraph and Cable Corporation. As the base for their far-flung operations they organized International Telephone and Telegraph (ITT). In ten years they had built a primitive island telephone system into one of the world's largest enterprises. Their many "national companies" sold stock in those countries, and their employees were preponderantly natives.

When the Behns bought Postal, it had about seventeen percent of the United States telegraph business and fed its overseas messages to Commercial Cables, with 37,000 nautical miles of Atlantic and Pacific cables. The Behns retained socialite, dilettante Clarence Mackay as president of Postal, but the man who ran it, as executive vice president, was Lt. Col. Augustus H. Griswold, a Western Electric veteran who had helped build the transcontinental telephone in 1914, and the telegraph and telephone systems of the American Expeditionary Forces in Europe during World War I. John L. Merrill remained president of All America Cables in which ITT had a one-fourth interest.[27]

In 1929, Sosthenes Behn tried to buy RCA Communications (RCAC), a wireless telegraph business, from RCA. Thomas W.

Lamont and Nelson Dean Jay of J. P. Morgan & Co., bankers for ITT and Behn, negotiated in Paris with Owen D. Young and David Sarnoff, vice president of RCA, and agreed on a price of 400,000 ITT shares, then worth 100 million dollars.[28] However, Senator Wallace H. White, Jr.'s[29] Radio Act of 1927 prohibited any merger of wire, cable, or wireless. Sosthenes Behn testified before congressional committees in a vain effort to obtain approval. When the deal with RCA was finally cancelled in 1931, however, the ITT stock offered was worth only fourteen million dollars.[30]

ITT's Postal Telegraph was starving, and its transatlantic cable and wireless companies were having rough competition. Believing it would make ITT dominant in the international field, Behn again sought approval for the RCAC purchase, and a merger of domestic and international companies, saying it was as wasteful to maintain separate offices in cities as to have two post offices.[31] He pointed out that Great Britain, Italy, France, and Germany had merged their overseas systems. Obviously inspired by ITT and RCA, stories appeared in the *New York Times* and other papers, expressing concern over the "failure" of the United States to merge its cables and radio to compete with Great Britain. One story quoted Chairman Owen D. Young of RCA criticizing the White Act as placing the United States at a disadvantage.

Several times during 1938, Western Union and RCA officials informally discussed the possibility of a merger, but leaders in Congress said the law could not be changed. Carlton said the White Act "seems to combine every prohibition against any association between wires and wireless that would benefit the public."

Behn declared fifteen million dollars a year would be saved if only Western Union and Postal were merged. Carlton replied that Western Union was not interested in buying Postal. Western Union officials marked significant sentences in news stories "planted" by ITT, claiming advantages to the public would result from a Western Union–Postal merger. They remarked that all of the "advantages" would be to ITT if it could get any company to pay Postal's deficits.

Walter Gifford just wanted AT&T's radiophone left out of any merger; his company already had everything going its way. On August 8, 1928, Western Union entered into four contracts to lease AT&T voice facilities, enabling it to expand its telegraph capacity without heavy capital investment.

In 1940, Commercial Cables, All America, Mackay Radio, and Sociedad Anonima Radio Argentina were acquired by American Cable and Radio Corporation, formed by ITT in 1939 with Warren Lee Pierson, former head of the Export Bank, as president. Fifty-seven and eight-tenths of its stock was owned by ITT. The Behn brothers passed on, but the giant ITT they created continued to grow in the hands of shrewd and able successors.

Notes

[1] Sir Oliver Lodge said it was reported in a paper Maxwell published in 1865.

[2] Orrin E. Dunlap Jr., *Marconi the Man and His Wireless* (New York: Macmillan, 1938).

[3] Anna Jameson, of a well-known Dublin whisky distillery family, and related on her mother's side to the Scottish Haig distillery family, went to Bologna to study music. There she met and married Giuseppe Marconi. Their second son Guglielmo was born April 25, 1875.

[4] "History of U.S. Army Signal Center and School," a 1967 manuscript by Helen C. Phillips, museum director and Signal Corps historian. In 1917 the Signal Corps established training courses in seventeen colleges and four camps, one of which was Camp Alfred Vail in Monmouth County, New Jersey, named for Morse's partner. Camp Vail offered telegraph, telephone, and radio engineering in nine- to twelve-month courses. During World War II, Camp Vail was renamed Fort Monmouth, and trained thousands of officers and men. See also Donald McNicol, "Communications in World War I and World War II," *Communications Magazine* (December 1941) 5-7.

[5] Arthur W. Page, *The Bell Telephone System* (New York: Harper & Brothers, 1941).

[6] In 1919 government communications in connection with the Peace Conference raised Western Union's cable total to nearly ninety million words. Growth of international trade kept the volume high after the war: Western Union cables handled 87,372,278 words in 1929, compared with 37,972,000 in 1913.

[7] *Saturday Evening Post* (July 26, 1930) 117.

[8] David Sarnoff, an immigrant boy from Uzlian, Minsk, Russia began work with Commercial Cable as a messenger at five dollars a week. He bought a telegraph set, learned to use it, and got a job as office boy for the Marconi Company. He became an operator at age seventeen, contract manager in 1914, assistant traffic manager in 1915, and commercial manager in 1917. Sarnoff became RCA's president in 1930 at age thirty-nine, and led in building RCA as a manufacturing, broadcasting, and communications company. He kept a Morse key in his desk with which he talked in dots and dashes with officials of RCA subsidiaries.

[9] Following World War I, Marconi devoted his life largely to scientific developments under Fascism. After his first wife divorced him, Premier Benito Mussolini was his best man when he married again in 1927. Mussolini appointed him president of the National Council of Research in 1928, and president of the Royal Academy of Italy in 1930. In 1932 he provided a microwave telephone to the Pope, connecting Vatican City with the Pope's summer castle Gandolfo.

[10] Two Americans, H. H. Dunwoody and G. W. Pickard, invented the quartz crystal and "cat's whisker" receiver that replaced the coherer. In 1908 Hugo Gernsback founded *Modern Electrics* (later *Popular Science*), the first magazine to contain many articles on radio; and, in 1913, *Electrical Experimenter* (later *Science and Invention*), about one-fourth devoted to radio. By July 1919 radio experimenters were so numerous that Gernsback established *Radio Amateur News*, filled with articles and diagrams; in 1929, *Radio-Craft*; and in 1930, *Short Wave Craft*.

[11] Amateurs used waves of 100 meters or so that were not crowded. Commercial stations used waves of several thousand meters.

[12] During those years I provided numerous ideas, acts, and even whole programs to promote telegraph services.

[13] Graham McNamee, one of the early announcers, appeared on WEAF as did the Happiness Boys, Ethel Barrymore, and singer John McCormack. The station was moved to 195 Broadway in 1923, and talented telephone employees who worked there performed on many programs.

[14] A.T.&T. *Bulletin* no. 4.

[15] Sarnoff letter to G.E., June 17, 1922.

[16] NBC paid A.T.&T. the same amount for

service during its first year.

[17]WJZ, the first regular broadcast station in the New York area, was established in 1921 by Westinghouse at Newark, sold to RCA, and moved to New York May 15, 1922.

[18]In 1898 Eldridge Reeves Johnson, who founded the Victor Talking Machine Co., built his first model in a tiny shop in Camden, New Jersey. His first record was the popular song of the day, "I Guess I'll Have to Telegraph My Baby." The Victor trademark was a dog "listening to its master's voice" in a Victrola horn.

[19]Lawrence Lessing, *Man of High Fidelity* (Philadelphia and New York: J. B. Lippincott Co., 1936).

[20]Sources: Personal conversations with Armstrong, F. B. Bramhall, Western Union transmission engineer, and others; also Lessing, *Man of High Fidelity*.

[21]Known as 24 Walker Street until Western Union moved its operating rooms from it to 60 Hudson Street in 1930.

[22]Thanks to I. S. Coggeshall, authority on communications traffic, for this information.

[23]Now most companies communicate direct via telex (landline-connected teletypewriters) and now fax or facsimile, and do not use that much code.

[24]*U.S. News & World Report* (June 22, 1978).

[25]Because the conventional coloring of maps to denote British Commonwealth territories was red, routes touching only British soil were called "All Red."

[26]Instead of permalloy, Mumetal, a new copper-nickel-iron alloy developed by T.C.&M., was used because it was less brittle.

[27]See chap. 11, above.

[28]Young's testimony before the Senate Interstate Commerce Committee, on December 11, 1929.

[29]White was a member of Congress then; later he was senator from Maine for many years. He was considered the best-informed man on communications on Capital Hill.

[30]*Telegraph and Telephone Age* (March 16, 1931) 134.

[31]"Behn Brothers," *Fortune Magazine* (December 1930).

Early to Modern Operating Progress

Before 1900, communications progress seemed slow, but several adventurers into the electrical unknown opened the door for today's multichannel automatic telephone and telegraph operating methods.[1] They were steps toward the computer, data processing, networking, message and circuit switching, coaxial and optical fiber cables, facsimile, lasers, radio beam, microwave, TV, and satellites—in short, the exciting communications world in which we live.

The path to multichannel automatic printing operation was discovered away back in 1855 by David Hughes, music professor at Bardstown College in Kentucky, who realized machines could be synchronized with his music tuning fork. He built a transmitter with a typewriter-like keyboard and a receiving printer with a typewheel. When a key on the transmitter was pressed, it raised a pin, and when a brush synchronized with the typewheel of the distant receiver passed over it, it flashed an impulse that printed the letter at that instant on a paper tape.

Because Hughes's twenty-key transmitter conflicted with House's printing telegraph patents in America, George M. Phelps of Western Union combined the Hughes and House printers in 1859, and it became the most successful printing telegraph machine in Europe.[2]

Until 1900 automatic telegraphy was only used for ten percent of the traffic because Morse instruments were much cheaper and easier to repair. Many operators used an "automatic" sending key, known as a "bug" because it made a fluttering sound as it produced any desired number of dots and dashes in rapid order. The major use of printing telegraphy was in stock tickers.

The grandfather of multichannel telegraphy, Moses G. Farmer, divided use of a wire between two operators. He had synchronous brushes revolve over divided rings at the two terminals, and sent the signals of the two operators in sequence—the first letter of each message, the second letter of each, and so forth.

In 1868, Joseph B. Sterns, president of the Franklin Telegraph Company, made the

18.1 UNIVERSAL KEYBOARD TRANSMITTER AND TICKER.

duplex (two message circuits in one wire) more practical by adding a condenser to the artificial line. He sold his patent to Western Union in 1872, and it provided more than 500,000 miles of duplexed circuits. As we noted earlier, Edison created the quadruplex.

With a tape perforator patented by John P. Hummaston of Connecticut, an operator could press a single key and bring into position the dots and dashes to transmit any letter of the alphabet. A foot treadle was pressed to punch the holes through a paper tape.

That led Wheatstone to develop his perforator, automatic transmitter, and recorder. He used three keys: one to punch dots; the second, a row of holes down the center of the tape for spaces; and the third, holes on the other side of the tape for dashes. The motor-driven transmitter sent the plus and minus signals as contacts penetrated through the perforations. The recorder brought a pin in contact with the moving tape for long or short periods of time. That system was used from 1883 to 1901 on some trunk circuits, duplexed at eighty to 150 words a minute in each direction.

Multichannel Automatic

A giant step in the rise of printing telegraphy was a synchronous system invented by Emile Baudot, a French government engineer, who used synchronous distributors, running at a constant speed. The operator set up a five-unit code combination representing any desired letter by depressing one or more of five piano-like keys, and the corresponding signals passed over the line. A set of five magnets in the receiving instrument was actuated by the signals, and the desired letter was printed.

Later, printing telegraphy with sliding, notched permutation bars or discs, was generally based on the Baudot principle. It could transmit over five channels simultaneously in 1878, and six in 1881. One-way Morse operation could handle 1,500 words an hour; Morse duplex, 2,700; and Hughes simplex, 3,600. But Hughes duplex and Baudot quadruplex systems handled 5,760. Baudot five-channel multiplex was used in European and other countries in 1914. The word "baud," meaning the number of code elements per second a circuit can handle, is derived from his name.

The Kingsley Printing Telegraph, using perforated tape transmission, chemically decomposed letters of the alphabet on a tape at the receiving end. Five magnets at the receiving end controlled five markers. To make a "1" on the tape, a vertical marker would receive current; to make the letter "B," a vertical line, three horizontal lines, and another vertical line.

In 1903, a system that printed telegrams in Roman letters was introduced by Charles L. Buckingham. It was slower than Baudot's but was used on a number of Western Union lines. Donald Murray, a newspaperman in Sydney, Australia, read about the monotype that cast and set type automatically by means of perforated tape. He produced an automatic printer and tape perforator, took his invention to New York in 1899, and exhibited it at the Astor House.

Postal hired Murray, and in two years his system was ready, but the company would not adopt it. It looked like a sewing machine with a typewriter mounted on it, and was jokingly called "The Baby," "Mur-

ray's Coffee Mill," or "The Sausage Machine." It was used on lines in Europe, Siberia, and India until it was replaced by the Baudot Multiplex in 1909 and 1910.

Using the best features of Buckingham, Wheatstone, and Edison's automatic telegraphs, Western Union chief engineer John C. Barclay produced a printing system in 1903. He used the Wheatstone automatic transmitter, which perforated tape in a five-impulse code. His page printer had five selecting relays, actuated by impulses from a revolving unison wheel. After the five code signals were received, a sixth impulse printed the selected letter. Called the Buckingham-Barclay, by 1910 it was used on 31,171 miles of sixty-three trunk lines to handle about 100 words a minute in each direction. A strike in 1907 speeded Western Union's conversion from Morse to automatic operation on 24,888 miles of line.

A duplex printing system by John E. Wright, a telegrapher, was used by Postal from 1909 to 1911, but Postal became discouraged with printing telegraphy and reverted entirely to Morse.[3] That could not compete with Western Union's printers, and Postal began using a Morkrum start-stop page printer, developed by Charles L. Krum, a cold storage engineer, and his son Howard, with the financial backing of Jay Morton of Chicago, founder of the salt company bearing his name.

"Start-Stop" meant the transmitting and receiving mechanisms made one cycle to select and print each character, and then stopped. The perforated tape transmitter sent the character, and the receiving printer's revolving typewheel printed the letter. Siemens and Halske Company of Berlin also produced a direct keyboard printer eliminating the perforated tape.

Western Union Multiplex System

18.2 MULTIPLEX OPERATOR IN OPERATING ROOM SENDING
MESSAGE BY PERFORATED TAPE.

Western Union's main New York operating rooms, at 195 Broadway since 1875, were moved in 1914 to the building at Sixth Avenue, Lispenard, and Walker Streets. Installed on seven floors were 1042 trunk lines, 581 loops, and eighty-four time-service circuits without interrupting service. Traveling belts and pneumatic tubes sped messages between locations in the building.

Because of the rapid growth of traffic, Western Union's laboratory produced a multiple-circuit printing telegraph system to avoid a large investment in additional lines. Since both Western Electric and Western Union were a part of AT&T then, Western Electric engineers joined in the effort. The Baudot, Murray, and Barclay inventions were used in developing Western Union's Multiplex system.

For two years Western Union engineers, led by W. A. Houghtaling and George Benjamin, tested models and made changes, and in 1915 an eight-channel multiplex was installed on twenty-five four-channel multiplex circuits.

World War I Speeds Operations

When Western Union's World War I telegraph traffic multiplied the demand for multiplex equipment beyond the capacity of Western Electric, Western Union engineers worked with the Kleinschmidt Electric Company to produce a direct, keyboard type-bar printer, based on Barclay's. To speed production, parts bought from typewriter and other companies were assembled in a Western Union shop. The Morkrum Company supplied a new typewheel printer and an improved perforator.

Those printers, installed in multiplex systems during World War I, placed many cities in direct communication with one another and aided the war effort. By 1920 Western Union had 368 automatic circuits handling seventy-five percent of the traffic. Its use of terminal sets had grown from 120 in 1910 to 3,000 in 1926; Postal had 450; AT&T and associated companies, 1,400; and railroads about 100. Western Union had 650 duplex multiplex circuits on trunklines in 1935 with a capacity of more than ten million words an hour, or ninety-five percent of its long distance traffic.

On Associated Press wires, the number of Morse operators reached a peak of 1,300 in 1928. A.P. began replacing Morse keys with printers in 1916, and in 1935 had 3,000 costing $2,500,000.[4]

Utilizing the absence of inductive interference in cables, in 1924 and 1926 Western Union placed the multiplex system in operation on its loaded transatlantic cables. An elderly woman visiting one of the cable stations had the operation explained to her. "I understand all of that," she replied. "What I don't know is how that paper tape crossed the ocean without getting wet!"

18.3 PART OF THE OVERSEAS TELEPHONE SWITCHBOARD IN THE LONG LINES DEPARTMENT OF THE AMERICAN TELEPHONE & TELEGRAPH COMPANY IN NEW YORK IN THE 1930S.

In its early years multiplex printed telegrams directly on blanks, because it was believed the public would not want strips of printed tape gummed to them. However, a tape-printer error could be corrected by merely retransmitting the word and pasting it over the error, instead of resending the entire message. That is why tape use was adopted in 1925.

Since as many as eight telegrams were sent simultaneously by multiplex, its maximum speed was 640 words per minute over a single wire. At each end of the wire four operators used perforators with typewriter-like keyboards. When the key for a letter of the alphabet was struck, a combination of one to five holes was perforated across a narrow paper tape. The tape then passed into a transmitter, called an "iron horse," and the five pins in the "iron horse" slipping into holes in the tape flashed the combination of impulses over the wire. A rotating distributor switched each combination of signals in sequence onto the line. At destination, a printer typed the letter.

A revolving brush contacted the segments on the distributor face plate at each end of the line, so that one transmitter and one receiving printer were connected with a segment of the face plate. As the brushes rotated at the same speed, a brush at each end of the line touched the corresponding contact, and one sending operator and a receiving printer were momentarily connected.

18.4 IN THE 1930S–1950S, AS TELEGRAMS ARRIVED ON TELEGRAPH PRINTERS, OPERATORS GUMMED THE TAPE ON TELEGRAPH BLANKS. THAT ALLOWED THE OPERATORS TO CORRECT ANY ERRORS BEFORE PASTING THE TAPE ON THE BLANK. IN LATER YEARS, ELECTRONIC SWITCHING AND OPERATION ELIMINATED ERRORS AND TELEGRAMS WERE PRINTED BY THE MACHINES DIRECTLY ON THE BLANKS.

Carrier Current Systems

Multiple channels used time division and frequency division multiplex. Following Western Union's installation of time division on trunk lines, AT&T introduced frequency division in 1918 to handle two or more conversations over one pair of wires.

While the Bell System installed 65,000 miles of carrier channels in its 7,500,000-mile plant, Western Union delayed until it could adapt it for telegraph use without infringing AT&T patents. To study AT&T's system, Western Union leased a Western Electric four-wire carrier system in 1930 between New York, Scranton, Elmira, and Buffalo. In three years, Western Union developed an eight-channel carrier system.

The carrier applied a radio technique to wire. Each frequency was, in effect, a different broadcasting station, and each receiving machine a home radio, tuned by electrical filters to receive only on one station's frequency. To avoid interference, the frequencies were spaced 300 cycles apart.

At first the tone generator of a Hammond organ was used to provide twenty-two tones. Each tone was interrupted rapidly in accordance with a code. The higher tones were far beyond the range of the human ear. A metallic circuit of two wires was used, with the current going over one wire and returning over the other, instead of through the ground.

A portable carrier system also was developed, weighing about seventy-five pounds. It could be installed on a pair of wires in ten minutes to increase their message capacity at major news events where wires were scarce.

The carrier conveyed the code by amplitude modulation (AM). A wideband AM system with twenty-two carrier channels was placed in service between New York and Chicago in 1936, and between New York, Washington, and Atlanta in 1937. Later, Armstrong's wideband frequency modulation (FM) was used, in which the receiver responded only to frequency change. A slight change in frequency conveyed the code impulses. Amazingly immune from fading and outside electrical disturbances, FM was just what was needed for carrier and radio beam. It provided a "super highway" on which there were a large number of voice bands, on each of which a number of telegraph channels could be operated.

When World War II began, the FM telegraph carrier was used to quickly increase facilities at Washington, D.C. The carrier provided millions of miles of circuits without adding a mile of wire. One user was the Civil Aeronautics Administration private wire system. By 1940 carrier was used to send as many as 288 messages simultaneously over a single pair of wires, and later more than 2,000 over a microwave radio beam system.

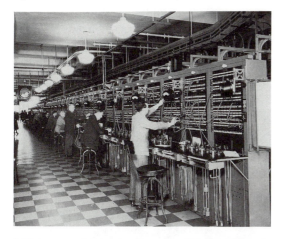

18.5 A BLOCK-LONG SWITCHBOARD IN WESTERN UNION'S NEW YORK HEADQUARTERS BUILDING HAD THOUSANDS OF CONNECTIONS WITH TELEGRAPH LINES THROUGHOUT THE COUNTRY. THOSE CONNECTIONS NORMALLY WERE UNCHANGED, BUT WHEN STORMS BROKE LINES OR MAJOR NEWS EVENTS OCCURRED, CONNECTIONS ON THIS BOARD WERE CHANGED TO ROUTE MESSAGES ON LINES AROUND THE BREAK OR TO PROVIDE ADDITIONAL LINES TO HANDLE MILLIONS OF WORDS OF NEWS.

Switchboards and terminal repeaters were used in main telegraph offices to test and regulate circuits, and to shift the flow of traffic from one line to another. In major cities dozens of "T&R" men worked at long switchboards, in which thousands of wires were interconnected.

Usually wires were permanently connected through the boards to provide for normal circuit assignments. Cords and plugs were used to test circuits, route traffic around a line broken by a storm, or connect circuits for a major news event. Orders for circuit switching were flashed to other cities by the chief dispatcher and his staff at New York. A national political convention, a baseball World Series, or a major disaster required setting up numerous circuits, and sending equipment and operators to transmit millions of words from the scene.

Birth of Teletype

As the telegraph business grew in the 1920s and 1930s, thousands of direct wires were added to link main offices with branch offices in hotels, railroad stations, and other public places. Also, firms using tielines to send and receive telegrams multiplied, as did companies leasing private wires. The tielines radically reduced messenger pickups and deliveries of telegrams. It didn't make sense to flash a message across the country and then wait until a messenger could pedal his bicycle to make delivery. A simple machine was needed that employees of all firms could use to send and receive messages.

The result was the teleprinter that became the "work horse" of communications. Western Union had rented start-stop, typebar printers, from the Morkrum Company in 1912, to meet the need for small-volume operation between small places, but their magnets and contacts often needed attention. In 1917, an improved printer was developed and installed on 105 circuits.

When the Associated Press had trouble distributing the swollen volume of war news to the New York newspapers in 1915, A.P. general manager Melville E. Stone called Kent Cooper of his Chicago staff to solve the problem. Cooper had Howard Krum install printer lines to the New York papers. Cooper succeeded Stone and was A.P. general manager for many years.

The Kleinschmidt Company was unprofitable and in 1917 was reorganized. Its real owners were Albert Henry Wiggin, president of Chase National Bank and longtime Western Union director; Charles B. Goodspeed of Buckeye Steel Casting Co.; and Edward Moore, son of Judge Moore of American Can. Faced with tough competition, Morkrum elected Sterling Morton, son of Jay, as its president. Goodspeed offered to sell the Kleinschmidt company to Western Union for $412,000, but President Carlton refused.[5] Morton then tried to sell it to Western Electric, without success.

Still lacking a simple direct keyboard printer for branch office and customer lines, Western Union worked with Morkrum again in 1920 and produced a Simplex Printer, a "typewheel tape teletype," but parts failed, and speed was only forty words a minute.[6] Working with Western Union engineers, Morkrum built a new model in 1923. Kleinschmidt Electric at Long Island City, New York also produced a Simplex printer.

The Simplex printers of both companies performed well, but became involved in patent suits that could have been disastrous to both. They merged in 1924 in the Morkrum-Kleinschmidt Company, and combined the best features of each printer. It was the printer Western Union had aided in testing and developing for a dozen years, and the company installed the first 100 in 1925.

Two families of printers were made by Morkrum-Kleinschmidt—the typewheel and typebar. The typewheel printer had the alphabet spaced around the rim of a wheel, mounted on a shaft attached to a gear wheel. Each time the armature of a magnet was attracted, the typewheel revolved until the desired letter faced a ribbon of paper. Then a reverse or increased current impulse caused another magnet to push the paper against the type and print the letter.

The typebar printer had wider use. Imag-ine a super typewriter with its keyboard in New York and its typebars in San Francisco, printing a letter the instant you touch the key. The signals traveling over the wire select the correct typebar, make the shift to figures, and make spaces between words and lines. Codes, like Baudot, were used to send thirty-two combinations of impulses. The typebar printer was mostly used between main branch offices. Multiplex was used on trunklines where the perforated tape could be checked for errors before it passed through the transmitter.

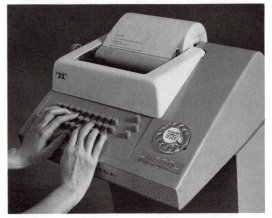

18.6 PRESSING A KEY ON A TELETYPE PRINTER NAMED TELEX CAUSED A TYPEBAR OF A SIMILAR MACHINE IN A DISTANT CITY TO PRINT THE SAME LETTER. WESTERN UNION BOUGHT THOUSANDS FOR USE IN BRANCH AND CUSTOMERS' OFFICES TO SEND AND RECEIVE TELEGRAMS. THEY ALSO WERE USED BY BUSINESSMEN FOR "TELEGRAPH CONVERSATIONS."

An important policy decision was made when S. M. Barr of Western Union's Plant Department prepared a memorandum that W. C. Titley, plant vice president, sent to President Carlton. It proposed that the company buy control of Morkrum-Kleinschmidt, capitalized at $1,500,000.[7] Carlton discussed the matter with Engineering Vice President G. M. Yorke, a former telephone man. Although Western Union planned to invest many millions in printers, Carlton had his secretary pass a "no" decision to Titley. It meant that Western Union's policy, as a communications carrier, was not to engage in manufacturing.

Morkrum-Kleinschmidt changed its name to Teletype Corporation, and from 1925 to 1930 Western Union bought Simplex printers from it in lots of 500 to 6,000. The company continued to make improvements in the light of operating experience. Although it refused to buy control of its manufacturer for less than $1,000,000, in 1930 Western Union bought 10,000 of the printers for $7,500,000, and the associated 100-wire concentrators for $1,500,000.[8] In 1932, Western Union had 18,500 printers. Teletype split its stock fifteen for one and paid dividends as high as twelve dollars a share.

The Bell System provided more printing telegraph circuits to customers each year: 70,000 miles in 1925; 175,000 in 1927; and 280,000 in 1928. The number of telegraph printers Bell used grew from 1,000 in 1925 to 16,000 in 1929. Their varied uses forecast trouble for Western Union. Until 1925, customers bought machines from Teletype Company and leased the circuits. Then AT&T paid Teletype $60,000 to redesign and simplify the page printer. Printer models 15 and 19 resulted.

Those were boom times for Teletype, and Morton decided it was a good time to sell. Colonel Behn of ITT refused it. Clarence G. Stoll, vice president of Western Electric, knew his company needed thousands of printers. On October 1, 1930, AT&T bought Teletype for $30,000,000 in a one-for-one share exchange of common stock, plus $1,467,795 to retire Teletype's preferred stock. The foreign rights were sold to ITT for about $1,250,000. AT&T's purchase placed Western Union at the mercy of AT&T when the two companies were competing for private wire contracts. A delay in obtaining printers could be fatal.

18.7 SCENE IN MAIN OPERATING ROOM AT WESTERN UNION'S BUILDING, 60 HUDSON STREET, NEW YORK CITY, IN THE 1940S.

Between 1925 and 1931, automatic operation reduced the number of Morse operators from sixty-three and one-tenth to fifteen and one-half percent of the total.[9] In 1930 printers handled eighty-six and four-tenths percent of 240 million Western Union messages; in 1931 it was ninety percent. Women operators, only twenty-two and three-tenths percent of the force in 1902, were sixty percent in 1931. The number of employees grew from 33,000 in 1910 to 60,000 in 1930; annual revenues, from $33,900,000 to $133,2000,000.

The thousands of printers placed in customers' offices provided direct-wire operation to telegraph offices. With direct wires from hundreds of companies to its local main offices, and many of them in use only a few times in a day, Western Union would have required thousands of printers in its operating rooms to connect one directly with each customer. Instead, it installed

18.8 Scenes in Western Union headquarters building, 60 Hudson Street, New York City. A pneumatic tube system linked twenty-three floors and departments, and underground to other offices in the city. Telegrams and other messages were placed in small cylinders and containers which sped through the tubes. More than 500 W.U. offices in other cities used 2,000,000 feet of tubes until electronic switching and teletype replaced them.

concentrators (switchboards) to which an operator could connect any of 100 customers and branch offices.

Teletypewriter Exchange Service

Following AT&T's purchase of the Teletype Corporation, as a subsidiary of Western Electric, it inaugurated Teletypewriter Exchange Service (TWX) on November 21, 1931. Central switching exchanges were established through which a subscriber could communicate by teletypewriter with any other subscriber. To make a call, the customer looked up the number in a nationwide TWX directory and called the operator to connect him with the desired party. Then the two subscribers could type their messages and replies. The basic charge was for five minutes, based on time and distance.

Morton had presented the TWX idea to the telegraph companies first, and received no encouragement, but a central office switchboard had been opened in London on July 1, 1910, to provide direct connections between subscribers, and Donald Murray predicted that telegraph offices would become automatic switching exchanges.[10]

Before starting TWX service, AT&T offered to include Western Union and Postal if they would guarantee thirty dollars a month to the telephone company for each installation. The telegraph companies refused. They saw little to be gained by helping the telephone company take over their business, destroy their identity with customers, delay delivery of messages when others were using the tieline, and wreck their own extensive tieline systems.[11]

To offset TWX, the telegraph companies inaugurated Timed Wire Service on December 1, 1931, placing "the telegraph facilities at the customer's disposal on a time basis." The sender would transmit a message by teleprinter, and the charge would be based on the distance and time required to send it. The TWX minimum charge was for five minutes, but for Timed Wire, three. This was a bargain because most messages could be transmitted in three minutes for about the cost of a fifty-word Day Letter. A five-minute telephone call from New York to San Francisco cost fifteen dollars; TWX, five minutes for four dollars; Timed Wire, three minutes for two dollars forty cents.

In three years, however, annual TWX revenue was $3,800,000, and Western Union's Timed Wire $1,700,000. The reason was that TWX provided two-way communications, while Timed Wire was one-way. When its TWX installations fell from 2,960 in 1932 to 2,447 in 1933, AT&T cut its rates twenty-five percent, reduced the minimum time to three minutes, and provided printer-perforators so that customers could accumulate tape until they had enough for three minutes. That resulted in rapid growth to 7,000 machines by the end of 1935. It became apparent, however, that much of the TWX business had been diverted from long distance telephone during a depression in which many telephone employees were on part-time, and all were told to push sales. The Bell System lost 290,000 telephones in 1931, and during the next four years did not earn its nine-dollar dividend, but paid it in part out of accumulated surplus.

For large companies, AT&T provided cost-saving combinations of TWX and leased private wire, accumulating and relaying messages at strategic points, instead of sending them direct to all points, and offering conference calls. Telegraph officials complained to the government that AT&T was providing TWX at less than cost and subsidizing it with high rates paid by local telephone users. Their protests fell upon deaf ears.

In the 1930s typebar printers that made carbon copies seemed destined to replace the typewheel printers, and Western Union developed a better typebar printer. Then it ordered the parts from Thomas A. Edison, Inc. of Orange, New Jersey; Underwood Elliott Fisher Company of Hartford, Conn., and others. It assembled 7,500 printers in its own shops at a saving of $2,000,000.

Another move to compete with TWX was the Varioplex. Noticing the difference in water pressure when more than one faucet in his home was turned on, Philo Holcomb, a Western Union engineer, got the idea of the Varioplex. He divided the capacity of a four channel multiplex system among as many as thirty customers, who did not have enough traffic individually to lease a channel and keep it busy. Telemeter Service, made possible by the Varioplex, was inaugurated on June 8, 1936, and gave each customer the advantage of low-cost direct-wire transmission with one correspondent, such as between headquarters and plant. The idle-

18.9 ROUTE AIDES CARRIED TELEGRAMS FROM ONE PART OF A LARGE TELEGRAPH OFFICE TO ANOTHER.

channel time of each customer was automatically at the disposal of the others. A meter counted the words, and charges were based on distance and a minimum of 25,000 words a month.[12]

Used also on ocean cables, Varioplex systems provided leased direct printer circuits to overseas offices of brokers, banks, and others. In April 1943, Varioplex began providing direct wartime circuits between the British Foreign Office, British Admiralty, and British agencies in the U.S., and between the U.S. State, War, and Navy Departments as well as their overseas offices.

In 1935, David Sarnoff of RCA predicted that printing telegraphy would be dead in five years, replaced by facsimile, but the thousands of printers AT&T and Western Union installed proved how wrong he was again.

How little competition Timed Wire and Telemeter gave AT&T was indicated by the rapid growth of TWX stations: 1932, 2,521; 1942, 16,595; 1952, 33,338; to 1962, 58,530.

High-Speed Message-Switching Network

The Multiplex system provided direct long-distance transmission between large cities without manual relays, but did not eliminate the several hundred million manual retransmissions of messages to and from smaller places.

As F. E. d'Humy,[13] engineering vice president of Western Union for many years, pointed out in March 1940, the telegraph business consisted of the transmission of multitudinous short messages from a large number of sending points, crisscrossing in every direction to an equally large number of receiving points.

Each telegram required handling by several people at each city where trunk lines met. There, arriving messages were sorted by destination. Fast-moving V belts and drag belts then carried them to sections of the operating rooms, where girls on roller skates glided gracefully to take each message to the operator sending to that destination.

To solve that problem, Western Union mechanized its printing telegraph network in the 1940s by installing fifteen high-speed "reperforator" message-switching centers, each serving an area of two or more states. Those big centers were at Richmond, Atlanta, St. Louis, Dallas, Oakland, Philadelphia, Cincinnati, Boston, Kansas City, Minneapolis, Syracuse, Detroit, Los Angeles, New Orleans, and Portland, Oregon.[14]

The centers were interconnected by trunk lines, and each was linked by direct wire with all telegraph offices in its area. To send a telegram, the originating operator typed two "call letters," indicating the switching center in the destination area, and then the message. The two call letters, arriving at the area center, caused an "electrical brain" automatically to select a trunk line to the destination area center and send the message. The message arrived in that center on perforated tape with the words printed above the perforations. Seeing the destination on the tape, a clerk pushed a button bearing the name of that city or town, causing the mes-

sage to be retransmitted to destination. The telegram was typed once at origin, and sped to its destination with human aid needed only to press a button.

Postal Telegraph also developed a message-center system. Employees tore incoming perforated tape off, message by message, and carried it to operators for automatic retransmission.

At first AT&T also had long-distance connections made by operators at each main point along the route, but in the 1930s AT&T set up a trunk-line system linking eight regional centers. Those centers reached 150 "primary outlet" cities, and they, in turn, had direct wires to 2,500 toll centers. Thus a maximum of four operators was required for a long distance call.

Walter S. Gifford, who succeeded Thayer as AT&T president in January 1925, was a Vail protege. While Gifford was president from 1925 to February 18, 1948, and chairman until December 31, 1949, the company's assets increased from three to more than ten billion dollars, and operating revenues from 655 million to two billion dollars. A friendly man with great ability, Gifford served from 1950 to 1953 as ambassador to Great Britain.

Wearing a bow tie and smiling, Gifford participated readily in public and industry celebrations, and encouraged cordial relations by his staff with Western Union people, including myself.

At a 1927 public utility conference in Dallas, Gifford stated that the basic Bell System policy was "to furnish the best possible telephone service at the lowest cost consistant with financial safety," putting public service ahead of profit. If earnings exceeded what was necessary to provide good service, he said, rates would be reduced. Gifford's speech was written by

Arthur W. Page, AT&T public relations vice president, who said all ideas in the speech were from Vail's speeches. The son of Walter Hines Page, a Doubleday, Page & Company partner who was President Wilson's ambassador to Great Britain, Page persuaded the telephone company to place public relations in the forefront of management responsibilities, staffed with public relations professionals.[15] AT&T was a far cry indeed from the Jay Gould era robbery of the public.[16]

Time Service[17]

18.10 James Hamblett was called Father Time because he led the establishment of Western Union Time Service, providing U.S. Naval Observatory time to radio stations, businesses, schools, and government offices nationwide. A time ball on the roof of Western Union Building at 195 Broadway signalled 12:00 noon to New Yorkers who checked their watches and clocks by it.

Until the 1880s every city had a different time, often two different times. One was by setting clocks at 12:00 when the sun was exactly overhead, and the other by the railroad, which used the time of its principal city. In 1877 Western Union began receiving the noon signals from the U.S. Naval Observatory in Washington, which automatically started a large iron ball sliding down a tower, visible for miles, on the roof of its tall building at 195 Broadway, New York. People in the streets and sailors on ships in the harbor set their watches by the time ball until the forty-seven-story Singer building obstructed the view in 1908.

James A. Hamblett of E. Howard & Company, Boston clockmakers, was hired by Western Union to provide Naval Observatory time over its network. Hamblet first established a time service to jewelers and other stores in Maiden Lane and John Street, New York, and developed a master clock with a cog wheel to transmit time signals automatically to subscribers. He gradually extended the company's time service to thousands of towns. With his snow-white beard, bushy hair, and glasses, Hamblet became widely known as "Father Time."

Charles Ferdinand Dowd, a Saratoga, New York schoolmaster, presented a plan to a railroad convention at New York in 1869 to divide the United States into four time zones.[18] In 1878, Sir Sandford Fleming, chief engineer of the Intercolonial Railway between Halifax and Montreal, proposed to divide the earth into twenty-four time zones, beginning with zero meridian (which is marked by the Greenwich Observatory, just southeast of London). His plan was adopted by an international time conference at Washington, D.C. on November 18, 1883. Congress passed the law on March 13, 1884.

The Self-Winding Clock Company of Brooklyn produced master clocks in 1885 that could be electrically synchronized, and Western Union installed them in its offices nationwide. At 11:55 a.m. daily, the seconds beats flashed from the U.S. Naval Observatory over the company's time network until, on the stroke of twelve, each master clock was checked for accuracy. Then for twenty-four hours the master clocks sent an hourly synchronizing impulse to all subscriber clocks at railroads, government buildings, schools, theaters, stores, factories, banks, stock exchanges, and on buildings. In later years radio stations and others replaced the master clocks with electric clocks.

Weather Maps and Reports[19]

Every morning after the 1860s the Western Union chief operator in every large city received telegrams from other cities reporting the weather, temperature, and direction of wind. He then set up his circuits to work with the least interference from the weather.

Cincinnati Manager Frank A. Armstrong realized these reports would be useful in business and agriculture, and prepared a weather map. On July 28, 1868, he persuaded the Chamber of Commerce and the Associated Press to support a daily weather map service.

Armstrong issued a daily weather digest to the A.P., and a map to local subscribers until October 10, 1870.[20] The Cincinnati Chamber of Commerce sent a representative to New York to see Western Union's president, who said his company would collect the reports without charge if meteorological instruments were provided for its offices. For three months the Chamber paid for the publication of a daily weather bulletin, based on Western Union reports from many cities.

Congress then appropriated funds, and on November 1, 1870 the U.S. Signal Corps began issuing a tri-daily weather map.

For the next sixty years reports were telegraphed from 250 stations of the U.S. Weather Bureau (now the National Weather Service). On the walls of Western Union's New York and Chicago operating rooms was an array of pigeon holes. As each report arrived, it was mimeographed on a sheaf of blanks ready-addressed to all of the 140 National Weather Service stations taking the report of that city, and transmitted to them. Nearly all of the 250 reports were taken by the Weather Service Offices at Washington, Chicago, New Orleans, Denver, and San Francisco, where the official forecasts were made for their sections of the nation. After 1930 these forecasts also were telegraphed over a leased National Weather Service network.

In recent years weather satellites have transmitted photographs at scheduled intervals. When a tropical depression might develop into a hurricane, airplanes check it for the Hurricane Bureau at Miami, Florida, and frequent satellite photographs track the storm's movement and development. Warnings are transmitted to Weather Service offices for the media.

The National Meteorological Center at Washington provides analyses and forecasts to National Weather Service offices and private subscribers over an 1,100-station, 23,000-mile facsimile network. Area forecasts are transmitted four times a day to about 650 government, military, and National Weather Service stations in more than 300 cities.

The Federal Aviation Administration used a nationwide teletype network to collect and transmit weather data to its offices, airports, and airline agencies. The Strategic Air

Command gathered weather data from its flights and military stations around the world, and transmitted maps over a leased facsimile network to its operations centers. These maps, arriving around the clock, not only included information needed in commercial flying, but high altitude data required for such purposes as ballistic missile tests.

In the 1980s the National Weather Service used a billion-dollar system that utilized the radar method developed during World War II of reflecting radio waves off objects in the atmosphere, like raindrops, and measuring light-wave frequencies to produce an outline of the weather. Weather warnings save lives and, it has been estimated, save forty-five billion dollars a year in property damages.

Stock Quotations

As business and trade grew in the small 1624 Dutch settlement that surrendered to the British in 1664, and was named New York, traders met daily under a buttonwood tree on Wall Street. On May 17, 1792, they agreed to create what became the New York Stock Exchange. They rented a room at 40 Wall Street and traded railroad, bank, insurance, canal, and other stocks.

After the Civil War, in 1865, other exchanges were established to trade in oil, cotton, corn, and gold at 10-12 Broad Street. The gold exchange there installed the glass-domed Edison ticker in brokers' offices. Its 215 to 225 character-a-minute speed with various improvements was adequate. In the 1870s the tickers were installed in principal cities as far west as Cleveland, with Morse operators retransmitting quotations on to some other cities. After ticker circuits reached Chicago in 1913, the number of

stocks listed grew from 15 to 191 in 1967.

Tickers lagged behind trading on the first "big" day, April 30, 1901, when 3,270,000 shares were traded. There were only eleven three-million-share days before 1929, when the tickers lagged behind trading, but Western Union installed 194 new glass-domed, self-winding tickers, called "300" tickers, because they handled five characters a second.[21] The characters were printed on paper tape that spilled from the ticker.

18.11 CALAHAN TICKER WHICH EDISON IMPROVED.

After a long speculative boom, William C. Durant of General Motors warned President Herbert Hoover in April 1929, that government clamping down on credit would be disastrous. The long bull market crumbled October 22, when six million shares were traded, and climaxed October 23, 1929, "Black Thursday," when 12,890,000 shares changed hands.

The tickers, hours behind, were blamed for confusion that fueled the stampede. It destroyed the jobs and savings of millions. Eighteen billion dollars in values vanished. Some brokers jumped from windows, and sad-faced men sold apples and pencils on street corners and stood in bread lines.

18.12 The Earliest Stock Quotation Tickers Often Got Out of Alignment, but Edison Invented This Universal Stock Ticker Which Could Be Brought into Unison with All Others on the Line from the Central Office Without Sending a Man to Each Location. (Courtesy U.S. Dept. of the Interior, National Park Service, Edison National Historic Site.)

Before the 15,000 miles of ticker tape recording the October 23 transactions were cleared away, Western Union, the New York Stock Exchange (NYSE), and Teletype were developing a 500-character-a-minute ticker system. At a cost of $4,500,000 Western Union added 5,000 miles of circuits, equipped eighty-two automatic repeater stations, had 10,000 "500" tickers manufactured, and trained 800 maintenance men at ticker schools. When all of the new tickers were installed, the system was placed in operation in September 1930. A new trans-

mission table automatically combined the flow of quotations sent by two operators.

For thirty years the "500" ticker met the need of markets trading less than two million shares, but fell far behind on the 13,726,200-share day on May 29, 1962. Again a faster ticker was installed in 1964 that could be stepped up as needed from 500 characters a minute to 600, 700, 800, 900 on the network, then 24,000 miles. Linked with an IBM computer and a new floor-reporting system, the "900" tickers were designed for 10,000,000-share days, after which lights would shine late in a myriad of Wall Street area windows.

In 1963 Western Union speeded up transmission to the 2,600 tickers on its 18,000-mile NYSE system to 355 cities, and, on January 1, 1968, speeded up the American Stock Exchange (ASE) system to 900 characters a minute.

Average daily NYSE trading increased to 12,970,000 shares in 1968, and ASE to 6,400,000. On April 3, 1968, NYSE records were broken with 19,290,000 shares, and seven days later 20,410,000 shares, with the ticker forty-one minutes late. In the 1970s 35,000,000-share days were not unusual; on October 10, 1979 it was 81,600,000; and January 8, 1981, set a new record of 92,890,000 shares.

Again the need for a faster ticker was evident, and alternatives were tried to reduce the number of quotations. A computer was used to provide a summary for sales of less than 100 shares and deals away from the exchange floor.

The telegraph company used the replaced "500" tickers to speed up the Philadelphia-Baltimore-Washington, Midwest, and Pacific Coast Stock Exchange systems.[22] Trading in commodity futures contracts grew to trillions of dollars a year. The NYSE only traded stocks, but on August 7, 1980, it opened a New York Futures Exchange in competition with the Chicago Board of Trade and Chicago Mercantile Exchange.[23]

The Securities and Exchange Commission proposed in 1971 that a National Securities Market be formed to include the comparatively small Boston, Cincinnati, Detroit, Midwest and Pacific Exchanges. The National Securities Dealers Association sponsored, and Congress passed the act in 1976, requiring that each trade be routed automatically to the exchange from which the best bid was received. An eight-million-dollar electronic quotation system—the National Association of Securities Dealers Automated Quotations (NASDAQ)—was provided, and over-the-counter trading grew to about half the volume on the NYSE. Many brokers computerized their back rooms to handle the greater load. A Merrill Lynch computer handled 15,000,000 shares in one day.

In 1979 the Cincinnati Stock Exchange pointed to the big exchanges in future doing away with face-to-face trading. Brokers' orders were transmitted to a computer in Jersey City, New Jersey which matched buy and sell orders and flashed confirmations. The trading system was owned by Control Data Corp.

For many years the telegraph company installed and maintained Trans Lux machines on brokers' boardroom walls to project the ticker tape moving across a long screen about eight inches high. The first automatic posting of stock quotations boards in brokers' rooms was by Teleregister Corporation in 1929. For a decade, Teleregister inventor, president Robert L. Daine, was aided by Western Union which became owner of the company. The automatic electric boards replaced boys writing quotations on blackboards in brokerages. An operator at the central office transmitted the quotations simultaneously to all Teleregister boards, and discs opposite the abbreviation for each company revolved to show the figures.

Western Union sold Teleregister to a group of bankers,[24] and it became a division of Bunker Ramo Corporation, which was acquired in 1980 by Allied Corporation, formerly Allied Chemical.

Teleregister, Scantlin Electronics, Ultronics Systems,[25] a General Telephone subsidiary, and Allied developed deskside or desktop machines on which brokers punched keys or dialed the symbol of a stock, and the latest quotation appeared on a revolving disc or tape. Scantlin followed with its Quotron I,[26] Ultronics with its Stockmaster, and Teleregister with its Telequote, providing high, low, and number of shares. Thousands of computer-operated news machines were provided by Dow Jones to supplement the tickers.[27] A larger General Telephone and Electronics (GTE) machine added customer portfolios, financial and research reports, and news bulletins. In 1975 GTE established a network providing computerized quotations to 400 cities around the world, and added trades on the London, Toronto, Tokyo, Hong Kong, and Singapore exchanges.

The National Association of Securities Dealers Automated Quotations System (NASDAQ) was placed in operation to provide quotations on 1,100 over-the-counter securities. More than 100 major security traders telegraphed hourly reports for NASDAQ to Bunker Ramo's computer center in New York, and brokers dialed their desktop Telequote III, Quotron, or Stockmaster machines for quotations. For more information, brokers used Telequote machines and screens. Some 20,000 over-the-counter stocks were handled by 4,300 firms and 200,000 registered NASDAQ members.

Electronic quotation systems in 1982 reported 136,000 deals for about 19 billion shares. Some people used software programs to plug computers into quotation circuits.

In a five-year bull market beginning August 12, 1982, the Dow Jones climbed from 776 to 2,700, and the market crashed on "Black Monday" October 19, 1987. A half-trillion dollars in paper values disappeared, and the public feared a disastrous depression like the one after the 1929 crash. The Federal Reserve opened credit lines to prevent a banking crisis. Business continued, but it was two years before the high prices and 100- to 200-million-share days returned. This was aided by the purchases of foreign institutions and pension funds. The market then broke the 3,100 Dow Jones record high. It surpassed that record on December 27, 1991, with 3,101.52, and finished the year with 3,178.83.

To reduce quotation volume, the automated daily opening report paired off market orders for 5,099 shares or less that arrived overnight. A Tandem mainframe computer with software provided a "designated order turnaround" that routed orders of 1,099 shares or less to a specialist who summarized them according to price. The "00" on numbers of shares and even large trades with only small price changes by institutional investors were omitted. A batching system to lump together trades in the most active stocks also was planned.[28]

Little known to many people, the ticker plays an important part in financing the companies that provide our daily needs. The prices for bread, butter and eggs, poultry, potatoes, sugar, and clothing are determined by bargaining on the floors of commodity, futures, cotton, coffee, sugar, and cocoa exchanges. Millions of yards of used ticker tape once were thrown from skyscraper windows to applaud heroes riding up Broadway, but its use in tickers affected our personal and daily lives.

Quotations and other financial news sped over leased wires from the Associated Press, United Press, Dow Jones, and Reuters to broadcasters and newspapers.[29] Telegraphed financial reports reach brokers, newspapers, magazines, radio, and television stations on Dow Jones and other broad tape news printers. Large brokers in turn digest the news and send their opinions over leased wire systems. In the future there may be one computerized communications network reporting NYSE, Amex, and over-the-counter trades. The rise and fall of securities prices concerns about thirty million stockholders, and the pension funds of 100 million people.

Since fortunes are built on receiving financial information, the business of providing it will grow. *U.S. News and World Report* predicted it would reach $1,100,000,000 in 1990.

The importance of communications to the financial world was illustrated when an AT&T repair crew mistakenly severed a cable in Newark, New Jersey, and put the busy New York Stock Exchange out of business for hours until an emergency cable could be substituted.

Another business use, initiated by telegraph men, was flowers by wire. Western Union Superintendent Valentine at St. Louis noticed the rapid increase in telegraph money orders, and suggested that florists organize an exchange of flowers by wire. The service spread nationwide, with the Florists Telegraph Delivery (FTD) clearing house and headquarters in Detroit. Western Union added millions to its revenues. Along with others, I have attended the annual conventions in different cities, and I assisted with advertising and publicity until Vice President

Willever called a halt on the expense. AT&T gradually took over FTD communications.

Notes

[1]This chapter may be of special interest to engineers. Sources: mimeographed compilations by Western Union's H. W. Drake (equipment engineer) and Hobart Mason (engineer of lines); "Development and Improvement in the Telegraphic Art" (1900 to about 1934) 37 pages; F. E. d'Humy (vice president–engineering), "Brief Outline of Technical Progress by Western Union 1910–1934," 562 pages; P. J. Howe (assistant chief engineer), "History of Technical Progress, 1935–1945," 505 pages; and articles in *Western Union Technical Review.*

[2]It was adopted in France and Italy in 1862; England, 1863; Russia and Germany, 1865; Austria, 1867; Holland, 1868; Switzerland and Belgium, 1870; for European international circuits, 1872; and Spain, 1875.

[3]Donald McNicol, "Big Pole, Little Pole," *Telegraph and Telephone Age* (January 1944).

[4]Newark *Sunday Call,* November 3, 1935.

[5]"$30,000,000 Worth of Teletype," *Fortune Magazine* (1932).

[6]*Telegraph and Telephone Age* (May 16, 1922) 227.

[7]Barr gave me a copy of the memorandum.

[8]See my article in *Telegraph and Telephone Age* (April 16, 1930) 175-76.

[9]In 1926, Western Union had 20,000 Morse operators; in 1944, just 2,000. In 1965, 122 nationwide, but practically all worked teleprinters, and were near retirement.

[10]*Journal of the Institution of Electrical Engineers* (London, March 1925).

[11]Memorandum by Charles E. Davis, Western Union official who analyzed the proposal and recommended the refusal.

[12]This is as described to me by the inventor.

[13]A man of great vision, d'Humy also was the inventor of the drive-up ramp garage.

[14]Of this I have firsthand knowledge: I managed the inaugurations at all fifteen centers.

[15]John Brooks, *Telephone* (New York: Harper & Row, 1975) 173.

[16]Prescott C. Mabon, "A Personal Perspective on Bell System Public Relations," a Bell booklet. (Mabon succeeded Page as A.T.&T. public relations vice president.)

[17]The following information on time service was checked by the Time Service Division, U.S. Naval Observatory.

[18]See also above, chap. 13, on "Old 195 Broadway."

[19]This section was checked for accuracy by the National Weather Service.

[20]W. F. McClure, age 90, told on May 10, 1935, how he placed the temperature and direction of the wind on the maps Armstrong provided to firms handling perishable goods.

[21]The exact number varied because the alphabet was in sequence around the typewheel, "AB" could be printed in $1/_{30}$ of a revolution of the wheel, while "BA" required $29/_{30}$. American Can, originally abbreviated CAN, required about one-and-a-half revolutions but it was changed to AC, only two steps on the typewheel. More than 200 company abbreviations were changed for that reason.

[22]The Midwest Exchange in Chicago was a consolidation in 1949 of the Chicago, Cleveland, St. Louis, Minneapolis, and St. Paul exchanges. The Pacific Coast Exchange was formed in 1957 by combining the San Francisco and Los Angeles exchanges. The New York Coffee, Sugar, and Cocoa Exchanges merged in 1979.

[23]The Futures Market traded in Treasury bonds and bills and foreign currencies.

[24]Source: personal knowledge. I attended the closing for Western Union.

[25]In 1970 Ultronics received quotations in twenty countries on six continents through a

transmission center in London. It reported transactions on stock exchanges in North America and Europe to computers serving brokers in Paris, Frankfurt, Geneva, and Amsterdam.

[26]Quotron, a unit of Citicorp Bank, was acquired by Phoenix Technologies.

[27]*Forbes Magazine* (May 16, 1962) said 330-pound Boston tycoon Clarence Barron bought the Dow-Jones ticker company and its *Wall Street Journal* in 1902 from Charles Dow and Edward Jones and expanded it. He left his estate to an adopted daughter, Mrs. Jane Bancroft. In 1962 her elderly daughters, Mrs. Jessie Cox and Mrs. Jane Cook, their children and grandchildren owned 58.3 percent of the great financial news company that also included *Barrons Magazine* and *Financial News Service* and a book publisher, all worth nearly one billion dollars.

[28]*Forbes Magazine* (October 8,1984).

[29]The AP used two computers to process and transmit stock quotations over its leased multiplex-telegraph system. The quotations moved at speeds up to 2,100 words a minute and arrived in newspaper offices nationwide in the form of typesetter tape, minutes after the New York and American Stock Exchanges closed. The computers replaced forty tabulators, fifteen typesetter operators, and twelve women who had checked the tape for errors.

Facsimile, World War II, and Disastrous Merger

Roy B. White, president of the Central Railroad of New Jersey, was elected president of Western Union on June 1, 1933. He was brought in during the great depression as a "hatchet man" to cut costs. On the day of his arrival he added a touch of comedy. At the door to the presidential office he stationed a plump ex-Pullman porter who wore a sky blue, swallow tail jacket with two rows of shining brass buttons as bright as the teeth exposed by a wide grin. The general effect was a bullfrog about to jump.

The new president wanted no suggestions. Sometimes he would send for me or another employee, talk for an hour in generalities, refuse to hear any questions or ideas, and dismiss the visitor, saying, "See what you can make out of that." Of course, nothing was.

Although Western Union's wealthy board included Vincent Astor, W. A. Harriman, George W. Davison, William Fahnestock, Edwin G. Merrill, John M. Schiff, William K. Vanderbilt, Albert H. Wiggin, Frederick H. Ecker, and Donald G. Geddes,

it had become a "beer and pretzels" business.

In the prosperous 1920s Western Union had liberal employee benefit plans and distributed a $16,700,000 bonus among its 75,000 employees. Carlton, now board chairman, had invested $415,694,458 to build a modern automatic-operated plant, but it could handle three times the business it had. Dividends of eight dollars shrank to zero during White's eight-year regime. Operating revenues dropped more than fifty percent, to $85,191,000 in 1938.

All of the carriers suffered. In 1932, even AT&T, which had paid nine dollars a share for many years, earned only $5.96, and lost 1,650,000 subscribers. However, until 1936, AT&T paid the nine dollars as a "moral obligation" to its stockholders. In 1932 Western Electric lost about ten million dollars and laid off nearly eighty percent of its employees. Putting dial telephones into use, Bell eliminated 70,000 operators in 1935.

Postal lost $1,233,122 in the first nine

months of 1932, and cut expenses drastically after losing $1,379,948 in the same months of 1931. On January 21, 1933, *The Magazine of Wall Street* mistakenly said AT&T and the telegraph companies probably had seen their greatest expansion, and that ITT was purely a gamble as an investment. By 1938 AT&T's telegraph revenues from leasing wires grew to $16,834,000, and teletype to $6,803,000. Like the irresistable movement of a glacier, AT&T continued its invasion of the telegraph business.

White had little knowledge of the telegraph. He gave each department quotas to cut costs, and then more quotas. To meet them, thousands of employees were fired or furloughed. After White's third year only 51,683 remained, and they had long vacations without pay and lower salaries.

There was a reign of terror. When employees made outstanding successes, Senior Vice President J. C. Willever would order, "No publicity. He (or she) would want more money." Employees said, "Yes," when a superior was wrong, and obedience to orders produced industry legends. One episode will illustrate.

Willever wired Kenneth W. Heberton, Western Union manager at Syracuse, New York to buy a certain prize bull for $700 for his farm at Washington, New Jersey. The animal was to be auctioned at the state fair. Heberton located the auctioneer, entertained him royally, and said his (Heberton's) job depended on the purchase. When Heberton called out "$700" at the sale, the auctioneer banged his gavel and yelled "Sold," amid startled protests from bidders.

The young manager pieced the $700 together from personal and local company funds, bought a blanket, and sheltered the animal overnight at the fair grounds. The next morning the bull was so bloated from oats and water that a veterinarian had to be called. The medicated bull took off across the fair grounds with Heberton in wild pursuit. When he caught the bull, it would not budge. Heberton twisted its tail, and a sudden blast covered Heberton's new suit! Finally the bull reached the railroad, where a crate was built. Forcing the bull into the crate by more tail twisting completed the ruin of Heberton's suit. After desperate effort and hundreds of dollars of personal expense, he finally got the bull off to Willever.

A quarter of a century later, when Heberton was the company's executive representative at Washington, D.C., Willever told him the bull earned $14,000 in "stud fees," but fear of Willever still prevented Heberton's telling about the purchase.

White had American District Telegraph, the burglar and fire alarm company controlled by Western Union, declare an extra dividend in 1933. That $1,200,000 and stringent economy enabled him to show a profit of $4,400,000. He planted spies in the company, and officials, learning their identity, adopted their suggestions no matter how foolish. They included such money losers as the sale of stamps to be used to pay for telegraph services, and Kiddiegrams to be sent to children.

When White resigned on June 1, 1941, the company was on a profitable basis, but he was a railroad man, and seized an opportunity to be president of the Baltimore and Ohio Railroad.

Facsimile Telegraphy and Television

During White's regime, facsimile was hailed as "The Telegraph of Tomorrow." People always wished to see distant things.

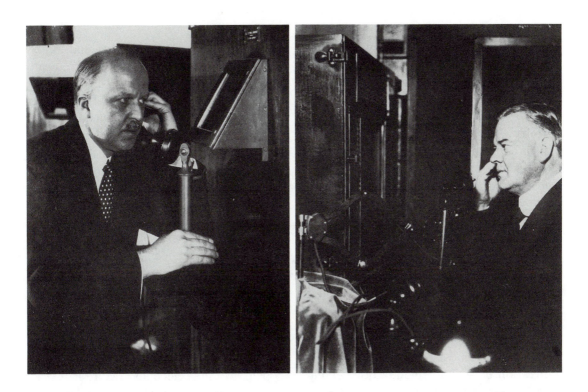

19.1 First public demonstration of long distance transmission of television in 1927, with AT&T President Walter Gifford in New York talking with and seeing Secretary of Commerce Herbert Hoover in Washington. (Courtesy of AT&T Archives.)

Back in 1847 F. C. Bakewell of London drew a design with an insulating fluid on a revolving cylinder. When a metal point passed over the design, current flowing over a line was interrupted, and made a corresponding mark on chemically treated paper on the cylinder. In 1851 he transmitted handwriting.

History repeats, and facsimile returned eighty years later. The advent of vacuum tube amplifiers around 1910 made it possible to detect minute electrical currents, and telegraph engineers realized pictures are composed of small bits that can be transmitted by converting them into electric impulses of varying strength.

In 1924 and 1925 a facsimile system, named "Telepix," was developed by Western Union in cooperation with newspaper inter-ests, and twenty papers were served over 8,000 miles of leased wire. It was costly, slow, and discontinued after one year. AT&T developed the more elaborate "Tele-photo" in 1925, but Western Union, as agent for it, found few customers. Later Telephoto was improved, and the Associated Press used it. Facsimile picture systems were developed also by Siemens and Halske in Germany, and by Belin in France. One line operated between Paris and Lyon in 1938.

Facsimile revived the idea of television. A German named Nipkow had sent images over wires in 1884 with a rotating disc that had small holes. The image was scanned by a light beam shining through the openings, and corresponding electrical impulses were transmitted. In 1925, Scottish inventor John Logie Baird transmitted the image of a doll.

Herbert E. Ives and his group in Bell Laboratories transmitted pictures in 1923 with a photoelectric cell and vacuum tube repeater, and in 1925, sent recognizable faces. In 1925 I witnessed a demonstration of television in New Jersey by Charles Francis Jenkins, whose scanning was much like Nipkow's. People walking in another tent were easily recognized. Using a scanning beam produced by Frank Gray, the first successful American TV demonstration over a distance was made on April 7, 1927. It showed Secretary of Commerce Herbert Hoover in Washington on a small screen, talking with President Gifford at New York.

On April 4, 1928, AT&T transmitted the first motion picture by wireless. V. K. Zworykin of RCA invented a cathode ray iconoscope camera tube in 1928, and Philo T. Farnsworth created a cathode ray receiving tube. The scanning disc was then discarded, and electronic line-by-line scanning began. In 1928 Baird demonstrated a color TV system in England, as the Bell Laboratories did in the United States.

Also in 1928, G.E.'s station WGY sent a feature act by television to a theater in Schenectady, New York. A two-way picturefone was demonstrated by AT&T in New York City on April 9, 1920. I myself later, at 195 Broadway, talked over a telephone line with a Bell Laboratories engineer miles away on West Street while watching his face on a small screen. By 1937 there were seventeen experimental TV stations and on July 1, 1941, New York's WNBT and the *Milwaukee Journal* were the first the Federal Communications Commission authorized to begin commercial operation.[1]

As more stations were licensed, the TV set in the living room became a status symbol. People tried to tune out vertical lines as they watched Hopalong Cassidy, Ozzie and Harriet, Howdy Doody, Amos and Andy, Lucy, John Cameron Swayze, and Edward R. Murrow. The twenty-four-line systems were gradually improved to 441 lines per frame and thirty frames per second.

TV convinced Western Union that facsimile also could revolutionize its industry, but it could not record messages without photosensitive films, requiring time-consuming processing. Then the head of its facsimile engineering group, Raleigh J. Wise, invented an electrosensitive dry recording paper in 1934 named "Teledeltos." At last messages could be received in typewritten, handwritten, or any other graphic form ready for use.

That "breakthrough" led to the fifteen-billion-dollar recording revenues copying industry today with more than five million copiers in use. Teledeltos was a black, carbon-impregnated paper, coated at first with vermillion-colored mercuric sulphide, but later in a light gray color, by using lead thiosulphate. Western Union sold millions of square feet of the paper for office copying use throughout the country, but was its own best customer because one facsimile telegraph recorder after another came from its laboratories.

The message to be sent was wrapped around a revolving cylinder that moved under a photoelectric cell. Reacting to the light and dark areas of each narrow horizontal strip of the message (about one-hundredth of an inch), the photoelectric cell's signals flashed over the wire. On the recorder, a wire stylus rode on the surface of Teledeltos paper on a revolving cylinder moving horizontally at the same speed as the transmitter drum. The current arriving in the stylus decomposed or burned away minute portions of the dry recording paper's coating.

Use of Western Union's "Telefax" began

on trunk circuits between New York and Buffalo on November 14, 1935, and between New York and Chicago October 15, 1936. Telefax sent telegrams, manuscripts, line drawings, maps, and page proofs for magazines. Since that required expensive long distance circuits, facsimile use grew mainly between local customers and central offices. The first branch office to use it was in the Graybar Building in New York in 1936.

Two years later, in 1938, Chester Carlson, a New York patent lawyer, made his first "Xerographic" image with what he called electrophotography. In 1944 Battelle Memorial Institute contracted to develop Carlson's process, and in 1947 licensed the Haloid Company, a photo papermaker, to develop and market a copying machine. Haloid coined the name Xerography from the Greek words for "dry" and "writing." In 1961 Haloid changed the name of the company to Xerox Corporation because of the success of the Xerox machine. Xerox became a huge global business.

Western Union's first fax machine was designed to send and receive telegraph messages; Xerox's were designed for office copying. Later they were coupled with telephones to send written-record messages (telegraphy) by dialing a telephone number and sending them over telephone lines to other fax users. About a million fax machines a year were sold in the 1980s in a technological explosion comparable with the computer's revolutionary change.

Those Telefax machines were too large and expensive for small-volume customers, and engineers spent months seeking a remedy. Then Garvice H. Ridings, a young telegraph engineer, had a sudden inspiration. He took a small, discarded tomato can to the cellar of his New Jersey home, set it up in an improvised frame to revolve, added an electric eye, and had the rough model of a small, low-cost machine. It delighted Vice President d'Humy; models were developed and production started in 1948.

19.2 AFTER WESTERN UNION ENGINEER RALEIGH WISE INVENTED THE FIRST FACSIMILE COPYING PAPER, ANOTHER W.U. ENGINEER, GARVICE RIDINGS, CREATED A FACSIMILE MACHINE TO SEND AND RECEIVE TELEGRAMS. THE MESSAGE WAS PLACED ON THE CYLINDER AND A BUTTON PRESSED TO MAKE THE CYLINDER REVOLVE. A FACSIMILE OF THE MESSAGE APPEARED ON A SIMILAR DISTANT MACHINE. I NAMED IT DESK-FAX. IT WAS USED IN BRANCH AND MAIN TELEGRAPH OFFICES, AND ON THE DESKS OF MANY THOUSANDS OF BUSINESSMEN.

The small machine was placed on the desks of businessmen, who sent and received telegrams in "picture" form by merely pressing a button. I named it "Desk-Fax." In a few years 38,000 Desk-Fax machines were in use, compared with 23,000 companies with teleprinters. Fifty million telegrams a year were sent and received by Desk-Fax with no possibility of error, and no messenger pickup and delivery.

In 1926 RCA started a picture service to London, and in 1933 to Europe and South America.[2] On June 11, 1936, RCA demonstrated an experimental facsimile circuit between New York and Philadelphia, using ultrashort waves.

Western Union's J. W. Milnor and G. A. Randall introduced a cable system in 1939

19.3 CABLE PHOTO TRANSMITTED FROM LONDON TO NEW YORK SHOWS ATLANTIC CLIPPER LANDING AT SOUTHAMPTON, ENGLAND AFTER ITS FIRST FLIGHT ACROSS THE ATLANTIC, APRIL 4, 1939. THE PHOTO WAS PUBLISHED IN AMERICAN NEWSPAPERS.

that transmitted six-by-seven-inch photographs in twenty minutes. The first one received at New York showed the arrival of the *Yankee Clipper* at Southampton, England, after its first transatlantic flight.

Improvement in quality resulted from the shielding that cables provide, and special electrical networks used to reproduce the gradations of light and shade of the original picture. When the electrical impulses grew weak and distorted in shape, intermediate stations amplified and corrected them. During a complete transmission, signals were amplified in voltage fifteen million times, and unwanted components of current were filtered out. The impulses received were converted to light of varying intensity on photographic film in a gas-filled tube. The film was on a cylinder rotating in darkness in synchronism with the sending cylinder.

Previously unknown quality resulted in strong demand, but only about 1,000 pictures were received before World War II government requirements for "secure" cable circuits

forced Western Union to give up international photo service.

Automatic facsimile machines resembling mail boxes were used at the New York and San Francisco World Fairs in 1939, and in hotels and other public places. The user pressed a button and dropped his message in a slot. The machine sent the message and a panel lighted up the words, "Thank you."[3]

Western Union also provided nationwide weather map facsimile systems to the Air Force Strategic Air Command, Weather Bureau, and Civil Aeronautics Administration. On the National Aeronautics and Space Administration system, stations could dial others and transmit letter-size copy.

Western Union demonstrated High-Speed Fax on March 13, 1951, and used it between New York and Washington. D.C. Pages, inserted in cylinders making 1,800 revolutions a minute were held in place by centrifugal force. A ninety-page magazine could be sent in an hour, the company pointed out, with "no processing required before or after reception."

David Sarnoff of RCA held a conference for the press and government officials at Washington, D.C. Evidently ignorant of the fact that the Morse key and sounder had been replaced long ago, Sarnoff declared that the dots-and-dashes telegraph was doomed and would not last longer, but would be replaced by the "epic advance combining the magic of television with the latest techniques in high frequency radio relaying and photography" that he would demonstrate. Simultaneously with his "demonstration," *Colliers Magazine* appeared with the "great historic breakthrough" proclaimed with photos in its leading article. Sarnoff said Ultrafax could take over a large part of the telegraph and the U.S. Mail.

19.4 Cable photos arriving at New York over Western Union transatlantic cable.

Somehow Sarnoff forgot to mention that Western Union created the paper and machines that made facsimile possible, and had sent facsimile by telegraph for years, or that the 1,000 pages of the *Gone with the Wind* novel he showed and claimed he just received from New York in "less than one minute and a half" required advance filming and development of the film for transmission. After transmission to Washington, the pages required enlarging and processing to make letter-page-size copies.

The receiving process used had been developed by the Eastman Kodak Research Laboratories for the U.S. Army, using chemicals and heat-resistant film. If Sarnoff had revealed the real process, it would have shown he could have hand-carried the book from New York to Washington in less time. After all its publicity, Sarnoff's Ultrafax was quickly forgotten. If anyone else had been guilty of that stunt, the media would have called it a disgrace.

Although facsimile was hailed in the 1930s as "the telegraph of tomorrow," the industry's later development was stronger in printer and data telegraphy.

World War II

After its experience in World War I, the Government did not "take over" the communications companies in World War II.[4] The same experienced people had to operate them, and they could be relied on for patriotic service. There were fifty million radio sets, and nearly everyone listened to the war news, heard the patriotic broadcasts of Roosevelt and Churchill, and aided the war effort.

The telephone and telegraph people again provided "all-out" service to the armed forces and war industries. Though handicapped by shortages of personnel and supplies, they handled an unprecedented volume of traffic. The international cable, radiotelegraph, and radiotelephone operators never forgot the feverish around-the-clock strain of full-capacity operation.

About 9,500 Western Union and 67,000 Bell employees volunteered to enlist in the armed services to handle communications. Others important to the war effort were instructed to remain on their jobs. A great array of telegraph, telephone, and radio equipment was developed. Military groups were trained regularly in company schools to install, maintain, and operate it.

When Italy entered the war against the United States in June 1940, it cut the five British cables to the Middle East, Far East, India, and Australia, and repeatedly bombed the cable station at Malta, killing and injuring the staff. The British then cut the Italian cable to Spain and Horta (Azores).

History repeats. Since its cables from Emden to the Azores were cut during World War I, Germany had laid new cables to the Azores and to Lisbon, Portugal. They were cut by the British on September 3, 1939, and diverted to Allied use. Some British cable ships were lost during the war while picking up, repairing, and diverting cables.

The Cocos Island cable station that was raided by, and caused the sinking of, the German cruiser *Emden* in World War I, was wrecked by a Japanese warship on March 3, 1942. Since the station on the route to Australia could not be defended, the company sent a radio message in plain language (knowing the enemy would intercept it), directing the staff to destroy all instruments because "Cocos has been permanently put out of action." It worked; Cocos was repaired, and the Japanese left that station alone.

Cable stations were prime targets of the enemy raids on Gibraltar, Crete, Port Said, Darwin, England, and elsewhere. Wrecked stations in England were relocated by digging tunnels and underground rooms; operators often worked around the clock in crowded rooms with little food, refusing to stop sending even while bombs fell and nearby buildings burned.

When a bomb hit Western Union's Electra House in London in July 1944, power was restored in twenty-eight minutes, and, except for two killed and others injured, 400 shaken men and women went on sending messages, surrounded by 300 tons of debris. At the Moorgate station operators worked on with crashing bombs and wreckage all around. An assistant called out, "'S all right chum. Don't worry. They're only dropping pamphlets."[5]

To provide telegraph service to new war plants, shipyards, military camps, naval bases, and Government agencies, Western Union's 1944 annual report said, "Since June 1940 over 7,000 projects for these war purposes have been completed." Western Union and AT&T set up direct lines to practically every war industry, airport, arsenal, ordnance depot, and plant, proving ground, airplane plant, Army camp, Naval, and Coast Guard office, ferrying command, medical depot, shipyard, Civil Air Patrol, medical or food plant, and Federal Reserve bank.

One 1941 Western Union project was a 33,000-mile printing telegraph network for the Civil Aeronautics Authority. It linked 180 airports, 400 Weather Bureau offices, Army, Navy, airlines, flying schools, and defense industries. Its purpose was weather reports, but, when occasion demanded, remote control switches were used to connect all points.

Censors delayed news early in the war, especially at London, and editors protested. The AP and UP moved their London bureaus to New York, and newsmen corresponded between London and Paris by cable via New York. At that time Germany had no censorship. U.S. correspondents in Berlin could wire stories to their editors and broadcast news. Germany started censorship later, and the Allies became more efficient in theirs.

Censorship, suppression of information and reports was designed to mislead the enemy, hurt its morale, or influence neutrals. The telephone-telegraph blockade between London and Paris contributed to wild rumors. Many big-headline stories in American papers were not true: A "great battle" with enemy planes over England was actually Allied batteries mistakenly firing on their own planes; the Siegfried line was not pierced; the Belgian dikes were not opened; and a second battle of Jutland was not fought.

Efforts of officials to escape blame for Pearl Harbor also resulted in untrue stories. One history of the war often copied another's errors. One said that a Western Union messenger was pedaling on a bicycle along a dusty road in Hawaii with a message from Washington, D.C. to the commanding general, warning that the Japanese would attack Pearl Harbor. The telegram had long been delayed in transit, the story said, and the messenger carrying it dived into a ditch and hid when the Japanese bombers arrived. I pointed out to the *New York Times* whenever it repeated that story, that it was false: the FCC had never permitted Western Union to have an employee or service in Hawaii. The *Times* ignored all requests for correction.[6]

American telecommunications carriers were notified on December 7, 1941, that the Navy would censor all international traffic, and field censors were stationed in offices of each company to check messages of the United States, British, and Russian Governments. All other international traffic was cleared with Radio and Cable Censorship Offices at New York, San Francisco, and Miami. Codes and registered addresses were prohibited, and amateur radio was banned.

A Defense Communications Board was created on September 24, 1940, by executive order "to coordinate the relationships of all branches of communications to the national defense." It included officials of the FCC, Signal Corps, Naval Communications, Treasury, and State Departments. Aided by various subcommittees, it governed the use of facilities, priorities of materials, and protection of plants and lines. President Gifford of AT&T was chairman of the Industry Advisory Committee.

New Signal Corps Telegraph System

When the United States entered the war in 1941, the Signal Corps handled one million words a day over its 57,000 radio-channel miles. Its only domestic radio channels were from Washington to New York, Boston, Atlanta, Chicago, Omaha, San Francisco, Ft. Sam Houston, Ft. Hayes, Wright Field, and Seattle, with connecting telegraph lines to major camps and stations. Its only overseas circuits were to Honolulu, Manila, Puerto Rico, Panama Canal Zone, and Tientsin, China. The network was operated largely by Morse operators in a small space in the Munitions Building, with out-of-date equipment.[7]

Faced with urgent need, the Signal Corps organized the Army Communications Service (ACS) in 1942 to take over the Traffic Division, Operations Branch, and Office of the Chief Signal Officer. Major General Frank E. Stoner was appointed Assistant Chief Signal Officer and Chief, ACS, and a big network was quickly developed.

Installation of a switching center for semiautomatic relay operations in the Pentagon in July 1943 made telegraph history. It was desperately needed to send command messages directing the conduct of the war, coordinating all military elements, and suppling millions of members of the Allied Forces. Scores of experienced Western Union men went to the Pentagon, and, faced with an "impossible" deadline, worked twenty-four hours a day in relays never numbering less than thirty-five. Scaffolds were built so men could work in layers above others, work to the point of exhaustion, and return after brief rest.

The entire intricate center in the basement of the Pentagon was completed in twenty-one days of furious effort, including delivery of equipment to Washington. The ACS network then could handle fifty million words a day over a modern worldwide network. Its center was WAR, the Communications center of the War Department.

Major General H. C. Ingles, chief signal officer, wired President Williams of Western Union that all known records for speed and quality of workmanship had been broken, and WAR had the fastest and most efficient communication with all the world.

Other ACS centers used in the emergency were at San Francisco, New York, Atlanta, Chicago, Dayton, and Dallas. They utilized old tape reperforator equipment Western Union had taken over from Postal. At those centers, uniformed men and women received messages in the form of perforated tape, and inserted it in transmitters connected by circuits to their destinations.

Other major installations were made for the War Department at Washington, at headquarters of theaters of war, Army groups, Armies, Corps, and Divisions. The networks were interconnected. Rapid communications was provided down to battalions, regiments, companies, and the smallest units of the armed forces. The portable telegraph key and sounder, telephone, and "walkie-talkie" radio

were used to maintain contact with units on the move. The semiautomatic Signal Corps system center at Edmonton, Canada, sent Army communications to Alaska and the Aleutian Islands over a 2,060-mile line Western Electric built in 1943.

The War Department asked AT&T, Bell Laboratories, Teletype Corporation, and Western Union to plan a system for Army field use. In thirty days the committee recommended a design, and Western Union produced 650 sets that included all wiring and relays. Soldiers could connect them for operation with a screw driver or coin in ten minutes. When a command post moved, the sets were shifted quickly—as many as twenty in a truck.

Over Western Union cables 314 million words sped between government agencies in Washington and London, where the high command coordinated Allied strategy. One of the best-kept secrets of the war was the invasion of North Africa. Western Union provided the cable equipment and trained the Signal Corps operators. Headquarters for the Africa Command was established in an office on the Rock of Gibraltar with nineteen teletype machines. On January 2, 1943, the Americans and British opened a direct cable, laid from Gibraltar to Casablanca, with a capacity of a million words a month. Another was established to Algiers, with American operators who had received only three months of training.

Communications in the invasion of Italy and Sicily were, in effect, warm-up exercises for the big event at H hour on D day, June 6, 1944. The communications system was ready on D day when ACS flashed the twenty-six-word communique from London that started the invasion of Western Europe. The news was immediately announced on the presses and news broadcasts. ACS provided channels for transmission of fifty radiophotos of the invasion, and prints reached American newspapers four hours after fighting began. A pooling arrangement of all carriers handled the news, and the American public received a dramatic description at breakfast.

The first assault waves of amphibious forces established FM radio communications to link their beachhead units. Wire-laying jeeps estabished telephone and telegraph communications between field and command posts, and portable switchboards were set up in foxholes. Post telegraph stations often were operated in trucks. Other trucks laid big reels of insulated wire as fast as they could travel. Each division had Signal Corps groups to operate by radio, telephone, and telegraph, intercept enemy radio messages, find directions, locate enemy stations, and construct, repair, and maintain communications.

It was a war with words. The civilian watcher on a lonely hilltop would seize a telephone and say, "Army flash. Planes coming from the east." Then messages would go to antiaircraft posts and fighter pilots. Air-raid sirens would sound, and lights would be blacked out. The Columbia Broadcasting System newsroom at New York, beaming propaganda news programs to Germany and Italy, where listening to American radio was forbidden, had a sign on the wall: "Will a man risk his life to hear the words I'm writing?" Radio broadcasts to Bataan in the Philippines gave American soldiers their only news. More than 15,000 radio "hams" provided service at home, reported any suspicious broadcasts, and joined the civilian emergency radio network to aid if bombs disrupted communications.

The first permanent communications with France after the Normandy landing was

provided by diverting the former German cable from Horta in the Azores.

Development of Telekrypton machines began in Western Union's laboratories in February 1940, five months after the start of hostilities between England and Germany.[8] They were used constantly on cables and secret broadcasts to Strategic Service agents in enemy territory. Only with duplicate copies of the "random texts" could anyone decode the messages transmitted.

Special switching centers were installed for the War and State Departments, FBI, Office of Strategic Services, Office of War Information, Maritime Commission, and other American and British agencies. Also, they were installed at such conferences as Hot Springs, Bretton Woods, Dumbarton Oaks, Chicago, Atlantic City, Montreal, and San Francisco (United Nations).

Newspapers frequently reported that a high American official "talked" with Churchill, Eden, Halifax, Eisenhower, MacArthur, or Clark overseas. What really happened was that General Marshall, for example, went to a Pentagon conference room and dictated a question. A telegraph operator sent it; General Eisenhower or another official read it on a telegraph printer in England and dictated his reply. The typed discussion was projected on a screen.

Another telegraph development was a film scene selector. As many as twenty military experts, viewing a battlefront motion picture, could press a button and designate certain scenes for study. Locations of the selected scenes were electrically recorded to speed the preparation of prints.

William Buckingham and his associates in Western Union Laboratories at Water Mill, Long Island, developed the first "point source" of light that projected parallel rays of brilliant light. A major scientific breakthrough, a step toward the laser, it was classified "secret," and used by the Army in light-beam telegraphing. The Navy used it to exchange intelligence between planes, and between planes and ground stations. The point-source beam was used also to "boresight" guns mounted on 6,000 Army and Navy fighter planes, and to train night fighter pilots by simulating combat.

AT&T's contribution to the war effort in personnel, inventions, equipment, and facilities was huge. One Bell Laboratories development enabled the Allies to spot enemy planes behind clouds or in darkness. It was a cathode ray tube used in conjunction with radar. A Bell school trained men to use the Western Electric-built radar and the M-9 gun director.

Bell Laboratories also developed FM combat radio equipment used on tanks, artillery, and command cars. Western Electric employees ran lines and installed equipment wherever needed. General Telephone and others carried out important defense projects.

Demand for telephone service broke all records. In 1945, twenty-two million Bell phones and sixty transcontinental circuits were added. However, AT&T's plant deteriorated because it was devoted so fully to the war. Two million orders for service were on waiting lists.

RCA centralized its research in laboratories at Princeton, New Jersey, and made 150 new electron tubes and 300 new types of radio equipment. It established radiotelegraph circuits to many countries, installed radio apparatus on ships, and beamed radio programs to foreign countries in collaboration with the War Information office. It also helped Army Communications Service standardize international radio circuits to connect with the telegraph and cables.

Communications people remember the armed guards at operating room doors and laboratories, bomb alarm drills, photo and fingerprint cards, areas and files locked to prevent espionage or sabotage, and emergency power generators for use if commercial sources failed. "Telegraph offices on wheels" accompanied soldiers on maneuvers. Since metals were needed for munitions, alloys and plastics were substituted, and communications employees donated their jewelry.

In addition to billions of words over government and war-industry private-leased wires, Western Union handled 190 million telegrams during the first year of World War II, compared with twenty-four million telegrams in the entire Civil War and 62,173,149 words in the Spanish-American War. At a cost of $3,000,000 it trained 9,344 operators in its schools in 1944 alone to replace employees in the Armed Services, and discontinued social telegrams and other nonessential services to concentrate on vital war communications.

Also in 1944, the company handled 15,897,802 telegraph money orders, transferring $715,972,000, much of it from war workers and military personnel to their homes. Within the three weeks after Japan surrendered, 850,000 government contract cancellation telegrams were sent to contractors and subcontractors, and they quickly wired their suppliers.

The unhappiest wartime job was the delivery of telegrams announcing death, injury, or illness, men missing in action, or prisoners of war. Such messages were not delivered between 10:00 p.m. and 7:00 a.m. Messengers chosen for this work were mature men and women, many of them gold-star parents who stood by to aid. Bright spots came when messages said the missing were found, or servicemen wired, "Am well and safe." For security reasons, they could not tell where they were, and many parents asked "where's that" about the dateline—"Sans Origine."

At war's end, Chairman Fly of the FCC said communications "have become woven into the fabric of our everyday life perhaps more than have the motor car, the train, the steamship, the airplane," and made possible the coordination of our land, sea and air forces. "Without them," he said, "there would hardly be modern mechanized war. Communications, like transportation, makes one-third of the battle. The last third is striking power." He cited a report that 12,000 telephone messages were necessary to build a single bomber.

World War II unleashed the full fury of technological change. Realizing that communications, atomic energy, and medical advances might mean the difference between victory and defeat, the government gave industry and universities contracts for research and development that continued after the war.[9]

The Singing Telegram

To encourage greetings back in 1914, telegraph officials provided a special "Holiday Greeting" blank with a spray of holly and red candles across the top. This was followed by Easter, Mother's Day, and other blank headings in four colors, and changed from time to time with paintings by famous artists. Suggested messages were provided to help pencil chewers compose their own, and millions were sent each year.[10] They were sent for business as well as social reasons. Congratulatory telegrams were sent on birthdays, weddings, births, appointments, Christmas, Valentine's Day, Easter, Mother's Day,

Father's Day, the Jewish New Year, Thanksgiving, and other occasions.

19.5 SINGING TELEGRAMS WERE DELIVERED BY TELEPHONE AND BY UNIFORMED MESSENGERS, GIVING PLEASURE TO MILLIONS OF PEOPLE FOR MORE THAN FIFTY YEARS BY WESTERN UNION, AND SINCE THEN BY NONTELEGRAPH MESSENGERS DRESSED IN CLOWN, GORILLA, AND OTHER COSTUMES.

One of the nonessential greeting services discontinued in World War II was the Singing Telegram. Since World War I, newspapers and magazines often had depicted the telegram as a death message, and most people stopped using them for social correspondance for fear delivery would alarm the recipients. I myself realized that was one reason ten million former birthday, anniversary, and holiday greetings, and social telegrams were not being sent during the 1930s depression. Western Union was losing a million dollars a month and was desperate for cash.

"What kind of telegram would be so

novel and amusing that it would convince people that sending and receiving telegrams is fun?" I asked myself in 1933. Then the idea dawned that no one ever sent a telegram in song. I looked for someone to receive the first one and learned that the birthday of Rudy Vallee—then a popular singer and whom I knew—was the next day, July 28. So I called an operator to my office, and a plump, jolly girl, surprisingly named Lucille Lipps, arrived. She called Vallee and sang, "Happy Birthday, Dear Rudy. . . . "

Mention of that first singing telegram appeared in Walter Winchell's and other newspaper columns. People called Western Union to send them, and good-natured operators obliged, though there was officially no such service. The idea spread quickly, and switchboards lit up like Christmas trees with calls.

J. J. Welch, traffic vice president, protested to First Vice President J. C. Willever that the company was not prepared to render such a service. Called to Willever's office, I was angrily informed that I was making a laughing stock of the company, and Willever refused to listen to the reason for the service. Public demand soon convinced him, however, that Singing Telegrams produced cash, and they did give pleasure to millions. Operators crooned them over the telephone and, later, uniformed messengers delivered them for an extra fee.

Every cartoonist, newspaper columnist, and radio comedian capitalized on the humorous possibilities of messengers singing at homes and offices, and the more they kidded the service, the more gleefully the public seized upon it. Fred Allen, Eddie Cantor, Bob Hope, and Edgar Bergen were among those who often exercised their sharp wits on the service on their network radio programs. One messenger with a singing voice

like Donald Duck was in constant demand by the public and broadcasters. The Singing Telegram continued on the wackiest, zaniest musical spree ever until World War II.

When the messengers returned from the war, they joined unions that banned uniforms and demanded high pay and professional singers, so personal deliveries were not resumed. However, telephone operators sang for birthdays and anniversaries, Valentine's Day, Mother's Day, and Father's Day. Mary Martin, of "South Pacific" and "Peter Pan" fame, launched the postwar revival at New York; the Andrews Sisters did on the Pacific Coast.

Because of the dwindling number of public telegraph offices, Western Union discontinued the Singing Telegram in 1974, but local enterprises sprang up in hundreds of cities. They provided personal delivery by dancers and singers in a wide variety of costumes, including tuxedo, clown, gorilla, belly dancer, and "Strip-a-Gram" strip tease. Thirty such companies were listed in the New York telephone book fifty-three years after I created the Singing Telegram. Western Union resumed accepting orders for personal delivery in 1980, using professional singers provided by Musicbox, Inc., but at prices too high to gain wide popular use.

Government Regulation

The Post Roads Act of July 24, 1866, provided that any telegraph company "shall have the right to construct, maintain, and operate lines of telegraph through and over any portion of the public domain of the United States, and over and along any military or post road of the United States." One phrase in the Act required telegraph companies to handle government messages at such rates as the Postmaster General might fix, and that cost the telegraph companies between fifty and one hundred million dollars a year.

In 1877 the Supreme Court decided communications between states is interstate commerce[11] and the Mann-Elkins Act of June 18, 1910, gave the Interstate Commerce Commission (ICC) authority to regulate communications rates, undertake property valuations, and prescribe uniform accounts and financial reports.

Radio communications was the subject of no less than twenty-six laws between 1910 and 1934. In 1912, the ICC and Post Office Department were given control over rates, and the Secretary of Commerce over radio station and operator licenses and wireless telegraphy. In 1920 the Secretary of the Navy was authorized to use government-owned radio stations to transmit press and private commercial messages between ship and shore.

When the number of radio broadcasting stations grew from 528 to 719 in 1926, chaos resulted from overlapping wavelengths, and the Radio Act of 1927 created the Federal Radio Commission to license and regulate radio stations. The Act prohibited purchase, lease, construction, or control of any telegraph or telephone system by a radio company, or vice versa.

Overlapping governmental jurisdiction caused so much confusion that President Franklin D. Roosevelt recommended the establishment of the Federal Communications Commission (FCC), and the Communications Act of July 1, 1934, created the FCC, with jurisdiction over interstate communications. Soon the FCC had 2,000 employees, a sixty-one-million-dollar budget, and was a burden on the carriers.

It had been feared that politicians without industry knowledge would be appointed as FCC commissioners and put pressure on news broadcasters and carriers to adopt voter-pleasing but ruinous rates. FCC political regulation was compared to a camel getting his head under a tent, with his body sure to follow. Critics said certain politicians thought stockholders were a small group whose rights could be disregarded.

AT&T welcomed regulation when the original Bell patents were expiring and regulation might prevent rate wars with independent companies. Regulation was needed to control "dog eat dog" competition, but the handcuffs were uncomfortable. Carriers were required to compile voluminous reports and obtain permission to change offices, hours, services, or rates. That forced the carriers to hire large staffs of lawyers, accountants, and officials who spent years tangled in FCC red tape. Appointments of FCC commissioners by presidents for seven-year terms were not based on communications experience.

The first major FCC investigation, from 1934 to 1939, was of the Bell System. It cost AT&T more than three million dollars, and the taxpayers about the same. About 350 investigators worked in the AT&T and Western Electric offices. Some 750 volumes of records, loaded on twelve handtrucks, were furnished daily. Some 8,800 pages of testimony with 2,100 exhibits submitted during the hearing resulted in twenty-seven volumes of reports totaling 15,000 pages. Telephone employees spent five years complying with 10,000 FCC orders. President Gifford testified over a two-year period, and called it a "one-sided investigation."

"To an unprecedented degree in the world's history," John Bickley, chief accountant for the inquiry, testified, "communications of intelligence by word or pictures, and, to some extent, printed news, is under the control and surveillance of a single private interest." Its complex intercorporate relations resulted in concentration of control over 200 companies, and the flow of income from all of the constituent companies up to the holding company at the top.[12]

When the hearings ended, Commissioner Paul A. Walker proposed that the FCC be empowered to consider the probable effect of Bell policies before they were even adopted. In effect, he wanted people with no experience in the business to prevent the telephone experts from making mistakes! He also wanted the FCC to have power to approve or disapprove all intercompany contracts, costs, prices, financing, and so forth, with the company paying the cost of a huge FCC organization to do so.

AT&T attacked that as the plea of people who wanted power badly. "The Walker Report," AT&T said, "blandly recommends that this Commission should be given the authority—without being required to assume the responsibility—of a super board of directors of the Bell System, with the power to review, and to approve or disapprove all Bell System policies."[13]

In its final report in June 1939, The FCC omitted many of Walker's wilder recommendations, but pled for great FCC power and a larger organization, supported by assessing the companies. It proposed many changes in patent licensing, regulation of Western Electric's cost accounting, and prices. Long distance telephone rates were cut again, while local telephone service, regulated by state commissions, remained high.

After another investigation, lasting ten years, the FCC concluded that it had insufficient evidence to decide whether the telephone rate structure was unlawful or discriminatory.

Western Union– Postal Telegraph Merger

The FCC adopted a policy of "continuing surveillance," resulting in informal rate and other adjustments. At one hearing Postal Telegraph-Cable Company officials testified it was unable to improve its service largely because of Western Union's exclusive contracts with the railroads. Actually, Postal's largest business was telegrams to places it did not serve, and had to pay Western Union to handle. The FCC went to the rescue of Postal in 1933 by asking Congress to authorize a merger of landline, cable, or radio telegraph companies.

Western Union did not want Postal, but in the depression, with the FCC investigating their real competitor, AT&T, they dared not antagonize the FCC. They also had a wholesome respect for the political power of ITT, which owned debt-ridden Postal and was determined to get rid of it.

Postal and Commercial Cables had been merged with ITT May 15, 1928, but ITT could not aid them financially. It had paid no dividends for years and was in financial trouble, but in Washington it had a charmed life. In every crisis some government officials rushed to ITT's aid. Its foreign enterprises were financed by New York banks, but they would lend no more after the bank crisis of 1933. Most of ITT's income was from foreign countries, where currencies were collapsing, and dollars were blocked from transfer to the United States.

ITT had paid $11,000,000 in 1931 to Krueger & Toll for a thirty percent interest in L. M. Ericsson Telephone Company of Sweden, a manufacturer, but the audit of the affairs of match king Otto Krueger led to a financial panic, suicide, and scandal. ITT needed money to pay interest on its own debts.

Then the Francoists revolted in Spain, fired 184 shells into the ITT building in Madrid, and announced the building would be blown up at 4:00 p.m. Col. Behn sent word that he was holding a meeting at 4:00 p.m., and, if the fourteen Americans at the meeting were killed, Spanish relations with the U.S. would be disrupted. The building was saved, but for years ITT lost one-fourth of its total income that had come from Spain.[14]

ITT had invested $44,000,000 in Postal, which went into 77-B bankruptcy in 1935 after losing $14,600,000 in four years. Alfred E. Smith and George S. Gibbs, trustees in reorganization proceedings, joined Col. Behn, in pressuring Washington to merge Postal with Western Union, and persuaded the Government to lend millions to Postal to meet its payroll.

To hurdle legal obstacles of the Sherman Antitrust Act and White Act that prohibited telegraph company control of radio, Senate Interstate and Foreign Commerce Committee Chairman Burton K. Wheeler of Montana conducted a survey authorized in 1939 by Senate Resolution 95. At his request, the FCC held hearings and recommended a domestic telegraph merger; however, it said the corporate history of Postal was "so complex and records so inadequate, that it is difficult if not impossible to determine its true financial condition at any time."

ITT persuaded the State Department and Federal Loan Administrator Jesse Jones to have the Export-Import Bank underwrite two-thirds of a $15,000,000 loan to Postal. In 1940, Spanish bankers offered $60,000,000 for ITT's eighty percent interest in the Spanish Telephone Company, $36,000,000 more than ITT valued it. Colonel Behn was delighted, but the State Department believed it was a German scheme. At its request Behn held off until the end of the war, and his cooperation so endeared his company in Washington, that ITT received another Export-Import Bank loan which enabled it to survive.[15]

Also to aid the "fair-haired boy," the FCC not only advocated a Postal-Western Union merger, but its consolidation with ITT's American cable and radio companies. That report was said to have been cleared with President Roosevelt in advance by FCC Chairman James Lawrence Fly. Fly also proposed a merger of Western Union's cables, which in 1938 had handled thirty and four-tenths percent of all overseas business. R.C.A.C. had twenty-three and four-tenths percent; Commercial Cables seventeen and four-tenths percent; All America Cables fifteen and two-tenths percent; Commercial Pacific one and nine-tenths percent; and Mackay Radio, Press Wireless, Globe Wireless, Tropical Radio, Southern Radio, and U.S.–Liberia, all together, eight percent.

Into this situation stepped a new Western Union president, Albert N. Williams, president of the Lehigh Valley Railroad, who succeeded White on June 17, 1941. Williams knew no more about the business than White, but was smart enough to know it and to welcome the advice of those who did. Williams was a likeable, good fellow who would put his feet on your desk and tell jokes. Probably he never knew the presidency had been offered first to Joseph L. Egan, whose title was public relations vice president. Egan was a lawyer, so busy with company contracts, stockholder, government, Congress, other carrier relations, and company policy, that he had to leave many public relations matters to me. Egan told me he refused the presidency because he feared the pressure would shorten his life—as it did later.

Albert Williams played the game well. War services caused revenues to increase faster than labor costs and taxes. About $5,000,000 in debt was paid off, and dividends were raised from one to two dollars. Testifying at Senate merger hearings, Williams attacked the proposed bill requiring the company to divest its profitable international communications, but not merge all record communications.[16] The latter included AT&T's TWX leased wire and teletype telegraph services, which were much larger than Postal's total revenues. "Any legislation to permit mergers should make it reasonably possible for the domestic telegraph company to take over the merchandizing of these Bell System telegraph services," Williams said.

The bill, approved by the Senate on June 23, 1942, required merger of the two telegraph companies, but not an international telegraph merger. The Navy claimed the merger "would disturb worldwide communications, and consequently be detrimental to the war effort," and asked that it be postponed for reexamination after the war. For twenty years the cable divestment hung like the sword of Damocles over Western Union's head.

The FCC and Congress would approve no sale to another cable company, but only companies in that business could operate it and offered to buy it. Finally, Western Union International was formed as a separate company, and operated the system until Xerox bought it on April 24, 1979. Xerox sold WUI to MCI on June 23, 1982 for $185 million.

Edwin F. Chinlund, board chairman of Postal, who, like Egan of Western Union, was responsible for the negotiations, testimony, and exhibits, told the FCC that his company's debt to the Reconstruction Finance Corporation was about $12,500,000. Federal Loan Administrator Jesse Jones pushed for the merger to bail him out of his loans to bankrupt Postal.

After hearings through a long, hot summer in 1942, the House of Representatives' committee reported, "the only alternative to a merger would be government subsidy to Postal, as proposed by its union" (American Communications Association).

The Army, Navy, and Federal agencies approved the merger, and the 78th Congress passed Public Law 4 on March 6, 1943. Western Union was not happy. The House provisions to permit cable and radio telegraph companies to merge had been refused by both the House and Senate. Also eliminated was a provision to repeal the Post Roads Act of 1866, which gave the federal government a forty percent discount. That meant a loss of millions. AT&T retained its very profitable $7,000,000 TWX telegraph business. The FCC refused a fifteen percent rate increase to help meet Western Union's merger costs.

When it took over Postal, Western Union was saddled with its thirty-five companies.[17] The bill guaranteed all Postal people hired before March 1, 1941, employment for five years at suddenly increased salaries, and all hired after that, a month's severance pay for each year of service. Their pension rights were to remain as before. FCC Chairman Fly hailed the merger act enthusiastically, and said the FCC would press Western Union to divest its cables.

Western Union issued 308,124 shares of a new class B stock to Postal shareholders. The FCC held a public hearing, lasting twelve weeks, at which Williams complained about the "number of onerous conditions" imposed by the act.

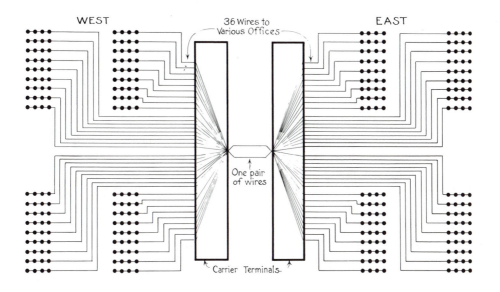

19.6 IN 1945, THE CARRIER SYSTEM, DIAGRAMMED HERE, ADDED A MILLION MILES OF TELEGRAPH WITHOUT ADDING ONE COPPER WIRE. ONE HUNDRED FORTY-FOUR MULTIPLEX OPERATORS SENDING EASTWARD AND 144 WESTWARD OVER A PAIR OF WIRES OR RADIO BEAM USING 26 SEPARATE MUSICAL TONES ACCOMPLISHED THE RESULT. THE SIGNALS WERE TRANSMITTED BY VARYING THE TONES ACCORDING TO THE TELEGRAPHED CODE.

Western Union had more than 208,000 miles of pole line, generally along railroad rights of way; Postal, about 31,000 miles along highways. Western Union had 18,677 offices and 13,500 agency stations; Postal, 3,948 offices. Most Postal lines, offices, and equipment duplicated Western Union's, were useless, and had to be junked. That required years.

Some 1,185 branch and tributary offices and 458 agencies were consolidated, and 3,500 of the 86,000 messenger call boxes were on duplicating teleprinter tielines and were removed. Because of the Act's onerous employee requirements, many former competitors with higher salaries were appointed to important Western Union jobs, including vice presidencies. It seemed that bankrupt Postal took over Western Union.

Egan carried a backbreaking burden through all of this.[18] Williams resigned to accept the presidency of Westinghouse Air Brake and Union Signal Switch. Egan succeeded him December 15, 1945, and continued running the company.

According to testimony before congressional committees and federal agencies, some officials and members of the American Communications Association (ACA), which represented former Postal employees, were Communists, stuffed ballot boxes at union elections, and engaged in violent tactics. That resulted in widely published charges that Communists handled government messages, and might relay sensitive information to enemies.

Having gone "all out" to aid the war effort, Western Union officials were upset. It had repeatedly urged Congress for a law to bar Communists from access to government communications, but the politicians refused because that would be "antiunion," and lose union votes.

The Commercial Telegraphers' Union (CTU, of the American Federation of Labor, AFL) applied to the National War Labor Board for an election by the company's employees. Despite strong protests by the company and CTU, the National War Labor Board (NWLB) ruled in 1944 that the employees had to be separated into a number of areas because the alleged Communist-led union had strength only in New York City, and would lose. ACA won the election only in New York City, and the CTU won everywhere else.[19] The company could only give employees a booklet quoting Federal laws and company rules about message secrecy and protection of the plant.

Five minutes before going out of existence at 5:00 p.m. December 19, 1945, ACA's friend, the NWLB, ordered a job classification program requiring thirty-one million dollars in retroactive pay and an increase of forty-five million dollars in annual recurring wages. That ruined the company's financial ability to comply with the merger requirement to submit a comprehensive plan to convert "existing facilities into a modern, efficient and nationwide system capable of competing with other communications carriers."

Hitler could not have devised a more effective way to wreck the nation's record communications. The NWLB order, two weeks after Egan became president, so impaired the company's finances that, for the first time in twenty-seven years, it asked the FCC to permit a fourteen-million-dollar increase in revenues. The company submitted a sixty-million-dollar modernization and

mechanization plan, saying it required the rate increase before banks would lend the money. Four months later a U.S. Department of Labor fact-finding board required another wage increase of twenty-four million dollars, regardless of the fact that the company was not earning the money to pay the previous award, and on April 1, 1948, it piled on still another wage increase of eight cents an hour, amounting to approximately eight million dollars.

During the war, the American Communications Association union gave AT&T trouble. Joseph Beirne, a former oiler, office boy, and Western Electric salesman, became president of the National Communications Workers of America (NFTW). His militant demands resulted in a wartime strike on April 7, 1947. There was mass picketing at the big Kearney, New Jersey Western Electric plant. Public opinion turned against the union, and in a month the Long Lines workers settled for a third of the union's demand, and all returned to work.

The awards by the liberal New Deal agencies were a great blow to Egan, who had worked ceaselessly to make Western Union a greater service company. His plans were in ruins also as the result of AT&T's TWX and private wire competition, and loss from acquiring Postal. FCC pressured it to divest its cables when antitrust laws did not permit their sale to any other carrier. In three years, government wage awards added about sixty million dollars annually on a company losing a million dollars a month. The company needed sixty-five million dollars for mechanization to cut labor costs and remain competitive.

A deeply religious man, Egan went to France to visit the grave of his Air Force son and namesake, Joe, who died in combat in World War II. There he died of pneumonia on December 6, 1948. The prediction he made to the writer two years before in a Washington taxi during the merger hearings had come true; the presidency he did not want did shorten his life.

AT&T also had severe financial and service problems resulting from government regulation. Like Egan, Gifford had a son killed in combat. He decided to retire at sixty-three when AT&T also was suffering from rate regulation. Its most effective vice president in political and governmental matters was Leroy A. Wilson, a neighbor and good friend of the writer. Gifford selected Wilson to succeed him as president in 1948 because of his Washington expertise, and he became chairman. Vice President William Henry Harrison left to be president of ITT; and Vice President Keith McHugh, to be president of New York Telephone Company.

A skilled fighter in Washington, Wilson pointed out that without money to expand the plant, service could not be improved, and he obtained badly needed rate relief. The government then sued under the Sherman Antitrust Act in 1949 to separate Western Electric from AT&T and open competition to other manufacturers. Knowing AT&T was needed to manage the Sandia atomic bomb project, Wilson declared that the ability of the Bell System to perform that major national service would be destroyed if the company was dismembered. He demanded that the suit be killed, and President Harry Truman did just that. The Sandia Corpora-

tion, a subsidiary of Western Electric, then produced the nation's atomic bombs.

In 1949, its union joined the Congress of Industrial Organizations (CIO). AT&T withdrew its recognition of CWA, and requested National Labor Relations Board approval, which was refused. In the next three years there were several CWA strikes, but AT&T stood its ground in hard-nosed bargaining.

In 1950 Wilson became ill with leukemia, and was told he had only a year to live. As he weakened, his close friend, Vice President Cleo F. Craig gradually took over the load. Wilson had saved the Bell System from Washington and big labor, and AT&T prospered. It sent many employees to colleges at company expense, on full salary, to broaden their minds, and better serve their communities. With long practical operating experience, Craig was the ideal choice as president.

AT&T and the government came to an agreement, approved by U.S. District Judge Thomas F. Meaney in 1956, that Western Electric would manufacture equipment exclusively for AT&T use, and license other companies to use its techniques.

The CWA struck against the Bell System again April 17, 1968. Although 178,000 walked out in 3,000 towns and cities, the Bell network then was so automated that the public only learned of the strike through the media, which overplayed it. After two weeks the union settled for $2 billion in wage and fringe benefit increases.[20]

Notes

[1]One TV station was operating in England in 1937, where the public owned about 1,000 sets.

[2]*Telegraph and Telephone Age* (March 1, 1933) 64-65.

[3]For later facsimile developments, see ch. 25.

[4]Likewise, Great Britain allowed Cable and Wireless Limited, with 155,000 miles of cables and ninety-one wireless circuits to function without government interference. The heroic role of CWL's 10,000 men and women in seventy countries and on the high seas was described by Charles Graves in *The Thin Red Lines* (London: Standard Art Book Co. Ltd., n.d.).

[5]Graves, *The Thin Red Lines.*

[6]RCA served Hawaii then. It was not until the 1970s that Western Union of Hawaii subsidiary was even permitted to provide intrastate teleprinter service there. The false "delayed telegram" story was repeated by John P. Roche in "The Jigsaw Puzzle of History," *New York Times Magazine* (January 24, 1971) 36.

[7]"Reins of War," a pamphlet describing the role of the Corps in the war, was the source here. I prepared the pamphlet for the Signal Corps.

[8]The telekrypton operation was described in more detail in ch. 17.

[9]Favorite TV programs after the war included Major Bowes's "Original Amateur Hour", the Arthur Godfrey show, the Ed Sullivan show, Jack Paar and the "Tonight Show," Steve Allen's "I've Got a Secret," the "Hall of Fame," "I Love Lucy," "Gunsmoke," and "Bonanza."

[10]Henry O'Reilly was first to promote the social and greeting telegram. Among his papers at the New York Historical Society is his notice of December 12, 1848 saying:

> The increasing use of the telegraph in the social as well as commercial and political relations of the community. . . . The Superintendents of the various lines are requested to cause the Telegrapheries in all the principal towns to be kept open for extra hours during the week including Christmas and

New Year—that "absent friends" may interchange the "compliments of the season" where they cannot enjoy the festivities of association at "the ole homestead."

[11]Pensacola Case 96, J.S. 1.

[12]*New York Herald Tribune,* March 25, 1936.

[13]Under the headline, "Craving for Power," the Seattle (Washington) *Post Intelligencer,* February 20, 1935, attacked the FCC for "seeking the sole power to decide when the antitrust laws shall be set aside in order to permit the establishment of a monopoly in communications."

[14]"I.T.&T.'s Nine Lives," *Fortune Magazine* (September 1945).

[15]Ibid. When the war ended, intervention by the State Department prevented Spain from expropriating the telephone company, permitting ITT to sell it to the Spanish Government in 1945 for $88,000,000.

[16]Source: personal knowledge. I assisted at the hearings in the Senate Office Building.

[17]Postal, a unit of ITT, controlled the "Associated Companies" (formerly the Mackay Companies), a Massachusetts trust which owned the capital stock of 35 corporations in various states known as the Postal System. Operation was conducted by Postal Telegraph-Cable Company, a New York corporation. In the 1940 reorganization under section 77-B of the Bankruptcy Act, the stock of the 35 Postal companies was transferred to a new holding company, Postal Telegraph System, Inc., separate from ITT's cable and radio companies. Thus, what Western Union acquired was one company owning 35 Postal companies.

[18]I know of this merger firsthand. My office was next to Egan's, and I assisted him during those years.

[19]It was not until 1966 that the metropolitan New York area was won by another union, the Communications Workers of America (AFL-CIO) which also represented telephone workers.

[20]*Reader's Digest 1969 Almanac,* 707.

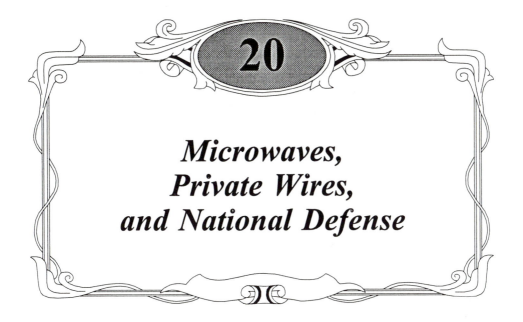

Microwaves, Private Wires, and National Defense

After World War II, rapid growth in public and business activity produced a severe shortage of communications facilities. More private wire systems, growing television networks, and long distance telephoning also demanded wideband capacity.

The shortage of facilities was a challenge to communications engineers. Western Union had only 2,300,000 miles of its own wire, and leased thousands of circuits from the Bell System to connect its 23,400 telegraph offices, 2,000 private wire customers, 55,000 teleprinter and Desk-Fax tielines to businesses. Expense of providing service had grown rapidly, spurred by sharp increases in telephone and telegraph wages. Millions of miles of circuits were needed, but copper was costly.

After much study, Western Union and AT&T arrived at different answers. Although Bell Laboratories had researched radio waveguides in the 1930s, AT&T placed its first commercial carrier systems on coaxial cables in 1941. The telegraph company placed its bet on the microwave radio beam.

As early as April 1, 1930, ITT had announced its engineers used "microrays" only seven inches long, and aerials less than an inch long at Herndon, England, in cooperation with its subsidiary, Le Materiel Telephonique of Paris, to talk across the English Channel. Around the same time Bell Laboratories tested ultra shortwaves less than ten meters,[1] and frequencies greater than 30 megacycles. In 1932 the British and French Air Ministries ordered the use of "microrays" to announce arrival and departure of cross-channel airplanes.

The ITT and Bell experiments called attention to the great area of unused frequency space in a realm virtually free from fading and static. They travelled down the stairway of wave lengths into an area of waves so short they were almost in the field of light waves, and had many of their characteristics. There they found a world of microwaves that oscillated a billion times a second.

That news led various oracles to gaze into their crystal balls and rush into print.

Journalism Professor Walter Pitkin of Columbia University[2] foresaw numerous radio stations costing less than $5,000 each, and radio listeners able to tune in programs on more than 200 different wave lengths.

Like many other inventions, the coaxial cable, with ultrahigh frequency radio waves transmitted through a hollow metal pipe, resulted from simultaneous, independent research at two places—the Bell Telephone Laboratories and the Massachusetts Institute of Technology. Both Dr. George C. Southworth of Bell and Dr. Wilmer L. Barrow of M.I.T. announced it at a joint meeting of the American Physical Society and the Institute of Radio Engineers on April 30, 1936.

The Bell System installed an experimental coaxial cable from New York to Philadelphia, containing a pair of copper tubes a little larger than a lead pencil. Each tube carried messages in one direction. A wire in the center of each tube was held by thin discs of hard rubber so that it would not touch the metal. It provided thirty-five voice channels, and repeaters were located at ten-mile intervals.

After two years of experiments, "talking movies," projected from New York, were seen in Philadelphia. It required a million cycle carrier current frequency range to transmit either a TV image, or 240 telephone messages.[3]

The first commercial service by coaxial began June 9, 1941, over a cable between Stevens Point, Wisconsin and Minneapolis. Using an amplifier every five miles along the 195-mile route, it could carry 600 simultaneous telephone conversations, or one two-way television channel.[4]

The Bell System then began laying short coaxials in Eastern states. Some contained eight transmission lines, and could carry

1800 telephone channels. Then a twelve-tube coaxial cable was developed, with a capacity of 9,300 voice circuits.

To meet the heavy demands of World War II, the Long Lines Department and the Northwestern, Mountain States, Pacific, and Lincoln Telephone Companies in 1942 laid a pair of 500-circuit transcontinental carrier cables from Omaha to San Francisco. These were not coaxial, but contained fifty-four pairs of wires. They were laid by a train of diesel-powered Caterpillar tractors, rooter, plow, and trailers, altogether nine pieces of machinery, 275 feet long, weighing more than 100 tons. In one continuous operation, it cut a narrow slit thirty inches deep, laid the two cables in the trench, and covered them.

When that was completed December 21, 1942, AT&T had four main transcontinental routes, with twelve-channel carrier current systems, where there had been only 250 telephone circuits as far as Denver and 200 west of that city.

Microwaves versus Coaxial Cables

Undeterred by the telephone company's faith in coaxial cables, Western Union installed the world's first commercial radio beam system in 1945.

Experiments with light beam telegraphy at its Water Mill Laboratories on Long Island had shown excellent results, but in the short distance between the Western Union Building at 60 Hudson Street and the Empire State Building, the dirt, dust and smoke of the city cut the beam signals to about one-tenth of their strength. In 1943 W. D. Buckingham's "point source" concentrated arc lamp invention was used to provide light beams for printing telegraph transmission from the Western Union Building to the *New York Post* seven-tenths of a mile away, and in 1944 to the *Brooklyn Eagle* one and eight-tenths miles away.

Even that powerful beam would not provide long distance facilities, and the engineers turned again to radio waves. Paradoxically, the lowest frequencies gave the highest performance, but the engineers were baffled by the refusal of the extremely short waves to bounce or bend over the curvature of the earth. They went straight off into space. In 1944 engineers found the cure was to move microwaves through automatic relay stations, and Western Union obtained licenses to use RCA, AT&T, GE, and Westinghouse radio relay patents.

First Microwave Beam System

Building a microwave beam system between New York and Philadelphia, Western Union erected a series of towers on hilltops about thirty miles apart, with each tower antenna in line of sight with the next one. Reflectors on each tower caught the narrow beam and sent the signals to a station below. There the signals were automatically strengthened, improved, and flashed up to another reflector, which beamed them on to the next tower. The beams were not subject to inductive interference from power lines and electric railroads, borealis, static, sunspots, lightning, fire, flood, ice, snow, or windstorm.

On October 22, 1945 Western Union announced a seven-year plan to extend the beam system and link major cities, replacing many of the familiar pole lines. In 1948 it completed a triangle between New York, Philadelphia, Washington, and Pittsburgh with twenty-one intermediate towers from

sixty to 120 feet high, at elevations up to 2,900 feet. The New York antennas were placed atop the company's twenty-four-story headquarters building by helicopter. A ninety-foot tower resembling a monument was constructed on 41st Street near Wisconsin Avenue, Washington, D.C. A tower on a bluff overlooking Pittsburgh relayed the beam to the roof of the Chamber of Commerce Building, and, in Philadelphia, to the roof of the Market Street National Bank.

20.1 WESTERN UNION CONSTRUCTED A TRANSCONTINENTAL MICROWAVE TRANSMISSION SYSTEM WITH MORE THAN 270 RELAY STATIONS BETWEEN BOSTON, NEW YORK, LOS ANGELES, AND SAN FRANCISCO. IT WAS THE FIRST COMMERCIAL RADIOBEAM SYSTEM. THE MICROWAVE TOWERS, SPACED 25 TO 30 MILES APART, CARRY EVERY FORM OF TELECOMMUNICATIONS. THIS IS A 300-FOOT TOWER AT ELLIS, NEBRASKA.

Microwave also was ideal for television. Everything in front of the TV camera is scanned line by line, like the lines on this page, and light variations are converted into a stream of impulses. When these electronic variations reach the home TV tube, they reproduce the scene 525 lines in each frame and repeat it thirty times a second to produce a continuous picture.

Western Union expected to place its FM carrier system on the beam to transmit 1,000 messages simultaneously in each direction, and also to provide video channels for TV broadcasting. Of course, the television networks needed to reach more points than Western Union's triangle provided, and a request was made to interconnect with the Bell System. When AT&T refused, Western Union appealed to the FCC, which instituted "appropriate proceedings" and gave no decision for eight years. That stymied Western Union while AT&T built an experimental microwave system from New York to Boston, providing service free until the FCC approved rates May 1, 1948.

AT&T extended its beam to Philadelphia, Baltimore, and Washington, and the national political conventions were telecast over 18 stations in nine eastern cities in July, 1948. In 1949 its midwestern coaxial cable and microwave network interconnected thirty-two TV stations for President Harry S. Truman's inauguration. Its TV facilities reached Chicago September 1, 1950. The political conventions of July 1952 were telecast over 107 stations in sixty-five cities, and an estimated seventy-five million persons saw Eisenhower's inauguration in January 1953 over 118 stations in seventy-four cities.[5]

After eight years, the FCC announced its decision in 1953. It refused interconnection on the ground that Western Union's plan would give the company a "blank check" to build facilities.[6] That gave AT&T a TV monopoly. The broadcasting networks still

urged Western Union to establish a video network because competition would lower costs, but they would not sign a contract to guarantee its use if it was built. That is why Western Union did not extend its beam system to Pittsburgh, Columbus, Cincinnati, and Chicago until 1957. By December 31, 1958, AT&T had 68,000 miles of TV circuits, not including "fall back" circuits for emergency use, and 19,860,000 miles of telephone circuits on the microwave system. In 1958 it also completed a microwave system across Canada. In the 1960s more than ninety percent of network TV was on microwave beam relay, and it was very profitable because the cost was the same regardless of the number of channels it carried. It handled television, telephone, and telegraph traffic, and by 1970, 699 TV stations grossed $1,400,000,000.

Television introduced the young to strange adventure, travel, science, and space world, and the old to a dream world of entertainment. It influenced opinions, clothing, food, pronunciation, sports, and our way of life. The telephone and telegraph business prepared for surges of activity when major news or entertainment programs were telecast. "Image making" became so important that candidates worked to improve their TV appeal.

Miniaturization revolutionized communications equipment after the transistor, a "semiconductor amplifier," invention in 1947 by Bell Laboratories engineers William Shockley, Walter Brittain, Robert Gibney, and John Bardeen. Brittain said it was lucky when they substituted a tiny piece of treated germanium with two gold contacts in place of an electron tube in a communication circuit, and it worked! The transistor had two pointed gold contacts less than two-thousandths of an inch apart.

That breakthrough in communications opened a new epoch in electronics. Solid state technology revolutionized progress in computers, hearing aids, medical instruments, electronic switching, miniaturization of room-size equipment, satellites, moon trips, and on and on. The transistor radio brought the outside world into desert tents, rain forest huts, and icy igloos.

The "Marshall Plan"

20.2 WALTER P. MARSHALL WAS THE 13TH PRESIDENT OF WESTERN UNION. AMONG THE MANY PROGRAMS CARRIED OUT UNDER HIS DIRECTION WAS THE CONSTRUCTION OF THE NATION'S FIRST MICROWAVE RADIO BEAM COMMUNICATIONS SYSTEM, COMPLETED IN 1962.

When Joseph L. Egan died in France, Walter P. Marshall was elected president of Western Union on December 21, 1948. Comptroller and executive vice president at Postal before the merger, Marshall had be-

come assistant to the president of Western Union, treasurer, and Contract Department vice president.

Damaged by Labor Board wage decisions, the company was losing a million dollars a month, had two bond issues of thirty million dollars to redeem in 1950 and 1951, and thirty-five million dollars in notes due in 1960. Although the company's ledgers were dripping red ink, the unions demanded more, went on strike from April 3 through May 23, 1952, and got it.

Tall, slender, with black hair and high cheek bones, giving credence to the report that he was one-eighth Cherokee Indian, Marshall had no pomp and puffery. There was no doubt whether his decision was "yes" or "no." He delegated authority in an atmosphere of friendliness and teamwork, and got results. The heavy deficits were replaced by profitable operation, and twenty million dollars was raised by selling a million shares of stock. He sold a number of Western Union buildings and leased them back. The headquarters building was sold for more than twelve million dollars to Woodmen of the World Life Insurance Society. He also sold Teleregister Corporation, the centrally controlled stock quotation board business, and the American District Telegraph Company, the burglar alarm, and fire sprinkler service systems company at a profit of millions. Such sales reduced indebtedness from seventy million to thirty-five million dollars.

Marshall invested in Microwave Associates, Technical Operation, Inc., Dynametrics Corporation, Gray Manufacturing, and Hermes Electronics, all with technological competence of special interest to Western Union. He carried to completion a $250 million automatic program, including the network of fifteen reperforator switching centers, described earlier, through which telegrams sped to destination after one typing at origin.[7] He also speeded Desk-Fax installations and leasing of private wire systems.

Looking to the future, Marshall proposed a national communications policy, called "The Marshall Plan," for separate telephone and record communications, with regulation to promote adequate, economical, and efficient service at reasonable charges, without wasteful or destructive competition.

Marshall proposed to President Truman's Communications Policy Board, created on February 17, 1950, that his plan be implemented by the following actions:

1. Elimination of the discriminatory twenty-five percent excise tax on telegrams, far higher than the tax on luxuries.

2. Acquisition by Western Union of Teletypewriter Exchange Service and other telegraph services of telephone companies.

3. An integrated system for the Government's record communications.

4. Merger of the international telegraph carriers.

Marshall repeatedly urged the unification of telegraph companies to compete with the unified oral service. He called AT&T's telegraph services "unfair competition," with low long-distance rates being subsidized by high local rates. Neither the FCC nor Congress would act. Fifteen years later, when the FCC obtained AT&T figures, the charge of subsidized competition proved true. AT&T had provided its telegraph services at a profit of only two and nine-tenths percent for TWX, and one and four-tenths percent for private wires, while local telephone services earned ten and seven-tenths percent.[8]

While Congress considered Marshall's proposed voluntary merger of international cable and radio companies in 1950, Western Union announced the development of the world's first cable amplifier, with three times

as many printing telegraph channels. Other international companies and the FCC constantly pressed for divestment of its cables, but the Merger Act prohibited sale to another carrier, and for twenty years no buyer made a satisfactory offer, and Western Union couldn't invest to modernize it because of the uncertainty. Telex service to 200 countries provided sixty percent of its international revenues, and leases twenty-five percent. Western Union finally divested its International Division on September 30, 1963, to a new separate company, Western Union International, Inc. (WUI). It received $5,400,000, but had to pay $5,750,000 to Anglo-American Telegraph to transfer its lease of cables to the new company. Western Union's net loss on the sale was $9,449,769 after tax credits.

Edward A. Gallagher, a Western Union lawyer, was president of "WUI," and the cable employees went with him. WUI's headquarters was at 26 Broadway. Years later it moved to 1 Western Union International Plaza, New York City.

TWX versus TWS

To compete with AT&T's Teletypewriter Exchange Service (TWX)[9], Western Union established dial-direct, automatic, two-way customer-to-customer timed wire Telex Service (TWS) in 1958. To communicate with another subscriber, the Telex user merely pressed a "start button" and dialed the number. He then pressed a "who are you" key, and his machine automatically printed the name and city of the distant subscriber to confirm the connection. The message was then typed, and a reply could be received on the same connection. It was two-way "written conversation" with typewriter-like keyboards. Rates were based on distance and time, with no minimum time.

Western Union installed Telex switching centers (using equipment of Siemens & Halske of Germany), at New York, Chicago, San Francisco, and Los Angeles, and connected them with a Telex system, linking 26 Canadian cities. It took business away from TWX and long distance telephone, and in November 1962, AT&T reduced the minimum charge from three minutes to one.

Radio Beam Nationwide

20.3 WESTERN UNION TECHNICIANS ARE SHOWN COMPLETING ADJUSTMENTS ON 300-FOOT MICROWAVE TOWER AT WARWICK, NEW YORK, ONE OF 267 STATIONS ON THE TELEGRAPH COMPANY'S 7,500-MILE TRANSCONTINENTAL SYSTEM. THE SYSTEM IS CAPABLE OF HANDLING ALL KNOWN FORMS OF COMMUNICATION.

In the 1960s, installation of Telex centers required trunk lines to interconnect

them. To avoid increasing its Bell System leases, already providing sixty percent of its needs, in 1959 Western Union began extending its New York to Chicago radio beam network nationwide. Aerial maps were made from coast to coast, sites were acquired, and towers—as tall as 300 feet—erected. One access road, up Plowshare Peak in California, cost $100,000. When a tower was built over a coal mine, the coal deposits under it were bought so no one could undermine it. Helicopters were used to place microwave reflectors on the roof of a 42-story building in Dallas, Texas, and on mountaintop sites.

The 7,500-mile system, costing eighty million dollars, began operating November 17, 1964, with 267 stations about thirty miles apart. For reasons of national defense, the route avoided critical target areas, but spur lines linked it with military installations and cities. One branch linked the Jet Propulsion Laboratory at Pasadena with the Goldstone deep-space tracking station in the California desert, to which spacecraft sent photographs and information about distant planets.

The system had a capacity of about 7,000 voice channels, but was equipped initially with 600 voice bands, adding eighty million channel miles to the company's network. Major cities along the route were Los Angeles, San Francisco, Oakland, Denver, Kansas City, Dallas, St. Louis, Chicago, Detroit, Cleveland, Buffalo, Syracuse, Boston, New York, Philadelphia, Baltimore, Washington, Pittsburgh, and Cincinnati. The system was later expanded to 8,400 miles, and interconnected with the Canadian National–Canadian Pacific microwave network, 3,200 miles from Montreal to Vancouver.

To assure continuity of service, signals traveled simultaneously over separate operating frequencies. Junction stations along the route merged the two received signals, and transmitted the resulting high-quality signal over dual channels to the next station.

The TAT Cables

Many people do not know that transatlantic calls and broadcasts travel along a cable on the ocean bottom. Many cables also are leased to large users, and carry a large volume of both telephone and telegraph communications for news services, governments and businesses.

The first transatlantic telephone (TAT-1) cable was laid in 1956 from Clarenville, Newfoundland to Oban, Scotland, as a joint project of AT&T, British Post Office, and Canadian Overseas Telecommunications Corporation. Complex 500-pound amplifiers were inserted in the fifty-one-circuit cable to boost the strength of voice signals at twenty-mile intervals. The route was by landline to Portland, Maine, radio relay to Nova Scotia, submarine cable to Newfoundland, landline across Newfoundland, and twin cables to Scotland. Each circuit carried the voice in one direction. TAT-1 carried ten million calls during its twenty-two-year life span.

The first telephone cable to Hawaii was laid by AT&T and the Hawaiian Telephone Company in 1957. TAT-II, the second transatlantic telephone cable, was a twin cable laid in 1959 from Newfoundland to Penmarch, France. It was jointly owned by AT&T and the French and German communications administrations.

On December 19, 1961, Queen Elizabeth II and Prime Minister John Diefenbaker of Canada inaugurated a cable called CANTAT from Oban, Scotland, to Grosse Roches, Newfoundland, to be a part of the British cable route to Hawaii, Fiji, New Zealand, and Australia. AT&T leased twenty-four of CANTAT's eighty circuits.

AT&T's first cable ship, *Long Lines,* was launched September 24, 1961, at Hamburg, Germany. The first task of the 511-foot, 17,000-ton vessel was to lay TAT-III, the first telephone cable linking the United States directly with England. It extended 3,500 miles from Tuckerton, New Jersey to Widemouth Bay, Cornwall, England. Placed in service in October 1963, it provided twelve simultaneous telephone conversations.

In six years C. S. Long Lines laid 17,000 miles of ocean cable, including the first Hawaii–Japan telephone cable, a second one from Hawaii to California, one from Guam to the Philippines, and one between Florida and the Virgin Islands. TAT-IV was laid between the United States and France in 1965; and in 1966, the Virgin Islands cable was extended to Venezuela. A 360-voice circuit coaxial submarine cable was laid in 1968 from Cape Town, South Africa to Lisbon, Portugal,[10] and a 720-channel cable (TAT-V) from Rhode Island to San Fernando, Spain in 1970.

The international record companies wasted no time in leasing channels in the new cables and fighting furiously for traffic to keep them busy. International communications grew with American activity in foreign affairs and trade.

The TAT cables were operated by Long Lines, AT&T's long distance and overseas operating unit that connected the twenty-two Bell System and about 1,600 independent telephone companies with the overseas gateway centers at New York City, White Plains, New York, Miami, Florida, and Oakland, California. More than 100 million telephones in the United States could call ninety-six percent of the world's 200 million telephones. In 1967 telephone subscribers began dialing their own calls from New York to London and Paris.

The FCC did not permit AT&T to provide international telegraph service, but it did lease channels in its cables to telegraph companies. Even an AT&T voice and record combination service overseas could have cost the record carriers millions in annual revenues. The FCC made one exception to that rule by allowing AT&T to provide record service to Hawaii, but treated that state as a foreign country by refusing to let Western Union serve it. Finally, on November 14, 1972, teletypewriter exchange service was established there by a subsidiary, Western Union of Hawaii, Inc., but only for intrastate service.

With its TAT cables, AT&T and the international telegraph carriers' revenues grew rapidly. International conversations via AT&T increased from 170 million in 1975 to 309 million in 1978, while telephone investment was 250 million dollars compared with 191 million dollars by the overseas telegraph carriers. In effect, international record carriers provided a "middleman" service between AT&T, the wholesaler, and the public. International trade and commerce is carried on with an unseen torrent of billions of written and spoken words, flashing along ocean bottoms, and bombarding the heavens over a spider web of circuits criss-crossing the world.

Private Wire Systems

As companies grew and became complex, their headquarters officials needed to receive reports from widely scattered points, and to send orders based on that data to plants, warehouses, and field offices. The volume of data required to manage large companies efficiently became enormous, and private wire systems were necessary.

When Western Union began providing private wires to business and government, only simple point-to-point lines were needed. Later, when a company required fast direct communication between its Eastern headquarters and a number of Pacific Coast offices, it would have been very costly to connect all offices with direct circuits to New York. Instead, they leased one transcontinental trunk line and short lines from it to their Pacific area offices. Messages were switched on to destination from one western center. Sometimes several small firms joined in leasing one long trunk circuit. The June 11, 1930, *Commerce and Finance Magazine* listed 465 financial houses, and 1,259 "establishments" using private wires.

20.4 VIEW OF SICOM (SECURITIES INDUSTRY COMMUNICATIONS) CENTER WHICH WAS AT WESTERN UNION'S TECHNOLOGY CENTER, MAHWAH, N.J. EQUIPPED WITH UNIVAC 418 COMPUTERS, SICOM WAS THE FIRST COMPUTER-CONTROLLED COMMUNICATIONS SYSTEM TO SERVE MEMBERS OF THE SECURITIES INDUSTRY ON A SHARED-USE BASIS. SICOM SPEEDS BUY AND SELL ORDERS, HANDLES EXECUTION REPORTS, MARKET NEWS REPORTS, AND ADMINISTRATIVE MESSAGES.

By 1930 AT&T's private wire leases totaled about 1,100,000 miles, and abuses began to appear. For example, the First Boston Corporation, an investment house, leased some circuits, and offered telegraph service to large companies at low rates, if they in turn would relay messages addressed to third parties. At FCC hearings in April 1935, the telegraph companies complained that such lessees should not be permitted to act as common carriers. That stopped the more flagrant abuses.

The first industrial user of Western Union's leased wires was one of U.S. Steel's predecessor companies, Carnegie Brothers and Company, Ltd., in 1881. U.S. Steel had the first system installed with switching centers in different parts of the country. It adopted high-speed switching, doubled its network in 1939, and handled 3,000,000 messages between sixty-one offices in forty-six states in 1941 to expedite production of steel for tanks, airplanes, shells, and other war materiel. During the war a number of switching systems were added, including the Federal Reserve System, Sears Roebuck,[11] and aircraft manufacturers. U.S. Steel's centers in eight cities, connecting 152 stations in eighty-five cities, were converted to pushbutton switching December 17, 1951.

William Blanton and other Western Union specialists converted early private wire systems to pushbutton, and then to automatic electronic switching. Charles E. Davies, Robert F. Dirkes and their assistants, surveyed the communications flow of hundreds of large companies and designed custom-made systems. News stories on the inauguration of these systems by large companies to improve their efficiency prompted hundreds of others to do likewise.

Companies leasing Western Union private wire systems became a "Who's Who" of American industry. In 1958 it provided 2,000 private wire networks with three million miles of circuits for about forty million dollars. The first section of an International Business Machines network was placed in

operation September 15 that year, with torn tape switching and 25,000 miles of circuits connecting 150 offices.[12] Automatic switching was substituted for pushbutton in the U.S. Steel, United Air Lines, and many other systems.

20.5 A SWITCHING UNIT IN A UAL CENTER.

The most sophisticated private wire systems were those of the airlines, because travel agents disliked the tedious job of searching through route and flight schedules, and calling airlines for reservations, when a quick question to a big computer-controlled private-wire system could get an immediate ticket confirmation. American Airlines, with its Sabre network linking 7,000 locations, and United Airlines, with its Apollo, were the largest. They recouped some of their multimillion dollar cost by charging travel agents for the service. Eastern Airlines also had a large computer reservations system that it sold to Texas Air for 100 million dollars.

Soon competition appeared from Trans World's Pars, Delta's Datas, Eastern's System One, and Tymshare's Mars. The airline systems grew until they were handling twenty billion dollars in reservations in 1983. Some airlines charged that American and United were using data from their systems to take unfair control and advantage of their agents. Such charges were denied, and the Aeronautics Board set up regulations to eliminate bias. Additional use of these systems was for hotel and car rental reservations.

Railroads also were large private wire users. Southern Pacific and Santa Fe had systems of thousands of miles in the Pacific Coast and southwest, and CSX had a 5,000 mile system named Lightnet linking the New York, New Orleans, Chicago, and Miami areas. Texaco and other oil companies used computers to process data on networks connecting their worldwide offices.

The private wire network transferring the largest amount of money and securities was the one serving the Federal Reserve Bank System. On an average day Federal Depository institutions transferred $500 billion, and bankers protested any delay, because it might force some corporation to borrow a large amount to meet a closing deadline.

National Defense Networks[13]

Having used all available equipment to meet government and armed forces needs during World War II, Western Union needed more to meet the peacetime needs of 55,000 business offices with facsimile and teleprinter tielines, the needs of FM carrier systems providing 1,000 channels on a single radio beam for the growing number of big private wire systems, the needs for interconnecting lines between its fifteen high-speed automatic switching centers and all telegraph offices.

At the same time the Blue Cross hospital and medical service ordered an 18,000-mile network to link 91 United States and Cana-

dian offices, and Liberty Mutual Insurance, Reynolds, and Boeing and Clark Instruments expanded their private wire networks to handle data processing.

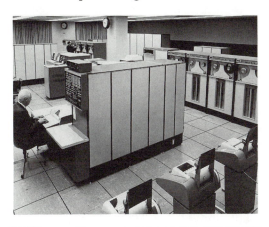

20.6 Western Union developed and installed the U.S. Air Force combat logistics Network in World War II to handle 60 million words daily. Shown here is one of the compound terminal centers installed at all military bases. It converted the coded data flowing into the network. It then transmitted the data over the network in punched tape at 20 characters per second. An IBM printed card in five channel (right) punched the information in 80-column cards.

Then the U.S. Air Force gave Western Union a contract for the world's largest private wire system—a new high speed electronic network to handle two and one half billion words annually. It was to require only 480 operators, compared with 1,800 then. The entire automatic system was to be completed in one year. The Air Force was to lease the system for $3,500,000 annually. In addition, a thirteen-million-dollar contract was received to establish a global network for overseas Air Force bases.

While that network was being installed, a follow-on, computerized system of switching centers, called COMLOGNET, was designed. The "log" in the name meant logistics. Faced with delays in shipments from suppliers, the telegraph company converted its old twelve-acre pole treatment yard in Chattanooga, Tennessee into a plant, where the complicated wiring and components were installed in large steel cabinets for the switching centers. Major General Dudley D. Hale, U.S.A.F. delivered a patriotic address, praising the 500 mainly local people for their loyal work on the project.

COMLOGNET was expanded, renamed AF Datacom, and placed under the direction of the Defense Communications Agency. New features and capabilities were added. It became the first data communications system to serve the entire Department of Defense, and was given its third name: Automatic Digital Information Network (AUTODIN).

Linking 550 tributaries and centers in the United States and worldwide, AUTODIN was the world's largest computer-controlled communications system. In 1964 still another contract was received to expand AUTODIN by adding four domestic switching centers, and increasing the capacity of the first five, to connect 2,700 substations. Simultaneously, the Defense Department announced plans to install ten more overseas centers, and bring AUTODIN's worldwide capacity to forty million punched cards, or the equivalent of 600 million words a day. In 1976 Western Union was awarded a ten-year contract to establish an AUTODIN II, with eight computer switching centers, using "packet switching" in which the data is broken into pieces, and routed on various paths until the entire message is delivered.

AUTODIN transmitted information in digital or binary form as a series of ones and zeros. Previously, different punched paper tape, punched card, magnetic tape, and computer devices were used by the four armed services, but AUTODIN used a common electronic language, and all could "talk" directly and simultaneously with each other. Each center used several large, solid-state electronic computers to control, store, and relay messages and data.

Information transmitted over AUTODIN was automatically encrypted and decoded. It was sent in four different levels of priority or urgency, so routine messages would not delay urgent ones. Western Union was systems manager, subcontracting the manufacture of equipment to RCA, IBM, and other companies.

When AUTODIN was formally inaugurated February 27, 1963, General Curtis LeMay, Chief of Staff, U.S.A.F., sent a message over the system that began: "The activation of this new data communications system is not, on the surface, so spectacular as, say the launching of a new ICBM; however, it is every bit as important to our security."

Lt. Colonel Frederick W. Schultz, Air Force project officer for AUTODIN, said: "AUTODIN is one of the greatest developments in the field of communications since the invention of the radio. . . . It is one whose concepts will be adapted to many future communications needs, not only within the military framework, but also in industry."

The Air Force alone had as many as 1,500,000 different supply items in warehouses in thirty-nine countries. With AUTODIN, numbers corresponding to supply needs were transmitted to a switching center, and flashed on to the Western Union supply stations or subcontractors' plants. The time required to get action was reduced to minutes, where previously it was days and weeks. AUTODIN also transmitted data relating to paying, housing, and moving troops, and collected and disseminated budgetary information.

Western Union also placed in service for

the Joint Chiefs of Staff in June 1961 an Emergency Message Automatic Transmission System (EMATS) to flash top priority instructions and orders to major military commands around the world. The push of a button would transmit one of a number of previously prepared emergency messages. A "hot line" provided by EMATS was intended to enable the United States and Soviet leaders to consult quickly to avert a war when an emergency situation developed. The messages, of course, were sent by telegraph and translated.

To meet the communications needs of nonmilitary departments, the Advanced Record System (ARS) was installed for the General Services Administration in January 1966. ARS, an integrated, common-user data and teleprinter information system, linked more than 2,000 teleprinter terminals of thirty agencies, including those of Social Security, Health, Education and Welfare, Food and Drug, Commerce, and Agriculture Departments in 600 cities. A Veterans Administration network, with hundreds of stations, was added in 1967.

In 1962 Western Union began providing a nationwide bomb alarm system to the Defense Department, with sensors that would be triggered by a nuclear blast and report their location on central display boards. The company also started a private wire leased "Hot Line," metered no-dial telephone service, used by merely lifting the receiver. The distant telephone called rang in one second. The voice circuits were on Western Union's transcontinental microwave beam system.

Providing "real time" transmission from user to user, ARS switching centers transmitted multiple address messages, and provided both circuit and message switching. It improved efficiency by sending and receiving messages by teleprinter, data on punched cards, magnetic tape or facsimile. An Electronic Data Systems' contract in the 1980s was "to integrate Army information processing worldwide, including hardware, software, technical support, user training, and documentation."

20.7 SCENE AT NORTON AIR FORCE BASE AF DATACOM CENTER, THE FIRST OF FIVE CENTERS OF THE WORLD'S LARGEST DIGITAL DATA COMMUNICATIONS SYSTEM, WHICH WAS TURNED OVER TO THE U.S. AIR FORCE BY WESTERN UNION. AF DATACOM WAS DESIGNED TO HANDLE SEVEN MILLION PUNCHED CARDS OR THE EQUIVALENT OF MORE THAN 100 MILLION WORDS DAILY.

A 38,000-mile telegraph network was provided to the computerized National Crime Information Center of the Federal Bureau of Investigation. It enabled Federal, state, and local law enforcement officials to obtain information on crime, criminals, and stolen property from a central computer in fifteen seconds. Previously that required up to two weeks. Linked with a Canadian system, it connected 6,000 law agencies.

Important military networks[14] also were provided by AT&T.

20.8 In 1960, at Siegelbach, Germany, the U.S. Air Force inaugurated a 450,000-mile vital defense communications network. This global system, specially designed by Western Union for the Air Force, consists of ten automatic high-speed message centers. Shown here is a scene in one of the "switching aisles" in a typical center from which messages move at 100 words per minute between Air Force stations.

1. Strategic Air Command Control Systems (SACCS) which funneled information to SAC headquarters at Omaha, and to the Air Force commands. This primary alerting system featured a "hot line" red telephone the SAC Commander-in-Chief could pick up to establish immediate contact with the President and the worldwide strike force of missiles and manned bombers. Special lines also went to NORAD and the Joint Chiefs of Staff. The nerve center of SAC's communications complex was a huge underground room, one side of which was covered by a three-story-high world tracking map, showing the movement of air traffic.

2. The Air Defense Semi-Automatic Ground Environment System of the U.S. Air Force (SAGE), a Western Electric engineered surveillance and weapons computer control system, maintained a watch over the skies of North America.[15] Later a United States and Canadian backup interceptor control system replaced the SAGE system.

3. The Ballistic Missile Early Warning System (BMEWS) transmitted ballistic missile- and satellite-tracking data from radars in Alaska, Greenland, and England over dual routes to computers at the North American

Air Defense Command System (NORAD) headquarters at Colorado Springs, a land-based missile station. The Space Detection and Tracking System (SPADATS) computers, linked with the tracking stations, kept records of all objects orbiting in space.

As stated earlier, Sandia Corporation was created as a subsidiary of Western Electric in 1949 to develop and monitor production of components and systems, except nuclear weapons. Sandia's laboratories were at Albuquerque, New Mexico, and Livermore, California. This "weaponizing" of devices by the nuclear laboratories was only one of Western Electric's important defense contracts. Western Electric was the prime contractor, producing the Army's Nike Hercules and Nike Zeus guided missile systems, and, with Bell Laboratories, developing the antimissile system known as Nike-X, later named Safeguard.

ITT also carried out many major defense projects, including the operation and maintenance of the U.S. Air Force Distant Early Warning (DEW) system, with thirty-three "listening posts," forming a 6,000-mile electronic "fence" across the top of the world from Alaska to Greenland. About 2,000 civilians and airmen manned this Artic radar line as sentinels, to warn against enemy attack by bombers or intercontinental missiles. ITT also developed the Strategic Air Command's worldwide control switching network.

Usually the communications facilities of each military command were dedicated to its use, and when a circuit was broken, idle capacity from another command could not replace it. Conferences between different commands were difficult. To remedy this, Secretary of Defense Robert S. McNamara ordered a network that would make all circuitry available at any time.

The result was AUTOVON, a direct-dial voice network, provided primarily by AT&T.[16] It made instantaneous communications available to all military commands to control their forces. Four levels of precedence were used. In the United States, AUTOVON switching centers were leased by the government, and owned, operated, and maintained by the carriers; those overseas were owned and operated by the government.

20.9 WESTERN UNION'S TRAINING REPRESENTATIVES (WOMEN AT LEFT AND RIGHT) MAKING LAST-MINUTE CHECKS OF THE OPERATION OF THE WESTERN UNION SWITCHING SYSTEM INSTALLED AT NASA'S SPACECON CENTER AT GREENBELT, MARYLAND. THE RECEIVING UNITS ARE IN THE BANK OF EQUIPMENT AT RIGHT, AND THE TRANSMITTING AND SWITCHING EQUIMENT ARE AT THE LEFT. NASA COMMUNICATIONS PERSONNEL ARE IN THE BACKGROUND.

AUTOVON provided special features to insure flexibility of use and a high degree of survivability. The number of switching centers grew rapidly to meet volume requirements; access lines doubled in 1966 to 12,701 miles, and the trunk lines from 2,120 to 4,599 miles. Sophisticated routing plans and concepts are used to "have available real time information in order to take the right control actions at the right time."

"There are four major differences between AUTOVON and AUTODIN which should be emphasized," Col. Lee M. Paschall, assistant director, Programs and Requirements, DCS, DCA, wrote in the March 1966 *Signal Magazine*. "These essentially represent capabilities of AUTODIN that are not available in AUTOVON. . . . First, AUTODIN is a secure system in that not only the text of the message is encrypted, but message and traffic characteristics are concealed. Second, AUTODIN switches can convert transmission to and from dissimilar terminal devices (card to tape, data to cards, and so forth), whereas AUTOVON subscribers must have identical devices for every other subscriber with whom they desire to communicate. Third, AUTOVON is like any telephone system. In order to communicate, the called party must be at his phone, or, if the called party is busy, or the call interrupted by a higher precedence call, must start over again. AUTODIN is a store and forward record system. Traffic can be held in temporary storage at the switch when the subscriber is busy with other traffic, and lower precedence traffic can be held at the switch while higher precedence traffic is processed to the subscribers."

AUTODIN and AUTOVON were a part of the Defense Communications System (DCS), with which the president, as commander-in-chief, the Secretary of Defense, and the Department of Defense "agencies" carried out the integrated management of the nation's forces. The DCS formed a belt around the world, with terminals in 96 countries, about forty-two million miles of communications. It was a two-and-one-half-billion-dollar plant investment, with 36,500 people, and operating costs of $600 million.

The Defense Communications Agency realized that "real time systems" were necessary for maximum efficiency, and had AT&T and Western Union representatives appointed as members of the switched networks management team named "the Switchman Group."[17] A Federal Telephone System (FTS) was placed in use by the General Services Administration in 1964 for use by 700,000 government workers in seventy agencies. The Bell System also provided separate data networks to the U.S. Weather Bureau, Federal Aviation Agency, National Aeronautics and Space, Veterans, and Social Security Administrations.

AT&T provided many large telephone and data private wire networks. One was a system that routed calls between 700 General Electric locations.

The national defense, F.A.S., Department of State, G.S.A., NASA, and other systems were a part of the National Communications System (NCS), established by Presidential Memorandum August 21, 1963. Its creation followed the Cuban Crisis of 1962, when the military, civil air, space, data, diplomatic, and other systems were incompatible, and could not work together in the emergency.

The NCS Emergency Action Group established direct links between the various federal networks to coordinate them, and provide communications for NASA's Apollo space project. AT&T and Western Union had important roles in the nation's defense.

Notes

[1] One meter is 39.37 inches.

[2] This was one of my professors; he wrote *Life Begins at Forty*.

[3] *Telegraph and Telephone Age* (January 1, 1938) 19.

[4] Arthur D. Hall (of Bell Telephone Laboratories), *A Methodology for Systems Engineering* (Princeton NJ: D. van Nostrand Co., 1962) 26.

[5] *Broadcasting Magazine* (May 11, 1953).

[6] F.C.C Opinion and Order 53-313, 1953.

[7] See chap. 18, above.

[8] Testimony by AT&T officials before the FCC in 1965 and 1966.

[9] See ch. 18.

[10] The cable was laid by Standard Telephones and Cables, Ltd., an ITT British associate, for South Atlantic Cable Company, Ltd., a South African company, to connect with a British cable to be laid to South America.

[11] The Sears company was founded by Warren Sears, a Midwestern Morse telegrapher, who bought a shipment of watches a merchant refused and sold them by wire to other telegraphers. He then started the company.

[12] I myself arranged and publicized many of the switching center inaugurations, but IBM refused "because AT&T is a good customer." Officials of many companies saw the news stories about their competitors' new systems, and decided to increase their efficiency by leasing one also.

[13] Source: personal planning and participation.

[14] These pages were edited by an Air Force communications expert.

[15] In 1966 SAGE was integrated with AUTOVON, described later.

[16] Article on AUTOVON by Lieut. Col. Albert Marks, Chief, AUTODIN Division, Defense Communications Engineering Office, DCA, in *Signal Magazine* (March 1966).

[17] A talk, "The Vital Link," by Col. John F. Gerstner, Chief, Communications Services Division, Headquarters, DCA, at John Carroll University, Cleveland, Ohio in April 1966.

The Computer World
We Live In

Few inventions have changed the world and people's daily lives as radically as the printing press did centuries ago and the telegraph and telephone did during the nineteenth century. But in the twentieth century the computer ushered in an electronic-technological revolution that changed our business, finance, production, insurance, banking, and other customs.

To succeed and prosper today we must learn how to live with computers. They make available what we need from vast stores of information. The computer world's dynamic growth has epitomized the nation's change from a smokestack economy to services and information management. The computer, it has been estimated, saves people the equivalent of a day a week in looking for information. It helps businessmen make decisions confidently in an increasingly complex world. Such technology becomes meaningful when it changes our lives for the better, and doubles our productive power.

The computer is not only a vital tool of industrial managers in production, sales, and service, but also of people in homes and small businesses. A new generation of students has grown up using the personal computer as a learning tool in schools. Students using computers take a greater interest in acquiring knowledge to amplify their accomplishments. Computer automation has ended some familiar jobs in old industries, but created more new ones. Computers have met the need to process the growing volume of business, finance, and mass production data sent over thousands of private wire telecommunications systems.

Like other important inventions, the basic technical knowledge was developed long ago. Two thousand years ago the Chinese moved beads on an abacus to do fast and versatile calculating, and millions of small computers are carried in men's pockets and women's handbags today to add, subtract, multiply, and divide quickly.

Not until 1642 was a gear-driven adding and subtracting machine invented, by nineteen-year-old Blaise Pascal of France to help his father, a tax collector. In 1801 Joseph-

Marie Jacquard of France developed the automatic weaving loom that used instructions punched into cards or paper tape to control the loom's operations—an idea used later in computers. Charles Babbage, an English genius, designed a calculator in 1822 with the elements of a digital computer, using punched cards to provide its instructions, do complex calculations, and set up its results in type. He built a small working model but ran out of money and could not continue.

When a conductor punched his train ticket in 1887, Herman Hollerith, a Buffalo, New York statistician, got the idea, produced a punched-card system to do statistical work, and used it to compile the 1890 U.S. census statistics. (Hollerith also, in 1896, organized the Tabulating Machine Company, which in 1911 was succeeded by the Computing-Tabulating-Recording Co., and in 1924 became International Business Machines.)

George Stibitz of Bell Laboratories demonstrated a machine in 1939 that calculated complex numbers and electrically tabulated the figures. That provided the basis for punched-card data transmission used during the 1960s, when the telegraph and telephone multichannel carrier, coaxial, and microwave channels were available to carry the data.

The first general purpose computer was built in 1930 at the Massachusetts Institute of Technology by a team headed by Vannevar Bush.

During World War II, an IBM-sponsored team led by Howard Aiken and Grace Hopper of Harvard, developed the Harvard Mark I (1944), the first automatic sequence-controlled calculator. Western Union engineers designed its input/output circuits, adapting telegraph reperforators and printers. The Mark I computers were a complex of adding machines and calculators, controlled by instructions in perforated telegraph tape. (Also during World War II, a secret British computer, the Colossus, broke the German codes, and was of great aid in winning the war.)

The Electrical Numerical Integrator and Computer (ENIAC) using 18,000 electronic vacuum tubes was built in 1946 by John William Mauchly, a physics instructor who conceived it in a memo, and John Presper Eckert at the University of Pennsylvania. That monster—the first electronic computer—weighed thirty tons and required 15,000 square feet of floor space. ENIAC performed simple addition in $1/_{5,000}$ of a second. This seemed fast at the time, but it was 100,000 times slower than present-day personal computers (PCs). They then believed six ENIACS could serve the entire nation.[2]

Western Union had used director-translators in its private wire and reperforator centers since 1948. Like a computer program, its director connected with a translator that stored routing information for messages, disconnected when the code was translated, and then controlled switching to the proper outgoing equipment.

The transistor[3] replaced the electron tube and became the heart of the modern computer which stored the program of instructions to be carried out. Texas Instruments and Hewlett-Packard did a big home calculator business by making them programmable.

In the 1960s many companies established computer centers that processed and transmitted data over their private wire systems. For example, all Reynolds Metals offices sent daily reports to a headquarters computer that processed them overnight, and officials started the day with the data before them. Large savings resulted, especially when important orders arrived and management knew which plant was producing what was needed.

In 1980, General Electric bought from United Telecommunications for 100 million dollars, a company named Calms that provided computer-aided design and manufacturing systems to save drafting and testing costs. Crown also developed a voice synthesizer calculator that "talked."[4]

A forerunner of information technology data networks was the Bank Wire, established by Western Union in 1950. It moved a billion dollars daily among 229 major banks in thirty-three states. Each bank transmitted its orders to the Bank Wire's center in the Western Union building in New York, where telegraph employees pressed buttons to switch arriving orders to destination banks. That avoided any bank's orders being seen by employees of another bank. Fifteen years later in 1965 the growing Bank Wire was converted to computer control with 40,000 miles of duplexed circuits handling 3,000 words a minute. That was one step toward our throwing check books away and becoming a checkless society.

In the 1980s, bank-teller machines were installed in thousands of public places. People used plastic cards to draw cash or pay bills at point of sale in stores. The first big maker of automatic self-service bank-teller machines in 1974 was Diebold, the bank vault company. An identification card was scanned by reader machines that sent the number through a central processing center to the bank which transferred the amount from the customer's to the store's account. Many home TV sets were linked with bank computers so customers could do their banking and pay bills in the comfort of their homes. By merely pressing buttons, bills were paid and reported on the home TV screen like an electronic checkbook.

In 1986 Check Robot, Inc. introduced a system to eliminate cashiers and checkout lines in grocery stores. A laser registers the prices of items on a conveyor belt and shows a running subtotal. The customer transmits his checking account number at the bank to pay the bill.

In 1954, Western Union provided the first big private wire system engineered for data processing for the American Steel & Wire Division of U.S. Steel, and also, for Sylvania Electric, the first nationwide private wire system designed exclusively for automatic transmission of control data. Sylvania's center was in a building erected by Sylvania for that purpose at Camillus, New York. Sylvania later became a division of General Telephone and Electronics (GTE), which later sold the center to North American Phillips.

Realizing the computer would permit the greatest labor saving and productivity in history, in the 1950s and 1960s hundreds of large companies set up processing centers, using computers of various manufacturers, but they lacked a universal language. Many had a programming code incompatible with and thus unable to exchange information with others. International Data Corp. estimated that from their birth in World War II until 1977 a half-million computers were sold in the United States.

It was necessary for Western Union to develop, build, and install conversion equipment for the various codes used in customers' networks. Another growing pain was companies ordering million-dollar computers without realizing their large capacity would be useless until private wire telegraph systems were installed to bring data to them. Wise companies hired communications specialists and put the horse in front of the cart.

The Silicon Chip

The tiny silicon chip, thin enough to pass through the eye of a needle yet hold millions of bits of information and multiply the efficiency of modern industry, is made from one of the cheapest and most available substances in the world—sand.

Making a chip is an intricate and complex process, and semiconductor companies had to invest hundreds of millions of dollars. The chips were floated through a ten-foot machine on a column of air to avoid any human contamination. The circuit was then etched on each tiny chip from glass masks. Transistors smaller than two-thousandths of an inch in diameter were placed on a quarter-inch square chip to make more than 300,000 memory cells, each containing a transistor and capacitor. At that magnification a speck of dirt would look like Mount Everest. (The $750,000 chip-making machines were made by Ultratech Corporation and GEA to improve on an earlier printer, made by SLJ Technology Laboratory, that projected circuitry on the wafers.)[5]

The giant—15,000-square-foot, thirty-ton—pioneer computers were replaced by one or several tiny silicon chips which had the same capacity. Texas Instruments advertised in 1988 it had provided twenty-four *billion* chips since Jack S. Kilby pressed a switch in 1958 and proved all parts of an electronic circuit could be integrated. Charles Sporck built National Semiconductor into a $1,100,000,000 company.

The silicon chip restructured communications and changed our way of life and business practices. It enabled the telephone to speak to you and say "The number you have called . . . " and provide voice, data, and television signals. There are chips the size of a postage stamp that can store the contents of an entire telephone directory. Chips give computer power to auto engines, copying machines, microwave ovens, video games, toys, and many other things.

There are too many ways chips are used to serve people to list here, but one will illustrate. Atlantic Richfield opened a computer-controlled gasoline service station with a credit card actuated dispenser system. The customer inserts the card, its validity is automatically acknowledged, and an invoice is printed. There is a slot for dollar bills for cash customers.

A typical 1992 personal computer has one microprocessor chip that serves as the central processing unit (CPU). It also includes perhaps seventy to 100 or more random-access memory (RAM) chips (all identical), several read-only memory (ROM) chips, and a number of other chips to connect things together and provide the disk control, monitor control, keyboard control, printer drivers, perform floating point mathematics, and other special purposes.

In the early 1980s, the Japanese concentrated on the memory chip (RAM), reduced prices, and took over a majority of that market. Many American semiconductor companies left that market to the Japanese and concentrated on microprocessors and other special-purpose chips with a higher profit margin.

In 1985, the market for computer chips was estimated at thirty-three billion dollars. Intel, Motorola, and Texas Instruments developed microprocessor chips, to be discussed later. In 1986 NEC of Japan was the world's largest producer of chips. The Japanese and Koreans had cut the price of their mass-produced commodity chips, but Texas Instruments was second and Motorola third. Apple, UNISYS, and National Cash Register

(NCR) were among the first thirty computer makers that decided to use chips in all their computers.

Needs of the defense network for faster chips caused constant improvements. For example, A-to-D converter chips were needed to convert sound waves and analog signals from different sources instantly into digital signals the data networks of the military could process.

Despite major chip inventions by American companies, Japan increased its global silicon chip business from eight billion dollars in 1984 to $17.6 billion in 1988, while United States sales grew from $11.6 to $13.2 billion. Likewise, America's share of world electronic production dropped in three years from fifty and four-tenths to twenty-seven and one-tenth percent while Western Europe lost little.

Research and development computer engineers found it stimulating to work in the same areas as other high-tech specialists. In the early 1980s, about 1,500 plants were located in "Silicon Valley," in Santa Clara County, California. Other aggregations of high-tech plants were on Highway 128 circling Boston; "The Research Triangle" near Duke University, Durham, North Carolina; "Automation Alley" near Michigan University; Naperville, Illinois, near Chicago; "Silicon Prairie" north of Dallas; "Silicon Forest" near Portland, Oregon; "Silicon Mountain" in Colorado; and a long computer company strip on the southeast coast of Florida. There also were "Brain Street, U.S.A.," "Silicon Glen" in Scotland, and "Samsung Center" in Korea. Inventions from these high-tech complexes made earlier computer paraphernalia obsolescent in a few years.

After one hundred million dollars and years of effort, Intel's Laboratory in Silicon Valley produced a microprocessor that included one and two-tenths million transistors on one one-square-inch chip. In 1984 Bell Laboratories developed a faster semiconductor for the big high-speed computers AT&T was building for a new government voice, data, and video network. In 1989 IBM built the first sixteen-megabit (sixteen million bits)* computer-memory chip only one-third by one-fourth inch in size. The sixteen-megabit chip will hold 1,600 pages of double-spaced typewritten text, four times the capacity of previous chips. AT&T joined with Mitsubishi of Tokyo in making and selling chips. (By installing a chip in the meter box at each home, water companies can read the meter by telephone without sending a meter reader.)

The early computer industry giants— International Business Machines (IBM, the standard-bearer in the twenty-nine-billion-dollar personal computer market), National Cash Register (NCR), Burroughs (now, joined with Sperry-Rand, known as UNI-SYS), and Honeywell—were joined by Hewlett-Packard, Xerox, Digital Equipment Corporation (DEC), Apple, and hundreds of others who spent heavily for research and development to speed technological progress. The usefulness of computer data increased as improvements were made with microchips, word processors, minicomputers, optical scanners, floppy disks, hard disk drives, and robots and changed the way people lived and worked. In 1980–1982 an estimated two million people lost jobs in manufacturing companies due largely to technological change; more such large job losses are to be expected. However, the telecommunications industry grew and provided new work for more people.

The industry began employing a new breed of studious young men who spoke a strange language about hardware, software, peripheral equipment, programming, bits, bytes, and bauds, modems, interfaces, masers and lasers (*m*icrowave or *l*ight *a*mplification by *s*timulated *e*mission of *r*adiation), eight-level ASCII (*A*merican *S*tandard *C*ode for *I*nformation *I*nterchange), BCD (*b*inary *c*oded *d*ecimal), and languages like Basic, C, ADA, C++, COBOL, PASCAL, FORTRAN, and PL/1. Most communications people hired during the Depression 1930s reached sixty-five, retired, and were succeeded by graduates of engineering schools. Education or experience as a programmer, engineer, analyst, or other computer specialist was sure to provide a good job. Companies recruited computer students in colleges and raided employees with such skills from other companies. Turnover was heavy and salaries of communications employees continued upward. The Bureau of Labor Statistics forecast that by 1995 the American high-technology companies would generate one and seven-tenths million jobs.

In 1963 President Marshall of Western Union foresaw the revolution in computer-data communications, looked for a man qualified to lead Western Union in a data processing world, and selected Russell W. McFall as executive vice president. McFall, forty-four, a well-built, crew-cut electrical engineer, looked as though he had just come from a university football field, but at General Electric had managed a staff of 3,500, including 700 engineers and scientists. He was then Litton Industries' vice president responsible for four organizations engaged in the design, manufacture, and sale of radiation and communications equipment for military and commercial use.

McFall was elected president on January 12, 1965, and reorganized the company. The multicolumn field organization was replaced by a single line in which area vice presidents were responsible for all activities in their geographical areas. All reported to an executive vice president. Morse's original dots and dashes were the same basic digital "1-0," "Yes-No" units used in computer data transmission. Using some of the computer techniques it developed in its switching centers for Telex, AUTODIN, and Advanced Record System, Western Union analyzed customers' needs, designed the private wire system to meet them, selected the best computer equipment to do the job, installed and tested the system, and trained the customer's employees to operate it. McFall saw a big new market for real-time computer services on a time-sharing basis, and concentrated on building systems to provide them.

Large companies with private wire systems need communicators skilled in electronic data transmission and processing to manage them.[6] A vice president of Merrill-Lynch-Pierce-Fenner & Smith, stockbrokers, managed that firm's 750,000-mile leased wire system, utilizing computers. A system called COMPASS enabled employees of E. F. Hutton[7] brokerage to communicate between its hundreds of offices via a central computer.

Data Networks, Banking, Reservations

Western Union established a computer-controlled system for customer use with centers in New York, Chicago, San Francisco, and Atlanta. Its multiple-access capabilities permitted several Telex users to work simultaneously with a Western Union computer. The Telex Computer Communications Service (TCCS) switched Telex (teleprinter exchange), TWX (teletypewriter exchange), and public messages through the national telegraph network. InfoCom, another shared-computer service, offered many advantages of a private wire computer network to customers without a major investment in big centers, equipment, installation, space, programming, and manpower. Another time sharing company, On-Line Systems, was acquired by United Telecommunications. There was even a PAX (private automatic exchange) to interconnect computers as the PBX (private branch exchange) did telephones.

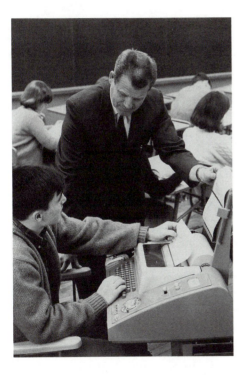

21.1 EARLIER IN THIS COMPUTER AGE, INFORMATION RETRIEVAL BY TELETYPEWRITER OVER LEASED TELEPHONE LINES MADE COMPUTER FACILITIES AVAILABLE TO STUDENTS IN MANY SCHOOLS.

Electronic banking is simply an electric connection between homes, offices, and banks so that banking services can be performed from a distance. Honeywell's Inco-

term subsidiary linked Seattle's First National Bank's 176 branches and 105 automatic-teller machines to provide instant status of every customer account. In the 1960s a score of companies, like CompuServe, established networks serving small business and home computer users on a time-sharing basis.[8]

In 1957 the telegraph company started a data processing system to handle its own payroll, financial reports from distant offices, operating performance, and inventory. When it established Telex switching centers in the 1950s, Telex subscribers could dial TWX users, and TWX could dial Telex. It was then the world's largest interconnected network for record communications, with subscribers in 168 countries. Its computer switching center at Middletown, Virginia handled international messages.[9]

Telex Corporation, the Telex telegraph-printer maker, had heavy losses in making computer peripherals, and lawsuits by IBM and others in the 1970s, but recouped by buying, for two hundred million dollars, Raytheon's ninety-million-dollar annual business providing computer terminals for airline reservations.[10]

For the securities industry, Western Union developed the Securities Industry Communications (SICOM) computer-controlled system for shared use. With teleprinters at wire and order rooms, branches and correspondents, and posts on the New York and American Stock Exchange floors, each SICOM broker could provide high-speed computerized handling of orders, reports, market news, and administrative traffic.

Mead Data Control provided a computer memory bank, used by Barrister Information Systems to enable 140,000 lawyers and judges to radically shorten their time spent researching the law. Another shared-time computer project was MEDINET, a hospital and medical information system planned by General Electric. It was designed to provide data by specialists, from a central library, locate empty hospital beds, and prepare bills. GTE provided the network. Security of SICOM messages was protected by special codes and controls programmed into the computer.

Like the cellular telephone, computer networks were vulnerable to invasion of privacy. With a modem that lets a computer talk with others, personal computer users have access to databases and other retrieval services. The fourth amendment to the Constitution was believed to protect the privacy of telephone conversations, but a court ruled that they were protected only if the telephone was plugged into a wall. That left the cordless cellular telephone unprotected. The Supreme Court refused to hear an appeal.

Officials began to realize that when information they need is packaged in databases and readily available by computer and communications networks, it becomes infinitely more valuable. It provided information they needed to make quick, well-informed decisions and meet or keep ahead of competitors. Officials with yesterday's sales and inventory reports on their desks the next morning could make important savings and speed production and shipment to customers. Purchases and sales underwent major changes both in business and the techniques officials needed to learn to succeed in business.

An EDP (Electronic Data Processing) Auditors Association official testified before a congressional committee that hundreds of thousands of "viruses" or "worms" had invaded U.S. corporations and government networks, and Congress needed to make it unlawful. "Hackers" who were pranksters but not criminals managed to break various computer network codes. One college student sent orders in code that pretended to come from an official source ordering computers to destroy stored records. That produced nationwide chaos until the culprit was found and emergency replacements made. When a computer technician ordered a bank to transfer eleven million dollars to his accounts, the Defense Department ordered contractors to encrypt all classified data sent over telephone systems. One group of four "hackers" was indicted on charges of a fraudulent scheme to disrupt the 911 telephone service in southern states.

Failure of a computer programmer to change routing numbers on a network handling the 900 service has twice virtually closed down AT&T's long distance service in the southern states. It was the worst crisis in the history of AT&T's service. The company spent millions to replace the equipment with digital transmission and expressed its apologies for the inconvenience by offering Valentine's Day and other calls at a discount. Its competitors tried to take advantage of the incidents.

Western Union's data processing technology center at Mahwah, New Jersey began operating with Shields & Co. as the first user. In 1976 SECTOR, a brokerage communications association, leased 240 voice/data circuits in Western Union's satellite system to handle the messages of its members. Then GTE set up Financial System One, serving more than 400 brokerage offices and their clients.

The Personal Computer

In the 1980s the personal computer changed the way people worldwide work and live. In 1954, Harry D. Huskey, later a professor at the University of California, developed a computer for personal use for

Bendix Corporation, using punched telegraph tape, but it cost $50,000 and required a truck to move it years later to the Smithsonian Institution for display at a telecommunications technology exhibit.

In January 1975, *Popular Electronics* published the first installment of an article on the Altair 8800, a "personal computer" developed by Micro Instrumentation and Telemetry Systems (MITS), a small electronics company in Albuquerque, New Mexico, founded in 1969 by Edward Roberts and three others. The Altair 8800 had no keyboard or display screen, but it was the first, full-fledged personal computer on the market.

Earlier, in 1971 John Blankenbaker, an engineer, put together a machine in his garage in Albuquerque, New Mexico with no keyboard or screen. It was programmed by turning switches and had only 356 memory characters, but his Kenbuck I was called a "personal computer." In 1986, a panel of judges at the Computer Museum in Boston declared Blankenbaker was first with a personal computer.

Chuck Peddle designed a personal computer in 1977 with keyboard and screen. Another who did was David Allison, curator of the Computer Exhibit at the American History Museum.

The first *commercially successful* personal computer was produced and marketed by Stephen Wozniak and Steven Jobs. Wozniak and Jobs were among the many computer hobbyists inspired by Roberts's Altair 8800 to design their own computers. In late 1975, Wazniak's and Jobs's "Apple I" emerged from Jobs's garage, which for a time was Apple Computer Company's world headquarters, and marketing and distribution center. The Apple I's success sparked rapid growth of the entire personal computer (PC)

industry (the Apple II was introduced in 1977). Apple, IBM, and Radio Shack became the largest producers. Their officials believed a million people would buy a personal computer when they realized how much it would increase their productivity.[11]

21.2 THE PERSONAL COMPUTER. (COURTESY OF IBM.)

"Computer Literacy," Apple said, "is becoming as fundamental a skill as reading or writing," and, in fact, today many universities require students to take an introductory computer course and be literate on the subject before graduating.

Encyclopedias and other information sources were being recorded on computer disks or tapes as databases that could be accessed by computer and shown on home screens. Personal computer users nationwide could press buttons in a remote-control retrieval system and have desired information transmitted from a central library. An autodial access system at Ohio State University permitted students to dial 8,000 programs from seventy-five courses. There were even two magazines sent by mail that had magazine covers but no printing inside, just disks that could be put in the computer. They were named *Big Blue Disk* and *Diskette*. Another system permitted storage of 500,000 pages on one reel of video tape, and numerous students could "read" the same or different articles at the same time. The system also shortened the time required for research in business. People buying computers only needed software and a modem to use them for investing, store operation, and so forth.

Starting with a Woolworth order for 8,000 terminals for its stores in 1979, IBM attacked the strong position National Cash Register (NCR), Burroughs, and others enjoyed in banking and retailing. Burroughs proudly set up "the biggest network in the world to connect about 700 of the world's largest banks." Its network centers were in Brussels, Amsterdam, and Culpepper, Virginia. Computer Sciences Corp, provided time-sharing systems and computer technology worldwide. One large early data processor was Commerce Clearing House which handled income tax returns for professional clients.

Reduction in size of the half-acre pioneer computer to the big mainframe computer, of course did not meet the needs of smaller companies and individuals. Digital Equipment Corporation (DEC) recognized the need for a minicomputer in 1960 and built it into a five-billion-dollar business by 1984. As minicomputers grew larger with advances in technology and began to overlap the capabilities of mainframe computers, DEC competed strongly with the mainframe computer makers. When later PCs began to overlap the capabilities of the mainframes, some companies replaced their mainframes with a network of PCs.

Growth of the personal computer (PC) business was phenomenal. It gave homes more brain power. It is a tool so easy to use,

so useful at homes and businesses and its uses so many, everyone should understand it. PC sales increased from 150,000 in 1979 to two million in 1982. National Science Foundation predicted forty percent of U.S. homes would have computer workrooms and two-way video services by cable TV in the early 1990s. The result should be a tremendous explosion in human knowledge.

Faced with high wage costs and featherbedding by the typographers' union, newspapers adopted computerized typesetting. The Associated Press provided a photo-offsetter with which papers receive telegraphed news photographically. Autologic, a Volt Information Services' subsidiary, provided high-speed digital typesetters to large newspapers.

News can bypass printing and delivery, appear on home TV screens, and be recorded if desired. The first such daily service was begun in July 1980 when the Columbus, Ohio *Dispatch* began sending news through the CompuServe home computer network to subscribers.

The use of personal computers to obtain information from databases grew rapidly. Dun & Bradstreet, Mead, McGraw Hill, Readers Digest, H.&R. Block, Lockheed, and others did a $1.9 billion business publishing databases, and the service grew thirty percent a year. Business Computer Network, a database distributor, spent $1.6 billion to establish and promote the system. It provided the service to subscribers for a monthly fee. It distributed 450,000 disks to attract home users. Others providing database information were CompuServe, General Electric Information Service, Dialog, News Network, and Western Union's Easy Link computerized service by telephone or Telex. About 3,100 large businesses used database service.

The brain of the PC is a silicon chip

with circuitry to remember and carry out the desired process. The software program tells the computer what to record on a floppy disk, or a hard disk for storage. Big PC makers were IBM, Compaq, Apple, Tandem Corp. (with a three-hundred-million-dollar personal computer business), Sun Microsystems, Synapse, Data General, the NCR Division of AT&T, Storage Technology, Hewlett-Packard, and Olivetti. Xerox unveiled a low-cost desktop computer that could be used also as a word processor. Wang also made word processor computers. Compaq had systems for multiple users, and battled AT&T to set the PC standards.

In 1985 GTE combined PC and telephone technologies to provide medium and large companies with voice, data, and electronic mail over standard telephone and local area networks. Thousands of travel agents, airlines, banks, and stores used minicomputers to provide instant two-way communication. Chains of stores were established across the the country to sell computers at cut-rate prices to individuals. Some companies moved their workers from crowded

21.3 PERSONAL COMPUTER KEYBOARD. (COURTESY OF IBM.)

New York City to New Jersey or elsewhere and retained their headquarters, but maintained the flow of data between workers and headquarters by computer.

Many companies connected their computer systems with homes of employees, transmitted work to them, and received the completed work from them by computer. Telecommuting saved the expense and time of travel to crowded cities. Employees and companies liked it, but the unions fought it. The advantages were so great that home work increased. Observers hoped home work would slow down the mad rush of population to big cities with their ghettos and crime.

In 1983 four million homes had computers for personal and business use, and a personal computer in every home became a hope of the industry.[12] Peripheral parts with strange names—mouse, track balls, tablets, light pens, joy sticks—controlled features of different makes and models. IBM developed a microchannel computer and hoped it would become the national standard, but a "gang of nine" PC makers developed a different computer channel standard and produced a new silicon chip to implement it.

Intel introduced the first microprocessors to power PCs on November 15, 1991. The "gang of nine" adopted their standard industry architecture and utilized the successively more powerful chips that were developed. It increased the usefulness of the PCs and their popularity. The nine PC manufacturers were AST Research, Compaq, Epson, Hewlett-Packard, NEC (Nippon Electronics Corporation), Olivetti, Tandy Wyse, and Zenith.[13]

Leading makers developed elaborate plans to automate and dramatically increase efficiency and productivity of the home and business office. Minicomputer sales numbers were twice as much as the big mainframes.

Makers planned software, word processors, "brainy" electronic typewriters, facsimile and printing telegraph terminals, CAD/CAM (computer-aided design / manufacturing). telephone switching, and work-flow diagrams to produce the automated, electronic, paperless office of the future.

The computerized office of the future will handle electronic mail, internal company telephone or written correspondence, banking, reservations, and so forth to save time. With such equipment, the workaholic businessman might even make the future office a workaholic instead of himself!

Hoping to take over part of AT&T's four-billion-dollar telephone equipment business, IBM made an agreement for Mitel, an Ontario-based company making telephone switching systems, to develop a family of telephone equipment for the big market it foresaw in automating offices and set a standard for efficient business operations. Systel added to that vision by using computer technology to turn the electric typewriter into a word processor. IBM hired employees of MCI (Microwave Communications, Inc.), AT&T's bitter competitor, to train IBM people in the telephone business. IBM also bought ROLM, a telephone switch manufacturer, which it later sold to Siemens.

Thousands of software programs were tailored to meet the needs of different companies and industries and increase the success of their managements. Computers were programmed to forecast the effect of management's decisions on everything from production to stock prices. Conferences were held with distant participants checking data

in their computers before answering questions. Since it was necessary to serve industries in which many members' computers were not compatible, the complexity of the work of software programmers increased. However, many personal computers had standard software packages for the more frequently used computer processes.

The trend was toward smaller, less-expensive computers. Apple made a laptop computer and Zenith Electronics Corporation produced a five-and-nine-tenths-pound microcomputer with a 3.5-inch-square diskette to store data instead of the usual 5.25-inch diskette. Zenith and others made a "notebook-size" computer for laptop use. Compaq's laptop computer was battery powered, 8.5x11 inches, and weighed only six pounds. The lap computers were for use in trains, autos, and hotel rooms. Canon, Hewlett-Packard, and others turned out low-cost, hand-held computers, and two toymakers offered a pocket computer terminal with a miniature keyboard.

The microprocessor—a silicon, integrated-circuit central processing unit—is the electronic brain of the personal computer. Major microprocessor makers are Intel, Motorola, Zilog, CPT Corp., and NBI, Inc.[14] Compaq made IBM-compatible PCs and, of course, they would run on software made for IBM's PCs. Both IBM and Compaq used the Intel series of miniprocessors, but Apple used those made by Motorola and ran on different software. In 1992, IBM and Apple announced they would be jointly developing a new generation PC workstation to use a chip to be made by Motorola.

Great Speed
Mainframe Computers

Big mainframe computers made by Cray, IBM, NCR, Control Data, Sperry, Burroughs, Amdahl, Honeywell, and National Semiconductor performed complex processes in small fractions of a second. All invested heavily in research and development (R&D) to stay competitive. R&D was estimated to cost ten *billion* dollars in 1984, of which IBM reportedly spent two billion and Apple almost one billion dollars. Repairing and servicing computers was a six-billion-dollar business.

IBM produced a new generation of big computers in the mid-1960s that set such a high performance standard that RCA, Xerox, and General Electric stopped making them after sustaining heavy losses.[15]

General Electric Credit Corporation and others bought big IBM computers, but IBM produced a new, improved series every few years, and companies selling or leasing "plug-compatible" equipment for use with IBM computers faced losses. Lloyds of London insured computer leasing companies against such loss, but suffered heavy losses itself. Comdisco, Inc., the largest dealer in used IBM equipment, benefitted.

One loser was Itel, so prosperous from leasing computers that it spent three million dollars on a party wining and dining 1,300 employees and spouses at Acapulco, Mexico in January 1979. Late that month a new IBM computer series burst the Itel bubble, and it had to struggle for survival. Itel had expected a billion-dollar year in 1979, but lost two hundred million dollars during the first nine months and sold some of its business.

The 1980 U.S. Census of Population and Housing used Sperry's Fosdic (film optical sensing device for input to computers) that digested four billion facts, filling 200,000 pages. Fosdic read the census questionnaires after they were photographed on microfilm and translated the answers into computer tape. It succeeded Sperry-Rand's UNIVAC (universal automatic computer) an early electronic computer, built for the Census Bureau thirty years before (1951).[16]

ITT established a data network with computerized automatic switching centers in the United States and Europe. Its time-sharing system, leased-line customers could use the computers while it processed data from its own offices. One customer was Tidewater Oil Company's world headquarters and western division in Los Angeles. Modular high-speed computers delivered quotations from the London Stock Exchange to brokers, and at the same time controlled ship traffic in New York Harbor and processed Voyager spacecraft pictures of Jupiter and Saturn.

Big on-line computers needed emergency generators. A brief brownout or blackout could cause million-dollar checks or orders to be sent in error. Power converters and controls to avoid such trouble were provided by GE, Inco, Teledyne, Emerson Electric, International Power, and others. Suits were brought by companies claiming malpractice for selling them computers not suitable for their use. Some large awards were made by juries.

IBM introduced the first commercial magnetic drive in 1979 and it revolutionized data processing. It stored information in computers, making data available quickly and was widely adopted for airline and car rental reservations, twenty-four-hour banking, and credit card transactions. IBM said its magnetic disk would hold all of the information in twenty-two newspaper pages in a space the size of a postage stamp. Successive developments increased the number of bits of information it could store. In 1989 IBM announced it stored one *billion* bits on one square inch of magnetic material.

An outstanding computer inventor, An Wang, a native of Shanghai, who came to the United States in 1949 with his father, founded Wang Laboratories as a one-man company in Boston, and invented the core memory that was the standard computer data storage until the semiconductor came into wide use. Then Wang sold his patent to IBM, but continued producing important equipment for the automated office, including a broadband cable network system linking word processors, computers, and telephones.[17]

Like An Wang, many others won fame and fortune in the computer industry. The most successful was Thomas J. Watson (1874–1956), who built and ruled IBM, the largest and most profitable computer company, from 1914 (when it was the Computer-Tabulating-Recording Company) until his death in 1956.[18] Watson had his salesmen dress like the executives they called on, with coats, white shirts, and ties, and the "IBM man" became a standard for well-dressed businessmen. He made the word "Think" his company slogan, and its color blue, so IBM was called "Big Blue." (The one-word slogan was only one of many things Watson learned from his former tyrant boss—who fired Watson in 1914—National Cash Register's philosopher-king John Patterson.)

IBM overcame Remington Rand's UNIVAC, the early leader, in large mainframe computers with great capacity for large users like the government. Cray's superconductor in 1986 performed an incredible 1,200,000,000 calculations per second.[19] Later, Digital Computer and Cray challenged IBM's mainframe computers, but IBM had seventy percent of the mainframe business, UNISYS 9.1 percent, Amdahl 8.3, and National Cash Register and others less. IBM accounted for thirty cents of every dollar spent for all computers.[20]

To compete with the burgeoning market leadership in personal computers of Apple II, Thomas Watson, Jr., who had succeeded his father as chairman, assigned to IBM's laboratory director, William C. Lowe, the task of making a superior personal computer. Lowe organized a committee of planning engineers called "the dirty dozen" to handle "Project Chaos." IBM's "PC" (IBM's name for the unit, but since PC has become generic, "personal computer") was so successful that it took the personal computer leadership from Apple. IBM had so many fine computer engineers that many left to join or form other successful companies.

Steven Jobs and Stephen Wozniak, mentioned earlier as cocreators of the first successful personal computer, founded Apple Computer Company. Jobs did a great job in promoting the company, but disagreed with John Sculley, later Apple's chairman, who wrote in his autobiography that he "regretably ousted" Jobs from the company. Jobs had brought Sculley over to Apple in the first place. Sculley reorganized the company several times and was criticized for his large salary and lavish lifestyle, even as chief executive of a company that grew in annual sales in four years from $1.9 billion to $5.3 billion.

As a lone wolf inventor, Jobs then created another personal computer at his home in Silicon Valley that did technological wonders and started marketing it as "the first computer of the 1980s," at his NeXT, Inc. company. He produced an easy-to-use workstation and software based on standards not compatible with or useable by Apple and IBM machines.

With many new products being created and rolling out of its doors, IBM's policy was to reveal nothing about any development until it was publicly announced, but Jobs at once announced his new computer and said H. Ross Perot, who founded Electronic Data Systems and became a billionaire, invested twenty million dollars in his company.

The family of Sirjang Lal Tandon sent him to America with $5,000 to get a college education. He worked at night as a busboy in restaurants and by day to earn a masters degree in Mechanical Engineering. In 1975, he started Tandon, his own business, producing low-cost recording heads to put data on floppy disks for storage, and in eight years had a $1.5 billion business and a fortune.[21]

Michael Dell, a nineteen-year-old whiz kid, founded the Dell Computer Company in 1984 before graduating from Texas University. Making and selling computers directly to customers, he did a $388.5 million annual business in five years.

Companies gaining a lion's share of the IBM-compatible microcomputer business, were Dell, Wang, Tandy, Commodore, Compaq, Kaypro, and Zenith. Many companies changed their names as they merged. Univac became Sperry-Univac, then Sperry combined with Burroughs and became UNISYS. Sperry was a leader in computer networking in which computers talked with each other and companies bought others with computer techniques they needed. The personal computer business was estimated by Basic Data to have grown from one billion dollars in 1981 to about thirty billion dollars in 1987.

There were joint computer ventures by IBM and AT&T. One by Sears Roebuck

with IBM was named Prodigy. It gave access to an IBM central computer by owners of personal computers to obtain news, catalogs, stock quotations, airline tickets, and hotel reservations. It sent orders by toll free 800 numbers to companies that advertised the products.

Moore Corp., Uarco, Standard Register, and a few others built a big business supplying tons of paper forms used by the computer industry, and Dataproducts Corp. found a way to make personal computers print forms by using lasers. In 1989 Judge Greene removed a ban he had placed on AT&T in 1982, barring it from the electronic publishing business. He permitted AT&T to originate, compile, edit, and provide an information database and transmit information on subscribers' request by computer.

Databases were expected to be a very lucrative business. Many publishers transmitted their pages to distant publishing plants to expedite delivery and the *Wall Street Journal* did so worldwide.

There were many other minicomputer makers like Fortune Systems, Osborne, Vector Graphic, and Victor Technology; floppy disk makers like Verbatim (Kodak), Dysan, and Xidex; and a hundred making parts. Many dropped by the wayside.[22] After the 1960s, IBM had a dozen manufacturing plants in Europe, and one year sold ninety percent of its products in Europe.

A key to success, said John T. Hartley, president of Harris Corp., maker of semiconductors, terminals, and other equipment, is "technology transfer" from defense developments to commercial products. For example, L. M. Ericsson Telephone Co.'s advanced computer controlled telephone systems resulted from a multibillion-dollar system it developed for Saudi Arabia.

The Russians apparently copied IBM

computers from blueprints, even to color, but did not mass produce, or make sophisticated use of them. The U.S. Department of Commerce reported that in 1977 Russia had only 20,000 computers.

Indispensable Software

Largest in number of employees were labor-intensive software companies producing elaborate instructions that each company needs for its computer system. Software (as opposed to "hardware," the physical machine itself) is the set of instructions that tells the computer what to do, and allows it to integrate the computer functions and generate reports like profit and loss balance sheets. Software is labor intensive but not capital intensive. The instructions must be precise and detailed. Software can even help people compose music, write a letter, or estimate a company's or project's chance of success.

Without a program a computer is just a dumb bunch of wires and pieces of metal and plastic. The program tells the computer precisely what to do with the data input and what process or formula to use in processing it. So many software packages were produced that the D.C. Software Directory listed 21,000 software packages for 200 kinds of personal computers, priced under $10,000.[23]

The big software service firms in 1990 were Microsoft, Lotus, American Management Systems, Novell, Ashton-Tate, Automatic Data Processing, Borland, Computer Associates, Computer Sciences, Consilium, Electronic Data Systems, Informix, Oracle, Paychex, and Software Publishing.

The largest and most profitable was Microsoft, with net revenues of $805.5 million in 1989. Backed by IBM, Microsoft developed the Microsoft System Disk Operating System (MS/DOS) designed to make

IBM personal computers as easy to use as Apple's. Microsoft's founder and cochairman Bill Gates kept making improvements and eventually introduced the "Windows" program that splits the computer screen into windows like individual screens. Users control the computer functions with symbols instead of words. Analysts said it would heighten IBM's competition with Apple's Macintosh computers.[24] Apple sued Microsoft for copyright infringement.

21.4 MUSIC ON THE PERSONAL COMPUTER. (COURTESY OF IBM.)

The agonizing complexity of creating a new software program, when there are changes in a company's business and products, frustrates even the experts. The programmer must understand a company's business completely. Mitch Kapor, who founded the Lotus 1-2-3 spread sheet that five million companies bought and made Lotus an eight-billion-dollars-a-year software giant, was so frustrated, he said, "Not a day passed that I did not want to throw my computer out of the window." One reason is that even one letter out of place in a software code can turn a million-dollar program into gibberish and the business into a nightmare.[25]

Software excellence is demanding but its future seems secure. It is needed to improve productivity as equipment becomes more expensive. Customers of a software company cannot easily change to others. Software firms require lengthy study and preparation to meet the special needs of a company. The future of Microsoft and Lotus seem as secure as McCaw is in cellular telephones.

Among the leaders in the software business in the 1980s were Computer Associates International that bought Cullinet Software; NOVELL and 3 COM, a maker of software for personal computers; Microsoft that had $1.1 billion in revenues in 1989; Lotus Development Corp. with its 1-2-3 software program that had a best-selling 1-2-3 Spreadsheet used by many personal computer companies.

Philippe Kahn immigrated from France in 1982 and in 1990, at age thirty-nine, made $1.6 million in salary and bonuses as chairman of Borland International, maker of Quattro and Quattro Pro, a computer spreadsheet program, and Paradox, a database manager. In 1991 Borland bought Ashton-Tate, its database rival, for forty million dollars in a stock swap and became number four in the software industry after Microsoft, Lotus, and Novell.[26]

Cray Research spent about fifty million dollars on research for its big mainframe computers and split into two companies, one headed by its founder, Seymour Cray, who continued research, and the other which marketed its latest model. McDonnell-Douglas engineers wrote a software program designed to help American hospitals using their computer system cope with the laws of all states and cities. The program was de-signed to be changed and kept up to date when laws and the economics or computers of the hospitals changed.

As more computers "talked" over telephone lines with each other at 1,200, 2,400, 4,800 and 9,600 "baud" (number of bits per second), and over local area networks at rates of millions of bits per second, the need for communication channels multiplied. Western Union had established a broadband exchange service in 1964 with a push-button telephone to select (1) the connection desired, (2) voice or data, and (3) a bandwidth of one-half, one, two, four, or twelve voice bands. One bandwidth permitted transmission of letter-size pages by facsimile. Twelve voice bands (48 kilohertz) handled 50,000 words a minute to a distant computer. Charges were based on distance, bandwidth, and length of time used. Microsoft, the software company which prepared the software operating system for IBM's PC and PS/2, advertised that it was first to propose putting a computer on the desk in every home and office.

Japanese Competition

After ENIAC, and other early American electronic computers were built, officials of Sony Corp. of Japan studied the American computers on a visit to the United States, agreed to pay royalties for the use of patents, and started making semiconductors. The Japanese Government provided the finances.

Fujitsu announced Japan's first computer in 1954. Soon Americans were buying Fujitsu's RAMs and small data machines. Hitachi and other Japanese companies bought an interest in or established plants in the United States. Texas Instruments and IBM built plants in Japan to utilize its more productive workers and lower labor costs, but Fujitsu's computer sales outstripped IBM's Japanese and Commodore International's subsidiaries in Japan. When IBM offered a new series, Fujitsu officials said they could match it in six months.

When the silicon chip changed the computer in America, making it faster and more reliable, Japan decided to compete in the most widely used RAM chip area. They obtained some American chips and began producing them. A battle between the Japanese and American semiconductor industries followed in which Japan's Toshiba, NEC, Mitsubishi, Hitachi, Fujitsu, and Matsushita Electric took over one billion dollars of the annual chip business in the 1980's by selling chips at less than American production costs. In 1984 unionized American companies owned more than half of the six-billion-dollar chip-making equipment, but only a small part of the chip business. The remaining American silicon chip makers, Intel, Motorola, and Texas Instruments, competed vigorously and had thirty-six and one-half percent of the world market in 1989. National Semiconductor still produced chips, but only for use in its own equipment in checkout terminals of grocery store chains. Making chips is expensive because a chip plant costs hundreds of millions of dollars, and millions of chips of each new design have to be sold to pay the cost.

Hitachi and General Motors' EDS unit acquired National Semiconductor's mainframe computer business, and General Electric sold its semiconductor business to Harris Corp. of Florida. Because of intense Japanese competition, National Semiconductor laid off 2,000 workers; Advanced Micro Devices dropped 2,400 in January 1988; and the U.S. pressed Japan to open its markets to American goods. In 1982 IBM gave Intel of Japan the job of providing design and process technology for IBM's random-access memory chip that stored more than 64,000 bits of data.

It was charged that Japan obtained details of American inventions important in fighter plane and submarine technology and sold them to Russia. A 1984 book, *Techno-Bandits: How the Soviets Are Stealing America's High-Tech Future,* by Malver, Hebditch, and Anning, described Soviet efforts to acquire militarily significant American technologies in the 1970s, without the large research and development expenditures to create them. Internal "moles" were planted in the Pentagon, Silicon Valley, and other sensitive spots.

American companies claimed the clever Japanese cloned their electronic and small appliance products and sold them at low prices, destroying American earnings. Semiconductor maker Intel[27] introduced the first microprocessors in 1971 and later rivaled Motorola for the lead in that business. It sued to stop NEC of Japan copying its microprocessors. To compete with the growing Japanese semiconductor business, leading American semiconductor makers and users formed Memories Inc., but could not agree on building a large plant or the division of prospective profits, so Memories Inc. became a memory. Japan required American companies selling products in Japan to be majority owned by Japanese and manufacture a large percentage of the products in that country. Japan made a show of opening the door by allowing American products to be sold in some cities, but not in the lucrative big markets like Tokyo.

Americans bought such a huge amount of Japanese products that the balance of trade left Japan with billions of United States dollars that the Japanese used to buy great amounts of American auto and optical fiber plants and land. One purchase was Rockefeller Center in New York City, a symbol of American technology and communications. Another purchase was the big computer company Gene Amdahl founded, that bears his name. Amdahl made mainframe computers, semiconductors, silicon chips, and laser equipment. IBM's competitor Fujitsu backed Amdahl and owned twenty-eight percent of its stock providing large computers and parts in the United States. Amdahl joined with Siemens of West Germany to supply mainframes and parts to Europe. Japan had an unprecedented grip on the American economy.

Chester Carlson invented the plain-paper copier (electrostatic "xerography" process, 1938), which was then further developed by Battelle Memorial Institute, Columbus, Ohio. The patents were sold to Haloid Corp., which later (1947) changed its name to Xerox Corp. When the patents expired, the Japanese copied it. In 1982 Xerox was copying the copier that copied its copier. Semiconductor transistors, invented by Bell Laboratories, were used in millions of Japanese TV sets bought by Americans.

Motorola decided to invade the global Japanese electronics market by making a wristwatch pager to let users know they are being telephoned. Motorola Chairman Robert W. Galvin said, "We can beat the Japanese in quantity, quality, and cost," and his company did sell pagers, microprocessors, RAM, and other devices to Japanese companies and worldwide. Motorola, the old Chicago-based radio and TV maker, became one of the largest makers of semiconductors, pagers, and other electronic products and sold some of its products in Japan.

A page one story by E. S. Browning in the February 27, 1986 *Wall Street Journal* said that after years of effort N. V. Philips,

a big Dutch company, "figured out a potentially revolutionary technology: how to use a laser to play sounds and pictures recorded on a plastic disk."

Learning of this in 1970, Sony, Japan's most innovative electronic company, hadn't even started experimenting with it, but after years of development produced a minisized laser record player that swept the United States market. It was one reason for the classic Japanese export triumph of a record fifty-billion-dollar U.S. trade deficit with Japan in 1985.

Government-owned Nippon Telephone & Telecommunications Company (NTT), valued on the Tokyo Stock Exchange at $330 billion, was the world's most valuable telecommunications company, but bought its first digital switching system from Northern Telecom of Canada. The world's second largest semiconductor producer was Nippon Electronics Corporation (NEC) of Japan, a nine-billion-dollar electronics and communications giant. NEC established a plant to produce digital exchanges in Dallas, Texas, a silicon chip plant in California, and bought an interest in M/A COM, an electronics telecommunications and cable TV company, and other American companies producing a wide range of telephone systems. Japan's top computer makers were Fujitsu and Toshiba. Toshiba announced it would develop a line of low cost high-performance computers based on Sun Microsystems' microprocessor and Sun's Unix system.

Motorola also led the way in global semiconductor sales. The New York Times estimated that in 1990 Motorola would do $3.3 billion in semiconductors, Texas Instruments $2.9 billion, Intel $2.4 billion, National Semiconductor $1.2 billion, Advanced Microwave $1.1 billion, and AT&T $873 million.

SONY, one of the world's leading electronic corporations, founded in 1946 by Masaru Ibuka and Akio Morita, developed Japan's first tape recorder and magnetic recording tape and obtained a license agreement from Western Electric in 1953 to use AT&T's transistor patents. SONY then produced a stream of electronic products including Japan's first transistor radio in 1959, color television, and cassette recorders. SONY also established manufacturing plants in the United States.

Eighth-grade Japanese children reportedly had as much mathematics and science education as American college graduates, and Japanese companies invested twice as much in research and development per employee as Americans. Asian engineers and scientists trained in the United States were offered top jobs to return to their countries. In 1984 Japan announced plans to cooperate with AT&T in a five-year project to help Japanese companies develop their own software. In China, the sleeping giant, Taiwan, and Korea, American companies like Microtech and Umax Data Systems began producing electronic communications components.

Strong believers in racial purity and resisting union recruiting, the Japanese favored Southern U.S. states for their American plants. Stories with headlines "Whose property is this anyway?" said patent infringements by Fujitsu, Hitachi, and Amdahl cost America billions. IBM went to court to protect its copyrights ranging from PCs to mainframes and an arbitration decision required Fujitsu to pay IBM $833.2 million, but allowed it to copy IBM's software.

In summary, the clever and industrious Japanese bought American technology products, cloned them, and provided what Americans wanted at low cost. They developed a home market while sheltered by trade barriers and sold billions of dollars of camera, TV, radio, calculator, auto, VCR, tape recorder, and other products to Americans. Then they bought U.S. companies, real estate, services, and technology with fifty-cent, depreciated dollars. Ricoh and other Japanese companies built United States plants and employed 300,000 Americans.

The Japanese were conditioned to accept authority in their lives. Having no real social security system, the Japanese—unlike Americans—were accustomed to saving a large part of their earnings, and provided capital for long-term enterprises without borrowing from other nations. In the 1980s, even the Pentagon bought computer chips from Japan, while Americans spent far more than they produced. Japan's computer successes were a crowning achievement in turning its devastated industry in World War II into a modern industrial power.

The Robots Are Coming

Most people have the idea from reading space age science fiction that robots look like little metal men with gears like arms and legs, metallic skin, flashing lights for eyes, and synthesized, imitation voices. Most real robots, however, are machines that may resemble a sewing machine. A robot is an assembly of mechanical devices with an arm that moves as directed by the central processor unit of a computer to perform some repetitive motion: for example, to weld or spray an auto body or do other dirty or dangerous tasks in hazardous locations or at simple jobs, mechanically replacing human workers. The term "robot" in the Czech language means "involuntary servitude," "forced labor."

American plants had been using, for example, devices with moving belts and machines that filled and put tops on soft drink bottles for a half century to save labor costs and lower prices, but many were surprised to learn from the June 18, 1982 *U.S. News and World Report* that Japan used 75,000 robot-like devices to improve the efficiency of their plants and replace many workers.

By leading the world in labor-saving robot use, Japanese industries gained an important competitive advantage over American companies with their high labor costs. Toyota Motors proudly announced robots did seventy percent of its work. Fujitsu-Fanuc, a world leader in making numerical controlled machines, said robots did thirty percent of the work at one of its plants.

To compete, American companies used robots to reduce the number of employees. In 1983, Westinghouse Corp. bought Unimation Inc., inventor of industrial robots. General Electric and United Technologies then established robot companies. Black and Decker bought twenty-seven computer-controlled machines to automate the making of tools and saved $2.3 million in labor costs in three years. Texas Instruments and IBM jointly leased large-scale automation systems. The Robotics Institute at Carnegie-Mellon University predicted that robots would be used where toxic chemicals or danger of radiation might be a health danger.

The computer revolutionized operations and production costs of American factories. Computer-controlled robots increased the number of items coming off a production line, but reduced the variety because mass production efficiency required few models. Robot use had begun in some American steel mills. The Japanese automated electronic plants to lower product costs, and produced seventy percent of the world's robots.[28]

Many American companies automated plants or used robots to perform some time-consuming processes. Most common computer robot uses were for spot welding, casting, spray painting, loading and unloading conveyor belts, timing high heat processes, and packaging.

Robots duplicate the motions of wrists, arms, shoulders, elbows, and hands. Some are programmed to simulate human intelligence. Many were made by Westinghouse's Unimation. Check Robot Co. provided for customer use of checks on bank accounts to pay Publix Supermarket grocery bills at the store. After ten years of research, IBM produced in 1982 a "precision assembly" robot with "sophisticated tactile and optical scanning abilities." It threaded guideposts in a cartridge, detected mistakes in ribbon alignment, rejected defective ones, sensed when the supply bin was empty and pushed gears to fill it. Robots work all hours and days and never get sick or go on strike. An army of robots is coming.

Speaking computers tell bank customers by telephone a transaction cannot be completed right now, wish them a "good day," use artificial intelligence to "think" when playing games with people, reply to simple spoken questions, and help route air traffic. Projects to develop more abilities were carried on at Columbia, Stamford, Yale, MIT, Carnegie Mellon, and other universities. Once computers only did arithmetic but later were used to decide questions like which is the best choice among several proposals and which strategy to use. It made professional expert advice immediately available. Only time will tell how far future computers will go in simulating our mental processes. Even desktop computers are programmed to use many artificial intelligence programs.

We must learn to live with computers because they are a part of our daily lives, even if only as used by others to help us. A brilliant human brain devised a computer service named TCAS that will know—even if the pilot cannot see it because of rain, fog, or time limitations[29]—that his plane is on collision course with another. With a synthesized voice it tells the pilot what to do to avoid a disaster. TCAS was developed at MIT's Lincoln Laboratories and Bendix and placed in use on airplanes around the world. The lives it saves may be our own.

Some computer people think of their minds as machines, and computers can store enough information to solve very complicated problems in an instant. For example, if you give the computer a program called BACON, the same data that astronomer Johannes Kepler worked with, it will discover Kepler's third law that relates the distance of a planet from the sun to the time it takes to go around the sun. Or, if you give it the data Georg Ohm worked with it can immediately find Ohm's law that relates the electrical resistance in a wire with current and voltage.[30]

The computer's ability to break sounds into hundreds of segments and assign numbers to them enables it to "hear" and ask questions, and take over some work of operators. If a caller dials a disconnected or changed number, the computer asks what number was called and switches the call to a record that asks the caller's name and number. It asks if the caller will pay the toll and if the answer is "no," dials the number and asks the person called if he will pay. If the answer is "no," it says that the call is refused. If a coin-box telephone is used, the computer tells the person to stay on the

telephone after the call to pay the tolls.

About 100 million Touch Tone telephones are used to obtain customized information services. For example, pressing buttons can reserve a seat on an airplane, get a weather forecast, or learn about a tax refund from Teletax, an Internal Revenue computer system. Stored record disks or tapes tell a central computer to find the taxpayer's record, stored in digital codes. The computer also could select the weather report for an area or convert tones into written or spoken reply.

IBM and Voiceworks made machines that typed from spoken dictation, and National Semiconductor had a voice system that called out prices at computerized supermarket checkout stations. Computers can talk back to us.

Fujitsu, one of Japan's largest global computer companies, advertised in *Forbes Magazine* that U.S. Sprint had it install the electronic repeaters and switching equipment for a 340-mile link through the bayous and swamps of Louisiana in a nationwide optical fiber transmission system. Matsushita[31], Philips, and SONY agreed on a standard format for "a five-inch optical compact disk to carry text, still images, motion video, and computer graphics, which could be used interactively and simultaneously."[32]

To expand its position in the graphics market, Eastman Kodak bought Atex which had new technology computer systems and software to process texts for publications. Computer-aided design and manufacturing (CAD/CAM) used computer graphics to make designs on the computer screen and manufacture them faster. It was hailed as having great potential.[33]

Computer scientists at the University of California at Berkeley then found an experimental way to replace the gangs of big disk-drive "dinosaurs" with small, fast, cheap disk arrays. Size had been a serious problem for Amdahl, IBM, and Computer Technologies, the big mainframe storage companies, and the smaller Maxtor, Seagate, and Conner Peripherals.[34]

The computer is doing the work and taking the place of a large number of middle management officials of companies, in some as much as forty percent. Those remaining need to utilize computer methods efficiently. Companies will be bidding for employees with the greatest skills to achieve the company's goals. K-Mart installed computerized cash registers at all 20,000 of its checkout counters, all connected with its headquarters. Other chains followed suit.

Sun Microsystems and Unix

Sun Microsystems was formed by a group of brash, brilliant, young MBA scientists at Mountain View, Silicon Valley, California. In six years of spectacular success they built a billion-dollar company. In the midst of a complex reorganization, when Sun most needed revival of its spirits, AT&T bought a twenty percent interest and agreed to sell its microcomputers. That was a splendid AT&T investment because Sun made major developments for data network systems. Sun's rapid rise was strongly aided by the development in 1982 of a "wonderkind" of the Information Age: the Unix computer-operating systems with a desktop workstation and Unix software. An operating system developed by AT&T, Unix allowed a great advance in office automation.

Unix, first marketed by Convergent Technologies, Inc.[35], tells the computer what functions to perform, and interprets the software program. AT&T advertised that the Unix system could run hundreds of different software programs on mainframes and mini-computers, and wanted it adopted as a common open standard for the computer industry. Western Electric began producing its version of Unix. Other producers of Unix software were Itel, Motorola, and National Semiconductor. To help establish Unix as the industry standard, Sun invited computer companies in 1987 to clone its ultramodern workstations, and Matsushita's Solbourne Computer Company in America did. Others in the computer industry became so concerned about AT&T's growing role that it established a separate unit to handle Unix operations.

AT&T said the operating systems of various manufacturers were as different as New York and Washington traffic rules, but perform practically identical tasks. Major producers of Unix software were AT&T, IBM, Phillips, Digital Equipment, Intel, Motorola, National Semiconductor, and National Cash Register. Computer professionals describe each as "having different architecture."[36] AT&T produced a software package combining the best features of the most popular version of Unix, and said it would be the standard for the future.

After a frustrating effort, pouring hundreds of millions of dollars annually into an unsuccessful effort to establish a big global computer business, AT&T decided it needed NCR Corp., formerly National Cash Register, which was profitable, the fifth largest U.S. computer maker, and had a large European and global computer business. After a long takeover battle, AT&T won in 1991 and paid NCR share owners $110 a share, or $7.48 billion. Pending the purchase, it sold about $650 million in stock.

To help finance that purchase, AT&T also sold the twenty percent share of Sun Microsystems it had bought for $577 million in 1985, but soon bought back five million of the Sun shares. Earlier in 1991 Sun had refused an offer by Apple Computer to join it in a venture to develop a laptop computer and launch a new operating system. That refusal seemed a mistake. Sun had thirty-three percent of the workstation business, but Apple and IBM were the world's largest in the huge personal computer production in which Sun was not a factor but needed to be.[37]

UNISYS, a mainframe computer company, began providing open Unix operating systems that work with all others and in 1988 had the largest open systems business. It handled electronic mail, wrote new software programs, and developed more commercial uses for semiconductors. National Semiconductor, Itel, Motorola, Digital Equipment, Micron, and other American companies became big producers of tiny electronic microcomputers that enabled personal computers to guide robots in performing exacting jobs. Unix software, developed with Sun Microsystems, gave AT&T a head start in the competition to control the basic functions of the computer.

NEC, the world's largest memory chip maker, and other Japanese companies were anxious to escape dependence on American microprocessors. Fujitsu purchased a five-year contract from Sun for the right to sell its microsystem workstations, giving Japan an entry to the microprocessor market. Exportation of chips to the Soviet bloc was prohibited and smugglers were arrested by the U.S. Customs Service. Over the years, secret military and industrial plans reportedly vanished by the truckload from Silicon Valley, where 1,500 high-tech companies were concentrated. Americans, Canadians, and Israelis were convicted of selling advanced American military technology to the Soviet Union.

In 1985 AT&T obtained a $945 million contract to supply and service 250 Defense Department minicomputers and produced an entirely new class of personal computers, programmed with Unix software, to enable users to do several jobs simultaneously in voice and data. In 1988 AT&T built a $220-million plant north of Madrid, Spain to produce semiconductors.

Sun and AT&T multiplied the usefulness of microprocessors, much to the discomfort of IBM, which had spent ten years and billions of dollars on research and development to gain leadership in PCs and microprocessors. Enthusiastic engineers, scientists, and technical experts bought 150,000 of AT&T's workstations. Sun's star desktop workstation was smaller than many personal computers, but much more powerful, with three dimensional graphics and high-performance number-crunching ability. Utilization of such powerful features, of course, required skilled writing of general purpose software.

Alarmed by AT&T and Sun's development and licensing of the Unix operating system, IBM, Digital Equipment, Hewlett-Packard, Apollo, and four foreign computer firms formed an alliance to develop a rival version of Unix. Other Unix versions were developed by Digital Data, Microsoft, Amdahl, Apple, UNISYS, Wang, Prime, Tandem, and Compaq. IBM alone had three or more versions and a diskette system, all copyrighted. A variety of such operating systems was developed to direct the flow of information for scientific, engineering, and business purposes, and disputes arose. A meeting of computer companies finally produced a compromise agreement. AT&T announced the adoption of a uniform, standard Unix design. IBM, Digital, and others finally adopted the Unix as the standard operating system. In a revolt of the clones, many personal computer makers agreed to join in an effort to take from IBM the power to establish the standard for design and components of personal computers.

IBM had a giant share of the personal computer market, but AT&T's comparatively small computer sales were aided by the growing use of computers in the communications networks linking credit card companies with banks, stores, manufacturers, retailers, insurance, hotels, travel agents, and airlines. Health Systems International network, for example, enabled nurses to put patient data into computers that measure diagnoses against normal treatment and prepare for tests that would be ordered, producing major savings for hospitals. At fifty-six computer memory-access centers, AT&T's Accunet, Digital, and other services provided answers to complex problems for business managers.

IBM gained leadership in 1988 in the personal computer business with 34.7 percent over Apple's 18.4 percent, Compaq's 16.7 percent, and all others' (including AT&T) 30.7 percent. IBM seemed destined to continue its rule in PCs but faced tough competition from Sun Microsystems, Digital Equipment, and Hewlett-Packard, the leaders in workstation sales, that were growing thirty percent a year. IBM introduced a line of much faster workstations to increase its share.

Strenuous competition for leadership can be expected to continue in computer companies. A popular saying is that the leaders can be identified by the arrows in their backs. New officials in some companies ousted the pioneers who founded and built the companies, and hired them.

AT&T's Data-Phone service enables customers to transmit data from homes to processing centers, and receive answers on the same printers. Touchtone telephones are used to query computers. Stores send card numbers by simply putting the cards into a transmitter to check a customer's credit. A computer answers.

AT&T and Western Union private wire systems handle written matter, documents, reports, or images. Facsimile terminal equipment is provided by Honeywell, Burroughs, Alden, Control Data, National Cash Register, Electronic Communications, Telautograph, Seeburg, IBM, and Stewart-Warner.

Xerox introduced an electronic publishing system to send text and illustrations to an electronic printer that set type to make plates for publishers' presses. Xerox also bought companies like Diablo Systems. Shugart Associates produced printers, and Century Data Systems made rigid-disk drives.

Computer Logistics Corporation, a national information services firm specializing in warehouse and shipping operations, was acquired by Western Union in 1971. During the 1970s, the telegraph company installed six electronic switching centers in major cities to integrate circuit switching for Telex, TWX, and data services. Central telephone bureaus were at Reno, Nevada, Bridgeton, Missouri, and Moorestown, New Jersey with 800 numbers.[38]

To enable computers to transmit to each other, Digital Information Service (DIS) combined cellular technology with digital transmission by radio to homes and businesses that had a tower or dish antenna, and the FCC approved its providing the channels. Companies offering the equipment for those systems were Northern Telecom (a M/A COM subsidiary), Rolm, NEC, USACOMM, SBS, Contemporary Communications, DTS

Inc., Digital Telecom, Tymnet, NCI, and Local Area Telecommunications, Inc. The popularity of DIS prompted the U.S. Postal Service to establish E-COM, an electronic-mail ("E-mail") service which transmitted messages to post offices in destination cities. Most computer transmission was digital over telephone circuits.

In the early 1980s an integrated service data network (ISDN) was planned to connect digital systems around the world. People were to use it to talk to computers and the computers talk to each other to pass data on to destination. "Digital" operation handles voice and data simultaneously. ITT announced it would show a "Digivision" prototype to handle shopping, banking, security, and information.

Protected by trade barriers, Northern Telecom, supplier to Bell Telephone of Canada, became a tough AT&T competitor with digital switches ranging from small office to big central office size. Nothern Telecom was fifty-two and one-half percent owned by BCE Inc., owner of Bell Canada. Another Canadian competitor for the PBX business was Mitel, owned fifty-one percent by British Telecom. Bell Canada did not permit equipment made by non-Bell companies to be connected with its system until 1980.

Regional telephone companies and others developed information services to help businesses increase efficiency and productivity. The personal computer, word processor, facsimile, Telex telegraph, pushbutton telephone, internal switching, and flow of paperwork were prominent in such plans.

A growing trend is for computer service companies to provide the data processing and record keeping of banks. Systematics Inc., a unit of Alltel Telephone Holding Company, was a leader in that trend.

Originally the computer was used mainly to do complex arithmetic, but as more people learned to take advantage of computer inventions and services already available, their use increased for cellular telephones, information services, banking, buying and selling, financial records, shopping by home TV, and keeping schedules, dates, and plans. When people leave home they can use a cellular telephone in the car, and when they leave the car, a miniature facsimile machine or telephone on the wrist can keep them in touch with home and office. A pager in the pocket will let them know they are being called. Faith in the brave new technology has become unbridled.

21.5 THE COMPUTER GOES TO SCHOOL. (COURTESY OF IBM.)

Students of all ages are trained in school with computers and learn to speak the language of the computer, video, and music-recording worlds. It makes many elders feel like Rip van Winkle. This chapter should help us gain a better understanding of how to live in our computer world.

This chapter's purpose also has been to

provide a brief history of the computer's birth, development, uses, inventions, major companies, and their leaders. Numerous mergers, alliances, and agreements between companies to pool inventions and sell each others' products take place daily worldwide. *The Wall Street Journal* and other publications provide such information, and should be read to keep up to date. Recent developments of course cannot be added in a book.

Notes

[1]This chapter is designed to help everyone understand and adjust to life in our computer world. Because the writer is not an electronic or computer engineer using computer jargon and technical details, it should be understandable to all readers.

[2]Eckert-Mauchly Computer Corporation, the first computer firm, offered to sell stock. Investors were worried by rumors that computers were a Communist plot, refused to buy the stock, and the company failed.

[3]The point-contact transistor was invented in 1947 (December 23) by three Bell Laboratories scientists, Walter Brattain, John Bardeen, and William Shockley, who found that a tiny crystal of germanium could do anything the vacuum tube could do, and do it in a much smaller space, more reliably, and by generating very much less heat. It also reduced the size of all kinds of electronic devices. The point-contact transistor was followed by Shockley's junction transistor (1951), and then by the microcircuit and the microprocessor.

[4]Remington Rand was the largest pioneer computer maker, and its engineers, experts said, were best qualified to take over the entire business.

[5]*Forbes Magazine* (October 8, 1984).

*A *bit* or *bit*nary dig*it* is a (single) unit of computer information that is the result of a choice between alternatives, that is, either *1* or *0* (also *Yes* or *No*, *On* or *Off*)—the basic function with which every computer, no matter how complex, does its work. A *byte* is a fixed number of adjacent bits that a computer can process together as a unit; a byte may correspond to a single letter, number, or other character. (There are 8-bit bytes, 16-bit bytes, and so forth.) Of course, the more bits a computer can process together (bytes), the faster it is.

[6]James McNitt, president of ITT World Communications Inc., in *Signal Magazine* (January 1966) said many managers of data systems are members of the Industrial Communications Association and/or the International Communications Association. Other organizations include the American Federation of Information Processing Societies, with its constituent societies.

[7]E. F. Hutton became part of Shearson-Lehman-Hutton, and then Shearson-Lehman.

[8]H.&R. Block agreed to acquire CompuServe in 1979. An article by Jeremy Main, "Computer Time-Sharing, Everyman at the Console," *Fortune Magazine* (August 1967) described the development of time-sharing since the first public paper on the subject was presented by a British mathematician, Christopher Strachey, in 1959. Main said more than thirty research systems at universities and "think tanks" and at least twenty firms offered commercial time-sharing systems which usually handled from thirty to sixty users at once. IBM and Control Data expanded their own time-sharing centers.

[9]*Telecommunication* (January 1979).

[10]*Forbes* (August 27, 1983).

[11]*Forbes* (January 4, 1982).

[12]*Forbes* (July 25, 1983).

[13]*Wall Street Journal* 1989 advertising supplement "Personal Computing for the 1990s."

[14]*Forbes* (February 15, 1982).

[15]In 1961 Texas Instruments built the first integrated-circuit (IC) computer. IBM followed TI's lead by producing System/360, the first family of computers, in 1964. The 360 was in the forefront of the third generation of computers, characterized by ICs and unit-to-unit compatibility. The 360 was approximately as powerful as the first personal computer produced by IBM in 1981.

[16]Jourdan Houston, "Gearing Up for the Census," *Kiwanis Magazine* (September, 1979).

[17]*U.S. News & World Report* (April 4, 1986).

[18]Watson opposed going into the computer business, but his sons argued for it. After Tom Jr. succeeded him, IBM grew until it was the largest of the 10,000 computer makers in America and the world.

[19]*Forbes* (November 23, 1981).

[20]*Changing Times* (July 1987).

[21]*U.S. News & World Report* (July 4, 1983).

[22]*Forbes* (October 8, 1984).

[23]*Forbes* (July 25, 1983).

[24]*Wall Street Journal* (May 21, 1990).

[25]*Wall Street Journal* (May 11, 1990).

[26]*U.S. News & World Report* (October 14, 1991).

[27]Intel in Silicon Valley was said to have virtually invented (in 1971) the microprocessor, computer-in-a-chip, the personal computer's brain. Motorola and Texas Instruments also were early chip competitors.

[28]*Wall Street Journal* articles, July 18, 1982 and April 4, 1983.

[29]Two planes on a collision course, at 550 miles per hour each, could be one mile apart three and one-half seconds before impact.

[30]*U.S. News and World Report* "Conversation with Herbert A. Simon" (n.d. 1982 and January 7, 1985).

[31]At date, Matsushita is the world's largest consumer electronics manufacturer.

[32]*Wall Street Journal,* July n.d. 1989.

[33]CAD/CAM (computer-aided design / manufacturing) dispenses with the laborious triangle, pencil, and compass work by designing on the computer screen.

[34]*Wall Street Journal,* August n.d. 1989.

[35]Convergent Technologies was acquired later by UNISYS, which was using systems based on the Unix operating system. *Forbes* (November 28, 1988).

[36]*AT&T Magazine,* no. 1 (1984).

[37]*Wall Street Journal,* October 5, 1991.

[38]*Western Union News* (May 1975).

Satellite
Communications

In awe of the heavenly bodies, early Babylonians, Egyptians, Greeks, Romans, Aztecs, and Mayans personified, deified, sculptured, built temples to, and worshiped the sun, moon, and planets. They made gifts and even human sacrifices on the altars of those gods. To them the sun god rode across the skies in his fiery chariot, Mars was the god of war and Venus the goddess of love and beauty.

In recent centuries humans have wondered if intelligent beings live on the moon and stars. In *Somnium seu astronomia lunari* (published 1634, four years after his death) Johannes Kepler, a German astronomer, described an imaginary trip to the moon. Two centuries later (1821) Carl Friedrich Gauss, German mathematician and astronomer, proposed communications with inhabitants of other planets by planting broad lanes of forest in a design, using the Siberian tundra as a gigantic blackboard. Littrow of Vienna suggested burning kerosene at night in ditches on the Sahara Desert, forming a geometric design. Jules Verne's novel, *From*

the Earth to the Moon (1865), Edward Everett Hale's "The Brick Moon," Dick Calkins's "Buck Rogers" cartoon strip, and others directed public attention to space. Nikola Tesla in 1889 and Guglielmo Marconi in 1920 reported receiving signals that might have been sent from other worlds.[1]

Like those naive gropings for the stars, communications in space remained a nebulous idea until 1946 when the U.S. Army Signal Corps carried on conversations between Washington, D.C. and Hawaii by reflecting signals off of the moon's surface. That signal's high altitude and slow orbit, however, made it unsuitable for regular use.

In April 1955 John R. Pierce of Bell Laboratories published a paper on technical requirements for satellite communications. Since microwave relay systems were using relay towers, Dr. Pierce suggested using satellites in the sky instead of towers on mountains to relay signals, and headed a team that built three prototype satellites.

A microwave tower in mid-Atlantic would have to be 400 miles or more high because ultrahigh frequency radio waves travel in a straight line and do not bend. However, placing a satellite in space to orbit around the earth had become possible because of progress in rocketry. A Russian named Tesilovsky developed a mathematical formula for rocket propulsion in 1890.

The "father of modern rocketry," Robert H. Goddard (1882–1945), physics professor at Clark University, Worcester, Massachusetts, filled notebooks with ideas about launching a liquid-propelled rocket capable of going beyond the earth's atmosphere and exploring other planets. As a sixteen-year-old boy, his imagination was fired by H. G. Wells's science-fiction *War of the Worlds,* and dreamed of constructing a space-flight machine. Then on March 16, 1926, on his Aunt Effie's farm in Auburn, Massachusetts, Goddard launched the first liquid-fuel rocket, using gasoline and liquid oxygen.[2] It was a historic first. Though it traveled only 184 feet and forty-one feet high, it proved the rocket was practical. Generally ignored in America, Goddard's rocket experiments stimulated developments in Germany and Russia. German V2 rockets rained on London during World War II.

In 1939, Wernher von Braun's team in Germany launched the first guidance-controlled vertical flight of a rocket that reached an altitude of seven and one-half miles. Von Braun, Germany's leading rocket expert, said Goddard was his boyhood hero. Ironically, the United States refused Goddard's offer to aid in rocket development.

Arthur Clarke, a British science fiction writer, suggested in the October 1945 *Wireless World* that communications could be vastly improved by transmitting signals to satellites powered by solar energy from a point on earth and receiving them back at another point. He said almost anything imaginable could be done with a communications satellite and predicted it would be used to explore the nearest stars. Clarke's ideas inspired more satellite developments.

Imagine the amazement and chagrin of American engineers upon hearing the bleeps of the first earth satellite, the Soviet Union's *Sputnik I* (*sputnik* means "fellow traveler"), a 194-pound yellow ball, circling the earth. On October 4, 1957, Sputnik passed 560 miles overhead, transmitting the first signals from space. Millions of Americans watched its flight. Newspaper headlines called it "America Beaten" and "Pearl Harbor in Space." Some writers said the earth was headed for disaster, but it was the dawn of the space age, and spurred American scientists and government officials to action. Congress voted funds for satellites, but before they could be produced, Russia was preparing to launch an intercontinental ballistic missile (ICBM), regarded as a military threat to the United States.

There were two kinds of communications satellites: passive and active. The passive ones merely bounced powerful signals back to earth. They required a sensitive receiver at an earth station. The active satellite is, in effect, a microwave tower in space that receives signals from earth and sends them back. They arrive back on earth faint as a whisper from their long trip on a line-of-sight path to the antennas of earth stations that pass them on into distribution networks of cables, wires, and radio beams carrying television, telephone, telegraph, and graphic communications. The first United States satellites were passive.

Scientists tried sending radar beams to the craggy surface of the moon. The U.S. Army then sent Score, a horn-shaped object into space, carrying a tape that rebroadcast President Eisenhower's 1958 Christmas message to earth.

The first passive communications satellite, *Echo I,* a ten-story-high aluminum-coated balloon, was launched by the National Aeronautics and Space Administration (NASA) on August 12, 1960. Folded in a canister, it soared 1,000 miles on a Delta rocket and expanded to full size when solid state material in the balloon changed to gas. Radio waves transmitted to it bounced back to earth.

In 1960 *Courier,* the first active satellite, received messages from earth, stored them and sent them to earth stations when they came into view.

On January 25, 1964, *Echo II,* a 134-foot balloon, was orbited—the first joint U.S.-Soviet Union space experiment—and millions watched it pass 700 miles overhead in polar orbit. *Echo II* was also a passive communications satellite.

The Soviet Union gave two dogs eighteen orbits in space August 18-20, 1960, and Enos, a chimpanzee, two orbits on November 29, 1961. Yuri A. Gagarin of Russia was the first man in orbit, April 12, 1961, on *Vostok I.* On May 5, 1961 Alan B. Shepard, Jr. was the first American in space. The first woman in space was Valentina V. Tereshkova of Russia (on *Vostok VI,* June 16, 1963); and the first space "walk" was by Aleksei A. Leonov of Russia—for ten minutes on March 18, 1965, from *Voshkod II.*

The first nongovernment satellite was *Telstar,* built by AT&T in 1967. *Telstar* carried the first live television scenes across the Atlantic and was followed in the same year by *Relay* that handled data and facsimile messages.

The early satellites passed out of range of their earth stations in a short time as they orbited the earth. A mobile earth station to track the moving satellites would cost fifteen to thirty million dollars and numerous satellites were needed to serve the world, but would cost billions.

A solution to the staggering cost problem was suggested by Arthur C. Clarke, who in 1945 had suggested use of satellites in space to improve communications. He said satellites should orbit the earth at the same speed it revolves, and always be in the same position relative to their earth stations. That required an orbit 22,237 miles above the equator.

Syncom 2, the first communications satellite, was placed in geosynchronous orbit over Brazil in July 26, 1963. (*Syncom 1,* launched in February 1963, suffered the failure of its electrical equipment. The first satellite communication between the U.S. and Africa was via *Syncom 2.*) The first commercial international communications satellite, *Early Bird,* later named *Intelsat 1* (see INTELSAT, below), was placed in synchronous equatorial (or geostationary) orbit over the Atlantic Ocean and linked the United States and Europe June 28, 1965, but there were only five earth stations with which it could operate. *Early Bird/Intelsat 1* was the first step toward a global satellite system.

The first unmanned soft landing on the Moon's surface was made by the unmanned Russian *Luna 9,* February 3, 1966 and by *Venera 3* on Venus March 1, 1966. *Venera* also sent data on the atmosphere of Venus on October 18, 1967. Unmanned American *Surveyor 3* sent the first pictures from the surface of the moon and scooped and tested lunar soil on April 17, 1967.

A United States manned flight by *Apollo 8* in December 1968 orbited the moon, and unmanned *Mariner 6* passed 2,120 miles above Mars on July 31, 1969.[3] Neil Armstrong and Edwin "Buzz" Aldrin, aboard *Apollo 11,* were the first men to walk on the moon, July 20, 1969. Their landing was televised in fifty countries. When Armstrong stepped on the moon he said, "That's one small step for a man, one giant leap for mankind."

America had entered the race for supremacy in space. It was not confined to military killer weapons but used satellites for peaceful exploration, communications, and commerce.

The U.S. Navy established a Minitrack network to keep track of satellites. Its antenna reception pattern formed a very narrow fan pattern. When a spacecraft passed through that fan, Goddard Space Flight Center at Greenbelt, Maryland received the signal on perforated tape and processed it by computer to produce a mathematical description of the spacecraft's orbit. The result was to bind the Western nations together in an electronic unity.

AT&T Satellites

International communications were growing about twenty percent a year, and AT&T realized it would need thousands of satellite circuits in addition to its transatlantic cables. To meet that need, AT&T decided to launch active Telstar satellites. It had 400 scientists, engineers, and technicians concentrate on space technology, developing such items as supersensitive, low-noise receivers with ruby laser amplifiers, waveguides, and traveling-wave tubes. AT&T already had developed the transistor and solar battery. Finally, an active satellite, *Telstar I* was assembled at the Hillside, New Jersey Bell Laboratories. AT&T President Kappel contended it was for telephone service and private property, while NASA, the govern-

ment, and politicians in Congress claimed it should have government ownership.

Kappel had been picked by Craig as his successor because he was a fighter—forceful and determined, loud and blustering at times. He browbeat subordinates but respected associates who would face him and shout back. Kappel, square jawed, stocky, of German ancestry, had started work digging holes for poles and erecting them.

Kappel was primarily responsible for changing AT&T's long-honored nine-dollar stock dividend that had attracted more stockholders than any other company, and was so safe that AT&T was known as the "widows and orphans stock." On December 17, 1958, Kappel persuaded his directors to split the stock three for one. His judgment was correct because earnings and the number of stockholders increased rapidly.

During Kappel's ten-year administration, fancy telephones and services were introduced and most party lines were discontinued. Employees of companies with central automatic switching exchanges could dial each other.

Carrying on the public relations tradition of Gifford, Kappel appointed as his assistant Public Relations Vice President Prescott C. Mabon, who called public relations "the heart of the business." Newsmen covering the communications companies were not happy with Kappel's roughness, but Mabon and his professional staff turned that into an advantage by calling it strength.

A decade of development costing a half-billion dollars and thousands of man-years of work were devoted to the production of the electronic telephone exchange switching system. It speeded telephone service and saved untold millions of dollars. As Mabon said, "The value of the service is related to the number of customers who can be interconnected."

After AT&T launched *Telstar I* with a Delta rocket from Cape Canaveral, Florida on July 10, 1962, it had managed to send television across the Atlantic Ocean but the received signals were weak. To remedy that, AT&T built a horn antenna 177 feet long and ninety-four feet high, weighing 380 tons, and mounted it on a rotating wheel to follow the satellite's path. The horn was in a radome at Earth Station No. 1, Andover, Maine, and similar stations were built at Pleumeur-Bodou, France and Goonhilly Downs, Cornwall, England.

With these horn antennas, and operating at high, microwave frequencies, *Telstar* was the first reliable satellite to relay television overseas. People in America and Europe watched some fifty telecasts, witnessing the dawn of satellite television.

U.S. President John F. Kennedy refused to participate in the *Telstar* inauguration or telecasts because he opposed AT&T, the private enterprise company that developed satellite communications and spent hundreds of millions to help the United States catch up to Russia and gain supremacy. Kennedy opposed AT&T's having ownership or control of its own satellite! Evidently he wanted control of satellite use through his appointees, with government operation to follow.

The "live" overseas telecasts thrilled millions. The satellite also relayed phone and telegraph communications, data, facsimile, and telephotos. Public response soon indicated the major use of satellites would not be for TV but telecommunications.

NASA launched an experimental satellite called *Relay 1* December 13, 1962 but it became silent a year later on December 21, 1963. NASA replaced it with *Relay 2* that provided communications with Europe, Japan, and South America and carried many

world news scenes. At the end of 1964 ground stations were in Brazil, Germany, Italy, Japan, Spain, and Sweden, completing world coverage just two years after *Telstar I* was launched.

NASA launched *Syncom 1 and 2* over the Atlantic in 1963 and they failed, but NASA replaced them with *Syncom 3* in 1964. *Syncom 3* handled telephone, telegraph, and news photo facsimile.

Congress Creates COMSAT

The common carriers sought control over satellite communications but since foreign governments were involved, congressional hearings were held, and the Communications Satellite Act (CSA) in 1962[4] set a national policy to establish a commercial satellite system to serve the world in cooperation with other countries. CSA created Communications Satellite Corporation (COMSAT), with American carriers to own up to half of its stock. President Kappel was unhappy because AT&T needed the satellite system it had developed and built to serve the public. Some people asked why, since free-enterprise AT&T is the main user and provider of the service, a big and costly congressional creation like COMSAT was necessary.

COMSAT was incorporated February 1, 1963, capitalized at $200 million. Some 140,000 individual investors, eager to participate in the space enterprise, snapped up the public half of the stock. Of the fifteen directors, six were elected by public stockholders, six by the carriers and three appointed by the president subject to Senate confirmation.

The International Telecommunications Satellite Consortium (INTELSAT) was formed to own a world satellite system, with eleven countries as partners.[5] The United

22.2 THIS ILLUSTRATES THE COMMUNICATIONS SATELLITES ORBITING THE EARTH AS OF MARCH 1964. IN THE LOWER RIGHT-HAND CORNER IS *ECHO II,* THE GIANT MYLAR-SKINNED BALLOON. ABOVE IT IS *TELSTAR II,* AND IN THE LOWER LEFT-HAND CORNER IS *TELSTAR I.* IN THE UPPER LEFT-HAND CORNER IS *SYNCOM II.* THE OTHER TWO SATELLITES ARE *RELAY I* AND *RELAY II.*

States representative in the consortium was COMSAT, which managed the entire system.

COMSAT's first commercial communications satellite, *Intelsat 1,* was built by Hughes Aircraft and placed in synchronous orbit over the Atlantic Ocean at the equator. It had 240 two-way voice channels, or one two-way television channel between earth stations at Andover, Maine and Mill Village, Nova Scotia, and provided England, France, Germany, and Italy with their first regular transatlantic TV facilities. With 100 member countries INTELSAT soon owned a global communications system. It sold and leased capacity in its system to telecommunications and television companies.

COMSAT launched *Intelsat 2,* first

named *Lani* (Hawaiian for "heavenly") *Bird,* over the Pacific Ocean October 26, 1966. Live television between Hawaii and the United States was inaugurated November 27. *Lani Bird* did not reach its exact orbit and a second *Intelsat 2* was launched January 11, 1967 to handle communications between the United States, Hawaii, Japan, Thailand, the Philippines, and Australia.

An Atlantic satellite was launched March 22, 1967 to supplement *Intelsat 1,* and a Pacific satellite on September 27, 1967 to supplement *Intelsat 2.* Intelsat 2 satellites were built by Hughes Aircraft and launched on Douglas Aircraft Delta rockets. Four satellites were then in orbit carrying communications across the Atlantic and Pacific, and serving two-thirds of the world.

To serve the entire globe, COMSAT ordered six satellites from TRW, Inc. in 1968 for an Intelsat 3 series. Each had a capacity of 1,200 two-way voice-grade telephone circuits. One was stationed over the Indian Ocean in 1969.

An advertisement in the April 1, 1971 *Forbes* said:

> In 1958 the company [TRW] became the first industrial firm to build a space-craft—*Pioneer 1*. Vela nuclear test detection satellites, orbiting geophysical observatories, and INTELSAT III. Commercial communication satellites followed. TRW's mission analysis, propulsion, and electronic systems played key roles in landing the first men on the moon.

At the end of 1977 there were 197 antennas at 161 earth stations in eighty-seven countries. In 1982 INTELSAT ordered a sophisticated thirty-nine-foot-high, twelve-foot-diameter satellite weighing more than 8,200 pounds to carry 33,000 telephone calls and four TV channels.

Rebelling against increasing AT&T rates, the three TV networks threatened to set up their own satellite systems. In January 1970 President Nixon's staff urged the FCC to encourage competition by throwing the satellite race open to everyone. The FCC did and half a dozen companies rushed to file. The first, on July 30, 1970, was Western Union, which had applied unsuccessfully as early as 1966 for approval to launch three domestic satellites, built by Hughes Aircraft, and provide 9,600 voice-grade channels to carry telegrams, Telex, TWX, data, television broadcasts, and Mailgrams. Western Union also planned six earth stations and thirty-one new microwave relay stations.

Before communications service by satel-lite began, AT&T provided circuits to 107 broadcasting stations in 1950. The first transcontinental broadcast was a speech by President Truman in 1951 and the first global broadcast was in 1967.

First American Communications Satellite

On April 13, 1974, America's first do-mestic communications satellite—*WESTAR I*—was launched by Western Union. It could relay eight million words or 600 million data bits per second. Distance rates no longer mattered; a message from New York to Los Angeles traveled 45,000 miles via satellite. Service began July 15. The Corporation for Public Broadcasting also put its national TV network of five ground stations and about 300 broadcasting stations on *WESTAR I.*

One man who realized the great future usefulness of satellites was Robert Wold, an advertising man who leased one of the transponders of an early Western Union satellite and made a fortune providing the facilities to broadcast individual sports and other events to companies that did not need expensive annual leases.

Western Union leased satellite circuits to American Airlines, Ford Motor, McDonnell-Douglas, Federal Express, and others. It used satellite circuits also in some of its private line data systems leased to about 2,000 companies. Even those uses, and telegraph, voice, and alternate voice-data services between many cities kept the large capacity of the nationwide microwave system and the satellites so busy that Western Union urged Data-Phone customers to send written mes-sages and data over its TWX network to conserve satellite circuit space.

A graphic chart in the *U.S. News &*

World Report, October 4, 1982, listed these space firsts:

1971—*Apollo 15* crew drives seventeen miles on moon, July and August. (Intelsat launched four satellites over the Atlantic, Pacific, and Indian Oceans.)

1972—First use of Satellite to assess Earth resources.

1973—First space craft probe of Jupiter by *Pioneer 10.*

1973—U.S. crew of *Skylab 4* sets endurance record of 84 days Nov. 16, 1973 to Feb. 8, 1974, and *Mariner 10* investigates Mercury, the planet closest to the Sun in March.

1975—Apollo (American) and Soyuz (Soviet) test project links their crews in space in July.

1976—*Vikings 1 and 2* perform long duration mission to Mars.

1977—*Voyagers 1 and 2* begin journey to Jupiter, Saturn, Uranus and Neptune.

1979—*Voyager 1* returns photos of Saturn's rings, reveals active volcanoes on Jupiter's moon.

1980—Soviet's *Soyuz 35* crew sets new endurance record, 185 days.

1981—Space shuttle [*Columbia*], first reusable spacecraft, begins first of four test runs.

Anik (Eskimo for brother), the first geostationary satellite, built by Hughes Aircraft and launched by NASA in 1972, was the first domestic satellite to reach Canada's farthest outposts, and was followed by two Aniks to link Canada's population centers. TELSAT, a government- and private-carrier-owned company, filled its circuits with Canada's TV, telephone, and telegraph traffic.

AT&T leased seventy-five percent of COMSAT's satellite circuit space. Because it required six-tenths of a second for signals to travel to a distant satellite and back, the lag produced an echo in telephone conversations and doubled when two satellites re-

layed them. To avoid that, telephone calls from Europe to Asia were sent by cable to America and by satellite across the Pacific. A device to eliminate the double echo was found later by scientists of Tele-Systems, a COMSAT subsidiary.

Plans for a global communications network to send information anywhere with the speed of light were announced by AT&T April 20, 1974. The domestic part was to be operated jointly with GTE Satellite Corp., a subsidiary of General Telephone and Electronics Corp. For $46.8 million a year, AT&T then subleased a three satellite system named COMSTAR. The first of the three was launched from Cape Canaveral May 6, 1976 and provided 14,400 telephone circuits to all fifty states via seven earth stations. Cost of that satellite, equipment, and earth stations was about $77 million.

When COMSAT also announced its worldwide plans, the international carriers protested it might strangle them. The FCC ruled that COMSAT could provide circuits only to common carriers and not to the public. The National Communications System (NCS) awarded a contract to COMSAT to provide thirty alternate voice-data circuits it could use to provide communications between Hawaii and the Far East.

After the Apollo moon mission, AT&T, ITT WORLDCOM, RCA, and WUI agreed to provide satellite TV facilities for manned spaceship splashdowns on different weeks to insure fair distribution of revenues. WUI provided the ocean-going station that relayed the TV scenes. Floating earth stations transmitted news and color TV of President Nixon's historic visit to China.

WUI recommended the establishment of a Satellite Industry Liaison Committee between the international carriers and COMSAT. COMSAT wanted no "intrusion by such a committee in its operations," but the FCC formed it anyway to administer the earth stations.

COMSAT then bought AT&T's earth station at Andover for $4.9 million August 29, 1965, saying all stations should be "an inseparable part" of its system. It built stations in 1966 at Brewster Flat, Washington, and Paumalu, Hawaii, costing about $7 million each, and three in 1967-1968 at Etam, West Virginia; Jamesburg, California, and Cayey, Puerto Rico, with ninety-seven-foot diameter antennas, and higher than a 10-story building.

The carriers opposed COMSAT ownership. On July 7, 1966 President Gallagher of WUI proposed a compromise, and the FCC decided on December 7 that COMSAT must share ownership of the American stations with the carriers. COMSAT and the carriers then agreed on March 24, 1967 to share ownership and the operating cost equally. COMSAT was to manage the stations subject to policies and decisions of the carriers' committee. Some 120 countries were connected with the global INTELSAT system by building more than 170 earth stations or by establishing a link with a station in a neighboring country.

COMSAT's Paumalu Earth Station, built on Diamond Head, the famous peak on Oahu Island, Hawaii, was typical of those that followed. The Paumalu station had two dish-shaped antennas taller than a ten-story building. Growth of satellite facilities was so rapid that one of every seven people on earth could watch the first astronauts walk on the moon July 20, 1969 by using all communications satellites and earth stations.[6]

A laser scanner developed by Addressograph-Multilith International (AMI) for Satellite Business Systems (SBS) scanned a page in two seconds and Hughes Aircraft satellites and earth stations used it. That revolutionized the production of many publications. Facsimile-based technology made it possible to produce a daily newspaper and deliver it promptly from printing plants throughout the United States and world. The Gannett chain of newspapers recognized its importance and established printing plants costing about $500,000 each in ten major areas to serve the first general-interest national newspaper, *U.S. Today*.

The *Wall Street Journal* at New York City was the first financial daily newspaper to publish by satellite. WSJ transmitted images of its pages from its Chicopee, Massachusetts plant to sixteen worldwide printing plants to expedite delivery. The *U.S. News & World Report* used a computerized system to produce page proofs in its Washington, D.C. editorial rooms, and transmitted them as electronic impulses to its printing plants in Chicago, Los Angeles, and Old Saybrook, Connecticut. The *New York Times* also sent its pages to printing plants in other cities. At each plant a phototypesetter using laser beam technology set each page of text in less than ten seconds. In ninety minutes a printing press plate was produced and put on the press.

"By the end of this century," said *U.S. News and World Report* Publisher John H. Sweet, "a subscriber to this magazine may be able to push a button and call up the latest report on a home TV screen. Someone may come up with a way to eliminate the printing press." Well, in 1989 there were what looked like magazines from their covers on newsstands, but instead of printed pages inside there were disks containing all of a magazines' printed matter, to be printed by a personal computer.

Versions of facsimile publishing were adopted by the printing plants of *Newsweek, Reader's Digest, Chicago Sun Times, Louisville Courier Journal, National Geographic, Forbes, The Economist,* the Government Printing Office, and the Central Intelligence Agency. In 1979 *Time* signed a five-year contract for Western Union to transmit facsimiles of its color pages from New York to small earthstations at its printers' plants in Chicago and Los Angeles. *The Wall Street Journal, Time,* and the European *Herald Tribune* transmitted their Far East editions to Hong Kong, and the *London Times* sent its first page to Puerto Rico.

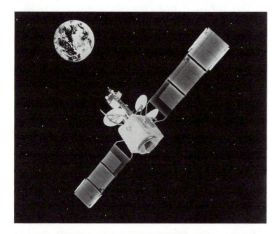

22.2 THE 12,000-CIRCUIT *INTELSAT V* COMMUNICATIONS SATELLITES WERE BUILT BY COMSAT FOR LEASE TO THE INTERNATIONAL TELECOMMUNICATIONS CARRIERS IN THE 1980S. EACH *INTELSAT V* HAD A CAPACITY FIFTY TIMES GREATER THAN *EARLY BIRD,* THE FIRST COMMERCIAL COMMUNICATIONS SATELLITE, LAUNCHED IN 1965.

The early satellites proved to be gold mines in the sky. In 1979 COMSAT's operating revenues were $262,635,000 and net income $40,185,000,[7] but as its number of launchings decreased, so did its earnings. AT&T and the international telegraph carriers leased circuits providing eighty-three percent of COMSAT's revenues. INTELSAT's

Atlantic, Pacific, and Indian Ocean satellites provided a substantial part of its revenues. In 1968–1970 COMSAT launched seven *Intelsat 3* TRW satellites, each with a capacity of 1,200 telephone circuits, and in 1971–1975 eight *Intelsat 4* Hughes Aircraft satellites were orbited with an average of 4,000 circuits, followed in 1975–1978 by six *Intelsat 4-A* Hughes satellites averaging 6,000 circuits.

A series of four *Intelsat 5* satellites was orbited for Ford Aerospace and Communications Corp. in 1980–1981, averaging 12,000 voice circuits and two TV channels. A McDonnell-Douglas Delta vehicle was launched on a General Dynamics Atlas Centaur as were many INTELSAT satellites.

In 1972 COMSAT, MCI (Microwave Communications, Inc.), and Lockheed Aircraft Corp. joined in planning a multipurpose domestic satellite system to compete with Western Union's, and asked FCC approval to serve all carriers except MCI's competitor—AT&T. They requested the FCC to "declare that the restrictions placed on COMSAT for services to Alaska, Hawaii, and Puerto Rico were no longer relevant."

The Soviet Union, France, and the United States, through NASA, provided satellites without charge to India in the 1980s. The United States also helped India establish television by satellite to help educate its illiterate millions. A nation with few telephones, India developed a 500-million-dollar space program with community television sets and programs in four languages.[8]

In 1979 the International Maritime Satellite Organization (INMARSAT) was formed by twenty-nine nations—there were forty-four members by 1990—and patterned after INTELSAT. In 1976 RCA, ITT, and WUI used Thor Delta rockets to launch three Hughes Marisat satellites—one over each ocean—to provide global communications with ships at sea, and passengers during transoceanic trips. Satellite communications spread around the globe in 1978. In 1980, there were 120 nations with 222 earth stations linked with the INTELSAT system that carried the majority of the world's growing international communications. Since satellites can pass signals on to others and each can send signals to a third of the world in its line of sight, three can reach the entire world. The British Royal Navy leased capacity in a MARISAT satellite launched in 1976 to provide high frequency communications with ships in the Atlantic Ocean. GTE's Sprint had its first satellite placed in orbit from a pad in French Guiana in Northern South America in 1984. GTE used it in its own operations and leased a part of its capacity to others.

By 1983 numerous communications satellites were in orbit carrying as many as twenty-four transponders to receive radio signals and automatically transmit them to earth stations. Their capacity was so great that many circuits were idle and competition between the carriers was intense, especially at major news events. When Western Union's sixth satellite did not achieve geosynchronous orbit, there was room for its pre-leased traffic on others.

American Satellite Company (ASC), a joint venture of Fairchild Industries and Continental Telephone, established a three-satellite domestic system in 1987, to transmit data, facsimile, and teleconferencing. It advertised that "When the leader in overnight delivery needed immediate, accurate communications for delivering the thousands of small packages and documents it handles, it turned to the leader in satellite communications—American Satellite." ASC recruited 300 satellite experts from Western Union,

COMSAT, IBM, and AT&T.

A common carrier demands complete continuity of service in meeting its responsibilities to the public. To do this, it must provide alternate routes by cable as well as satellite when sun spots or other radio interruptions occur. That was why AT&T applied to the FCC for permission to lay ten submarine cables, to maintain a balance between cables and satellites. Since overseas telephone use was growing twenty-five percent annually, AT&T said it would need 20,000 international circuits. COMSAT protested that more cables were not needed and would impede development of its satellite system, but the FCC finally authorized AT&T to lay the TAT VI and TAT VII cables. Five other companies then agreed to join AT&T in laying the TAT (transatlantic) cables in 1983 to provide 4,000 additional communications circuits.

The FCC required COMSAT to transfer mainland traffic with Alaska and Hawaii to domestic company satellites, but ruled that AT&T could not provide private wire service over its international channels. The company reacted to that by selling its 2,295,750 COMSAT shares.

In 1975 the Collins Radio Group of Rockwell International demonstrated the first mobile earth station transmitting commercial color TV programs. The fifteen-foot-diameter dish antenna mounted on a thirty-five-foot-long trailer performed well via Western Union's WESTAR satellites.

Many companies developed and supplied important satellite parts. Probably the most prolific was Hughes Aircraft Company that built satellites for Western Union and others and sold satellites outright instead of leasing them. Hughes had divisions with numerous technologies that provided such parts as fiber

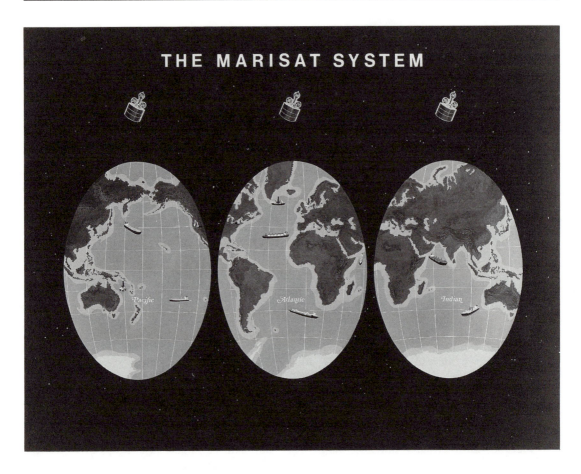

THE MARISAT SYSTEM

23.3 MARISAT SATELLITES ARE STATIONED IN SYNCHRONOUS ORBIT OVER THE ATLANTIC, PACIFIC, AND INDIAN OCEANS. THE MARISAT SYSTEM DEVELOPED BY COMSAT GENERAL CORPORATION, PROVIDES HIGH-QUALITY COMMUNICATIONS TO THE U.S. NAVY AND THE COMMERCIAL SHIPPING AND OFFSHORE INDUSTRIES. MARISAT SERVICES BETWEEN THE SHORE AND SHIPS AT SEA INCLUDE MODERN, RELIABLE TELEPHONE, TELEX, FACSIMILE, AND DATA COMMUNICATIONS.

optics, millimeter wave communications, microprocessors, lasers, solar cells, stabilizers, spacecraft flight controllers, and design work. NASA's *Pioneer* satellite that orbited Venus used a Hughes radar mapper for the terrain of Venus and sent back voluminous data. It was expected to burn up during 1992 in the atmosphere of Venus. Harris Aerospace provided parts for Telstar, Apollo, the lunar excursion module, data relay, and the space shuttle.

Westinghouse produced various critical accessories in satellite technology used in the 1961 Gemini-Titan, 1969 Apollo-Saturn, and 1981 Columbia (space shuttles) series of spacecraft. It provided a video camera, remote power controller, master timing unit to synchronize electronic and electrical systems, fluorescent lighting to reduce heat, devices to ease vibration and shock in landing, and power static inverters to provide electricity. Rockwell International built many satellites.

In 1984 Federal Express said it would launch a one-billion-dollar satellite system to expedite its package delivery business and start an electronic mail service. (FedEx's ZapMail plans were cancelled because potential customers were acquiring their own FAX machines and thus had no use for the proposed ZapMail FAX service. However, FedEx does have an extensive in-house satellite system for logistics and training.)

NASA collected many millions for launching satellites. COMSAT's subsidiary—COMSAT General—provided technical and operational services to countries with earth stations. Fairchild Industries subsidiary, American Satellite Corporation, and Scientific Atlanta did a booming business building smaller earth stations and antennas for homes and businesses. Use of solar cells that Bell scientists invented reduced the cost of the eight *Intelsat 5* satellites in the 1980s by increasing their lives to about seven years.

Under the leadership of Chairman Joseph H. McConnell and President Joseph V. Charyk, COMSAT strived to control ever larger areas of telecommunications. COMSAT even entered the home satellite television retail business, supplying television equipment through Amplica, Incorporated, a company it established. In 1974 COMSAT and IBM announced they would establish a domestic satellite communications business, handling voice, image, and data. They said COMSAT General planned to pay five million dollars to acquire the one-third interest Lockheed Aircraft and MCI each had in CML (Comsat-MCI-Lockheed) Satellite Corp. COMSAT seemed determined to be a monopoly itself.

COMSAT and INTELSAT had only one major competitor: AT&T, which foresaw a large increase in international communications as a result of growing foreign trade. In the 1980s and 1991 AT&T laid ten optic fiber ocean cables in partnership with cable companies of other nations. Each cable was designed to handle 500,000 simultaneous telephone conversations. Other foreign companies laid fiber optic cables separately from AT&T.

President Lyndon Johnson had a Task Force investigate "whether the present division of ownership ... best serves our needs." However, financial utilities writer Gene Smith of the *New York Times* wrote December 15, 1968 that copies of the task force 450-page "secret report" had been presented to Johnson, but was destroyed because it already had been distributed to many publications, presumably by COMSAT which the report favored to be the government's chosen instrument for both voice and telegraph international communications.

COMSAT revealed its intention in an application to the FCC for approval to build a network of 132 earth stations with four satellites at a cost of about $148 million. It intended to provide telephone and telegraph service over circuits to the television networks in all states.

Realizing COMSAT's intention was to take over all of the nation's communications services, the other carriers were up in arms, pointing out that COMSAT was created to be a "carriers' carrier." The FCC curbed COMSAT's more ambitious expansion efforts, but it tried strategies to get around FCC's roadblocks. In one end run it had COMSAT General organize Satellite Business Systems (SBS) in 1975, with COMSAT, IBM, and Aetna Life and Casualty as partners.

That end run worked. In 1977 the FCC approved SBS's plan to orbit satellites and provide a nationwide all-digital telephone system for a voice, data, video private wire switched network, using SBS antennas on customers' premises. SBS launched its first satellite (*Intelsat 5*) December 6, 1980, with twice the capacity of any before. In 1980, a court overruled antitrust protests by competitors and the Department of Justice. COM-SAT lost confidence in SBS when it lost $360 million in three years. It pulled out of SBS in 1985 and sold its earth stations business to AT&T for $150 million.

In 1980 the FCC authorized COMSAT to compete directly with the international carriers by serving broadcasters. It approved American Satellite, Satellite Business Systems, General Telephone and Electronics, Southern Pacific (railroad) Communications Company, and others launching satellites for domestic service, and reduced the cost of leasing a telephone circuit fifty percent in a few years. COMSAT leased three COMSTAR satellites to AT&T, each for $1.3 million a month for seven years.[9]

Satellites expanded the variety, volume, and flexability of communications facilities. In 1979 an international electronic mail system known as INTELPOST was developed, and TELESAT Canada and the U.S. Geological Survey demonstrated their use of satellite systems.

With hundreds of satellites planned for orbit in the 1980s, NASA considered charging passengers and companies for making science and technology experiments on space trips to recover some of its expenses. McDonnell-Douglas Company and Johnson & Johnson planned to make drugs and other products in space. Foreseeing a potential profit, Minnesota Mining and Manufacturing Co. and many others also engaged in research to produce products in space. An eight-billion-dollar space products business was predicted.

Over AT&T objections, the FCC allowed RCA to launch a satellite in 1979 to replace circuits RCA leased from COMSAT. RCA celebrated with a blitz advertising campaign. Twitting old Ma Bell, it said, "Good-bye Ma Bell," RCA is taking over the telephone business. However, the RCA satellite disappeared when it was launched and RCA had to take its hat in hand and beg AT&T for channels to provide those it already had leased to companies in the missing satellite. Ma Bell had the last laugh, and charged RCA a high price for the channels it desperately needed.

Encouraged by such interest, NASA launched the first space laboratory in 1984. It was thirty by fourteen feet in size, weighed eleven tons, and contained seventy-five experiments from nine nations. On board were particles of dust from interstellar space in an experiment by Washington University at St. Louis, seeking clues about the formation of the sun and planets by dust particles that had been bombarded for years by atomic oxygen. Other experiments involved cosmic rays and tomato seeds. When the seeds were returned to earth in 1989 after six years in space, the one and one-half million seeds were experimented with by NASA and Park Seed Company to raise "heavenly tomatoes." Studies indicated that NASA's space station could be used for experiments for thirty years.[10]

Astronauts had become so skilled in maneuvering space ships that when an eleven-ton "long exposure science laboratory" was about to plunge to a fiery death on earth in 1989, they maneuvered space shuttle *Columbia* beside it at 17,400 miles an hour, secured it with a robot arm and brought it to earth in its cargo bay. An American spy satellite providing highly sophisticated data on Soviet activities for years finally broke into four parts and fell, but a communications satellite that malfunctioned was "parked" by the space shuttle *Atlantis* in a higher orbit to postpone its fall so that its valuable instruments could be saved from incineration in falling into the earth's atmosphere.

Pagers, Cellular Telephones, Cable TV

When you are not near a telephone, a caller can dial a transistorized radio receiver in your pocket. It buzzes and you go to the nearest telephone. That "Bellboy" personal signaling service will find you anywhere in the area the transmitters reach. The first radio pager beeped a doctor on a New York golf course on October 15, 1950. Use of pagers soared, and in 1987 seven million pagers of several companies were in use.

They showed the need for a telephone wherever you are, and Systems Incorporated began offering cellular telephones to meet that need. The operating system enabled a cellular telephone in an automobile, truck, or restaurant to transmit or receive conversation by radio through a cellular control center, and local telephone office, and over the telephone network to the number called. Cellular companies obtained franchises to operate in municipalities. The value of the franchises was based on the "POP" (population of franchised area.) The RHCs, led by Ameritech and Bell Atlantic had the largest value per POP and largest number of POPs. They were followed by "pure cellular" companies owned by Alltel, Centel, Contel, GTE, LIN, McCaw, Cellular Communications, Metro Mobile, and U.S. West.[11]

When the cellular telephone was introduced, it was hailed excitedly by people who wanted to use telephones around their pools and public places, as well as homes, autos, trucks, and buses. Some chatted while standing in traffic jams or called the office to postpone appointments. Many carried them in their brief cases. In Los Angeles alone the number of cellular telephones climbed to 250,000. Car phone bills up to $500 a month were reported. The large number of subscribers resulted in crossed circuits and conversations overhead by unintended listeners. Bell Laboratories developed a cordless telephone that allowed a caller to stroll about home, office, or store and talk without worry about static from electric fixtures and flourescent lights. It used a radio transceiver that acted as a base station.

McCaw, the largest pure cellular operator, sought for years to acquire control of LIN, the second largest, with excellent management and very profitable. McCaw's offers climbed, as LIN repeatedly refused offers increasing from $6.5 billion to $7.5 billion and finally $154.11 a share for 21.9 million LIN shares in addition to the four million shares McCaw already had bought. LIN terminated an agreement to merge with Bell South's cellular service which dropped out of the bidding.[12] The purchase made McCaw in 1989 a great national cellular empire.

In the United States in 1980 there were 10,000 commercial TV, 4,600 AM radio, 5,200 FM radio, and 250 educational TV stations. However, the public was long since unhappy because of the lack of quality and variety of TV programs and motion pictures and because of their false representation of relationships between Americans to impressionable young people and foreigners.

Recognizing public dissatisfaction, entrepreneurs formed television cable companies (cable TV or CATV) and engaged in hot competition to obtain franchises from many cities. By 1974, 4,000 central CATV stations, using fourteen-foot antennas on their roofs, were receiving TV programs from seventeen orbiting satellites, and transmitting them via cable to millions of homes, businesses, and schools for a monthly fee. In

1975 Time, Inc. started providing motion pictures to CATV companies for broadcast to their subscribers for a monthly charge. That resulted in Home Box Office, Showtime, Cinemax, the Disney Channel, and the Movie Channel providing movies and other programs to millions of homes.

On April 10, 1981 President Ronald Reagan urged the adoption of a national telecommunications policy to approve direct satellite television service to homes. One of COMSAT and IBM's purposes in forming SBS had been to beam programs to homes. To avoid paying for CATV service, however, thousands installed large dish antennas in their yards or on rooftops to receive CATV programs direct from satellites. Battles ensued in countless city halls over whether to allow the big dishes on homes and taller buildings. Major direct satellite broadcasters, RCA, United States Satellite, Direct Broadcast Satellite, and Satellite Television Corp. scrambled their signals to prevent direct broadcast freeloading, but devices were soon produced to unscramble them.

General Instrument Company reportedly made the descramblers used in 2.5 million satellite-direct homes. Some carriers not only made descramblers but supplied programs for broadcast by satellite directly to subscribers' homes. Oak Industries had two subsidiaries: Oak Media provided programs with scrambled signals and Oak Orion provided descramblers to its subscribers, about 600,000 in 1977. Oak had agreed with Telsat Canada to broadcast its programs by 1983. Cable TV companies charged the dish antenna decoder was an illegal clone of their instrument. Oak Industries asked users to swap their clones for a descrambler (decoder) it made and pay the cost. Most did not.

Burnup & Sims (B&S), a twenty-billion-dollar company, did a large and very profitable business installing CATV cables. In 1982 B&S had installed a fifth of the 500,000 miles of TV cables in use. B&S Chairman James Robinson III saw cable TV as "the window to the world," to be used for shopping, banking, and investing, as well as entertainment.[13]

Other large providers of cable TV systems, equipment, and programs were ESPN, CNN, TBS, USA, MTV, Nickelodeon, and TeleCommunications, Inc., the Family Channel, TNN, C-SPAN, Financial News Network, Consumer News, the Business Channel, and Lifetime, with a total of about fifty million subscribers.[14] Understandably, the commercial TV companies denounced free use and sale of their programs as piracy, and Congress rewrote the copyright code of 1976 to make possession or use of a descrambler a crime. When possession of a descrambler became a crime punishable by six months in prison or a $2,000 fine, the FBI raided a Tallahassee, Florida firm, seized its list of customers and wrote to them to bring in their descrambler and fill out a questionnaire telling where they bought it.

Many former descrambler users became subscribers to CATV stations specializing in news, sports, stockmarket, education, weather, movies, or music. Cable TV pleased the public and its growth indicated half of all American homes would have it.

With cable TV people could learn from teachers, shop, buy, and obtain timetables, menus, catalogs, and other items. Each channel provided different subjects or services, such as local meetings, entertainment, education, or sports. There were outages when their screens were blank, and many programs were reruns of major network shows, but subscribers had lots of channels to choose from.

Viewers could applaud, denounce, or

"talk back" by using a response channel. By pressing a "Yes" button they could order a product shown on the screen, and a central computer would order shipment and send the bill. Warner Amex CATV invested $20 million in a two-way "supermarket of electronic service," called QUBE, at Columbus, Ohio, that expanded teleshopping to other cities and added automatic two-way police and fire alarm service. Zenith Radio even produced a "Space Phone" with which people could answer the telephone on the home TV set by pressing a button, and use the set as a speaker telephone.

With CATV companies offering as many as eighty channels, the future newspaper may be recorded by facsimile at home while we sleep, and we could vote on issues, talk back, ask questions, and get written answers on our TV screens. The FCC staff urged the establishment of low-cost mini TV stations to provide programs for blacks, farm areas, foreign-speaking, religious, and other special-interest audiences. Ma Bell was not happy to see communications wires other than hers entering homes.

Pressured out of the cable TV business by the FCC, the commercial networks began selling daily news packages and other programs to CATVs to broadcast. CBS formed a subsidiary named Viacom International to handle that business. Viacom joined with the largest cable TV company, Teleprompter, in a highly profitable service to 7.5 million cable TV homes. Westinghouse tried to buy Teleprompter from its crafty, stormy founder, Irving Kahn, but in 1980 the *New York Times* bought the reported $100 million business that had cost Kahn about $2.5 million.[15]

Reacting to public opinion, the FCC, which had imposed tough restrictions on the young industry, began deregulating cable

TV, encouraging more companies to compete. It did not want to regulate the small companies because, it said, they were not strong enough to engage in "predatory" practices. Then the Supreme Court decided cable TV was not a common carrier subject to FCC control, and that the FCC should deregulate the 8,900 radio stations because competition would guarantee public service.

The threat of stronger competition for CATV appeared in 1989. Cable Television Systems Corp., NBC, Hughes Electronics Division of General Motors, and newspaper publisher Rupert Murdoch's News Corporation agreed to put up one billion dollars to establish a satellite pay TV "Sky Cable" service with a satellite beam so strong it could be received by an inexpensive twelve-inch antenna dish on a home window sill or roof and provide sharp, high-definition video pictures from a selection of 108 TV channels. The small antenna could be used where the big satellite dishes were prohibited by zoning laws and in boats and recreation vehicles. Hughes was to provide the satellite with a powerful beam and NBC, the high-definition TV standard. Suppliers of TV cables and local stations were not happy at the prospect of being displaced but program producers looked forward to greater demand for their services in the 1990s.

Microwave Associates, a former Western Union controlled investment, recognized the interlocking nature of the industry's parts.[16] It made electronic components for many companies and grew, acquiring Digital Communications, Linkabit, and other companies. Local TV stations and newspapers began buying control of CATV and local telephone companies, and leased space for CATV cables on telephone poles. Microwave Associates changed its name to M/A COM, Inc. which acquired numerous telecommunications computer and printer companies and joined them in one system in 1980. M/A COM named its nationwide network Macomnet and produced sophisticated business communications systems.

The FCC had been formed to license broadcasters because many companies used radio frequencies and interference in radio broadcasting resulted. Though CATV used cables and not air waves, the FCC exercised authority over cable TV. It allowed independent telephone companies to establish CATV systems in rural areas and towns, which put AT&T's operating companies with pole lines partly in the CATV business. Dow Jones, Knight-Ridder and Gannett newspaper chains provided news to CATV stations. AT&T planned to advertise its Yellow Pages directory on CATV and send a list of plumbers, for example, on request. Similar services were provided by government telephone companies in Canada, France, England, and Japan. It was predicted that by 1999 electronic newspapers, services and entertainment on cable TV would serve half of all U.S. families.

In the late 1980s, a home TV shopping company, Prodigy Services, was established by Sears Roebuck and IBM. Its purpose was to provide twenty million CATV viewers access to a central IBM computer, with news, catalog shopping, stock quotations, airline tickets, and hotel reservations services. The monthly service for $9.95 and a start-up kit for fifty dollars were reportedly sold to about 60,000 subscribers. NYNEX provided Prodigy Services in the Northeast.

In 1984, a home videotext TV cable service in Orange County, California enabled subscribers to read parts of the next day's newspaper before it was delivered. In 1984 COMSAT merged its direct broadcasting satellite unit with United Satellite Communi-

cations Inc., 51.9 percent owned by Prudential Insurance Co. The principals in that deal were to include Douglas Rune, owner of United Press International.

"We view ourselves as video magazines," said Thomas Wheeler, president of the National Cable Television Association. When Ted Turner's all-news cable TV network began beaming news around-the-clock via satellite to cable systems nationwide from Atlanta on June 1, 1980, Wheeler called it "a historic telepublishing event."

Digital image enhancement, a technique developed in satellite photography, improved home TV reception and made weather forecasting more precise, especially for hurricanes. Videotapes and video discs were produced by North American Phillips, MCA, Magnavox, U.S. Pioneer Electronics, Matsushita, Toshiba, GE, RCA, and CBA. They were widely used in homes to record TV programs for later viewing. Motion pictures on discs were sold for home TV. Rights to record and sell many films were bought from producers by Magnetic Video Corp., a subsidiary of Twentieth Century Fox, RCA, MCA, and a CBA and Metro-Goldwyn-Mayer venture.

Satellite programming was the business of CompuServe, Times Mirror's Gateway, and Knight Ridder/AT&T's Videotron, which delivered information by telephone or two-way CATV. It opened a new world of entertainment and information. With a Naber adapter, the CATV subscriber could obtain stockmarket closing figures and other reports by using software programs. A signal selected the program, or information wanted from the central database, and transmission began in ten seconds. As many as thirty television stations collected the desired satellite programs at a central location for subscribers.

Cable TV grew because it provided a

wide variety of special programs to audiences with special interests like news, sports, business-finance, weather, movies, and foreign languages. The Home Box Office CATV alone spent $545 million for movies it broadcast in 1987. By 1989 the largest cable TV operator was Telecommunications Inc. and it bought half of the Showtime cable network from Viacom Inc., the leading movie service. Estimated Cable TV revenue from subscriptions and advertising grew from about two billion dollars in 1980 to twelve billion dollars in 1987.[17] The value of cable TV companies soared. In 1989 an American Express subsidiary paid $145 million for sixty percent of the cable TV subsidiary of Warner Communications. With broadcasts around the clock with more than a hundred stations competing with newspapers and magazines for attention, people were bombarded with much useless information in a modern Tower of Babel. It was time consuming for people to select information important to their lives and careers.

Cable TV continued its spectacular growth. In 1980 its advertising revenue was $35 million, but for the 1980s it totaled billions of dollars. Telecommunications, Inc. had become so large in 1989 in cable TV that it spun off three billion dollars of its assets into new programming and cable companies. The old television networks and newspapers were concerned about competition by the CATVs for the public's attention. Of course all was not perfect in the brave, new CATV world. Some channels offered mostly reruns of old public TV programs, and the selection of good old movies was extremely limited.

Since satellites have a life of only a few years, replacements are necessary. Three of Western Union's early WESTAR satellites remained in use in the 1980s handling voice, video, and data. AT&T replaced the three COMSAT cables it leased, using earth stations produced by Hughes Aircraft.

Faced with scores of changes in companies, services, and rates in the satellite business in 1984, the FCC stopped regulating the satellite services like those of AT&T, WUI, RCA, and SBI. At last they were free to make changes, but competition permitted few rate increases.

Satellites and Defense

A Satellite Early Warning System (SEWS) was established to warn the United States of attack, to get strategic fighters into the air, and to launch a missile counterattack. Signals from SEWS satellites to earth were relayed automatically by telegraph to Air Defense Command display boards in Cheyenne Mountain, near Colorado Springs. SEWS Command had 500,000 circuit miles linking eighty locations including thirty-four overseas bases. Over this system, tracking data was transmitted to a computer at Goddard and the Manned Spacetrack Center at Houston, where decisions are made like "go or no-go," and "retrofire." At other times, Houston handled instructions and reentry commands to astronauts. In the 1960s and 1970s the armed services had eighty-three communications satellites put into orbit, but also depended on those of the carriers. The Defense Communications Center in Reston, Virginia awarded Western Union a contract in 1979 to provide a satellite communication network employing digital transmission for up to three million bits of information a second. It implemented a single, integrated, switched-digital network to serve all armed services.

U.S. spy satellites routinely took closeup photos of anything suspicious with powerful lenses and transmitted them to intelligence experts at U.S. stations around the world.[18] The National Security Agency used satellites for photographic reconnaisance to gather sophisticated military data and monitor telemetry signals during missile tests by other countries.[19]

There were three basic parts of NASA's STADAN (Satellite Tracking and Data Acquisition Network). One used telemetry to track and control manned space flights. The second was used to track and command other unmanned space flights, and the third was a tracking and telemetry network formed by two "fences"—north-south and east-west—so that any satellite would pass over one of them. When it did, the data was transmitted to Goddard Space Flight Center, which operated the network, and its digital computer established the route, and transmitted it to stations in an average of 6.6 seconds. The stations received it on tape and sent it on to Goddard to tell what happened in space, and when and where. A parabolic "dish" antenna used by STADAN swept the sky a full 180 degrees to follow spacecraft, send commands to satellites, and receive their signals. The key station at Rosman, North Carolina had two eighty-five-foot-diameter dishes 120 feet tall, each weighing 300 tons. Other stations were at Fairbanks, Alaska; Canberra, Australia; Quito, Ecuador; Santiago, Chile; and Johannesburg, South Africa.

The Laser and Maser

Inventions of the laser and maser were two major advances in the history of electronics. Maser (1955) is an acronym for *m*icrowave *a*mplification by *s*timulated *e*mission of *r*adiation, and was created first. Laser (1957) means *l*ight *a*mplification by

*s*timulated *e*mission of *r*adiation, and is based on the same principle.

The idea of microwave amplification by stimulation had been discussed by Albert Einstein and V. A. Fabricant as early as 1940, and others tried to develop it. Charles H. Townes, a Bell Laboratories consultant and Columbia University professor, H. I. Zeiger, and J. P. Gordon, aided by knowledge of telegraph and later telephone development of microwave beam systems, and Western Union's powerful point-of-light beam invention, succeeded in isolating high-energy ammonia molecules, using their energy to emit microwave protons with identical wavelength and wavephase. They found that when the high-energy molecules were contained in a partially mirrored chamber they strongly interacted, emitting an extremely coherent beam from a small opening. The three scientists had laid the groundwork for the maser.

Theodore H. Maiman announced in 1960 he had built the first working laser and produced a beam of coherent light, using a ruby crystal to generate it.

Because masers operate at very precise frequencies, they serve as the basis for very accurate timing, like atomic clocks. Because masers can amplify a microwave signal with very little electrical noise, they found important use as amplifiers of weak microwave signals from distant sources, like stars or satellites.[20]

The first maser was built in 1954 by Townes, Zeiger, and Gordon, using the background of earlier developments. Townes, Arthur L. Schawlow, and others continued experimenting and adding improvements in its operation. In 1964, Townes, Nicholai G. Basov, and Aleksandr M. Prokhorov were awarded the Nobel prize in physics for their research in quantum physics that led to the development of the maser and laser.

Different types of lasers were developed. One was used to store large amounts of information on small computer disks, and used the reaction of the visible laser beam to encode bits of digital information on the disks as modulated signals the computer can use. Seiko Epson Corporation marketed a laser printer that printed six pages a minute as they arrived by telegraph. The laser also is widely used in surgical operations. One kind of laser is used to correct nearsightedness. Another—the excimer—was used to correct the eye focus by shaving off a layer of the cornea. Both were said to eliminate the need for eyeglasses. Three corporations—Summit Technology, Visx, and Taunton Technologies—developed powerful ultraviolet lasers designed to "sculpt" the cornea of the eye, and tested them on human beings.[21]

Charles Townes said, "The laser gives an intensity of light which is as much as a billion times the intensity of light on the surface of the sun. You could carry all the telephone conversations in the world on one light beam," and "There is enough bandwidth to give everybody his own radio station." Sent through a pipe to protect it from interference by rain or fog, the laser's frequency waves are in parallel, regular order. The laser significantly increased the capacity of cables.

When the frequency of electromagnetic radiation is increased to the microwave level, the beam's capacity becomes thousands of circuits. But since light waves are from a thousand to a million higher frequency than microwaves, the laser could provide enormous capacity when communications growth reaches a volume where all available frequencies are needed. Computer screens will use ultrafast laser light switches instead of

picture tubes. Future power plants may use laser-powered fusion instead of nuclear power to make steam and turn turbines to produce electricity.

However, Gordon Gould of Columbia University had a similar idea to Townes in November 1957, and recorded it with a candy store notary. Townes later argued that he could not file it for a patent because he asked the Pentagon for money to build it and the Defense Department classified it as secret. A court upheld his claim in 1982 and awarded the patent worth a fortune to Gould.

Satellites Explore the Universe

NASA gave Western Union Space Communications, Inc. a ten-year contract to provide in-orbit tracking and data relay service for manned and unmanned space satellites. The channels were to be used jointly by NASA and Western Union, which orbited four satellites from the space shuttle to carry out the contract. It still needed more circuits to reduce the $85 million a year it was paying AT&T to lease facilities in 1985, and more to monitor the unmanned satellites, so it added two WESTAR satellites.

Major improvements were made in the 1980s with big radio telescopes, rockets, satellite imaging, and other cosmic discovery devices. Satellite communications played a vital role in man's efforts to extend knowledge of our solar system and unravel mysteries surrounding the creation of the universe. Scientists had heard squeals and beeps from galaxies in deep space, and tried to learn if they were from intelligent beings on other worlds. NASA's "Search for Extraterrestial Intelligence" program provided funds for research, development, equipment, and engineering by universities, AT&T, and others. A tremendous advance in space science resulted from technical education to provide the skills needed to compete in a computer and satellite world.

A Soviet Union space reconnaisance spy ship crashed in the Canadian wilderness in 1978. The Soviet Union already had a spy ship that crossed the United States twice a day. All major countries had satellites carrying high resolution cameras checking other countries. However, the Soviet ship crash caused the United States to redouble its preparations for defense in a possible space war. The Soviets reportedly had developed an orbital nuclear bomb, so the United States started developing a network with powerful rays to destroy enemy bombs or intercontinental missiles approaching America.

The United States moved a satellite to the Near East to aid Israel in its Yom Kippur war in 1973. To show its adaptability as a hunter-killer, the Cosmos 970 satellite was launched in low orbit and shifted by ground command to only 600 to 700 miles above the earth, and exploded to show how such satellites could destroy others.

The Deep Space Network for lunar and planetary missions used eighty-five-foot parabolic antennas and was operated, as mentioned earlier, by the Jet Propulsion Laboratory of the California Institute of Technology at Pasadena. On the *Mariner 4* flight, communications and tracking were maintained on its 228-day, 325-million-mile trip, and photographs of Mars were sent 134 million miles back to earth. In the 1980s the *Voyager 1 and 2* satellites sent data and spectacular photographs to earth as they passed Jupiter, Saturn, Uranus, and Neptune many millions of miles away, dramatically increasing man's knowledge of the universe. On their deep space flights, they took photographs of all of the planets except one, and unveiled many mysteries.

The Manned Space Flight Network used thirteen ground stations, and tracking and support ships. The mission control center at Houston, Texas used a number of computer/communications control centers and relay stations. NASA Communications Network (NASCOM) provided a worldwide system of telegraph, voice, and digital data.

Manned and unmanned American satellites passed the planets as they moved around the sun at various distances in the 1970s and 1980s and used scientific instruments to obtain information about their atmosphere, temperature, volcanoes, climate, landscape, and rings and moons circling them. Both manned and unmanned satellite trips produced startling information that made people eager for more knowledge about our solar system, space travel, and the creation of the universe.

In 1610, Galileo (1564–1642) reported discovering with his telescope that, among other things, Jupiter had four moons revolving around it. Because he dared to contradict church teachings about the universe, he was arrested, tried by the Inquisition, and confined under house arrest (from 1633). In October 1989, 379 years later, a spaceship named in Galileo's honor began a six-year two-and-one-half billion mile flight to solve some of the mysteries about Jupiter, the largest planet in the solar system. It was expected to reach Jupiter six years later, in 1995.

Various other robot spacecraft with names like *Magellan, Pioneer,* and *Voyagers 1 and 2* were sent to pass Jupiter, Venus, and the other planets and use robot satellites to obtain atmosphere, temperature, and other data. Photographs from earlier probes showed huge rings surrounding Saturn and Neptune, and the moons of other planets. They transmitted an enormous volume of amazing facts back to earth, producing more mysteries that will have scientists studying them for years and advancing theories of the universe's creation.

Constant hurricanes blowing and a volcano were found on one of Jupiter's moons erupting and blowing sulphuric gas and frozen nitrogen five miles high. Other photos showed a cold spot on Triton, largest of Neptune's moons, and eruptions in the turbulent atmosphere of other planets. The two *Voyagers* made amazing discoveries in their ten-year grand tour of our universe, passing Neptune and its moon Triton, Jupiter, Saturn, and Uranus. After passing Saturn, *Voyager 1* went beyond the edge of our solar system into deep space. *Voyager 2* travelled nine and one-half million miles at seventeen miles per second, nearing the edge of the solar system.

The planets are no longer just points of light in the sky to us, but other worlds in space with various forms, colors, and temperatures that would freeze or incinerate human beings, and nothing to reveal living things. Some planets seemed to be great fuzzy balls of gas with no firm ground on which to land. Earth seems to be the only planet on which man could live in our solar system. On a dry lake bed in New Mexico, however, where a thirteen-mile row of twenty-seven giant antennas is tuned in and listening to the sounds of the universe, hope remains that there may be life beyond our solar system.

With the huge telescopes at Palomar Observatory in Pasadena, California astronomers discovered a quasar, a brilliant point of light, with a star-like body about eighty-two trillion billion miles from earth. That discovery raised a question as to whether the universe was formed earlier and faster than had been believed.

In 1981 many astronomers thought that fifteen to twenty billion years ago all matter in the universe collapsed into a very compact mass to very great density. They thought that triggered an explosion, a "big bang," producing hydrogen, helium, and lithium. The particles expanded outward at a great speed. Bits of the matter were attracted to each other by gravity and coalesced, forming galaxies and stars, all of which continued outward away from the big bang. A theory in 1989 was that cold, dark matter—lithium, helium, and hydrogen gas particles—gradually clumped together because of gravity to form the stars and galaxies over a few billion years. Astronomers believe the debris from the sun's formation and some of the stars are likely to have spawned the planets.[22]

Cameras on satellites photograph weather patterns on earth, providing data for daily forecasts and hurricane tracking. Medical scientists monitor the health of astronauts to aid space navigation. Satellites map ocean currents and photograph and chart the sea floor to probe the mysteries of the deep.

In the oceans, vast pools and currents of water determine climate and what the weather will be. A *U.S. News & World Report* article "Secrets of the Sea" (August 21, 1989) said:

> As cars and factories spew billions of pounds of carbon dioxide and other "greenhouse gases" into the air every year, the ocean will largely determine how fast the atmosphere will heat up and whether the climate will change in potentially devastating ways for all of the earth's inhabitants.

Satellites monitor the oceans, aid weather predictions and give warnings of climatic changes, like global warming.[23]

To make the benefits of INTELSAT's

global satellites available to the poor and illiterate of the world, mostly the equatorial countries, INTELSAT began in 1973 using a satellite so powerful that the most primitive TV could receive its programs. It was to provide lessons in reading, writing, farming, combating diseases, and improving their way of life.

After 1,500 satellites were launched from Cape Canaveral, Florida, people paid little attention to launchings. They were more concerned about the danger of so much "trash" falling on earth. Satellite communications were estimated to serve one of every four people on earth. They monitored natural resources like crops, forests, mineral deposits, rivers, and oceans. Phone calls from Washington to Peking, weather maps from 23,000 miles in space, and TV news of world importance promised communications benefits. Satellites connect private wire systems for large companies and industries like banking, insurance, medical, store, and restaurant chains.

A new dimension was expected to be added to man's knowledge of space by Edwin Hubble, a Kentuckian who earned a Ph.D. in astronomy at the University of Chicago with other postgraduate studies at Oxford University. Hubble's lengthy observations with the big telescope at Mt. Wilson Observatory in California, produced surprises

for scientists back in 1924 by proving that there are other galaxies beyond our universe. In 1929 he announced that the universe was expanding at a speed shown by his formula called Hubble's Law. He said all major objects in the universe were moving away from each other.

Hubble died in 1953, and the Hubble Space Telescope—named in his honor—was expected to have greater capability and cost than any other. When launched into earth orbit from a space shuttle in 1990, it failed to live up to expectations and will need repair in space.

Past telescope pictures from spacecraft were distorted through the earth's shimmering atmosphere, but the Hubble telescope, it was said, would be free of distortion and provide a clear view of the planets and a deeper look into space. It is expected to vastly expand man's knowledge of the universe before the new millennium year 2000.[24]

Americans have witnessed incredible changes in their lifetimes—from oxen and horses to trains, automobiles, airplanes, and manned space travel, typewriter to computer, water power to nuclear power, and Morse dots and dashes to satellite data transmission. However, planetary science is still in its infancy and more secrets of the universe are out there in space for man to learn. We are lucky to live in such a wonderful age.

Notes

[1]Two years before Marconi demonstrated his wireless telegraphy (in 1895), Nikola Tesla, a naturalized Yugoslavian, demonstrated radio transmission at the 1893 World's Columbian Exposition at Chicago. The U.S. Supreme Court ruled that Tesla's radio patents predated Marconi's.

[2]Milton Lehman, *This High Man* (New York: Farrar, Straus & Co., 1963) a biography of Goddard, who for fifty years developed many features of modern rocketry.

[3]*U.S. News & World Report* (October 3, 1977) and the 1969 *World Almanac.*

[4]Public law 87-624. 87th Congress. Approved August 31, 1962.

[5]"Agreement between United States and other Governments," 37 pages, August 20, 1964.

[6]*COMSAT at 15,* a 54-page anniversary booklet, provided many facts in this chapter. "Buz" Aldrin, one of the first two men on the moon told the writer, when they were on a TV network program, that he was too busy on the moon to think about how the moon landing would boost the satellite industry or how excited millions were about the landing.

[7]COMSAT pamphlet "Via Satellite," and COMSAT annual report.

[8]*Forbes* (March 26, 1984).

[9]*COMSAT Magazine,* no. 2 (1981).

[10]Ft. Lauderdale, Florida *Sun Sentinel,* December 17, 1989.

[11]Smith-Barney, Harris & Upham Co. *Analysis of Cellular Industry* (1988).

[12]*Wall Street Journal,* November 28, 1989.

[13]*Forbes* (May 25, 1981).

[14]*Chicago Tribune,* October 19, 1989.

[15]*U.S. News & World Report* (September 29, 1980).

[16]From conversations with the president of Microwave Associates.

[17]*Wall Street Journal,* June 23, 1987.

[18]*U.S. News & World Report* (November 16, 1981).

[19]*U.S. News & World Report* (June 26, 1978).

[20]*Academic American Encyclopedia,* 21 vols. (Danbury CT: Grolier Inc., 1983; 1989–).

[21]*Wall Street Journal,* February 6, 1990.

[22]*U.S. News and World Report* (September 11, 1989).

[23]*U.S. News & World Report* (September 24, 1989).

[24]"Journey to the Beginning of Time," *U.S. News & World Report* (March 26, 1990).

Our Communications World Disrupted

Manufacturing industries provided a majority of America's economic growth before revolutionary changes in the telecommunications industry during the 1980s added thousands of companies and millions of jobs. Telecommunications became such an indispensable part of our personal lives that we took it for granted like water and electricity. In the 1980s new communications products and services appeared almost daily to aid management control, production, sales, distribution, and credit.

Two companies—AT&T and Western Union—that helped build the nation and knit the intricate pattern of our personal and business lives were then shattered by—in my opinion—congressional blunderers and ideological bureaucrats.

To provide complete telecommunications to the growing nation required the combined voice and record services of AT&T and Western Union, and they merged in 1909. However, in 1911 the Supreme Court ordered the dissolution of Standard Oil as a violation of the Sherman Antitrust Act. Then in 1914, President Woodrow Wilson, lacking understanding that the communications industry was a natural monopoly in a country dedicated to free enterprise, forced AT&T to divorce Western Union.

Congress confirmed the basic policy of AT&T and Western Union as law by passing the Communications Act of 1934. It said the two companies were "natural monopolies that, if adequately regulated, could render a superior service at lower cost than that provided by competing companies." The Act also said it was "to make available, so far as possible, to all people of the United States a rapid, efficient nationwide and worldwide wire and radio communication service with adequate facilities at reasonable charge."

The Communications Act also created the Federal Communications Commission (more-inclusive successor to the 1927 Federal Radio Commission) to referee conflicting uses of radio frequencies, and the telegraph and telephone. In the next forty years, the twenty percent of American homes with telephones grew to ninety-five percent. The idea of having numerous companies digging up streets for cables or filling the sky with poles and wires was not popular, but the FCC and Congress designed their decisions to produce competition by a large number of companies.

AT&T Vice Chairman William M. Ellinghaus said on June 14, 1976:

> Where we once had a coherent national communications policy, the one established by Congress, we now have an inconsistant nightmare created by patchwork decisions of a federal agency, the FCC.

Ellinghaus correctly predicted that competition would result in wasteful duplication of facilities, increase in the cost of service, and regulatory decisions that scattered segments of the world's best services among many carriers.

Western Union and AT&T were wrecked by members of the FCC and Congress, influenced by companies eager to seize their businesses, but saying their purpose was to "promote communications progress."[1] The companies appealed to Congress to overrule the FCC, but nothing was done to preserve the world's best service.

The Wrecking of Western Union

When Western Union merged several telegraph companies to establish rapid communications "everywhere," it proved that numerous companies produced chaos, but one provided good national service. The telegraph had the advantage of providing a written record that could be read and reread, and convey a clear, understandable message. When Postal Telegraph was born and built lines along highways to principal cities, it had to turn messages over to Western Union to reach most destinations. That bankrupted

Postal, but congressional politicians—to "preserve competition"— kept it in business many years with tax money.

While Western Union was loaded with urgent communications during World War II, a strike was called by a 3,000-member labor union with alleged Communist leadership that had gained control of Postal. ITT owned Postal and persuaded Congress and the FCC, which controlled all-important rates and earnings, to require a merger and a twenty-one percent increase in wages and benefits. Western Union argued vehemently it could not afford to do so unless it was permitted to buy AT&T's TWX (Teletypewriter Exchange Service) and merge the other international carriers with its cable system.

The Merger Act that Congress passed was the reverse of what Western Union required.[2] It did not permit the TWX purchase or international merger, and forced Western Union to pay Postal's debt to the government, divest the nation's major cable system and hire all Postal employees at suddenly increased salaries for a minimum of five years. It was a prescription for disaster.

Minutes before it passed out of existence, the National War Labor Board awarded the former Postal employees many millions in "retroactive" wages.[3] Western Union people called it a Communist plot to wreck American business and defense communications in World War II, but the media denounced Western Union because the former Postal "Communist" operators were handling government as well as other messages. The press ignored the fact that the company had pled with Congress repeatedly to make it unlawful for Communists to send defense messages. The FCC allowed AT&T to continue its inroads in the telegraph business by subsidizing its low long distance with high local rates.

Union activity and increased labor costs were primarily responsible for the replacement of many public offices with agencies in stores with no telegraph employees and reduction of employees from 75,000 in the 1920s to 20,000 in 1971, just as unions killed their members' jobs in other industries.

Handicapped by an FCC ruling depriving it of its international Telex business, Western Union agreed with TRT Communications Corporation, a United Brands division, to provide twenty percent cut-rate service to large volume international Telex customers. To encourage competition, the FCC approved that.

To free itself from some of the FCC's straitjacket regulation, Western Union established a parent company—Western Union Corporation—with the Western Union Telegraph Company as a subsidiary.

The 17,000 Western Union employees represented by the United Telegraph Workers Union and the 3,000 in New York represented by the Communications Workers of America went on strike in 1952. That cost a twenty-one percent increase in wages and benefits over a two-year period, and forced the company to cut costs by reducing the number of employees. Union activity had reduced its 75,000 employees to 12,000. AT&T's answer to union pressure was direct dialing by the public, without operators.

The strike accelerated Western Union's transition from high-labor-cost, public-telegram service to an integrated-information services company, with nonregulated subsidiary companies providing data systems channels, engineering, computer facilities, and management services. Western Union acquired Distronics Corporation, offering management control, electronic teleprocessing, and systems services to the plumbing and electrical supply industry, and Teleprocessing Industries, Inc., which developed and managed computer communications systems. It also established Western Union Data Services Company, providing and servicing wire terminals.

Customers were given direct transmission access to the network through automatic Telex/TWX switching centers that required few Western Union employees.

In 1979 the FCC decided to end Western Union's international "monopoly" by giving to another company its inbound cable traffic from the "gateway city" terminals to domestic addresses. Not surprisingly, the FCC selected to handle that telegraph service the Graphic Scanning Corporation of Englewood, New Jersey, owned by the favorite of Washington politicians, ITT World Communications.

Of course, Congress and the FCC did not cause all of Western Union's troubles. Its employees had great loyalty, experience, and skill, but in the 1930s its "Wall Street" directors elected as presidents a series of former railroad presidents they liked but who had no experience in the telegraph business. Those new presidents approved proposals that already had been money losers, and appointed officials with no experience for their specialized professional responsibilities.

When Presidents Egan, Marshall, and McFall succeeded the railroad men, they recognized Western Union's physical plant was technologically obsolete and labor intensive, and that computer/data networks were needed. They spent about two billion dollars in fourteen years to modernize. News stories on inaugurations of hundreds of private companies' wire systems caused others to consider improving their efficiency also by leasing networks, and thousands did.

AT&T showed its friendly feeling for

Western Union for many years by giving a fifty-seven percent discount on its large bill for leased wires. New companies like MCI (Microwave Communications, Inc.) petitioned the FCC to force AT&T to give them the same discount, and AT&T raised Western Union's bill.

23.1 THE TELEGRAPH CAPITOL OF AMERICA FROM 1930 TO THE MID-1960S WAS THE 24-STORY WESTERN UNION HEADQUARTERS BUILDING COVERING THE BLOCK BOUNDED BY HUDSON, WEST BROADWAY, WORTH, AND THOMAS STREETS. WHEN WESTERN UNION BUILT IT IN 1930, IT WAS ONE OF THE LARGEST IN NEW YORK CITY.

Government pressure caused AT&T to sell its Teletypewriter Exchange Service (TWX) to Western Union for ninety million dollars. TWX provided direct telegraph communication through exchanges. Business growth during McFall's presidency aroused hope of a turnaround, but government rulings and regulations kept the company deeply in debt. McFall established a network to check on credit card credit, put the first

communications satellite in space and acquired Computer Logistics Corp. specializing in warehousing data operations in 1974.

In 1962 Western Union began providing a nationwide bomb alarm system to the Defense Department, in which sensors would be triggered by a nuclear blast, and report its location on display boards. Western Union also had a "Hot Line," metered no-dial telephone service in which the user merely lifted the receiver, and the distant telephone rang in one second. The voice circuits used in that private line telephone service were provided by Western Union's transcontinental microwave beam system, but it was scarcely a fly speck compared with Ma Bell's leases for telegraph use.

More awards to labor and regulatory decisions forced reorganization, expense reductions, early retirements, and sale of properties. Severance of 20,000 employees from the company in 1971 resulted from a 104-day strike and again destroyed the work providing the strikers' jobs.

One Western Union president had appointed a lobbyist in Washington, D.C. as vice president for public relations, saying his political friends could avert threatened crippling legislation, but they did not, and his appointees had no experience in public relations work. One vice president brought in "Yes, boss" pals as assistants and would hire no one with the professional experience he lacked. Expensive outside agencies were retained for projects only an employee with special knowledge could carry out. Another claimed credit in a magazine for a nationwide centennial celebration of which he was a last-day spectator. Officials played internal politics at three-hour luncheons in company-expense clubs.

Some Western Union officials opposed the singing telegram and public relations release wire, both of which I created, as an invasion of their territories. The public was delighted, however, and sent millions of singing telegrams that gave pleasure to their friends and provided desperately needed millions of dollars to the company during the Great Depression.

A British company later established a public relations release wire with 9,000 miles of circuits to 300 media locations. It was similar to a public relations newswire service I had started for Western Union years before when I was president of the Public Relations Society: New York.[4] The company did not want it then, but saw its value in 1969 and paid $9.5 million to buy the service from the British company.

FCC decisions designed to increase competition reduced income to $18.1 million in 1973, after awarding $10.4 million in severance pay to employees. A financial man was appointed to head Western Union's holding company, and an experienced telegraph man to operate the business.

The U.S. Post Office Department and Western Union jointly started a Mailgram Service January 1, 1970 which telegraphed mail to area points from which messages were delivered by mail. In 1975, 22.6 million mailgrams were sent. In 1972 the company built a large computer switching center to handle all telegraph messages at Middletown, Virginia, and phased out its fifteen reperforator switching centers.

At last, in 1977 the company elected a telegraph man as president: Walter Girardin, with forty-two years of Western Union experience, from messenger to executive vice president. He gave it a valiant try, but it was a generation too late.

In the 1980s, several companies bought large blocks of Western Union stock, hoping to gain control. Pacific Asset Holdings (PAH), a Los Angeles investment firm including the Ball brothers of Texas and H. Ross Perot, founder of Electronic Data Systems, planned to buy control and sell shares to pay the company's debts, but ran into government obstacles.

Curtis Wright, an aerospace and industrial equipment company, bought a large block of stock in 1981, but was acquired by Teledyne. Curtis Wright Chairman T. Roland Berner, then Western Union's largest stockholder, became its chairman and chief executive officer, but his takeover plan cooled when an administrative law judge ruled that Western Union had to refund $74.6 million in Telex "overcharges" to the international carriers, whose excessive rates of return already were a matter of FCC concern.

In 1982 Western Union took options on 4,240,000 shares of stock valued at $132 million in E. F. Johnson Co., maker of electronic parts for AT&T and GTE mobile telephones. Western Union installed three new Metrofone switching centers in its long distance telephone system serving about 80,000 commercial and residential telephones in 100 cities. To pay for the centers, Western Union sold twenty percent of its satellites to Fairchild Industries and Continental Telephone for $20 million, and nine transponders to Dow Jones and others for about $100 million.

More losses resulted when the FCC, with federal court approval, required Western Union to abandon Telex service, its largest source of income. The FCC had allowed Western Union to start Telex and other telegraph services to Hawaii in 1971, and Mailgram to Hawaii in 1976. The FCC also permitted the other international carriers to provide direct service to American customers. That diversion of Western Union's American haul to others cost the company four million dollars (forty percent of its foreign revenue).[5] To provide more telegraph competition, the FCC authorized Graphnet, Inc. to provide record services and the court of appeals upheld that in 1981.

With FCC approval, AT&T sharply increased the cost of Western Union's leases of telephone lines, largely through cables under city streets and in office buildings to reach customers. That cost the telegraph company $39 million in 1983. In 1984 the FCC approved the company's participation in cellular, mobile telephone ventures. To pay the cost, the company sold two of its satellite transponders to Vitalink Communications Corp. for $11.5 million and its interest in Vitalink. Competition reduced its electronic mail service earnings and it sold another two WESTAR satellite transponders to Public Broadcasting Service. Bonneville Satellite Corp. also agreed to sell two to WUI, Inc.

Western Union had bled much red ink in the early 1980s, lost $531 million in 1987, and was one billion dollars in debt. It was amazing that the grand old company could sink so low, after aiding the development of the nation for more than a century, and carrying news of milestones in peoples' lives, like birth, marriage, and death.

A new chief executive, Robert S. Leventhal, a former Harvard Business School student and Navy employee who had aided other troubled companies, was brought in to rescue the company. He soon had British Cable and Wireless, MCI, GM's Electronic Data Systems, Citicorp. Resource Holdings, MDC Holdings, Meridian American, and individual financiers considering purchase of Western Union. Leventhal struggled to avert bankruptcy by reducing employees and cutting salaries.

Bennett S. LeBow, a bearded New York investor seeking profits by turning troubled companies around, presented a plan to restructure Western Union in 1987. LeBow, Western Union, and MDC Holdings, owner of a large block of its stock, jointly announced agreement on LeBow's plan. It provided for LeBow to acquire control of Western Union for $25 million, merge the telegraph unit with its holding company, buy ITT's World Communications' international Telex subsidiary, sell it to Western Union, and combine it with its domestic Telex.[6]

Leventhal submitted LeBow's plan to a meeting of preferred stock and bond holders in 1987, but the debt holders were reluctant to swap their securities for a new issue of preferred stock. More stockholder meetings were held and bankruptcy threatened until approval was received. LeBow had borrowed $15 million from Canadian Imperial Bank and $10 million from partners to take control of Western Union with $25 million of other people's money.[7] He then bought ITT's Telex business and combined it with Western Union's. Realizing that Western Union's long distance telephone business was too small to compete, he wrote off a loss of $603 million and abandoned it. Though it was deeply in debt and losing money, Western Union paid five officials $1,196,099 in 1984. That is like giving a big bonus to the captain of the *Titanic*!

The company escaped bankruptcy but remained a great tradition, its spokesman said, like the five-cent cup of coffee and five-cent cigar. It still had a three-satellite communications system. In 1987 direct communications by Telex remained its most profitable service, followed by the electronic Mailgram, private wire systems, Money Transfer, and Opiniongram. Western Union transmitted many Mailgrams to post office computers in regional cities where computers received the messages for delivery. Many companies paid Western Union to send messages to the post office computers. Western Union also offered to deliver singing telegrams by telephone for twenty dollars, too high priced to restore volume use.

To aid finances, the company sold its interest in E. F. Johnson, Teleport Communications, several cellular telephone companies, and Airfone service for airplane passengers. Chairman LeBow engaged Western Union in more financial services, and in 1986 sold its profitable Government Services Division based in McLean, Virginia, to Continental Telecom Inc., later named Contel, for $155 million, and even sold its nationwide microwave network that provided the backbone of its facilities. He threw away Western Union's major asset by changing its name to New Valley Corp.

Company officials had so little regard for its great past they did not even retain its library and museum, with irreplaceable original publications and equipment gathered for years by myself and others before they retired.[8] Some of the historic telegraph and cable treasures the library and museum contained were donated to the Smithsonian Institution.

In a generation, Western Union's image underwent a metamorphosis. It was no longer constantly brought to the public's attention as an essential, personal, business, and defense service. The telegram was still used by business and government, but its big personal, social, and greeting services had faded. The telegram became extinct in England after 112 years and was no longer synonymous with rapid communications in America. However, Western Union still had the potential to rise from its ashes and regain an important role. Money orders, Teletype, data, and facsimile network communications

with everyone with a desktop telegraph machine remained a growing need. No other company had Western Union's experience in such telegraph operations and services.

Western Union sealed its fate in 1990 by selling its profitable Business Services Group to AT&T for $180 million. Telegraph was part of AT&T's name and it needed the record data (telegraph) communications of large businesses. It is ironic that the use of facsimile, invented and first used by Western Union, replaced most of its telegram business.

Ma Bell Loses Her Children

In three boom periods this century, some companies became giants by merger and rapid growth. Those whose growth exceeded earnings experienced financial problems especially during the Great Depression of the 1930s, but AT&T expanded steadily to meet the needs of a growing nation. In 1981 AT&T was fondly called "Ma Bell" by 1,042,000 employees, proud of their job stability, generous earnings, pensions, health, and other benefits. After its divorce from Western Union in 1914, AT&T's headquarters in Western Union's old building at 195 Broadway in New York's financial district was the mecca of the telecommunications world.[9] In 1977 AT&T had $36.5 billion in operating revenues ($18 billion of it from long distance calls), and $93.9 billion in assets.

AT&T had spun a web of record and voice communications around the world with landlines, satellites, microwaves, and cables. It's network control center was moved in the 1970s from Western Union's old long lines building at 24 Walker Street, New York, to three large buildings on a 422-acre tract in Bedminster, New Jersey. A big wall map of the world with thousands of lights indicating route conditions enabled controllers to shift the flow of 600 million calls a day over different routes as need arose.

23.2 IN 1982, AT&T MOVED ITS HEADQUARTERS FROM 195 BROADWAY TO A NEW 37-STORY BUILDING AT 550 MADISON AVENUE, NEW YORK CITY. IT HAS AN OPEN STREET-LEVEL PLAZA, A "SKY LOBBY" AND RECEPTION AREA SOME FIVE STORIES HIGH, AND A THROUGH-BLOCK, GLASS-COVERED ARCADE WITH A VISITOR AND EXHIBITION AREA.

AT&T then built a new thirty-seven-story headquarters building with a Chippendale top at 550 Madison Avenue between 55th and 56th Streets, in New York. It provided 500,000 square feet of space and cost $110 million. The company planned to retain its former headquarters building at 195 Broadway.

As one of the world's largest companies, mammoth AT&T had long been a target for politicians and lawyers. Certain politicians, and the media had built a public fear of monopoly since cartoons in the press in

1909 depicted President Theodore Roosevelt with a toothy grin smashing the trusts with his big stick. In 1913 the Justice Department filed an antitrust action against AT&T for acquiring Northwestern Pacific and other telephone companies. AT&T agreed to stop buying independents, divested two, and the suit was dropped. AT&T welcomed Government regulation, however, when its early patents expired and competitors appeared.

The Justice Department sued in 1949 to separate Western Electric from AT&T to end the "equipment monopoly." Secretary of Defense Charles E. Wilson said divestment of Western Electric would "effectively disintegrate critical defense projects." The Clayton Act in 1950 brought many mergers within government control. Justice Peter Stewart said, "The government always wins." In 1952, 1954, and 1956, the Eisenhower administration tried to get rid of the old Western Electric divestment case, saying that doing so would not greatly damage telephone service.

In 1976, 1978, and 1979, Congress considered acts to open the long distance market to more competition, as the Carterphone decision had in 1968 by permitting connection of non-Bell equipment with the Bell System. In 1974, however, AT&T's telephone equipment business had reached about four billion dollars. Western Electric made AT&T's switches and rented them to independents. It received one billion dollars a year providing private wire systems to many companies, while ITT, General Dynamics, Stromberg-Carlson, and others sold about $200 million of PBX and other telephone equipment to independents.[10]

Companies with numerous telephone extensions saved the cost of several operators by using a dial system that enabled all extensions to make "inside" calls. To call outside, "9" was dialed first. That gave the company PBX operators more time to handle incoming calls.

Many large companies used a Centrex System. Callers dialed a central office number plus the extension of the individual called. The PBX operators only handled calls for whom the caller did not know the extension number.

Many companies used a compact, desktop switchboard, handling up to 200 stations. The company operator could select any extension by pressing one of 200 buttons. Illuminated buttons indicated busy lines and call waiting. Another set provided a button to "hold" a call by pressing a button while calling someone for information or to switch a call. That reduced "callbacks" and saved time. Another button held a call for a busy line, and connected the caller when it was free. A preset telephone conference could be held with a half dozen associates by merely pressing a button.

Touch Tone service, in which calls are made by pushing buttons instead of dialing numbers on a rotary dial, was introduced in 1963. In 1965 the number of independent telephone companies had decreased, in twenty years, from 5,000 to 1,792, but the number of telephones doubled to near the total in Britain and France. The United States, with six percent of the world's population, still had half of the world's telephones.

In 1967, Haakon I. ("Hi") Romnes, a gentlemanly engineer, president of Western Electric, succeeded Kappel as AT&T's chief executive, after a period of rioting, arson, and looting in several cities. A strong advocate of equal opportunity, Romnes went to great extremes to hire, train, and employ hard-core welfare people nationwide. He established a plant in the Newark, New

Jersey slums to train and hire minority employees. After years of strong civil rights leadership, the thanks AT&T received was charges of racial discrimination by the Equal Employment Opportunity Commission to the FCC. Although AT&T had hired more minority employees than any other company in the world, the EEOC held hearings throughout 1972 and pressured AT&T into signing a consent decree in January 1973 to do just what it had been doing, and also pay seven million dollars to 7,000 women and minority employees as "back pay."[11] While President Romnes held the annual stockholders' meeting in 1970, 2,000 young people marched outside shouting antiwar slogans and demanding that AT&T stop serving the nation's armed forces.

Economies in plant building by Kappel, to maintain fair earnings in years the FCC refused rate increases, had resulted in the Bell System being unprepared for the large increase in telephone business in the late 1960s, and service deteriorated. Poor service in New York City in 1969 resulted partly from vandals and thieves wrecking 35,000 of the 100,000 public coin phone booths, and inability of some new employees to cope with their operating jobs. Breakdown of service was so bad that 1,500 qualified employees from other cities were transferred to New York.

In 1976 an administrative law judge recommended that AT&T be broken up, but the FCC issued a 535-page memorandum, saying tighter control was needed. The company was required to provide its wire facilities to cut-rate competitors. Howard Anderson, president of the Yankes Group, a consulting firm, said, "AT&T right now is like a battleship with a leak below the waterline. . . . If it continues unabated, you have to take that battleship to drydock and turn to something else." Romnes persuaded the FCC to approve a rate increase to finance plant expansion and improve service.

The old Communications Act of 1934 separating domestic from international service was no longer workable. Western Union was providing long distance telephone service to companies, and leasing twenty percent ot its satellite capacity to TV broadcasters. Some of its satellite circuits served news companies, but an FCC ruling prohibited Western Union from providing international Telex service. That hurt AT&T also because it was the major wholesaler of international circuits, and it terminated 56,000 jobs, 15,000 of them managerial. That was only one in a series of FCC rulings to "promote competition." Another decision allowed long distance competitors like Satellite Business Systems, Southern Pacific, MCI, and U.S. Sprint to interconnect with the Bell System, enabling them to lease AT&T's lines and use them to offer WATS and other services at cut rates to AT&T's customers.[12] Despite the FCC Common Carrier Division's efforts to weaken it, however, AT&T earned more than six billion dollars in 1980.

Prominent among telecommunications equipment competitors was United Technologies (UT), the big jet engine, elevator, helicopter, semiconductor, and integrated circuit company. UT also offered simultaneous digital voice and data telephones, PBX equipment, telecommunications, accounting, and management information systems. UT bought three other equipment makers in 1982 for about $100 million. General Electric also offered a line of telephones for sale, and Memorex International, N.V., bought Telex Corporation, which had about two billion dollars in annual equipment sales and was a large supplier of accessory computer parts to IBM. The equipment companies

received another boost when Canada removed its tariff on their products.

When John D. deButts succeeded H. I. Romnes as AT&T chairman in 1972, the demand for telephone facilities during the Vietnam War had resulted in delays of thousands of residential and business installations. A new breed of "Take it or leave it" operating personnel had been hired under political pressure, and service deteriorated. Monopoly criticism reappeared.

Having climbed the long AT&T promotion ladder via the presidency of Chesapeake and Potomac Telephone and Western Electric, deButts was determined to restore a high standard of service. He reorganized and revitalized the sales organization, and set up 1,500 centers to sell and service a wide variety of telephones, electronic typewriters, and personal computers. He achieved a 10.2 percent increase in earnings and increased dividends per share six times—from $1.60 to $4.60—in seven years. However, service was still below standard, and the FCC decided the cure was more competition.

MCI Enters the Battle

A new long distance telephone company, Microwave Communications, Inc. (MCI), based in Washington, D.C., obtained FCC approval in 1969 to build a microwave system between Chicago and St. Louis and provide private line service at low rates. AT&T protested that MCI's rates were unfair; it did not have the heavy expense of serving small towns everywhere, and MCI was "cream skimming." MCI's president, B. Orville Wright (no relation to the pioneer aviator), bought 62,000 miles of line from the United States unit of Northern Telecom, owned by Bell Canada Enterprises of Montreal. He concentrated on selling private wire systems to AT&T's largest customers and had Fujitsu Ltd. of Japan install a fiber optic cable network in New York City. MCI spent heavily for advertising and engineering to compete with AT&T and GTE technology and gained a million customers, including some of the 500 largest, but had to use Bell facilities to reach most places.

Ambitious MCI Chairman William McGowan lobbied against AT&T, and found support among some in the FCC, Congress, and the courts. He argued that MCI would never receive fair treatment dealing with the twenty-two Bell operating companies as long as AT&T owned them.

The son of a coal mine union organizer, McGowan was rough and tough, in contrast with the gentlemanly AT&T executives. He had the support of Senators Edward Kennedy and John Tunney. Ironically, McGowan had to depend on the network he was trying to destroy to carry on his business. Certain senators wanted to give MCI, with a few lines, the use of millions of Bell customer lines and telephones and use of the system AT&T built at a cost of billions.

DeButts and the state public service commissioners replied that such competition would deteriorate service and increase its cost. The FCC and politicians' friendly attitude to McGowan, and congressional opposition to AT&T were puzzling because AT&T had three million stockholders and a million employees, and everyone was a customer opposed to higher rates. AT&T was accused of spreading misinformation and said there was no reason "for a monopoly of the equipment that transmits the signals." Encouraged, MCI announced plans to provide its own network by orbiting a satellite system in cooperation with a company organized in 1970 by Lockheed Missiles & Space Co.

AT&T was so annoyed by pro-MCI statements of FCC Commissioner Nicholas Johnson that it asked him to disqualify himself in AT&T cases. In court later it charged that "an FCC commissioner" had been on the MCI payroll. It also was outraged by the hit-and-run tactics of MCI, offering below-cost rates to sign up a few big companies—rates AT&T would have to provide to everyone as a common carrier. AT&T appealed, but the FCC ordered AT&T to provide the cut rates to everyone. The Court of Appeals refused to hear an appeal.

Common carriers had been required to provide approved services and rates over all interstate routes, but now politicians allowed MCI to provide selected services over a few profitable high-volume routes between cities. AT&T responded in 1973 by proposing two-tier service with low rates on large-volume, low-cost routes, and higher rates on low-volume, high-cost routes, but the FCC refused. Witnesses warned that FCC refusal would weaken companies and produce such chaos in service that, like Humpty Dumpty, it could never be put together again. DeButts fought on Capitol Hill and in 320 speeches in five years, arguing against "pervasive regulations" and antitrust proceedings.

In 1981 the FCC awarded MCI $1.8 billion in treble damages from AT&T (later reduced to $931,880) because AT&T refused to provide long distance service at low bulk rates for MCI to provide a private wire network to General Motors. To finance its battle against AT&T and build a microwave system and fiber optic cables, MCI issued $1 billion in notes and warrants.

In 1975 the U.S. Court of Appeals upheld an FCC order requiring AT&T to provide interconnections to MCI, and the Supreme Court refused to hear an appeal. That encouraged Southern Pacific Railroad and others to enter the private wire business, and AT&T lost $125 million of it in the first nine months of 1976.

AT&T learned MCI's Execunet service was violating the law prohibiting switched long distance service. McGowan claimed it was an accidental violation, but AT&T cut off the circuits MCI was using to provide it. This violation had not been detected earlier because of the use of microwave circuits for long distance and Bell operating companies for local facilities.

Largest Antitrust Suit in History: U.S. versus AT&T

Bernard Strassburg, chief of the FCC Common Carrier Bureau, and young lawyers in the Justice Department Antitrust Division persuaded recently appointed Attorney General William Saxbe to file the largest antitrust suit in history—U.S. vs. AT&T—in Federal Court on October 1, 1974. Chairman deButts, who believed "You can't deal with tough problems in a gentle way," replied at a press conference that there was no need for the suit because AT&T was providing the highest-quality, greatest-available, and lowest-cost service in the world, and competition would disrupt service and increase cost. He predicted correctly that the FCC policy of allowing selective entry to the private wire field by companies that do not share its obligation to serve the entire public "would compromise the integrity of the national network and its quality of service."

U.S. vs. AT&T, charging anticompetitive behavior, apparently was partly the result of Strassburg's annoyance with AT&T representatives not showing what he considered proper deference to him, and complaints from MCI and congressmen. A typical 1977 news headline was "High Court Denies Ma Bell Immunity to Antitrust Suit," and a story that the court refused again to interfere with the government's effort to "break up the mammoth Bell System."

U.S. News & World Report said the FCC was supporting the right of "specialized common carriers to go into the long distance business." Not mentioned was the fact that nationwide communications requires a huge plant, capital, and debt. AT&T required large capital investment. At the end of 1976, AT&T had $35 billion of debt, $2.9 billion of preferred stock, and $34.3 billion of common equity. Therefore, total capital amounted to $72.2 billion.[13]

The Largest Antitrust Trial

One of the young activist lawyers in the Justice Department was Harold H. Greene, who had escaped from Hitler's Germany with his Jewish parents when he was sixteen. When young Robert Kennedy, a passionate advocate of civil rights, was appointed Attorney General by his brother, President John F. Kennedy, he became friendly with Greene, and the two cooperated in writing a strong civil rights bill. Greene became an activist promoting the Kennedy "Grand Design," and was rewarded with a lifetime federal judgeship in the prestigious District of Columbia Federal Court.

The new judge received exclusive jurisdiction over the U.S. vs. AT&T case in 1978 when Judge Joseph Windys, to whom it was assigned, retired. With an ambitious judge who seemed dedicated to overturning the American establishment, AT&T was doomed.

Greene lost no time setting a schedule for pretrial filings and ordered AT&T to turn its evidence in the MCI case over to the Justice Department. Numerous filings of statements, contentions, allegations, proofs, and stipulations followed in 1978, and Greene set the trial date for September 1, 1980, and then January 15, 1981.

Defense Secretary Caspar W. Weinberger urged the government to drop the suit because the nation's main military communications network was provided and maintained by AT&T and its breakup would weaken national defense. Weinberger said a single defense network with central control was required—not several smaller ones. Fearing that wrecking AT&T would allow Japanese and German companies to take over control of the American electronics and communications industries, Commerce Secretary Malcolm Baldrige urged Congress to act to prevent it. Even the Department of Justice requested an eleven-month delay to allow Congress time to act, but Greene was determined to hold the trial regardless of pending legislation and statements by both sides that they had agreed on a "concept" to settle the case.

William Baxter, antitrust lawyer in charge of the case, vowed to litigate the suit "up to the eyeballs." Philip Verveer, the crusading young Department of Justice lawyer leading the trial team, wanted interconnection with the Bell network for all competitors, and the end of "local monopolies." To him AT&T was a dragon to be slain by giving its business to others. George Saunders of the Chicago law firm Sidley and Austin, representing AT&T, was equally determined to save it. He stalled the case with motions demanding the millions of government papers relating to telephone regulation.

Chairman deButts retired in 1979, and was succeeded by Charles Brown, a friendly president, whose office door was open to visitors. DeButts had been a demanding, rough-talking, no-compromise fighter; Brown was a quiet, thoughtful executive leading a new generation of employees dedicated to friendly service. To finance a record-size construction program, the company raised seven billion dollars in 1980 by selling stock. Brown soon realized (1) the company would have to accept some losses to survive, (2) providing local service was expensive, and (3) long distance was essential to AT&T. He criticized government overtaxing, overspending, and overregulating.

Kenneth Anderson, a tough antitrust lawyer who replaced Verveer, wanted to break Ma Bell by cutting her limbs off. Protests inspired by AT&T flooded Washington from every state asking Congress to act, but Democrat Chairman Peter Robino persuaded his House subcommittee on communications to reject such legislation. Anderson decided to negotiate an out-of-court settlement, and the Justice Department replaced him with Gerald Connell as chief lawyer for the trial.

When Ronald Reagan was elected president, Justice Department officials, fearing replacement, softened their tone, and negotiations produced a tentative agreement. Both sides applied to Judge Greene on January 5, 1981 for a postponement while they finalized the wording. With his visions of presiding at the great trial fading, Greene ordered the trial to start in ten days. He would decide the fate of the nation's largest and most essential company without a jury.

In 1981 AT&T revenues had increased in four years from about $21 billion to $57.3 billion, and net income to $6.91 billion. AT&T handled 200 billion calls a day, 18.6

billion of them long distance. It had 142.5 million customer telephones in use, 1,042,000 employees and spent $18.1 billion for plant expansion. About one-third of the world's 472.1 million telephones were in the United States—175.5 million telephones. That was 8.4 telephones for every ten Americans.[14] The Bell system interconnected about 1,600 independent companies that had more than 13 million telephones and $13 billion invested in plant. Thirty percent of the revenues of the interconnected independents was from AT&T for their share in long distance calls.

In 1981 Congress had wrestled six years with a proposed law to deregulate and open all services to competition except local telephone. The new Attorney General William French Smith declared, "In an economy based on unfettered competition, efficient firms should not be hobbled under the guise of antitrust enforcement." He thought more competition was not necessary and bigness does not mean badness.

The suit was six years old when the trial began on January 15, 1981. The courtroom was crowded with officials, lawyers, politicians, and journalists. In his opening statement, Gerald Connell argued that AT&T was so big and powerful it crushed competitors. George Saunders then presented the company's opening, arguing eloquently for hours that free enterprise had enabled AT&T to compete and grow big by excelling in service, and basic American free enterprise policy should not be changed.

A recess followed to permit negotiations for a settlement to proceed. When agreement was not reached, the trial resumed on March 4. Connell presented witnesses who testified that AT&T was unfair to MCI and others and had monopolistic practices. Business witnesses told about disputes with the com-

pany. McGowan of MCI, the star antitrust witness, described his battles with AT&T, losses suffered, and the necessity for MCI to use the Bell network. Officials of other independents echoed McGowan's charges. On cross examination, Saunders put in the record various alleged McGowan activities, such as putting an FCC commissioner on his payroll.

Alarmed because they feared Judge Greene would destroy the nation's communications system, President Reagan's top officials formed a committee to persuade the president to dismiss the case. Hearing of that, Baxter went to the White House and protested to the president and cabinet. He convinced no one, but someone leaked the story to the press that the president might dismiss the case. Judge Greene declared from the bench he would disregard such stories, and Baxter announced he would press on with the case. The plan to get the president to act was dropped, apparently to avoid the possibility that Democrats in Congress would twist it into an anti-Reagan political hearing.

On July 19 the Department of Justice and AT&T requested a recess to give Congress time to enact the legislation, but Judge Greene refused, and AT&T's defense began on August 3. The judge was angry when AT&T introduced a Justice Department document criticizing the plan to get the president to dismiss the case, calling it an intrusion. Greene was irritated again when Saunders presented state utilities commissioners to testify for AT&T.

His temper really boiled when Saunders presented a list of 250 witnesses he would call: cabinet members, other influential government officials, presidents of leading national associations, and prominent business leaders. It showed the nation's leaders backed AT&T. Trying to prevent such testimony, Greene declared he would not hear or consider statements by people "without real knowledge" of the telephone business. Saunders replied that his witnesses could add important information in the case, but during five months of AT&T testimony he did substitute some who favored AT&T. Tens of thousands of pages of documents were presented.

Worried by the judge's antagonism, AT&T presented a lengthy motion to dismiss the case, while Chairman Brown called members of Congress to urge speedy action on the regulatory legislation. He also launched a campaign to convince the public that the law was needed to save their telephone service, and had company employees from all states lobby their Congressmen, but their pressure was so intense the lawmakers rebelled. In a seventy-three-page decision, Greene denied the motion to dismiss the case, and his wording clearly indicated his belief that AT&T was a monopoly violating the Sherman Antitrust Act.

Faced with opposition by the White House, the rest of the cabinet, and even the head of his own Justice Department, Baxter resumed negotiations. After more lengthy meetings, an agreement was reached on January 8, 1982.

Settlement of the Suit

After more days of argument and lengthy briefs, Greene issued a 178-page opinion approving the agreement but modifying it in ten areas. He rejected changes of his modifications proposed by the Justice Department and AT&T, and on December 16, 1982 issued a 471-page reorganization plan.

After more hearings, Judge Greene issued a 225-page proposed decree requiring divestment from AT&T of the twenty-two Bell System operating companies, the establishment of seven regional holding companies (RHCs), and AT&T's loss of advantages of its superior technology and size.

Greene's decree also restricted the RHCs from providing long distance service, by interchanging facilities with others or providing facilities for electronic publishing.

AT&T was required to divide its $149 billion assets, $65.7 billion annual revenues, and $7.2 billion net income[15] between itself and the seven regional companies. That reduced AT&T to about one-fourth of its former size. Greene called the "local" companies "the key" to AT&T's power, and severed them entirely from their parent. The decree made the complex communications world much more complex.

Each of the twenty-two Bell System operating companies became a subsidiary of one of the seven regional holding companies: Bell Atlantic for the mid-Atlantic states, based in Philadelphia; Bell South for the southeastern states, Atlanta; NYNEX for New York and New England, New York; Pacific Telesis for California and Nevada, San Francisco; Southwestern Bell for the southwestern states, St. Louis; U.S. West, for the northwestern and mountain states, Denver; and American Information Technologies Corp. (Ameritech) serving the Midwest, Chicago. Each RHC included cellular mobile telephone service licensed by the FCC. Ameritech was to publish directories and develop new products, like optic fibers. The twenty-two divested companies could continue providing their very profitable Yellow Page directories.

Because of their great benefits to the nation's trade and economy, "AT&T was to retain Western Electric Company for manufacturing and supply, its long distance network, international communications, information (data), and Bell Laboratories, one of the world's greatest research and development centers. The 3.1 million AT&T shareholders were to retain their AT&T shares and receive one share of each regional company for each ten AT&T shares they owned, with no reduction in total dividends.[16] The seven RHCs were listed on the New York Stock Exchange. AT&T was to select presidents to head the RHCs, and provide them with sufficient facilities, personnel, systems procurement, engineering, and technical information to perform independently.

The decree said Western Electric was not separated from AT&T because of its importance to national security. Sixteen of the world's top twenty-five telecommunications equipment companies already were foreign and provided one fourth of the PBXs in America. Also, AT&T had billions of dollars in telephone equipment in homes and business offices that sometimes was repaired, replaced, or provided to new customers.

Judge Greene was out of the country when the final agreement was ready, so the two sides got U.S. Dictrict Court Judge Vincent Biunno in Newark, New Jersey to dismiss the case and include an old Western Electric case. Biunno's court was where AT&T had settled the Western Electric case under the Tunney Act in 1956 by agreeing to stay out of the computer/data business, and both parties wanted to connect it with the current dismissal to modify the Tunney Act.

AT&T's public relations was not just a department, but its business, and handling of the settlement showed its professional skill. Working in secrecy, in very limited time, AT&T prepared book-length background information, a case chronology, and a series of news releases. It planned first to inform Bell officials. The public then was to receive the news through press conferences in all states and mailings to magazines. All statements were designed to reassure the public that partial destruction of the company was necessary and would benefit everyone. It recognized that times had changed, with new technology and data processing and said the public opposed monopoly and wanted competition.

Brown revealed the settlement confidentially at a meeting of headquarters officials and the twenty-two Bell System company presidents. Upon returning to their states, the presidents were to announce the divestment at press conferences and provide the prepared information. When the news reached employees nationwide, they were shocked and dismayed. It was like being told their world had turned upside down. The three million stockholders were puzzled about the economic implications of the unprecedented divestment. AT&T had been considered the safest investment for widows and orphans.

As modified by Judge Greene, the decree required AT&T to provide the operating companies with all support services they needed, and not engage in any "non-monopoly" business to gain "an improper advantage" over competitors. Many wondered how it could meet all of the requirements and survive. Judge Greene's decree stated that the FCC was incapable of enforcing the provisions of the decree and he would hold periodic hearings to do so. Any state law or decision that conflicted with his decree was invalid, he wrote, because the Federal court had supremacy and jurisdiction over state decisions.

His decree sought to "embody a congressional desire to put an end to great aggregations or capital because of the helplessness of the individual before them," and "since communications is the keystone of the economy, it was dangerous to entrust it to one company."

Just before the settlement was to be announced, Judge Greene heard of it, flew back to the United States, and declared the agreement was not official because he had not approved it. He was angry at not having the honor of closing the case, but it had been settled and closed. At a press conference in the National Press Club in Washington, Baxter announced the agreement, saying it met the Justice Department's requirements. Chairman Brown then stressed public advantages of the settlement. Some wondered if Judge Greene would hold a grudge against AT&T, which had spent about $300 million to defend itself. Seventy lawyers had made court appearances on its behalf and hundreds worked on the case.

After hearing the onerous terms of the decree, many stockholders sold their shares thinking a smaller AT&T would earn less and pay lower dividends, but those who did not sell were wise. The RHCs prospered, split their stock and increased dividends.

The agreement required the RHCs to share the cost for 8,000 AT&T employees to provide engineering, manufacturing, and administration services. A central task force was formed to manage the changeover, including representatives of the RHCs and an AT&T advisory staff to meet national requirements. Meetings of stockholders were held in eight cities to answer questions. Bargaining began with unions representing AT&T and Western Electric employees. MCI and GTE divided their long distance services into regions like the RHCs, and launched advertising campaigns.

Results of the Reorganization

Strong public concern was evidenced by thousands of news columns about the trial and its aftermath. Large piles of clippings, releases, and reports of companies, Justice Department documents, magazine articles, and books are summarized here briefly.

On D (divestment) day, January 1, 1984, nearly ten years after the suit was filed, AT&T named presidents of the seven RHCs and divested itself of ownership in local exchanges and directories. MCI and GTE asked Judge Greene to also strip AT&T of two of its greatest assets—the name "Bell" and its legendary blue Bell telephone trademark—and Greene gave both to the RHCs. It was a terrible price to pay, but the court's decision was final. AT&T was allowed to use the Bell name for Bell Laboratories and overseas operations only to avoid confusion. In 1984 AT&T sold its thirty-one-story Western Electric headquarters building at 222 Broadway, New York City for about $110 million and transferred thousands of employees to Berkeley Heights, New Jersey. The sprawling Western Electric Hawthorne Works near Chicago and some others were phased out to cut costs.

Though exhausted from years of all-hours work under pressure, directing the battle, Brown and the other Bell officials had the enormous task of segregating parts of the Bell System, serving for years on committees on services, dividing personnel with the RHCs, finances, legislation, public information, and so forth. An interchange compatability forum was established where carriers could discuss their technical compatibility problems. Decisions by the committees made changes in the lives and careers of millions of Bell and other industry employees and in everyone's rates and services.

The personnel committee siphoned off 3,000 Bell Laboratories employees to staff Bell Communications Research, called "Bellcore," a separate research center based in Livingston, New Jersey to aid the RHCs. Bellcore soon had 7,000 employees and a $200 million budget. Problems arose when each RHC wanted to own Bellcore's major developments. AT&T engineering was required for the RHCs to keep pace with technological progress. Seeking public approval, AT&T announced September 23, 1983 it was asking FCC permission to slash long distance rates $1.8 billion.

In the midst of their turmoil and all-hours pressure, a congressional committee called the AT&T officials to Washington to assure it that federal and state authority "to protect the public" would not be challenged! Fifteen competitors spent about $200,000 in 1982 on a public relations campaign to make the settlement tougher on AT&T, and help them seize AT&T's business. The attack gained so much public support that AT&T, the victim, was forced to spend two million dollars on a campaign to defend itself.

Eager to seize AT&T's four-billion-dollar business-equipment sales while the telephone company was "on the ropes," IBM made an agreement in 1982 with Mitel, an Ontario-based switch maker, to develop a family of business-office automation equipment such as digital switches for the automated business office of the future. IBM's idea was to be a major competitor of Western Electric's equipment. IBM also hired MCI, AT&T's enemy, to give some of IBM's personnel courses in the telephone business.

The work of state public service commissioners multiplied. Regulation of the RHCs presented problems. For example,

U.S. West's operating units continued using its operating company unit names—Mountain Bell, Pacific Northwest Bell, and Northwestern Bell—in dealing with customers who often knew little about U.S. West. Infringement by AT&T, RHCs, and independents on each other was another problem. In 1984 MCI's $1.8 billion antitrust judgment against AT&T was overturned on appeal. Tired of long antitrust suits, the Justice Department dropped its thirteen-year-old case against IBM. The Communications Workers of America Union had to negotiate contracts with all of the new companies.

Never before had such a huge business as AT&T's $44 billion equipment and long distance been thrown open for others to grab, and frantic efforts followed. In addition to U.S. Sprint, MCI, GTE, and Western Union Metrophone, there soon were about 4,000 competitors, some with services like Skyline and Allnet, and foreign telephone companies who required connections with new American carriers.

In 1984 MCI announced plans for a one-billion-dollar expansion, and an agreement for American Express to market its electronic mail, 800, and WATS services. About 300 discount carriers, called resellers, were born, leased AT&T circuits, and offered cut-rate services. With its long distance business declining, AT&T asked the FCC to end control over rates and services and give it freedom to compete by redesigning rates. AT&T's volume of calls also declined because the RHCs handled many calls within their regions that formerly were AT&T long distance calls. Competition for long distance calls was so strong that AT&T's share decreased to about seventy percent, and AT&T dropped thousands of long distance operators.

Since Republican President Ronald

Reagan had appointed a majority of the FCC commissioners, there was hope that controls on rates and services would end, but Judge Greene made a career of the AT&T case by retaining jurisdiction over enforcement of his decree. The antitrust lawyers' reputations were established as giant killers and some became high-priced Washington advisers to telephone companies. In their rush to dismember Ma Bell, bureaucrats, politicians, and independent telephone officials showed little concern for the nation's vital communications needs.

Judge Greene's philosophy had been expressed in his approval of the agreement: "Power that controls the economy should be in the hands of elected representatives of the people, not in the hands of an industrial oligarchy." Many communications people, however, feared service by numerous companies would be as disastrous as many railroads with tracks of different widths, and trains operating by different rules.

Fierce Competition by RHCs and AT&T

At Congressional hearings the RHCs repeatedly urged the passage of a law removing Judge Greene's restrictions, and allowing them to provide the content as well as the facilities for services and to manufacture telecommunications equipment. Naturally AT&T opposed such a law and also RHC invasions of its long distance, international message, and computer/data business. The RHCs were unhappy because some large companies leased or built private wire systems bypassing their lines and exchanges. For example, Westinghouse established its own microwave system linking its twenty-two buildings in the Pittsburgh area.

NYNEX and U.S. Sprint showed international ambitions by obtaining FCC approval to buy a stake in transatlantic optic-fiber cables. Southern New England Telephone bought equipment from GTE. The RHCs sought FCC and Greene's approval to extend their telephone and data networks into other regions and shopped among AT&T's competitors to lease long distance circuits.

RHCs also invaded each other's territories to sell equipment, mobile phones, and information services. They claimed Judge Greene's restrictions were tying their hands but increased their rates about fifty percent, and their revenues and earnings set new records. A federal appeals court prohibited the FCC from ordering refunds when RHC earnings exceeded permitted profit margins.

The RHCs bought control of cellular and radio paging franchises that provided services in each others areas. Ameritech and U.S. West led the way in publishing Yellow Page directories, used more than Bibles. Southwestern Bell even teamed up with Dun & Bradstreet's Donnelly Directories Co. to publish them on the East Coast in competition with Bell South. Because directories were very profitable, the RHCs ignored criticism by the media, unhappy with their competition for the advertising dollar. Some RHCs published them for large cities in other RHC regions. The Yellow Page directory business became a battleground, on which Bell South led with one billion dollars in directory revenues, followed by other RHCs, GTE, and United Telecommunications.

Americans had become highly mobile by 1974 and the FCC awarded air space for mobile cellular use in thirty large cities to AT&T and others. Radio calls were established for about 100,000 vehicles with transmitters and receivers serving small areas, and calls were interconnected with the worldwide telephone network.

Cable TV boomed, and in 1983 General Electric did a billion-dollar business in cellular radios. Soon nearly half a million American homes subscribed to cable TV companies that became the darlings of Wall Street. In 1984 the RHCs and United Telecommunications, a large telephone and a computer holding company, also bought control of many cable companies, competed in providing optic-fiber cables underground along rights of way in nineteen states, and planned a 22,000-mile network.

Judge Greene ordered changes in AT&T's, the RHCs', and the independents' operations, rates, and services, resulting in overregulation. Under the 1974 Tunney Act, a judge could determine whether a settlement was "in the best public interest," but if the law did not say precisely what that meant, the judge could interpret it to mean almost anything he wished.

Despite pleas by AT&T, Judge Greene allowed the operating companies to provide home and business equipment and refused to permit the combination voice and telegraph services FCC had approved for Pacific Telesis and Bell Atlantic. Sometimes the RHCs called the denials a blow to consumers and Greene forcefully renewed the court's authority, berating the FCC for not approving services. Eventually Congress will learn two rulers are too many.

Judge Greene finally permitted RHCs to provide electronic mail with computers relaying recorded voice messages, and they also engaged in computer software and foreign ventures. The Pacific Telesis Group, with $16 billion in assets, 79,000 employees, two million shareholders, and 2,000 high-technology customer companies in Silicon Valley, was a leader in fiber optics and information technology. More than sixty percent of its access lines were serviced by software-controlled, computerized switches. It was strong also in cellular and paging services.

Instead of favoring Ma Bell, her children bought equipment from companies like Northern Telecom, Digital Data, Hewlett Packard, TIE Communications, Memorex, International N.V., NEC (Japanese), and others, mixing it with their familiar Western Electric equipment. They also bought computers from Wang and other AT&T competitors. ITT proudly advertised it sold digital technology equipment to Bell South.[17]

Entering the telephone and computer repair and maintenance business, Bell Atlantic advertised it provided "the world's largest independent depot repair service for Digital Equipment Corporation computers and peripherals, and serviced 3,000 DEC, Amdahl, and IBM products." The federal government was Bell Atlantic's largest maintenance customer. Pacific Telesis, Southwestern Bell, and GTE led in cable TV and cellular mobile service.

The Bell operating companies formed alliances with independents that provided local access for long distance calls. New England Telephone and CSX Railroad planned a 20-state network. Bell Atlantic proposed a service that would display the caller's name and number while the bell was ringing, and it was attacked as an invasion of privacy. It also installed a dual system for Telefonica, Spain's national telephone company. Southwestern Bell joined a Japanese venture to design and install data systems.

NYNEX, aggressive in promoting computer networks, established a chain of ten centers with 1,500 hardware and software items to meet customers' needs, regardless of whether they used IBM, Apple, COM-

PAQ, Digital, Tandy, or other computers. NYNEX's engineers solved artificial-information and speech-synthesis problems for customers. Restricted to its own northeastern-states area, but wanting to go nationwide, in 1986 NYNEX acquired IBM's product sales centers. Outbidding AT&T and MCI, NYNEX won a Security Industry Association contract to provide an optical-fiber network linking thousands of brokerages in the New York City area.

When Judge Greene relaxed his rules on information services, Southwestern added them at once. NYNEX agreed to buy AGS Computer's professional services and software products business to increase its financial services. It expanded across the country with offices in major cities.

Ameritech installed "teller machines" at airports, shopping malls, and street corners in the midwest, enabling people to push buttons and withdraw money from their bank accounts, transfer funds, and pay bills. It provided data systems for brokers, banks and securities markets and the health care industry. It also built a private cellular network in Argentina. Bell South International (BSI) was created as a subsidiary to compete with Japanese and ITT international businesses, and was a partner in a consortium to build and operate, in Argentina, South America's first cellular network. BSI also was a consultant to India and Guatemala, agreed to modernize Spain's telephone system, and bought paging companies in Australia.

Separately from Bell South, NYNEX contracted to provide a monitoring system for Shanghai telephones. Pacific Telesis and NYNEX planned digital, fiber-optic cables for Europe and Japan, but lost a chance to enter the transatlantic telephone business when Britain's Cable and Wireless PLC withdrew from the venture. NYNEX invest-ed more than $1.7 billion in improving its local exchange network in 1984, expanded its optic-fiber network by enough miles to circle the earth, and installed digital switching on customer lines. It also tested a service for surgeons and engineers to hold video-voice discussions over its optic-fiber cables.

Bell South, U.S. West, and the U.S. Air Force bought large central office switching centers from Northern Telecom of Canada, a leader in digital technology and the second largest manufacturer of communications equipment in North America. One of Telecom's products was mail-voice message systems. Wisconsin Bell bought central office electronic switches from Siemens. Naturally, many AT&T people who were transferred to the RHCs felt some loyalty to Ma Bell and did not favor their strong-arm tactics in seizing AT&T business.

With so many companies interconnected, the telephone network was in effect a giant computer in which each customer's needs were programmed on a memory core to provide automatic, customer-tailored service.[18]

Industry leaders had long advocated a national policy, and decisions after the AT&T breakup showed some progress toward a combination of voice and digital telegraph data, wide-open competition between numerous companies, both narrow and broadband operation, and new services to save time and effort for the individuals, and speed to information and decisions for business.

AT&T's divestiture had increased competition, and reduced its assets and employees, but Judge Greene, the FCC, and Congress still seemed determined to prevent its regaining much of its former power. The public had been persuaded that competition by many companies could provide good service. But those who underrated Ma Bell were mistaken; she arose from her wheelchair, leaped into the battle, and gave her strong young competitors, U.S. Sprint and MCI, a great fight. In the decade after AT&T's divestment of the RHCs, competition from Sprint and MCI reduced its share of the long distance business from eighty-two to sixty-nine percent.

The "Baby Bells" went far beyond the areas assigned by Judge Green, with enterprises too numerous to list, but the following summary is suggestive.

Most of them established cellular and cable-TV services in foreign countries, worldwide. Ameritech and Atlantic Bell each bought half of the New Zealand telephone company. Bell South had paging and Yellow Page publishing companies, and sold tele-communications equipment. Pacific Telesis had cellular systems in Germany and Thailand and a credit-card system in Korea. U.S. West had cable interests in Britain and France and engaged in real estate developing, as NYNEX did.

For two years after their birth, Judge Green unhappily watched the Baby Bells expand beyond the areas he assigned. Then he exploded in an opinion blasting them "for their ascent into the ranks of conglomerate America, rated far higher on their list of priorities than the provision of the best and least costly local telephone service." Though he was appointed to preside over a suit to enforce the antitrust law, Judge Green assumed authority over telecommunications policy, equipment, services, and joint enterprises. Only Congress can change that.

Notes

[1]Since I retired before Western Union and AT&T were disrupted, this chapter's principal sources were the *Wall Street Journal*; documents provided under the Freedom of Information Act by the U.S. Justice Department; Steve Colls, *The Breakup of AT&T* (New York: Athaneum, 1986); and correspondence, company news releases, and stockholder reports. The familiar yellow blank and its uneconomical messenger delivery to homes was fading into American memory, replaced by services to business.

[2]Source: personal knowledge. I prepared statements for hearings and news releases, and attended hearings in Washington.

[3]See chap. 20, above.

[4]I developed the first release wire to transmit rush news releases to meet deadlines. The idea of wider use began in the 1930s when some friends and I formed the National Association of Public Relations Counselors. As the board member with ready access to telegraph service, I arranged with Western Union superintendents in Detroit, Chicago, Atlanta, San Francisco, and Los Angeles to form the first local chapters of NAPRC. The NAPRC then merged with a smaller public relations society in San Francisco to form the now big Public Relations Society of America.

[5]Western Union report for 1980.

[6]Telex Corporation, Tulsa, Oklahoma, maker of intercommunications equipment and telephone systems, was not involved.

[7]The *Wall Street Journal,* August 14, 1989, said that since 1985 LeBow "has been buying into troubled companies and trying to turn them around. Mr. LeBow has succeeded in two cases and has made millions for himself and other insiders. But the public shareholders have gotten poorer."

[8]For example, I found the historic dining room chairs and table around which Cyrus W. Field and associates organized the company to lay the first transatlantic cable. I obtained the descendants' approval, dug the set out of the hay loft of their barn in Connecticut, and trucked it to the museum in New York.

[9]This was where I started in the industry.

[10]*Forbes Magazine* (February 15, 1974).

[11]John Brooks, *Telephone.*

[12]WATS (wide area telephone service) enabled companies to list their numbers in other cities and permit distant customers to use an 800 number to call them toll free. The company also could use its 800 number to call others.

[13]Data provided by AT&T Financial Media Relations.

[14]*U.S. News & World Report* (October 5, 1981).

[15]AT&T 1982 *Annual Report,* 5.

[16]Information regarding final agreement, modifications, and review of the case as provided by the Justice Department.

[17]ITT became a large hotel owner in 1992, and left the communications business except for a thirty percent interest in Alcatel N.V., a European communications equipment maker. *Wall Street Journal,* February 21, 1992.

[18]One of the first electronic switching systems was installed by Bell Telephone of Canada at "Expo 67" in Montreal. The five-million-dollar system had central control, temporary memory and permanent memory. It was said to represent a "breakthrough" in communications technology and resulted from thousands of man-years of research and development.

In 1967 also, ITT's first integrated circuit and computer-controlled electronic telephone switching system was installed in Belgium to handle telephone and Telex switching.

Divestment Sparks Revolutionary Change

In 513 B.C. a philosopher named Heraclitus wrote, "Nothing is permanent except change." That certainly was true of telecommunications in the 1980s.

During the 1980s, Congress dismembered the world's finest communications system. Hundreds of smaller companies were born and the fiercely competing newcomers produced such radical changes in telephone and data equipment, operations, service, and costs that the lives of the industry's millions of workers, stockholders, and everyone as customers will never be the same.

D-year 1984, the year of divestment, opened the door to so many revolutionary technological developments and uses that both business and residential customers were confused. Millions began using beepers, mobile telephones, and cable TV. Thousands of new computer data networks linked companies in the bank, broker, credit, retail, law, hospital, hotel, travel, transportation, and other industries. Such systems enabled business to provide faster and more efficient management, operation, and service. Domes-

tic and global trade and competition grew rapidly, and problems multiplied.

Divestiture, restructuring, reorganization, and refinancing were only the first steps in AT&T's procession of changes. A large part of telephone company business became written-record telegraphy. AT&T divided its organization into two areas: long distance services and technologies to provide electronic digital data information networks.

Bell Laboratories inventions helped all communications companies compete. Many merged, bought an interest in others, or joined in ventures with foreign companies. ITT divested its $1.7 billion private wire and interconnect businesses and sold its international Telex, but retained its electronic mail, news terminal, telephone, switching equipment, insurance, financial services, and high technology companies. So many companies made major changes or merged that desperation became a way of life for a million employees whose jobs were threatened or lost. AT&T's 3,100,000 shareowners and those of the other companies also worried.

AT&T's divestiture cleared the way for services delayed for years by the decade-long antitrust suit. AT&T had developed two cellular mobile telephones in the early 1970s but was busy fighting for its life while others blanketed the nation with them. The FCC stalled off action on AT&T's application for wave space for the service until 1977, and delayed until 1980 a service to record and store telephone messages until the person called was available.

The 1,600 independent telephone companies in the United States all connect with the Bell companies. On long distance calls it is not unusual for three companies to share the "haul" and the tolls. Many independents have electronic switching central offices which convert a telephone network in effect into a giant computer in which each customer's special needs are programmed on a memory core to provide automatic, custom-tailored service.

Birth of numerous competitors produced operating problems. Video-telephone conferencing, for example, was blocked by differences between technical standards of companies. Machines of a company carrying 1,500,000 bits of information per second could not talk with those of companies with other standards.[1] Even Bell companies had to make changes in installation, maintenance, call processing, and network planning to connect with others. However, AT&T, Datapoint, and others unveiled a desktop picture telephone in 1986, and that may provide the spark to popularize video conferencing.

Major Competitors

ITT, one of the world's largest telephone equipment companies, had a seventeen-billion-dollar business with 150 affiliated telecommunications and electronic manufac-turing companies, and more than 200,000 employees in more than eighty countries. It had a large role in worldwide commerce and trade. AT&T bought one billion dollars of ITT's digital switching systems and other equipment to settle an antitrust suit claim that it only used Western Electric products and would not buy from other suppliers.

As we have seen, ITT was founded by the Behn brothers in 1920 and became the foreign equivalent of AT&T. It bought overseas manufacturing interests in 1925, and diversified into a huge conglomerate of many companies under Harold Geneen's leadership beginning in 1957. Geneen was succeeded in 1980 by Rand V. Araskog, who fought off corporate raiders. In 1981 ITT's engineers developed a spread-spectrum technology in which each message was chopped into small segments and transmitted over a number of radio channels for defense communications, so eavesdroppers would only hear static.

Only three percent of ITT earnings then were from communications. Because most of its operations were in other countries, and one merged with French government-owned CCE, many Americans regarded ITT as European, while Europeans considered it American.

In 1960 ITT and Bell and Howell demonstrated a system to transmit mail by facsimile in the same way Western Union's Desk-Fax sent telegrams between telegraph and business offices. In 1967, President Lyndon Johnson appointed retired AT&T President Frederick R. Kappell to study the Post Office Department operation, and Congress then authorized the establishment of an electronic-originated facsimile postal service. Disappointed because the service was expected to save labor costs by handling most of the mail service but had handled only 25 million letters, the U.S. Postal Service cancelled the facsimile service.

A large amount of ITT's income came from its telephone switches and, in one decade, it spent one billion dollars to develop a new switch system for its worldwide sales. The breakup of the Bell System gave ITT an opening to take over the American switch business, and it launched a 105-million-dollar drive in 1984 to sell its System 12 switch in America and Europe. It was a bold gamble because a large sale was needed to pay the heavy costs.[2]

In 1985 ITT agreed to acquire the American Broadcasting Company TV and radio networks, but the Justice Department feared it would use programs to promote political and economic deals with foreign governments, and ITT terminated the agreement.

Another worldwide equipment supplier was Siemens AG, the Munich, West Germany-based telecommunications equipment giant with manufacturing and assembly plants and offices at 350 places in the United States. Siemens won contracts for big U.S. Air Force network switching centers and installed private and central office switching systems for Western Union and other companies.

Siemens established an integrated voice/data information management system linking its coast-to-coast American offices, and provided cellular and other systems through Tel Plus Communications, its American equipment dealer subsidiary. In 1984 IBM bought Rolm, a leading supplier of PBXs, but in 1988 Siemens offered to buy Rolm for more than one billion dollars to become one of the largest suppliers of PBX exchanges.

With more than $31 billion revenues and $3.1 billion income, Siemens was a competitor with AT&T. It bought an interest in GTE's international telephone business in 1985 for $420 million, and bought control of several telephone switch and equipment companies.

The telephone switch business from small companies up to 150 employees was estimated at $2.1 billion. Siemens developed a hybrid cross between a PBX used by an employee to switch calls and an automatic section switch to do so.

In the 1980s various telecommunications companies scattered from Europe to the Far East, with about $73 billion in global sales, formed an alliance that in the future could lead the entire industry. Only AT&T and Alcatel N.V. had more equipment business than Siemens which showed an interest in allying with AT&T.

In 1987, the Department of Commerce feared that joint ventures of American carriers with German, Japanese, and other foreign companies would deprive the United States of the capacity to produce central office digital switches. It noted that U.S. Sprint was running out of switches to operate its lines, and AT&T and Northern Telecom of Canada produced seventy percent of the $3.7 billion of central switches in America. Northern Telecom was partly owned by Western Electric, and provided many of AT&T's switches. The Carterfone decision had given a boost to Digital Switch, an American maker of sophisticated large switches for the long distance companies, but MITEL, NEC, and Fujitsu made more switches.[3]

Earnings from cellular telephone, real estate, and Yellow Page Directory businesses doubled regional holding companies' (RHCs) incomes. The independents had lost their substantial revenue from AT&T's long distance tolls, and tried to replace them by merging, cooperating with others, or duplicating AT&T and RHC services at cut rates. Like the RHCs, AT&T's 300 long distance competitors were delighted to have less government regulation.

Before the AT&T divestiture, the largest independent was General Telephone and Electronics (GTE), serving all states, British Columbia, Quebec, and the Dominican Republic. Started in 1883 as Mutual Telephone Co., now Hawaiian Telephone Co., it acquired several other companies including Almon B. Strowger's company that owned his 1891 patent for the automatic dial telephone. With the Richland Center Telephone Co., it formed the General Telephone Company in 1918. In 1926 it was named the Associated Telephone Company, and reorganized a decade later as GTE.

A leader in optical fiber technology, GTE installed a 2,400-telephone-circuit optical-fiber system for the Belgian government that required switching devices every six and one-half miles with lasers to provide the light beams. Two optical fibers the size of a violin string were said to have the capacity to carry all of the telephone calls in New York City.

In 1987 GTE had $15.4 billion revenues, $1.1 billion net income and large electrical communications product sales, but its largest business was fifteen telephone companies. Its mobile cellular telephone franchises served twenty-one million people. It offered a variety of terminal and transmission equipment, electronic, microwave, satellite earth-station, and data-processing systems. GTE's twenty-five-plant equipment business competed with AT&T's packet switching, time sharing channel systems developed for the Armed Services. In time sharing, short segments of a message, or packets, were sent interleaved with other packets, and directed to their destinations by computers at each switching center.

GTE bought the Telenet data network linking 250 cities and forty countries.[4] Telenet's electronic mail via 200 switching points produced a large profit. Backed by Time Inc., Bessemer Securities, and the Palmer Organization, Telenet's founders had dreamed of becoming the businessman's telephone system in the 1980s by using Bell System facilities and providing packet service. Although it had large customers like General Motors and Hughes Aircraft, Telenet lost millions when GTE took it over for $55 million. As a GTE subsidiary, Telenet provided unregulated data services to business and teamed with Siemens in providing equipment to the U.S. Army and others. GTE profited handsomely from advertising in 1500 telephone directories it published in nine languages.

The next largest independent was U.S. Sprint, a long distance voice, video, and data telecommunications carrier, formed on June 15, 1983 in a 50-50 partnership of GTE and United Telecommunications (UT) of Kansas City. GTE had bought a Southern Pacific Railroad's long distance telephone subsidiary named Sprint for $150 million and renamed it U.S. Sprint. That was a brilliant deal because Sprint owned Spacenet, and GTE did not have to start its long distance service from scratch. Also included in the deal was Southern Pacific Satellite Company, a subsidiary that owned two space satellites, cable TV, electrical equipment and dial-up stock-data broker service.

U.S. Sprint spent three billion dollars in three years, and installed the first all-digital optical-fiber network in the nation. AT&T, deregulated and modernizing, was tough competition, however, and GTE's long distance business also suffered heavy losses. GTE then sold a majority of its interest in

U.S. Sprint to United Telecommunications for $600 million in 1985, wrote off a loss of $175 million, and formed a joint venture with AT&T in the digital network switching business.

GTE sold its remaining shares in U.S. Sprint to United Telecommunications in 1988. U.S. Sprint provided voice and video service to 150 countries and data transmission to ninety, using channels in an optic-fiber transatlantic cable. GTE then acquired Contel in the largest merger in telecommunications history for $6.6 billion in 1992.

U.S. Sprint and MCI (Microwave Communications, Inc.) continued as AT&T competitors. Before the sale, U.S. Sprint inaugurated a "FONECARD" that customers could use to make collect long distance calls, and customers of 800-service businesses to make free long distance calls to them. The 800 calls grew and generated large revenues, as did telephone and data equipment, cellular, and paging services.

McGowan maneuvered MCI into second place in long distance revenues in 1988, and himself into a fortune. (AT&T was first with 76.85 percent, MCI 9.78 percent, U.S. Sprint 7.84 percent, and United Telecom 1.02 percent.) MCI had been initially unprofitable because of the high cost of leasing AT&T circuits and trying to match its technology and services, but its owners poured more millions into it. After a dozen years in the computer industry, Bert C. Roberts, Jr. joined MCI in 1972 and became its president and CEO in 1986, determined to double its business. MCI announced in 1987 that it would pay $160 million to buy RCA's old Global Communications, Inc. which had Telex and data circuits to 200 countries.

After an absence because of a heart attack in 1987, McGowan shared the chief executive power with V. Orville Wright, who had returned as vice chairman. MCI's business grew, aided by Judge Greene's decisions and FCC rulings that enabled it to lease Bell lines at bargain rates. Earnings surged and MCI announced plans to buy $699 million of its stock back from IBM, its largest stockholder.

William Esrey, president of United Telecommunications Inc., based in Kansas City, Missouri, boldly committed United Telecom's three-billion-dollar revenue company to a two-billion-dollar investment in a 23,000-mile optic-fiber cable network to be completed by 1988. His purpose was to compete for the $47 billion long distance telephone business, and the result was lower cost, improved service, increased business, and its fiber network in 1988 was to be larger than any other along the East Coast, through Chicago and parts of the South.[5] The company faced tough competition because of AT&T's 11,000-mile fiber network and rate reductions. Acquiring Atlanta-based Insurance Systems of America, a holding company for telephone and computer service firms, for $41.5 million, United Telecommunications broadened its markets and prospered.

After the divestiture, AT&T and the independents had to pay access fees to the RHCs for long distance calls, and often the RHCs shared the "haul" and the tolls. The FCC set a fee of several hundred dollars for each interconnected line, providing a very substantial increase to RHC incomes.

Still another competitor was Chicago-based Allnet that built no lines but leased circuits from the large carriers, and established a national network while AT&T was fighting divestment. After three years of undercutting MCI and Sprint rates, Allnet had $143 million in revenues and $5 million in earnings in 1983. Its cut-rate policy was succeeding, but the requirement a few years later of uniform rates by all carriers deprived Allnet of that advantage.

An integrated services digital network developed by Rolm Services Division of IBM enabled telephone lines to carry data, voice, and TV video simultaneously in a stream of one-zero signals called "bits." Telephone people dreamed of using it in huge networks to meet many personal, business, and government needs.

To increase equipment sales abroad, AT&T formed a 50-50 European partnership with N.V. Philips, the European telecommunications and electronics giant, to make telephone exchanges and transmission equipment. After four years AT&T gained control of the venture by buying another ten percent. To double its sales of data communications equipment, in 1988 AT&T acquired Paradyne Corp. for $150 million. It also obtained a toehold in Europe by agreeing to acquire a twenty percent stake in Ilatel, owned by Italy. Ilatel's parent, Stet SpA was to acquire twenty percent of AT&T's International Network Systems.

In 1983 AT&T bought twenty-five percent interest in Ing C. Olivetti of Italy, and began selling its products in the United States. AT&T hired an Olivetti official to head its computer/data business, but he was replaced by a man familiar with American policies, and the Olivetti alliance vanished.

Satellite Business Systems (SBS) entered the leased private wire system competition in 1981, providing service with a satellite/data system designed primarily for large companies. SBS had been organized in 1975 by IBM, COMSAT, and Aetna Life and Casualty with an investment of $600 million.

Chicago-based Centel, one of the largest independents, a leader in digital switching and fiber optics, was started by an electrician who bought small telephone, electric, and gas companies and named the combination Century Telephone and Utilities Corp. A veteran AT&T operations official, Robert P. Reuss, became chairman and CEO when it was a half-century old, renamed it Centel in 1972, and restructured and diversified its business. He had combined about 200 small telephone companies in 1987. Centel had 1,300,000 customer lines in ten states and became a big supplier of business telephone and intercom business and data systems, financial and electric services. Recognizing cable TV as a billion-dollar wave of the future, Centel bought a number of cable TV companies—as did Teleprompter, AT&T, Time, and Warner Communications—put together six cable TV companies, and agreed to sell them for more than $1.4 billion in 1989. A subsidiary, Centel Cellular, with Honeywell and Field Enterprises as partners, provided an electronic service that enabled cable customers to receive stock quotations and other information on their TV screens.

Like Centel, Contel was a large, growing independent. Originally named Continental Telephone, it had 1,492,189 telephones, more than one billion dollars in assets, and was the third largest non-Bell company in 1969. In 1986, Contel had 2,500,000 telephone customers in thirty states and Canada, five billion dollars in assets and three billion dollars in sales. Contel was a pioneer in digital switching and provided extensive data processing and mobile cellular services. In 1984 Contel joined with France Cables and Radio, Alltel Telephone, and California Microwave in organizing AGRO to "end long distance service by bouncing signals off a satellite," for twenty-six cents a minute to all places, "because all cities are the same distance from a satellite."[6]

In 1987 AGRO merged with COMSAT

to compete in international communications via satellite. As mentioned earlier, Contel bought Western Union's national defense division in 1986 for $155 million and provided data networks to the United States and other countries.

Another competitor was Telephone and Data Systems (TDS) based in Chicago. In 1985 TDS operated sixty telephone companies serving 200,000 customers in twenty-two states, radio paging service in twenty-eight cities, and cellular mobile telephone service in the ninety largest cities. It was affiliated with and served other companies, and had $119 million revenues.

A fast growing competitor was Alltel with many telephone lines from New England to Texas. Alltel acquired other companies with digital switch operations, optic-fiber cables, data systems, mobile cellular, and equipment supply services. Telephones with push buttons instead of dials had been introduced in 1963, 800 service in 1965, picture photo in 1970, and cellular service in 1978.

Other equipment and long distance companies were organized in the 1980s to provide special information for an industry or profession. For example, MACOM provided legal and medical data networks. Some long distance companies quit after learning how difficult it was to compete with AT&T and the RHCs. In 1984 Ford Motor and American Network bought a seventy percent stake in Starnet, owner of five small long distance companies, but after two years Ford sold its interest to American Network.

After AT&T's divestment, the products of its principal competitors for the growing electronic switch business ranged from small PBXs, to direct incoming calls to a company's offices, to computer central office digital switches that route calls over telephone networks. Leaders in the four-billion-dollar switch business in the United States were AT&T, Northern Telecom, Rolm, NEC, and Siemens. Others were United Technologies, Republic Telecom Systems, Plessy of Great Britain, ITT, Alcatel of Italy,[7] Fujitsu, Iwatsu of Japan, Ericsson of Europe, and TIE Communications. Northern Telecom introduced the digital switch that provided call forwarding, call waiting beeps, and other services.

AT&T long lines still carried the network programs from radio and television studios to broadcasting stations, and a major share of international communications. It also leased domestic, satellite, and international circuits to other carriers. The other big wholesaler of international circuits was COMSAT, but AT&T leased seventy-five percent of COMSAT's satellite circuits. In 1978 there were 360 million telephones and 364 million TV sets in the world.

In 1984 Bell Laboratories invented the optical fiber that changed the world's wire and cable equipment. Harold Geneen of ITT said at the 1976 annual meeting that pulse-code modulation (PCM, multichannel, encoded, pulsed transmission of data—see below) opened up the application of low-cost mass transportation of information. Judge Harold Greene ruled in 1989 that AT&T could enter the growing electronic publishing business in which PCM via fiber optics could be used in transmission of news. Bell engineers were delighted when a laser in an experiment pulsed an incredibly fast two billion times a second through eighty-one miles of glass fiber, reducing the number of boosters needed in long distance light wave communications. That was especially important to undersea cables. It was reminiscent of Western Union's lightbeam invention and transmissions between New York buildings two generations earlier. AT&T announced it would lay an optic-fiber cable to England and France.

Despite all efforts to curb it, AT&T remained a worldwide symbol of American free enterprise, capitalism, and technological progress. Eventually, the Commerce Department decided there was enough competition, and advocated deregulation and freedom for AT&T and the RHCs.

The RHCs and independents bought billions of dollars of big digital telephone systems and other equipment from General Dynamics and other American, West German, and Japanese companies instead of Western Electric. That caused AT&T to drop 321,000 employees, close huge Western Electric plants, and sell equipment in Japan's six-billion-dollar telecommunications market. With the RHCs, U.S. Air Force, and independents buying central office switching systems from foreign companies, AT&T announced in 1988 it would join GTE in developing and marketing a new generation of switches.

As worldwide business growth demanded more facilities, AT&T established twenty-four offices in cities around the globe to sell its products in ninety countries. It sold eighteen million shares of stock and raised rates to finance global business, trying to recoup the $1.9 billion its divestment had cost.

Changes Confuse Customers

For years telephone customers were losers from the breakup of the world's most efficient communications service. Once a rarity, complaints became commonplace, often because of long distance noise or call delays. Noise resulted from the use of cheap foreign telephones and calls passing through the equipment of two or more local companies.

Changing long-established habits was confusing to customers. For many older people, the telephone was their major contact with the outside world and a lifesaver in emergencies. When asked if they wanted a "pulse" or "magnetic field" telephone, most did not know rotary dial telephones send a series of pulses that select the numbers, or that magnetic field pushbutton telephones send musical tones or beeps and can magnify sound for the hard of hearing. People also were asked if they wanted call waiting, call forwarding, three-speed calling, three-way calling, nine-number memory of frequently called numbers, last number redial, phone call screening to eliminate unwanted sales calls, or speed-a-code service, at an extra two to seven and one-half dollars a month. Did they want a portable or stationary phone? What color and style? Alexander Graham Bell would not recognize the fancy new instruments as telephones.

Companies like Telephone and Data Systems with 235,000 customers in twenty-eight states offered radio paging, cellular mobile radio telephone, and cable TV. Many companies claimed their services were best and cheapest. No wonder people were confused!

Service deteriorated because of confusion, snafus, regulatory and union requirements, and restrictive laws introduced by politicians to please groups of constituents. Homes and businesses with special problems experienced installation delays, and high installation and repair costs. Since communication is the lifeblood of stockbrokers, delays were nerve-wracking; one brokerage had to delay the opening of forty offices. Delays often resulted until services or operating facilities of long distance companies were coordinated with local companies.

Some financial firms with big private

wire systems left AT&T. Merrill Lynch, Chase Manhattan Bank, and Drexel, Burnham, Lambert went to MCI, and Sears Roebuck and Dean Witter went to U.S. Sprint, then a unit of United Telecommunications.

Fighting to retain its technological leadership, AT&T made major advances. It replaced electromechanical and analog technologies with digital-integrated circuits, using pulse-code modulation (PCM) to convert speech samples into digital codes. PCM samples a speaker's voice many times per second, encodes it into binary digital form as a series of pulses, and sends them to the destination where they are decoded and turned back into continuous analog speech. The signals are sent sequentially, interlacing the codes of many different voices. Each signal occupies the entire bandwidth for only several billionths of a second.[8]

ITT invented a technique for enabling a light beam to carry 6,000 PCM channels. Since most telegraph traffic consists of binary digital data, Western Union used PCM methods for data transmission. The data/information explosion demanded a redoubling of telegraph channels every few years. Two scientific marvels invented to help meet such demands were maser (*m*icrowave *a*mplification by *s*timulated *e*mission of *r*adiation) and laser (*l*ight *a*mplification by *s*timulated *e*mission of *r*adiation). Like the World War II point-source light created by Buckingham of Western Union, referred to earlier, the laser projects a narrow intense beam of coherent light in a single channel.

ITT's QUAME company developed a "Laser Ten" business printer in 1985 that it said would compete with all business computers and word processors when plugged in with interface modules. Hewlett-Packard became the leader in laser printers, and IBM introduced a laser printer.

To remedy long distance delays and noise level problems, AT&T created a five-class network and replaced its electronic switches with "smart digital switches" that routed calls over the best and most economical routes available. The average number of phones per home increased to two, and daily conversations to one and three-tenths *billion*. AT&T was so busy with problems it did not hold a big celebration of its hundredth anniversary on March 3, 1985. However, it did pay $3,000 a day for two days to Don Ameche, then seventy-three and still jokingly called Mr. Bell because he had played the part in a movie biography of Alexander Graham Bell in 1939. Ameche appeared at a dinner in New York to start a year-long promotional campaign in which he gave inspirational talks to AT&T pensioners.

Subscribers had to choose a long distance company and pay three dollars monthly for access to it. Fear of change, and memory of the good old days when one-company service was as simple and easy to use as apple pie, made AT&T the favorite selection.

The FCC had ruled June 1, 1983 that telephones must be sold or leased to subscribers. Homeowners were baffled by notices to buy their home wiring and telephones they thought they already owned, and pay monthly for repair insurance, telephone attachments, directory assistance, city franchises, and so forth. When a family or company moved, it had to remove its home telephone equipment and return it to the company, or pay for it.

About 500 American companies made telephones, but many people bought telephones made in foreign countries where labor costs made prices low, and returned millions of old phones to AT&T. Of course, instead of paying two dollars a month to rent the phone they could have saved money by buying it. Many residential telephone manufacturers suffered heavy losses. The major remaining makers and their percentages of the business were AT&T with twenty-one percent; Tandy, fourteen percent; Dynascan, eleven percent; Uniden, ten percent; ITT, eight percent; Comdial, seven percent; GTE, six percent; and others, twenty-three percent.[9]

People calling from hotels, hospitals, or airports were enraged by high surcharges for their calls because of added operator-assistance charges. Independent companies also installed pay telephones in many hotels, dormitories, barracks, and hospitals, and charged a fifteen or twenty percent commission. For example, *Consumer Reports* reported:

> An Albany, N.Y. man called a friend 125 miles away after 9 P.M. on a pay phone. Three times there was no answer. He talked three minutes on the third try and the bill was $10.00: $1.00 for the call, plus $3 for each "no answer." A son called his home from an army base and in a month the home bill was $692, because the home telephone did not have access to the long distance company the son used. It cost more to call another city in Florida than to New York City because of Federal government regulations. One caller exclaimed, "At least Jesse James used a gun!"

Many people mistakenly thought the long distance company access charge of $3.50 a month was paid to AT&T. Even the chairman of the House Commerce Committee, California Congressman Dingell, mistakenly launched a bitter attack on AT&T for its "atrocious" access charges. AT&T did not receive the access charge; it was paid to the local telephone companies whose exchanges handled long distance calls. Month-

ly telephone bills soared. Their bills listed federal, county, and city taxes, franchises, long distance and information calls, leases, maintenance, "dial-a-joke," horoscope, time, weather, political good wishes, get well wishes, offtrack betting, and other services.

A personal identification service the RHCs offered for local calls showed the caller's number on a display screen and was denounced as an invasion of privacy. Stores automatically recorded credit card information for future sales efforts and resold the lists. The result was sales calls and mail advertising from salespeople who even knew the model of the car people were driving.

Various services were developed to try to pacify unhappy customers. When the doctor, dentist, lawyer, or businessman is out, important calls may arrive. Thousands of "live" telephone answering services (TAS) began operating over extensions of subscribers' office or residence lines. They gave caller's numbers or messages to their subscribers. Hundreds of answering services were provided by Western Union International, formerly the cable department of Western Union, now owned by Xerox and named WUI/TAS.

Incoming wide-area telephone service (Inward-WATS) permitted distant customers to call a company without cost by using an 800 number. Many companies also used wide-area telephone service (outward-WATS) to make an unlimited number of calls in a certain area.

One new service aroused protests: "Dial a Porn," offered by Sable Communications of California. A suit to stop it resulted in decisions by two federal appellate courts that telephone companies have the right to refuse such services. The U.S. Supreme Court then ruled that "indecent" but not "obscene" communications are protected by the First

Amendment, and telephone companies can treat its providers as lawbreakers.

Companies minimize customer waiting time by using recorded voices that say calls will be answered by the first idle operator. The calls were "stored" and automatically connected in the order received as soon as an attendant was available.

To make telephone service faster, AT&T spent $100 million developing an electronic switching system, without which it said expanded service could not be provided in later years. It had been tested at Morris, Illinois, improved, and placed in operation in the central office at Succasunna, New Jersey, on May 30, 1965. Hundreds of other electronic switching systems followed and the company planned to provide electronic switching nationwide by the year 2000, replacing electromechanical equipment at a cost of more than $12 billion. Electronic switching, also used in the GTE telephone network, operated in millionths of a second—12,000 times faster than former apparatus.

Many company headquarters had telephone in addition to telegraph tielines to their sales offices, plants, and warehouses to permit unlimited calling at a fixed monthly charge, and it aided coordination of business activities with business-machine data, teletypewriter, and facsimile transmission.

The public was not happy with the changed industry and escalating costs, but telephone officials said customers eventually would benefit. For years Americans were adapting to the changes. They missed the convenience of "one-stop shopping" for services. Local companies increased charges because of higher labor costs, and were criticized by the Consumer Federation of America. The RHCs complained that regulation was sluggish and splintered and a coherent policy was needed. Even Charles F.

Rule, head of the Justice Department antitrust division, expressed doubt in an address at the 1987 Brookings Institute Telecommunications Conference. He said divestment results had not all been positive and likely served to slow "technological progress of American telecommunications markets."

A new breed of poorly informed, indifferent operators was hired whose mumbled answers were hard to understand. AT&T spent six million dollars a year to educate its new kind of employee in basic reading and math to enable them to do their jobs. To remedy the loss of the famous "voice with a smile," the telephone companies replaced a large number of operators with a system of computers that completed all kinds of calls and gave well-worded instructions in a clear recorded voice. It reduced union labor costs. The recorded voice often said, "You must dial one" (and make a long distance toll call) to a home or business a short distance away. The nation was divided into local and transport areas (LATAs) within which a call was local; it was a long distance toll call if it went over the line into another LATA.

Use of RHC credit cards at pay phones resulted in most pay phone long distance calls being routed over AT&T lines. Apparently still determined to help MCI and hurt AT&T, Judge Greene ordered the RHCs to give charge card revenues to MCI and other competitors. That required all cards to be called in by the RHCs and separate charge systems established for all companies at a cost of millions of dollars. Newspaper headlines said, "Phone company does less, charges more, and pleads poverty." Like residential users, communications managers for large companies had to decide which hardware offered by hundreds of manufacturers to buy.

People who have telephone calling cards

should not give their numbers to strangers. If they do, they may receive bills for thousands of dollars in calls to foreign countries that they did not make. Some people posing as telephone company inspectors and others have obtained calling card numbers and sold them to large users, who make numerous expensive foreign calls in a few days.

Many people regarded Ma Bell as an old friend that had deserted them, and wished Congress would restore AT&T as a symbol of American leadership and high-quality, low-cost service, because calls then were handled from origin to destination over one compatible system, with no transfers.

If Americans were confused, foreign telephone officials were more so. They had to decide which of many new American telephone and telex companies to connect with. They feared chaos would result from the switching installations required. Government owned, they had no wish to deregulate and foster competition, and limited the number of foreign countries whose connections they would accept.

AT&T launched a massive sales campaign, and spent $200 million to launch its computer business. It transferred 28,000 employees to American Bell, which first offered the Advanced Information System—a shared-data network service in which computers "talk" with each other. It provided anything from a basic telephone to a computer network with keyboards, printers, data screens and message switching centers linked by an applications processor.

Although American Bell had a billion-dollar business, thousands of stockholders, and advertised, the public knew little about it. People dealt with their state telephone company and many did not know the name of their RHC like Ameritech, NYNEX, U.S. West and Pacific Telesis. Several telephone operators were asked, and none knew of an American Bell Company. Seeking identity, the RHCs considered eliminating the name "Bell" from their state companies.

American Bell's first job was to establish service between New York and Washington, D.C. as the first leg of a digital-optic-fiber cable to carry messages in light pulses. In 1988 AT&T began operating a transatlantic optic-fiber cable with live TV camera and voice transmitted by laser beam to audiences in New York, London, and Paris.

After the FCC freed MCI and Sprint from virtually all federal scrutiny in 1983, the RHCs could change rates and services and enter new markets. AT&T appealed for equal treatment and obtained deregulation. The industry was in a turmoil with hundreds of competitors bombarding the public with sales letters, calls and ads. The chairman of the Telecommunications Subcommittee in Congress led a drive to strengthen the operating companies and reduce AT&T's size again by separating Western Electric from it. Congress, however, declined to reopen that Pandora's box of trouble.

AT&T changed the old three-minute minimum charge for calls to one minute and announced that anyone with a facsimile transmitter could send (telegraph) a written message by telephone to anyone with a facsimile recorder. Other companies followed AT&T's lead. AT&T also offered a WATS discount whether transmission was by voice, fax, or data. Western Union advertised that anyone with a computer terminal on his desk could transmit a message by digital facsimile to as many as 999 locations at the same time and receive automatic confirmation of delivery. MCI also announced a facsimile-network service.

To enable their divisions to work as a

team without leasing a private network, some large American companies like General Motors, Boeing, McDonnell-Douglas, and Westinghouse established their own networks, using satellites and microwaves to bypass the long distance companies. Some sold their surplus capacity to others, bought or leased computer networks, and reached their local offices by connecting with the RHCs.

One of the largest private wire networks had been leased from Western Union for many years by Merrill Lynch, a stockbroker with hundreds of branch offices. After the AT&T divestment, Merrill Lynch established a $100-million network but had communications companies manage it. Charles Schwab & Co., other brokers, and Fidelity Investments, had large private networks. AT&T created a System Integration Division to sell and maintain private networks. Tandem- and packet-switching systems, databases, processors, carrier systems, and Centrex lines were used on networks handling large-volume traffic.

Access to private networks was provided by long distance companies and the RHCs. In effect that extended RHCs nationwide when long distance companies switched their calls to other RHCs. Ameritech soon provided connections to about 250 long distance companies, and such switching added about twenty-five percent to all RHC earnings. One example of this was the first integrated-services digital-data network (ISDN) provided by Bellcore and Illinois Bell to McDonald's (hamburgers) Chicago world headquarters in 1985.

To retain its leadership against aggressive MCI, U.S. Sprint, and GTE competition, AT&T offered a ten-to-fifty-percent long distance discount plan to any size business, on every direct-dialed, interstate, or interna-

tional call using its wide area service (WATS). MCI had experienced many frustrations because of erratic service, technical troubles, and angry customers. At times MCI was near failure, but it had government help, survived, and grew strong in the 1980s. MCI and U.S. Sprint provided strong competition and AT&T fought back with new services, lower rates, and more sales people.

After repeated appeals, Judge Greene freed the RHCs of regional limitations, and they could provide databases that customers could access by telephone. The RHCs also wanted to offer nationwide data systems for travel reservations, shopping, weather, stock quotations, security, alarm, banking, cable TV, patient monitoring, energy management, and information and bank systems with terminals in customers' homes and offices. And they wanted to make their own equipment. The judge ruled that the RHCs could only lease systems with which companies could provide reservations, banking, and other services. He did allow them to offer information services, voice mail, and selection of pay TV cable programs by telephone.

For example, Bell Atlantic advertised services ranging from international communications equipment and nationwide private wire systems to software, counseling, and telephone services in large states.

When AT&T's long distance system was being digitalized, Wang and Comp-Systems provided voice and telegraph data service over Western Union lines at higher speed than Western Union had dreamed. Instead of sending line by line on a revolving cylinder, whole pages were sent. Telephone companies soon provided voice mail through PBXs that recorded messages and relayed them to destination cities for telephone delivery when the addressees were available.[10] Among large makers of mail PBXs were

Northern Telecom, Rolm, AT&T, Intercom, Siemens, Mitel, and NEC of Japan.

Another PBX competitor was Alcatel, a synergy of four companies providing communications and information processing systems. The four provided PABX and key systems, computer support and maintenance, optical-fiber and cable systems, and transmission and switching products. Formerly the four companies had been units of ITT and Thomson CSF Operations.

A federal appeals court ruling allowed the RHCs to acquire small telephone companies outside their regions, but they bought few because it was politically dangerous. They diversified by acquiring unregulated cellular mobile telephone, paging, and cable TV companies.

In 1985 IBM announced it was buying MCI stock, indicating more competition for AT&T's long distance computer business; but IBM soon sold its MCI stock and the RHCs reached settlements with MCI to avoid suits based on alleged monopoly practices. The RHCs established more exchanges, and increased rates because a larger number of customers could be called.

Electronic Data Systems (EDS), a spectacular enterprise in computer/data technology, was founded in Dallas in 1962 by H. Ross Perot with $1,000 and no employees. Perot began serving customers, trained computer experts, and gained service contracts with NASA, New York City Welfare, Technological Services System, and General Motors Acceptance System. In twenty-five years EDS was a $4.9 billion worldwide business with net income of $323 million. Its largest job was managing the U.S. Army and Navy networks, but Perot's strong criticism of GM's management policies caused GM to buy his eleven million shares and control of EDS for more than $700 million. It made

Perot a billionaire. The agreement Perot signed to sell EDS resulted in a bitter lawsuit by General Motors because Perot continued carrying out one billion dollars in electronic data contracts for Medicaid and others. GM's purchase of EDS put it in the telecommunications business.

Another General Motors subsidiary, G.M. Hughes Electronics, was a leading satellite system producer and operator. In 1987 it acquired M/A-COM Telecommunications, Inc., a leader in small aperture terminal earth stations, and renamed it Hughes Network System. It installed private satellite networks for large companies.

Pulse Link, a company spawned by the data transport business, enabled people using their telephones to obtain access to home information like shopping, banking, news, sports, education, and Dow Jones Financial News. The service required a special computer and standard television set. Pulse Link introduced a line of desktop computers, assigned 800 employees to sell data networks, and built up two billion dollars in computer/data sales, but lost one billion dollars doing so. However, no company matched the astute technology, pricing, and sales strategy of IBM, king of the mainframe and personal computer world. AT&T reduced its computer force.

In 1991 Comcast, Time-Warner, and other cable TV and cellular telephone company owners, eager to invade the telephone business by linking their systems, asked the FCC to permit a test.

Another company that showed great promise was Telenet Communications Corp., the first FCC-approved commercial packet-switching system, a government-developed technique using minicomputers and software to break up data from customers' computers and terminals into small packets of electronic information, sending each to the fastest available circuits (satellite, microwave, or cable) then reassembling the packets for the users. Telenet, established in 1978, had an eighty-city network mainly over Bell System lines and seemed ripe for acquisition by AT&T or another communications giant.[11]

Companies require changes in a constantly changing world, and computer analogs of data provide management with guidance for decisions in making changes. Sometimes analysis shows the need to diversify but some companies delayed too long in acting because officials opposed policy changes, while others made the correct moves and prospered. Singer, the sewing machine company, went into the manufacture of electronic, aerospace, and military equipment; Slumburger, Sears Roebuck, and others also diversified successfully. Thousands of small enterprises specializing in a product or service for which they found a real need, prospered when large plants and labor costs were not needed. Some utilized information available in data bases to find what most people in an area needed and wanted. In the 1980s more venture capital than ever backed small enterprises.

To challenge Reuters Holdings PLC's near-monopoly communications service to foreign exchange traders, AT&T and Telerate, Inc. bought an interest in Global Transactions Services Co. AT&T sold its interest to Global, and Dow Jones bought ninety percent of Global Transactions and operated it with about 600 terminals around the world.[12]

Weary from years of struggle to save his company, AT&T Chairman Brown retired in 1986 and was succeeded by James E. Olson, a dynamic leader who had started work in the northwest clearing brush from pole lines and trash from manholes. College educated, he had climbed with bulldog tenacity through the AT&T ranks to be president, chief executive, and vice chairman in 1979. A determined man other officials were careful not to cross, he began turning the company's finances around by reducing forces, regaining the company's momentum and prestige. After only one-and-a-half years, however, he underwent an operation for colon cancer and died April 18, 1988.

To succeed him, AT&T elected Robert E. Allen, fifty-three, who had acted for Olson during his illness. Allen had joined Indiana Bell in 1957 and moved through various telephone management positions until 1965 when he entered Harvard Business School for Management Development. In 1969 he became general commercial manager, then vice president, rates and revenues. He was president of Pennsylvania Bell and Chesapeake and Ohio Bell before becoming a vice president of AT&T in 1978 and president in 1981.

Allen, more sophisticated, reserved and low key than Olson, announced that he would follow the policy set by Olson, who said:

> Our focus is on three strategic priorities. The first is to strengthen and enhance our core businesses—long distance service, network telecommunications equipment for telephone companies, and equipment for business and home use.

Allen planned to develop new computer networking capabilities and lay transatlantic optical-fiber cables.

Impressed by American telecommunications advances, the European Common Market planned in 1985 to modernize its communications over a ten-year period at a cost of $200 billion. It decided to build a multination high-tech broadband network, using optic-fiber cables, satellites, and digital switches to meet such needs as electronic banking and credit verification.

In America the cost of replacing old wires and cables with optic fibers, installing a new digital data network, scrapping outdated equipment, competing with IBM, and retiring thousands of employees was a heavy drain on AT&T's income. It cut expenses but had its first annual loss in 1988: $1.67 billion. With its improved plant, however, it was a stronger competitor with a wider variety of popular toll-free 800 and other services.

Some politicians and others argued that bigness was bad, monopoly dangerous, and thus more dismemberment was required. The FCC proposed to limit AT&T's income to three percent less than the inflation rate. Since its 1984 breakup AT&T had reorganized several times but tried again in 1989 to appease critics by breaking its major businesses into between fifteen and twenty-five smaller units and giving their heads power to make basic decisions.

The company's goal—set by Theodore Vail long ago—was to provide the finest service at the lowest possible cost, but it was not easy when there were many competitors. After her weight reduction, Ma Bell still had to fight off attacks by bureaucrats and politicians. Judge Greene's rulings often placed AT&T at a competitive disadvantage, and Congress, wavering between socialistic do-goodism and American free enterprise, did not help.

In 1908, AT&T had advertised that its goal was to have "a phone in every home." Eighty years later its goal was universal service, with integrated voice, data, and image facilities adapted to customer needs, and universal access through adoption of compatible standards, equipment, and interconnecting networks.

Network for Entire Federal Government

The prohibition against RHCs providing nationwide services was criticized anew when the U.S. Air Force needed a software system for its $3.5 billion data network. In the 1980s the General Services Administration (GSA) approved a proposed network using the Unix software system to make the hardware carry out programmed instructions by computers, suggest a better way to express thoughts, and direct computer network traffic to the proper storage discs. Motorola provided silicon miniconductor chips for the Unix operating system at the base level of the Air Force's mainframe and minicomputer operations.[13]

The General Services Administration's purpose was to put the major technological advances of the preceding twenty years to use in a mammouth network serving all government departments and improve government efficiency, as Western Union and AT&T defense networks had done for the military. AT&T's Autovon military network alone used nine solid state switching centers, linked about 2,000 stations, and had a daily capacity of twelve million punched cards, equivalent to 160 million words.

The GSA plan was to replace all federal government communications systems with one network named FTS 2000. Called "the computer contract of the century," it was to cost ten to fifteen billion dollars in ten years. That was great news to the industry. AT&T and Sun Microsystems,[14] and also a group with Martin Marietta, MCI and the seven RHCs, made bids. Then IBM, EDS, and Digital formed a coalition with the RHCs, MCI, and U.S. Sprint to bid against AT&T

and Sun, but EDS withdrew when Sprint would not agree on its share. GSA then split the contract, making switches a separate item, and AT&T won four switches and the RHCs seven. The IBM group formed an Open Software Foundation and AT&T countered by establishing Unix Foundation to market a global data network. The project became snarled in controversy and came to a standstill when members of Congress objected to AT&T providing a major part of the system. It proved again that politics is no way to achieve efficiency.

Abandonment of the project was threatened when some RHCs were charged with obtaining AT&T's secret bid from a GSA engineer they reportedly entertained. AT&T revealed it also had received confidential information. GSA then suspended bidding and awarded the contract, sixty percent to the team led by AT&T and forty percent to U.S. Sprint's team with GTE, United Telecommunications, Martin Marietta, the RHCs, and MCI. U.S. Sprint also had a joint venture with Siemens of West Germany. MCI appealed to the GSA for years to void AT&T's big contract.

GTE cooperated with AT&T to provide an integrated network with advanced Bell technology features grafted onto GTE's digital telephone switching system. Bell Atlantic later received a contract to connect Federal agencies in the Washington, D.C. area. The 1988 contract split between AT&T and U.S. Sprint for the network to serve the entire Federal government was the largest communications system ever awarded and could reach $25 billion.

AT&T continued its leadership in developing improvements in Unix for twenty years with the aid of Sun Microsystems and Unix International, an advisory group. Another group, the Open Software Foundation,

developed a different version. The two were the survivors among about 200 versions, and companies found selection of one for their use far less confusing. The big central office switching systems they planned were designed to permit high speed call connections and data transmission.

While AT&T spent $1.6 billion in 1988 on Bell Laboratories research and development of its plant and services, and to fight stiff U.S. Sprint and MCI competition, Ameritech alone spent about $1 billion to improve fiber-optic cables and electronic-digital switching. With all companies striving to excel, telecommunications equipment and services approached state-of-the-art perfection.

Canadian Telecommunications

Telecom Canada, a consortium of Dominion telephone companies, pooled their long distance revenues. Canadian long distance revenues had been used to keep local rates low, but cut-rate U.S. Companies like MCI tried to extend their cut-rate services north of the border. Canadian National and Canadian Pacific railroads merged their telecommunications subsidiaries and that improved service, but a cry for more competition and cut rates arose in Canada when companies in the states began invading the Dominion. Canada Bell Enterprises, operating the local companies in Quebec and Ontario, announced it would require higher rates to compete with the invaders. Canada Bell owned fifty-two percent of Northern Telecom which profited from an upsurge in equipment sales and its purchase of Trans-Canada Pipelines.

In five years from 1982 to 1987, Northern Telecom, the telephone equipment unit of Bell Canada Enterprises, grew to have the largest corporate income in Canada. It offered $385 million to help Trans-Canada Pipelines try to buy Dome Petroleum, and had large real estate, printing, and oil interests. Northern Telecom also produced a network that provided simultaneous integration of voice, data, text, and graphics transmission over a twisted pair of telephone wires.

Telecom Plus International, a GTE subsidiary and Canada's largest independent supplier of telecommunications systems, experienced operating difficulties, as others did from cutthroat competition after the AT&T split up in 1984. In 1985 Telecom Plus acquired Compath National, greatly enlarging its operations and revenues. Siemens then acquired a thirty-three percent interest in Telecom.

Cam-Net Communications Network, a Vancouver-based telephone company organized in 1984, provided a discount telephone service from Canada to the United States and overseas. In two years Cam-Net seized a large chunk of British Columbia's ninety-million-dollar toll revenues and moved its network facilities and operations to Seattle.

After the AT&T breakup, the RHCs and other independent telephone companies installed computer-controlled switches in their central control offices. Northern Telecom was the beneficiary of very large operational changes necessary to handle the sophisticated new services. It paid off: profits from those services were large and their cost

was low because they required few or no operators. Northern Telecom advertised proudly in 1990 that the U.S. Air Force selected its telecommunications systems for 100 bases worldwide.

Many private wire network customers also added PBXs and Northern Telecom fiercely competed with Rolm for the PBX business. Rolm's earnings lagged and IBM paid $228 million for fifteen percent of Rolm which soon introduced a new generation of PBXs and became a factor in IBM's office automation equipment plans.[15]

Government-owned Cables and Wireless PLC increased its ownership in Mercury Communications from fifty to 100 percent. Mercury was the only telephone company not government owned in Great Britain, but did not prosper.

Judge Greene watched with growing annoyance as the Baby Bells rushed into new businesses like Yellow Pages, cable, TV, and cellular telephone, and into other countries, expansions that added billions to their revenues. His patience exploded in a blistering opinion in April 1991. The opinion said AT&T was imbued with a service mentality, but the Baby Bells devoted too little attention to providing the best and least-costly local telephone service.

Who will win the battle between the judge and industry remains to be decided, but in effect, Judge Greene is a one-man regulatory agency, instead of Congress and the FCC.

Notes

[1]*Forbes* (December 16, 1985).

[2]*Wall Street Journal,* December 19, 1985.

[3]*Wall Street Journal,* March 22, 1989.

[4]Apparently to increase competition with AT&T, the FCC approved GTE's acquisition of Telenet, which was strong in packet-switching technology and electronic mail.

[5]*Forbes* (September 23, 1985).

[6]Contel was merged into GTE August 7, 1989. The result: $25 billion combined market value, largest U.S. local exchange and second largest cellular telephone operator.

[7]Alcatel also made PBX, optic fiber, and metallic cables.

[8]AT&T 1966 *Annual Report.*

[9]U.S. News & World Report.

[10]A. N. Wang, founder of Wang Laboratories at age twenty-eight invented the magnetic memory core essential to the computer industry, and built a fortune of $1.6 billion. Wang was still producing major advances in the 1980s.

[11]*Forbes* (April 17, 1971).

[12]*Wall Street Journal,* November 29, 1989.

[13]Motorola, a seven-billion-dollar electronics equipment systems and components manufacturer, also was a large supplier of pagers to the United States and Japan.

[14]Sun Microsystems is the brilliant engineering group described in chap. 21, above.

[15]*Forbes* (February 2, 1984).

Living in
Our Information Age

The motto on the National Archives Building—a quotation from Shakespeare's *The Tempest*—reads

What's Past Is Prologue.

Without history there is no future.

When Johannes Gutenberg invented the first moveable-type printing press around 1450, it was a major advance in communications that compares with the invention of the telegraph and telephone and, like them, changed the world. Digital-data telegraphy, digital telephone, and laser publishing in distant cities are remote descendants of Gutenberg's moveable type.

Great nineteenth-century advances in telecommunications technology aided in building the nation and mechanizing American industries. The twentieth century will be recorded in history as one of unprecedented growth, adding a blanket of wires, channels, cables, and beams around the earth, like the nerves of the human body. Many convenient, easy-to-use services produce a higher level in everyone's lifestyle. Indeed, our under-standing and utilization of telecommunications are important both to our work and our success in life.

Even the confusion and complications in global network operations seem headed for solutions. In 1988 computer equipment and software leaders in the United States, Europe, and Japan agreed to form an open system in which equipment and software will conform to an international standard.

We humans have doubled our total knowledge every fifty years, but telecommunications progress was faster. Senior citizens remember the Morse telegraph key, glass-domed stock ticker, and crank telephone—antiques compared with today's modern electronic equipment. Some may even remember the Wright brothers' first heavier-than-air flight, Lindbergh's first solo nonstop transatlantic flight, and the huge zeppelin *Hindenburg*'s tragic explosion on arriving in New Jersey from Germany.

People may well wonder how we ever survived before the invention of modern telecommunications services like radio, TV, computers, pushbutton and cellular telephones, laser, electronic banking, cable TV, home shopping and banking, and national and global news.

Edward N. Cole, General Motors president, said:

> We have shortened the time from scientific breakthrough to market applications. Photography took 113 years from invention to application. The telephone took fifty-six years . . . radio thirty-five . . . radar fifteen . . . television twelve . . . and transistors five . . . but laser rays made it from laboratory to application in only ten months.

Computer inventions resulted in more-efficient business management, and met our needs for immediate information on almost any subject. Now we should prepare for life and work in an automated, checkless credit society, in which computers talk with computers, and giant business and government networks store the facts of our lives in their memory banks.

It seems easy to examine modern advances and trends of telecommunications progress and predict the future, but the speed at which the communications world turns soon makes most predictions outdated. The pace of change in telecommunications and computer technology has been so rapid that even Wang and other leaders in it have been unable to keep up with it.

Inspiration is the beginning of the inventive process, whenever a need is recognized. Ford did not invent the automobile, but applied the principle of mass production to its manufacture. Morse did not invent the Morse key, sounder, or dots and dashes, or build the lines and organize companies. He had the inspiration to invent the telegraph but others produced the first successful one.

The Canadian educator Marshall McLuhan (1911–1980) wrote that New York is a city to which people travel fifty miles each day to talk to each other on the telephone. McLuhan added that homes in the future will be designed, wired, and insulated to provide a computer communications workroom. Homes and business offices also will be equipped with computers, word processors, display screens, disk storage, temperature control, cordless, multiuse telephones, and other electronic aids. Just think of the comfort and convenience of telecommuting and working in an office where you feel—and are—at home, and can choose your hours and dress as you please.

About five million telecommunications workers in construction, sales, engineering, consulting, computer programming, law and accounting, arts, music, and other work do not commute daily to crowded drug- and crime-infested cities.[1] They stay at home at workstations doing assigned work with computers and flash results to headquarters. Homework helps decentralize business and slow the undesirable rush of population to large cities. Companies also are establishing many offices in areas where the employees live.

Link Resources, a market research firm, found in 1992 that more than thirty-two percent of American people work at home offices using personal computers, and that more than thirty-nine million people appeared to do all or a part of their paid work at home in 1991.

Computers will eventually control and enhance the safety of 250-mile-per-hour trains, solar-fueled electric autos, and superspeed airplanes. Passenger aircraft already have computers that analyze and monitor internal and external operations and report any problems. The instrument landing systems and altimeters send digitally encoded signals to the flight-control computer, which in turn sends signals to the digital engine controller and the control surfaces of the airplane. Meanwhile, the central maintenance computer interrogates each instrument to discover any problems, inform the flight engineer, and communicate any problems to the next airport so service personnel can prepare for any landing service required. If a collision between two planes is threatened, another instrument orders the pilots to climb or dive to avoid a crash.

Since failures are life-threatening, critical commercial airline instruments have a high degree of redundancy. Often instruments incorporate two different computers, which check on each other. Critical instruments are often in triplicate, so that if one misperforms, the flight computer will still have two which agree, and take the majority opinion.

In retrospect the progress of telecommunications seems more evolutionary than revolutionary, because complex new technologies present many problems, and humans react slowly to change. However, we will have to adjust our lives to a paperless office and a checkless credit society. Our records are pushbutton-available to banks, stores, and others. Some wags say that the paperless office will arrive at the same time as the paperless bathroom, but the paperless office will actually arrive when computer disk retrieval replaces the enormous paper files.

Sales procedures have changed; the sales clerk inserts our bank credit card in a check robot machine to record a purchase and deduct the price from our bank account; grocery purchases are passed over a low-power laser beam to record the item and price on tape; and sales records are transmitted to a computer to show what items to reorder. If the charge is to a national credit card, the clerk passes the card through a small slot to transmit the data to the credit company. A computer there automatically checks thousands of accounts in an instant and flashes approval or disapproval.

Once an error gets into your national data record, it may be years before you can get it corrected. Charles Weidner, forty-three, of Dryden, New York, with a wife and eight children, fought for years to be officially alive because a typographical error in the government's social security files recorded him as dead. Federal and state tax bureaus mistakenly made dual files for me by misspelling my name, and it required years to convince them to correct their error and stop demands for payment of taxes for both files.

The Federal Reserve Bank estimated that in 1978 thirty-eight billion checks were written annually on 100 million accounts. Check writing was a great national pastime. But electronic transfer of funds with computer networks grew thirty-fold in a decade. As a result, the number of checks used was reduced by twenty million when Social Security and insurance payments were deposited by electronic transfer to personal bank accounts. Credit card companies and store credit cards also reduced the number of checks written, because one monthly check paid for a number of purchases. Fewer checks enabled one big bank with branches to do without 10,000 employees. Factories also needed fewer people when they automated payrolls and other records with computers, but the growing telecommunications industry and its suppliers hired more millions.

Our economy is based on the distribution of information. It is a major commodity to which our social fabric is tied. And yet, although our modern society cannot be maintained without telecommunications technology, American schools have been graduating only 100 engineers to each 1,000 lawyers, and American technology fell behind for years. That loss is being remedied gradually, as children grow up taking compulsory courses in computers and mathematics in most schools. But unless they also learn how to use modern telecommunications services skillfully, they may never have a meaningful occupation.

Satellites, McLuhan said, will make the human family one tribe again. But that would require centuries because the differences between national populations, races, and individuals are so great. More understanding may develop, however, because global trade using telecommunications breaks down borders.

Automated World Communications

Nearly all countries use some modern American equipment in their communications, but there was one exception. I once spent a day showing the Shah of Iran how American communications were equipped and operated. The Shah, who spoke excellent English, said he would modernize Iran's communications, and gave GTE a $500-million contract to install a telephone system, but Iranians objected so strongly to change in their medieval ways that the Shah was unable to make the improvements, and lost his ancient throne and his life.

Despite technical problems, there was rapid growth of facsimile use in the 1980s, speeding page-size messages to distant business offices and homes over the telephone network. "Fax" also carried electronic mail and revolutionized personal and business communications. However, the volume of unsolicited advertising and appeals for donations kept many facsimile machines so busy they could not be used as intended. People protested, and the first law to prohibit junk mail by facsimile was passed in Connecticut in 1989. The Governor signed the law while his fax machine was tied up for days with messages urging him to veto it. Portable facsimile machines were taken on trips. Business people sent messages and documents by simply dialing a number and starting the transmitter.

A half-century earlier, Western Union engineers invented facsimile paper and machines, and the company sought to place a Desk-Fax[2] transponder on all business desks. Many thousands were installed and used, but the service was not advertised and pushed. But unlike old soldiers, that idea did not fade away. Laser copiers and minifax machines were produced by Xerox, Canon, Kodak, Mita, NEC, Ricoh, Toshiba, and others. Stores and libraries provided them for customer use in making copies.

Development of sophisticated facsimile equipment continued in the 1980s. Brook Trout of Wellesley Hills, Massachusetts, developed a circuit board to connect personal computers in different business offices. Brook Trout soon had four million dollars in service revenues. Federal Express also established a facsimile service to help its customers speed their express service, but few would pay the extra cost.

Most present transmission of information uses telephone lines, ocean cables, and satellite channels, along with telegraph, telephone, facsimile, or video transmissions. Confidentiality is preserved by packet switching, scrambling, speed variation, and other technologies.

The best use of all this information,

however, will still be left to the individual. The computer can analyze, but it is the person who makes the decision based on the information at hand. Languages of differing computers present major obstacles, and global satellite communications might require an international language.[3] A start was made in 1986 on establishing a global telecommunications system.

Major advances have occurred in personal computers through the use of super-microprocessor chips, electronic-digital processors which can compute numbers a thousand times faster than older units. Advances in programming techniques have also provided clearer pictures from space on television, and aided artificial speech recognition. Intel, Motorola, and Texas Instruments spent millions in the 1980's to develop equipment like the digital signal processor for the purpose of analyzing, rearranging, and strengthening signals to provide greater picture detail.[4] High-definition television was being installed in 1991.

Microsoft's Windows computer operating system and the "mouse" were designed for use with IBM, IBM-compatible, and Apple/Macintosh personal computers. Mouse is computer terminology for a hand-size remote control tool which has a ball with sensors to show on-screen the direction and distance it is moved. It is used to position an arrow on the computer monitor. When the arrow is over the on-screen picture of the proper item, clicking a switch on top of the mouse selects that item (program, maintenance file, clock, or other). Windows and the mouse are intended to simplify computer operation even for those who are not at all familiar with computer engineering and design.

Another trend is to smaller equipment. More elaborate computer workstations will be widely used. Also, more automatic teller machines, and more computers that recognize handwriting and protect against fraud will be used by banks. More companies will sell personal computers at $500 to $1,500. These will be easy for nonprofessionals to use in homes to handle household management, budgeting, and bookkeeping, and to keep records of small businesses.

In 1989 The NeXT Co. advertised the first computer in the world to be made exclusively for desktop publishing. It said the user "could use read/write eraseable optical storage." It had a single optical disk that could store 256 megabytes (MB), "removable for portability and added security." NeXT also said the computer made the power of the Unix high-performance operating system much simpler to use, provided display computer printing, used VLSI (very large scale integrated technology), as well as sending and receiving multimedia mail and was much faster than other desktop machines. It promised new technology in computer networking, desktop publishing, and workstation performance.

A new day in electronic mail was forecast by Motorola's announcement in 1992 that it had agreed for Embarc, its wireless message network, to be used by GE's Information Services, and IBM's Information Network for E-mail users to broadcast messages to one or thousands of people simultaneously, to be received at their desks or on a portable personal computer while traveling.

In 1992 also, AT&T, which for decades had been gradually taking over Western Union's written-record telegraph service, had its engineers developing a small, portable, personal digital computer with which typed or handwritten messages could be sent by means of wireless telephones. Reception was to be on portable small screens.

Another computer trend of the future was IBM's development of a portable-pen-based computer. With a pen or stylus, the user, whose handwriting the computer was programmed to recognize, could enter or retrieve data by writing orders on the screen.

Some experts predict a computer will be invented that will take and transcribe dictation. It will recognize the boss's voice, type what is dictated, convert it into digital code, and transmit it over a global printing-telegraph network.

In the 1980s Sun Microsystems and others introduced the first workstations that were very powerful computers with displays and keyboards much like personal computers, but with extended graphics and computational capabilities.

IBM developed a workstation in 1989 based on its version of Unix. It competed vigorously with Sun Microsystems, Hewlett-Packard and Digital Equipment that had eighty to ninety percent of the five-billion-dollar workstation market. Over the years, mainframes, workstations, and personal computers have become more powerful and tend to overlap in capabilities.

Information is as pervasive in our lives as the air we breathe, and the information industry is a $600-billion business in the United States alone, growing at a rate of twenty percent, NYNEX estimated in 1989. The communications networks are the super-highways over which voice and data information will travel in growing volume.

Computer-controlled robots with human-like programming will replace more telecommunications employees. The basis for that prediction was the comparison of a slice of a guinea pig's brain with the far-larger human brain that led Roger Taub of Columbia University, Richard Miles, and Robert K. S. Wong to experiment with simulations

of human brain-wave sequences. The joining of biological and computer techniques led to a great discovery.

Simulating the activity of human brains, IBM scientists produced brain waves so nearly like human ones that they applied computer techniques to them. A spectogram of the human voice was made to translate word sounds into digital impulses. The computer read the digital impulses and made corresponding sound waves to give verbal replies. Computers can talk, and telephone services with electronic instead of human operators will answer questions, giving verbal replies and instructions with a recorded synthesized voice.

The New York Stock Exchange began using synthetic speech in March 1965. When brokers dialed their computer for a quotation on any of 16,000 stocks, the computer's voice answered with a vocabulary of 125 prerecorded words.

In a related area, the human voice is as distinctive as a fingerprint, and recognition devices will enable banks and others to identify callers.

A computer with a display screen might translate foreign words into English. Spoken words, however, would have to be precise ones which the computer-translator would recognize.

AT&T introduced a telephone interpreter service in 1990 to enable people speaking 143 languages to talk with English-speaking persons for $3.50 a minute. General Motors had a car for sales purposes which answered questions by combining a computer with discs activated by certain words. It also warned if oil pressure was low or a seat belt was not fastened. Texas Instruments had a chip with a vocabulary of 200 words that provided a basis for communication and image-processing services.

Fone Base System of Vienna, Virginia developed a voice-message service. Subscribers called the 800 number of a data base, and a central computer, using voice-imprint technology, recorded callers' voices and numbers, and carried out instructions such as forwarding calls. Western Electric made equipment that remembered and performed many computer functions.

As its name indicates, the Recognition Equipment Co. made electronic devices that could read printed material and convert it into code which could be stored, edited, and manipulated by computers.

The creation of a new wonder of the future—a single machine combining the four basic functions of telecommunications—has been discussed by Ricoh, Canon, Xerox, and Eastman Kodak.

Motorola will be a large factor in the growing computer chip, desktop computer, and cellular telephone equipment business. Its progress in engineering, manufacturing techniques and foreign company alliances indicate large future production of data network equipment. Its vest-pocket telephones will enable busy people to keep in touch.

General Electric agreed in 1989 with AT&T, British Telecommunications Ltd., and France Telecom, the French telephone monopoly, to build a network linking GE offices on six continents for under $300 million.

A Link Resources graph shows that Apple Computer provided seventy percent of the computers in grammar schools in 1987, about sixty percent in 1987 and 1988, and fifty-five percent in 1989. Those years also showed that IBM provided about five, twenty-two, twenty-five, and thirty percent. IBM stepped up a campaign to "supplant Apple as teacher's pet" in United States primary and secondary schools, and Apple cut the price

of its Macintosh computers for schools.[5]

Like the saying "The show isn't over until the fat lady sings," the computer added singing and dancing to its show. It used laser diskplayers to provide music, and dancing with color graphics like television cartoons on a screen. With software programs to guide it, the personal computer currently can provide a multimedia show in addition to its basic data processing function.

American Business Information of Omaha produced the American Business Disk with 8.3 million business names and addresses. It included the lists in 8,000 Yellow Page directories. Personal computer users with a CD-ROM drive used these lists, and "let their fingers do the walking."

The twenty-six-volume *Compton's Multi-Media Encyclopedia* talks. It was recorded by Britannica Software of Chicago in 1989 on one compact disk that it sold to schools and libraries for $750. It provided the printed words, voices, and music of famous writers, speakers, and composers. It took seventy workers eighteen months to make the recording.

For quick reference, publishers also record dictionaries on disks to replace some printed volumes. Spelling checkers and thesauruses have been added to word processors.

Major Trends Will Continue

Rapid increase of optic fiber use in landlines and ocean cables forecasts its use in other telecommunications equipment, and a bright future for optic fiber makers. The fiber is also used in bodies of motor vehicles and boats. Optic-fiber cables are under the streets of many American cities to carry private telephone services. In the 1980s, new companies provided high-speed data systems

25.1 A HOUSEWIFE CAN PICK UP HER TOUCH-TONE TELEPHONE AND TAP OUT CODED NUMBERS TO ORDER GOODS FROM HER FAVORITE STORE THAT SHE SAW ADVERTISED IN THE LOCAL NEWSPAPER.

to corporations, sending numerous simultaneous communications over each optic-fiber cable.

In 1988 AT&T laid a transatlantic optic-fiber cable from New Jersey to England and France, with those nations sharing the cost. Named TAT-8 because it was AT&T's eighth transatlantic cable, it provided thirty-six channels to carry many simultaneous communications. AT&T planned a similar cable to Japan. Apparently convinced that optic fiber will be used in computer operations, IBM bought an interest in Corning Glass Works, an optic fiber producer. GTE demonstrated in 1988 that a video camera

could send TV pictures of a sleeping baby by telephone. It showed that people at home could videoshop by cable TV, and "attend" classes, lectures, and video conferences with as many as fifty-nine conferees "attending," by telephone.

Telephone-video conferences can be held with officials in multiple cities to make deals and hold policy discussions. Architects, builders, banks, and others can plan the building of houses. A video screen at each place would show conferees presenting figures and plans. Distant speakers will train personnel groups in production, service, and sales.

Use of cordless telephones and walkie-talkies grew so rapidly after the FCC approved them in 1981, that forty-five million were predicted for 1995 with twenty million in the United States, ten million in Europe, and three million in the United Kingdom. McCaw and Oracle announced they will broadcast data on a cellular system. McCaw bought LIN Broadcasting and Contel bought McCaw's Southeastern cellular properties.

Bell South, McCaw, the RHCs, GTE, and LIN were the largest cellular telephone companies. Bell South was the largest with about 500,000 subscribers, but McCaw was so profitable it paid its chief executive Craig O. McCaw, son of its founder, $54 million in salary and bonuses in 1989, and had already given him one and one-half million shares of stock in 1983 and 1986.[6] Many cellular subscribers were in large city areas crowding the frequencies and causing conflicting calls to be blocked. The FCC came to the rescue with approval to use higher frequencies for digital systems and tripled the number of calls they could handle.[7] Motorola began producing minimobile telephones in 1990, and pocket-size telephones may replace the full-size cellular ones.

If past trends continue, "big-brother" federal government could assign a call number at birth and have a small transponder attached to an arm of each person. Thereafter, wherever people go they could be called. Growth in use of portable radiotelephones will continue. People can carry a pocket telephone and make calls anywhere in the world.

Major cellular telephone makers were Motorola, Graphic Scanning, GTE, Western Electric, Ericson Telephone, and Japanese companies. Used in autos and trucks, these cellular phones keep service people in touch with the office to report and receive instructions. Many people will use infrared attachments on earphones to receive calls from small transmitters. Use of pagers by seven million doctors and others in the 1980s will continue to increase until cellular telephones replace them. Motorola devoted an entire Florida plant to pager production, with a computer-controlled assembly line and a series of robots programmed to test and insert components and package them. It estimated the robot line tripled the plant's efficiency. Millcom Inc. made a pocket beeper that reportedly received three-word printed messages in four cities. A dime-size telephone "teleset" device to fit in the ear was made by COMPASS International.

Satellites carry solar cells to convert sunlight into electricity and recharge the batteries required for communications control of operations. Development of long-lasting solar cells will make longer probes into deeper space possible.

Many viewers of network TV programs sent protests that misleading impressions are given of lives and relationships of Americans. Public and cable TV tried to correct this with stations and programs appealing to a wide variety of tastes and interests.[8] About half of American homes will subscribe to cable TV. Many homes will use a yard or rooftop dish antenna to receive programs directly from satellites. Dish antenna production became a several-billion-dollar business for Burnup and Sims, Scientific Atlanta, Cox, Teleprompter, Telecommunications, Viacom, United Cable, UA Columbia, Metrovision, Capital Cities, Continental Cablevision, Cris-Craft, Premiere, Comcast, and others. Pay TV stations will continue scrambling transmission signals to make people with rooftop dishes pay, but many viewers will decode the signals. A higher-power satellite technique will permit the use of a much smaller dish. Digital computer techniques will provide sharper TV definition, and people will use larger wall screens. Public use of high-definition digital TV in 1992 was forecast by General Instrument Corp., Zenith, and AT&T.

25.2 AUTOMATIC DIALERS ARE ONLY ONE OF MANY SERVICES OFFERED BY THE TELEPHONE COMPANIES IN THE BELL SYSTEM WHICH HELP BUSINESSES SAVE VALUABLE WORKING TIME.

Since the seven regional telephone companies enter all American homes, there is no need to build TV cables to them. Surely telephone companies will lease circuits to carry cable TV programs, home shopping, and banking services. They also could provide equipment to connect cable TV in homes, and collect for cable TV on telephone bills.

Floors of the stock, commodity, and futures exchanges will become quiet, orderly scenes, using an electronic trading system in which computers receive bids, match them with offers, confirm deals, and report them over quotation services to brokers, media, and public.

Increasing use of integrated circuits will keep semiconductor makers like Intel and Advanced Micro Devices Inc. busy. With a fifteen-billion-dollar annual business, in the 1980s major makers of big high-technology computers were Cray, IBM, Apple, Prime, Cado, Minicomputers, Wang, Digital Equipment, Advanced Micro, Control Data, Scientific Genentech, Datametrics, Barry Wright, Comserv, Computervision, Analog Devices, Auto-Trol, Gerber Scientific, Datapoint, Tymshare, Electronic Data Systems, Data Terminal Systems, Dataram, Floating Point, National, Syks, CPT, Adage, Paradine, Rolm, Tandem, UNISYS, and NCR. In the 1980s, as personal computers gained power, far surpassing the capabilities of mainframes of only a few years earlier, there were fewer orders for big mainframes.

Software Essential

As mentioned earlier, the largest and most profitable software company was Microsoft with net revenues of $803.5 million, for example, in 1989. Backed by IBM, Microsoft developed operation systems designed to make IBM personal computers as easy to use as Apple's. Microsoft's founder and cochairman Bill Gates kept making improvements, eventually introducing the Windows program—a computer operating system—that splits the computer screen into "windows," with individual screens for each program installed on the computer. With Windows, users can control the computer functions with symbols instead of words. Analysts said it would heighten IBM's competition with Apple's Macintosh computers, which used similar pictures.[9]

The agonizing complexity of creating software programs, when there are changes in a company's business and products or to improve efficiency, frustrates even the experts. The programmer must understand a company's business completely. Mitch Kapor, who founded Lotus' 1-2-3 spreadsheet that five million companies bought, making Lotus an eight-billion-dollar-a-year software giant, was so frustrated, he said, "Not a day passed that I did not want to throw my computer out of the window." One reason is that even one letter out of place in a software code or command can turn a million-dollar program into gibberish and the business into a nightmare.[10]

Software excellence is demanding, but its future seems secure. It improves productivity as equipment becomes more expensive. Customers of one software company cannot easily change to another. Software firms require lengthy study and preparation to meet the special needs of a company. The future of Microsoft and Lotus seem as secure as McCaw is in cellular telephones.

Unix, described earlier as the wunderkind of the Information Age, matured and improved with constant development for fifteen years and became a major AT&T triumph, as it was used widely by computer companies. Unix was a pillar of strength during AT&T's complex reorganization.

Artificial intelligence technology firms created computers that think like humans and perform work people formerly did. They already are used to simulate problems and report logical solutions. Some programs can even distinguish fragrances, diagnose diseases at hospitals, read printed matter with optical scanners, help design buildings, help prepare tax returns, and monitor industrial production. Robots will perform many business tasks, gain more general use, and revolutionize manufacturing and communications work.

We have witnessed a great increase in complexity. To avoid technofright, students need to learn about artificial intelligence, mathematics, and uses of computers.

Aid for the Handicapped

Transceptor Technologies, Inc. of Ann Arbor, Michigan produced a personal computer to aid the handicapped. It recognized the voice of its master and carried out orders automatically. It would turn on appliances and lights at a certain time daily, take telephone messages, and dial telephone numbers to return calls.

Computers will aid the deaf, dumb, blind and those who have lost their limbs to maintain communication with others. The deaf will use display screens; the dumb will use a computer keyboard to "talk"; and the blind will listen to recorded stories and information from speaking computers and databases. The deaf will enjoy TV programs because words spoken on network shows will be printed on the screen when they are said. TV shows recorded on disks can be played by the "Filebrailer" for the blind to hear. A special keyboard will be used by the blind to transmit in braille over telephone lines and communicate with blind friends.

Mute and immobile patients can "talk" with "Code-Com" sets, tapping out messages in code. The vibrations are transmitted through a sensor disk, and an "Opto-Com" measures the stare of an immobilized person at a series of letters of the alphabet to indicate words. GTE produced a "companion" telephone with automatic sensors that detect an interruption in a person's normal activities and dials for aid.

Obeying a telephoned voice command, a robot will open a refrigerator, take a TV dinner out, start the oven, light a candle, blow out the flame on the match, remove dinner from the oven, and put it on the table. One such robot was developed for quadriplegics at the Veterans Administration Center, Palo Alto, California.

Raymond Kurzwell, an MIT graduate, created the Kurzwell reading machine "that scans words printed on a page and transforms them into spoken language at the rate of 250 words per minute." Through a special control panel, a blind operator can command the machine to spell the printed words aloud. Kurzwell also invented a computerized machine that reproduced sounds of various musical instruments. His inventions met a need by many and Xerox bought his company for six million dollars in 1980.[11]

In 1983 Digital Equipment Corp. introduced a "speech synthesis product that reads text stored in a computer to listeners who dial for it with touch-tone telephones."[12]

IBM also produced a "Phone Communicator" for the handicapped. This used a voice synthesizer to convert messages, typed on a computer, into a man's voice. The person receiving the message replied by typing an answer and continued the "conversation."

Sophisticated Modern Services

The birth of numerous optional services helped American telecommunications business soar to $185 billion in 1988.[13] In "the good old days" one call to your AT&T state operating company solved any problems. Now new services make our lives easier and safer, if we understand which ones we need and how to use them. New services with

memories provided by silicon chips in telephone company computers will remember the number that called while you were out, forward the call to designated numbers, or dial to return the call for you. Also, it will signal that a call is waiting when your telephone is busy. Abbreviated dialing enables you to push two buttons and call a list of people you call frequently.

Companies buy lists of possible customers and use a flat rate 900 service to call and offer services or products. Often the sales talks are lengthy, recorded, and waste time. The answer to that is for a person to use an answering machine or use Caller ID service. This informs subscribers by showing the caller's number in a small box when the telephone rings. ID service was criticized as selling personal financial information about the caller without his permission. You also can have your telephone blocked from receiving sales calls from horoscope, insurance, home improvement, brokers, and other services. Caller ID service will grow because companies want it, and regional telephone companies profit from it. Objections may end when telephone subscribers can dial a number code to block all sales and other calls not preceded by the letter "p" for "private call."

Repeat Dialing keeps calling a busy number for you, and with Automatic Callback you push a button and the telephone dials your last caller. Ringmaster provides each family or office member with a different number of rings. Of course, such services add to the bill. The call-interception and call-waiting services were challenged by civil rights societies as an invasion of privacy.

Business people can call their homes and instruct a computer-controlled superphone at home to turn on the oven, heat, or lights.

Telephones can be programmed to call the fire department if there is smoke, or the power company if the electricity goes off. Telephones and computers will recognize voices and obey a simple order like "Call the police" or "Call an ambulance." Touch-tone telephone use grew rapidly in the 1980s. By pressing its buttons quiz questions could be answered and other new services used.

A "party line" service allows lonely people to chat with each other, and picture phones enable them to see the faces of callers. A Maryland company had a "Dial-a-Sex" service at nine dollars a call, with a record of a woman cooing her passionate love to the listener. It would make old Ma Bell blush. Parents were annoyed when the bills listed such calls by their teenage children. One "Dial-a-Porn" service was provided in Houston by a company based in Seattle.

Knight Ridder, a big newspaper chain, spent about $40 million on Videotron, an information banking and shopping service for homes with personal computers, but had only 3,000 subscribers and dropped it in 1985. Users of cellular pocket telephones and personal computers needed to identify distant callers. This led to a marriage of telephones with computer data bases stuffed with information about customers, the *Wall Street Journal* reported.

Transtech provided Automatic Number Identification or ANI, "that operated within the so-called Integrated Services Digital Network that enabled a computer and telephone user to send and receive voice and data simultaneously over a telephone line." The companies to which Transtech and others provided these services were called "technological Peeping Toms." Laws were proposed to curb the invasion of privacy.

The telemarketing venture by IBM and Sears Roebuck for personal computer users, named Prodigy (see chapter 21, above), was financed and heavily advertised, and became a $200-million annual revenue business in 1990. When a written order or note was sent by computer, the synthesized voice of an IBM central computer answered and told what the charge would be for the service ordered, including news, stock market trading, catalog shopping, and recipes from a cookbook data base.[14]

GE, American Express, and many other companies set up telemarketing operations that became popular. Millions of people used the 800 telephone service to answer advertisements of the 800 companies. Many operators were employees of the telemarketing companies and not well informed about the 800 companies and their advertised products. Customers calling to order products were alienated when they were put on hold or switched to other operators who also could not answer questions. However, we can expect better training of employees and more telemarketing in future.

Global telegraph, facsimile, and voice networks, telemarketing and data-base uses were still being perfected, but a shortage of radio frequencies resulted from the proliferation of wireless networks. No doubt more frequencies will be added by utilizing super-high frequencies and refinements of other technologies.

Long distance telephone companies offer competitive rates for different hours, days, plans, and groups to gain advantages, but the customer seeking the most inexpensive service is confused because dialing habits, hours of work, and individual needs determine the cost for each person or company.

Governments of nations have had different regulations, services, and equipment, but in recent years many have adopted equipment compatible with world standards, ending their isolation and connecting them with the global trade and commerce networks. A global satellite network will provide circuits for cellular telephones. In 1991 Motorola announced plans to build such a network at a cost of two billion dollars. There were many agreements and joint ventures between American long distance telephone companies seeking more international calls and foreign carriers needing an American ally.

Airline reservation computer networks were pioneers in network efficiency and were improved repeatedly. The traveler could insert a credit card and push buttons on a keyboard to make a reservation, and receive a boarding pass. The credit card number flashed to a distant credit company where a computer checked thousands of numbers in seconds and flashed the answer. At the airport the pass was scanned by a laser that ordered a computer to issue baggage tags. Telecommunications continued its role in operation and service throughout the trip.

United Airlines had a big computer network with switching centers at Los Angeles and other major cities. Consulting firms advised a combination with American Airlines' SABRE reservations network to provide 60,000 video terminals with six mainframe computers. Other industries improved services, using efficient computer networks.

There were about 2,000 computer service bureaus in 1978 doing a $2.4-billion business by sharing the use of their big mainframe computers with numerous customers to process data. When the comparatively low-cost business computer and mini-computer came into wide use in the 1980s, many companies used them for their standard business accounting and routine business and personnel uses, but continued using service bureau mainframes for sophisticated analyses. The computer service companies still did a business of hundreds of millions of dollars, and larger service companies were acquiring smaller ones.

Higher Speed, More Wonders

Service in New York City was complicated by interference and noise when the city became blanketed by a maze of high frequency radio waves serving its concentration of financial, banking, insurance and international businesses. Special engineering cured that.

Rapid growth of data and other telecommunications produced large-volume traffic requiring higher-speed circuits. Once a microsecond—one millionth of a second—was the goal of circuit speed. Then came the nanosecond—one billionth of a second. Illiac, developed by Burroughs Corporation and Illinois University, is said to make computations in the nanosecond range—at the rate of one billion bits per second. Some advanced circuits reportedly switched in a picosecond—one trillionth of a second.

A tiny laser beam, developed by GTE's laboratories in 1978 blinked on and off forty-five million times a second, and its pulses of light carried telephone voices and TV signals. Sent through 200 strands of optical fiber in a pencil-size bundle, the laser beam could handle 80,000 simultaneous telephone calls. It was used to telecast a football game, and the Winter Olympics at Lake Placid, New York in 1979.

Maser and Laser Beams' Great Future

"Optical Maser," a paper published by developers Charles Hard Townes and Arthur Leonard Schawlow on December 15, 1958, described their invention, and they filed a patent for it for AT&T. The first operational "optical maser" or laser was built in 1960.

Once featured as a "HI-SCI war-in-the-skies" death ray, the laser became recognized as "the great healer." High intensity laser beams are used in place of the scalpel in delicate surgery. Optical fibers carry images and allow doctors to see inside the stomach, bladder, intestines, uterus, and other parts of the body. Computers will be used increasingly to solve problems, and with robotic assistance may perform operations, supervised by a surgeon. The combination may eventually be sufficiently accurate and precise to make incisions and repairs along cell boundaries, without leaving unsightly and uncomfortable scars. Optic fiber has the capacity to transmit tremendous volumes of information, modulated onto laser beams. Many more uses of laser beams will be found, as were laser surgery and dentistry.

Since computer memory chips already are tiny and have thousands of minute transistors crowded on them, the laser beam may be used to increase the speed and capacity of communications circuits. The laser would provide parallel beams that optical computers could use, for example, to search police fingerprint and photo files in an instant.

An indication that laser beams will replace electricity in superfast data processing was revealed by AT&T at an optical computer conference in Kobe, Japan in

1991. The experiments by Bell Labs in new "photonic" technology combined a network of tiny lenses and mirrors with a quartz glass disk an eighth of an inch thick. It reputedly carried more information to computer chips faster than ordinary electrical connections.[15]

Another indication of the trend to smarter and smaller equipment was the announcement that the GO Corporation, based in Silicon Valley, California had developed an electrical fountain pen in the base of which a tiny sensor replaced a keyboard and allowed the computer to recognize the handwriting of most of its users.

In 1991, the NCR division of AT&T introduced a tablet input notebook computer, which allowed simpler data entry and could recognize handwriting.

In 1992, American and Japanese companies joined in a venture to develop a machine combining video, graphics, text, and sound. Toshiba was to provide the hardware, Apple the software, IBM a semiconductor it was to develop, and Sharp Corporation an electronic organizer. This wonder was to be created by converting data, pictures, and sound into digital form so they could be used on personal computers. With telephones carrying video, the next step could be computers receiving video on their screens. Sony, Japan's largest maker of electronic equipment, was not in the venture.[16]

Field Enterprises, Honeywell, and Centel joined in a Chicago venture to put electronic news in homes and offices in 1982. When the desktop laser printer was developed, it provided a way to do that and also mix graphics with text. Hewlett-Packard established a market for its eight-page-a-minute laser printer. Then IBM produced a ten-page-a-minute laser printer in 1989 priced at $1,600 to $1,700, slightly lower than Hewlett-Packard's. IBM predicted the total market for laser printers would grow to $3.9 billion in 1993.[17]

Another use of the laser beam is to "read" digitally encoded music on computer disks. It avoids the wear and damage to which conventional records are prone. Because they provide clearer sound, a large number are in use.

Sony and others made laser videodisk players for home entertainment. Their expanded use forecast the production of compact four-and-one-half-inch disks. One such disk contained the entire twenty volumes, 100,000 pages, of *Grolier's Academic American Encyclopedia.* The encyclopedia could be printed out on the screens of personal computers with a disk reader or from central libraries that offered on-line service.

In 1949, an exciting marvel was discovered by Professor Dennis Gabor of England. He produced the hologram by using interference of a pattern of light beams on a subject on a photosensitive medium, and produced the result by illuminating it, using a point source of light with a high degree of coherence. Laser coherent light sources have made holograms practical. By simultaneously exposing a photographic plate or film to an object and the light source, the interaction of light waves on it produced a three dimensional hologram image almost indistinguishable from the real thing. Using red, green, and blue laser beams, color with great fidelity is added.[18]

Bell Laboratory experiments with holography developed practical communications uses in picturephone service and photolithography to record holograms in great detail. The close space pattern required the subject to be absolutely still, but when that problem is solved the hologram will be used in live television with three dimensional scenes that make sports and other events seem to be taking place in your home.[19]

After a frustrating decade pouring money and manpower into the computer/data industry, making alliances with other companies to sell each others' products, and cross licensing patents, AT&T still suffered heavy loss from its computer business. However, its interest in Sun Microsystems success in establishing Unix as a widely used software operating standard, and takeover of NCR in 1991 aided AT&T. It even buried the hatchet in its war with IBM, joining it in a venture to share the financial risk of establishing an expensive memory chip plant, and began making some personal computers. In 1990 there were personal computers produced by all companies in about one of every four American homes.

More Wonders Predicted

Great thinkers and writers have made predictions that never materialized. As former President Herbert Hoover told me at the University Club in New York long ago, "It is never safe to predict what will happen in the future." Though the following predictions are based on trends, some may not happen soon. It may be a long time before astronauts explore the universe and land on other planets. That would require trips lasting for years, long-lasting solar batteries, and spaceships in which astronauts could live—with air, food, and water—and make landings.

The telecommunications companies will put more money and effort into the services that have the least rate regulations, such as cable TV, cellular telephones, and pagers. The Baby Bells bought control or established many of the early cellular and cable TV companies in the United States. Then they established, bought control of, or formed joint ventures with cellular and cable TV companies in Britain, South America, Australia, Canada, and other countries.

The picturephone will be used much more in the future. It seemed its time had come in 1964 when New York World's Fair visitors saw and heard people over a telephone line. Service was planned in Pittsburgh and twenty-five other cities, but the cost for a picturephone was $1,500. The monthly service charge was $100 because the picturephone required 250 times as much channel space as a telephone. So, cost prevented its wide use in the 1960s.[20] As wider bandwidths are available to homes wired with glass fibers, the picturephone will become affordable and commonplace.

In the future, you can be reached by anyone calling a cellular telephone in your pocket. You also can use it to make calls. Skillful use of modern services can increase your efficiency.

Biologists, chemists, computer engineers, physicists, and others have sought to create lifelike forms that could evolve into human-like beings. In the distant future, perhaps lifelike robots that now walk, talk, and answer questions with synthesized voices indistinguishable from human voices may receive chemical and mineral additions and become biological creatures like those on earth millions of years ago. The first living cell on earth may have resulted from a freak accident when lightning struck a rare combination of chemicals and matter. Man may find that combination after long years of trial and error.[21]

Instead of letting technofright delay them, people should start learning about the complex telecommunications industry and keep learning as more developments are made. There always will be competition, and knowledge will lead to success.

Since Galileo questioned traditional knowledge and insisted that the earth moves and revolves around the sun, man's great advances have been made by doing the "impossible." A few decades ago, we had none of our ultramodern services. In our information age, failure to learn the lessons of history dooms everyone to repeat failures, but if we learn what produces results, we can select the road to success.

In 1966, the *Wall Street Journal* predicted that men would land on Mars in the twentieth century, travelers would fly from New York to Tokyo in less than two hours, and commuters would strap rockets on their backs and jet to work. It was also predicted that 220,000 computers would be in the United States by the year 2,000, but that figure was far too low, with over fifty million in use in the U.S. in 1991. A safe prediction is that one billion dollars will be added to the telecommunications industry's assets in the next ten years in spite of much equipment becoming obsolete.

The size of the telecommunications industry and the numbers of its workers grew rapidly in the 1980s as it kept step with the needs of people to have direct communication with others around the world. As the last decade of the twentieth century began, about fifty million people used computers, nine million used facsimile machines, and five million in the United States used mobile telephones.

A final prediction is that the telecommunications industry will continue to be the wave of the future. It will provide millions of good jobs to people who understand it and good investments for savers. It will continue to grow because it provides a necessary service that people worldwide need and use regardless of depressions. Corporate traffic, voice, data, and teleconferencing between countries is increasing fifteen to twenty percent annually, and the rate in the United States, the largest user of telecommunications services, is still growing at least eight percent a year.

According to Michael Kennedy, telecommunications consultant at Arthur D. Little, Inc., in Cambridge, Massachusetts, "From five billion dollars in 1991, international revenues from corporate traffic will grow by at least twelve percent annually, to eight billion dollars by 1995." "By the end of the decade [Kennedy] expects such revenue to rise to $14 billion, out of a total world communications market of $750 billion."[22]

• • •

If you who have traveled through the centuries with this story of how man overcame the barriers of time and space, have gained knowledge that will help you avoid mistakes and lead more responsible, successful, and happy lives, my purpose will be achieved, and many years of research and writing will not have been in vain.

Notes

[1]*Wall Street Journal,* June 4, 1990.

[2]I named and promoted the Desk-Fax in the 1940s and 1950s after announcing it to the press in New York, with the personal assistance of Eleanor Roosevelt.

[3]"English has become the language of science and technology," said Roy J. Leffingwell, publisher of the *International Public Relations Review* in Honolulu. "One half of the world's newspapers and one-half of the world's scientific journals are in English. There are 1,068 different languages." (Leffingwell used George Washington University's 1961 classification of world languages in which the several Chinese dialects were counted as one—Mandarin, the chief dialect of China.)

[4]*Wall Street Journal,* June 20, 1983.

[5]*Wall Street Journal,* May 11, 1990.

[6]*Forbes* (May 28, 1990).

[7]*Wall Street Journal,* June 20, 1983.

[8]The Supreme Court aided rapid growth of cable TV in 1968 by ruling that CATV (*community antenna television*) companies do not have to pay royalties to broadcasters, precipitating a scramble for franchises in many cities. However, the FCC ruled that CATV stations can use signals from other areas only with permission of the station owners. A year later the Supreme Court reversed the FCC action, requiring public access for cable TV users.

[9]*Wall Street Journal,* May 21, 1990.

[10]*Wall Street Journal,* May 11, 1990.

[11]*U.S. News & World Report* (June 13, 1983).

[12]*Wall Street Journal,* December 14, 1983.

[13]From Associated Press reports.

[14]*Wall Street Journal,* November 9, 1990.

[15]*Wall Street Journal,* April 11, 1990.

[16]*Wall Street Journal,* July 6, 1992.

[17]*Wall Street Journal,* October 10, 1989.

[18]R. J. Collier, "An Up To Date Look At Holography," *Bell Telephone Laboratories Record* (April 1967).

[19]*Bell Telephone Laboratories Record* (April 1967).

[20]An experimental picture telephone at AT&T headquarters was demonstrated to me in the early 1950s. I and an engineer in the Bell Laboratory building miles away on West Street, New York, had a face-to-face conversation.

[21]Idea inspired by Associated Press article by Nancy Shulins, October 23, 1990.

[22]*Wall Street Journal,* October 4, 1991.

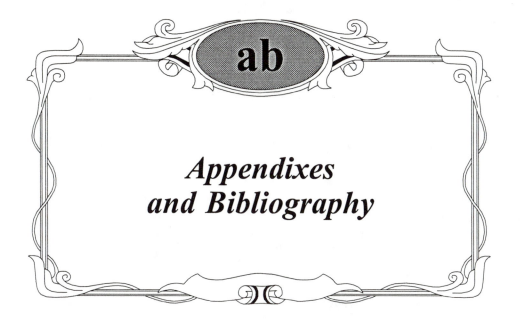

Appendixes and Bibliography

Appendix 1

Making History Live

Knowledge of the historical developments that made our nation great is a basis of responsible patriotism. Realizing how little Americans know about the great contributions that telecommunications made to our nation and our way of life inspired me to try to fill some of that void by planning, writing, and staging a series of five national historical celebrations.

The largest such celebration was the centennial of the telegram—man's first modern service enabling Americans to communicate quickly over a distance. The observance on May 24, 1944 began with a reenactment of the sending of the first telegram to Baltimore from the old Supreme Court room in the national capital. Congressmen who formerly were telegraphers used the original Morse instrument and installed a plaque of the first telegram in the room.

A large audience then attended the major public ceremony held in the great rotunda of the Capitol by a congressional committee and government officials. A Victory Ship was christened by Leila Livingston Morse, granddaughter of the inventor, and launched at Baltimore. A U.S. centennial stamp was issued and a banquet was attended by numerous national officials, congressmen, and industry leaders. Vice President Harry Truman was the principal speaker. Observances were held nationwide by schools and libraries.

The final nationwide observance in the series was on October 24, 1961, the centennial of Western Union's completion of the first transcontinental line. At the central ceremonies at Omaha, Nebraska, a pageant reenacted the start of construction on the line in front of the courthouse, with President Kennedy repeating Abraham Lincoln's reply

A.1 THE JOINT CONGRESSIONAL COMMITTEE FOR THE OBSERVANCE OF THE CENTENNIAL OF THE TELEGRAPH ON MAY 24, 1944 EXAMINES THE ORIGINAL TELEGRAPH INSTRUMENT, USED IN A REENACTMENT OF THE SENDING OF THE FIRST TELEGRAM, "WHAT HATH GOD WROUGHT!" THE COMMITTEE IS GATHERED IN THE OLD SUPREME COURT CHAMBER AT THE CAPITOL, SCENE OF BOTH THE OLD AND THE NEW CEREMONY. LEFT TO RIGHT: SENATOR WARREN R. AUSTIN, REPRESENTATIVE EDWARD G. ROHRBOUGH, SENATOR BURTON K. WHEELER (COMMITTEE CHAIRMAN), REPRESENTATIVE ALFRED L. BULWINKLE (VICE CHAIRMAN), AND SENATOR WALLACE H. WHITE, JR.

to the first transcontinental message. A civic luncheon and an academic procession by Creighton University followed, with the dedication of a library and the presentation of an honorary degree to Western Union President Walter P. Marshall. Observances were held at all towns and cities along the old line to San Francisco.

The three other anniversary observances were of the invention of the telegraph, the first demonstration of it at New York University, and the birth of Western Union with its first national communication services. The latter included a pageant at Rochester, New York, which I narrated.

Appendix 2

Presidents' Gold Telegraph Keys

General Edwin Seneca Greely, a relative of Horace Greeley, had a telegraph key made for use by President Cleveland to open the first Chicago World Fair in 1892. Presidents Harrison, McKinley, and Theodore Roosevelt also used it. Then it went to the New Haven (Connecticut) Colony Historical Society Museum.

To participate in distant ceremonies, five later presidents and one first lady used a gold telegraph key that George W. Cormack, who discovered gold in Alaska, presented to President Taft for use in opening the Alaska-Yukon-Pacific Exposition at Seattle June 1, 1909. The key and nuggets were solid gold—the first found in the Klondike rush. Taft handed the key to Edward W. Smithers, chief White House telegrapher, who brought it out for use on many occasions.

President Wilson tapped the key to blow away the Gamboa Dike, opening the Panama Canal on October 10, 1914. President Coolidge used it dozens of times: to open the Holland Vehicular Tunnel between New Jersey and New York City, the seven-mile bridge between San Mateo and Hayward, California, the seven-mile tunnel of the great Northern Railway through the Cascade Range in Washington State, and the Waterway connecting the Atlantic Ocean with Bay Mabel, Ft. Lauderdale, Florida, and so forth. Mrs. Coolidge opened two women's meetings at distant points.

President Hoover tapped the key to open the Long View Bridge connecting Washington and Oregon, the Chicago Board of Trade Building, the International Vehicular Tunnel between Detroit and Windsor, Canada, new plants of the Chicago Daily News and Detroit Times, and on other occasions.

"Doc" Smithers, as he was called, was the Western Union operator stationed at the White House during the McKinley administration. He aided so much during riots in the Cleveland administration that he was appointed chief telegrapher on the day the *Maine* was blown up. "In those days," he recalled, "there was only one telephone in the White House. It was installed in the War Room and anyone using it had to go to that room. . . . Naturally, the telegraph was very important."

AT&T installed the first telephone in the White House (December 1, 1878) in a corridor telephone booth, outside the executive office. One day in 1901 Smithers sat in his White House office. "Not a key was sounding," he said, "when suddenly they all starting clicking at once. I walked over, sounded the key, and it seemed like every Buffalo operator had gone crazy. When the message finally came through, it told of the president's [McKinley's] death."

When Smithers retired, he took the gold key with him. He was succeeded by Dewey E. Long, who served as chief White House telegrapher.

Learning about the historic keys from research, I arranged presentations of lucite and gold instruments in the Oval Office to Presidents Roosevelt, Truman, Eisenhower, Kennedy, and Johnson.

The first president to have a telephone in his office was Herbert Hoover on March 27,

1929. Telephone and telegraph men have enabled each president to maintain contact with Washington wherever he traveled, and provided wire facilities to newsmen accompanying him. For thirty-three years, Western Union's man handling these matters in the White House was Carroll S. Linkins, whose name in large gold letters was on the press car of presidential trains.

One of Roosevelt's uses of his gold key was to transmit a beam to the distant star Arcturus. The returning signal turned on the lights of the 1933 International Exposition at Chicago and started the twenty-five big bells of the carillon at the Fair playing, as they did every day thereafter. A musician in different cities played daily on a midget piano with twenty-five keys corresponding to the carillon keys at dusk to turn on the Fair's lights. Roosevelt used his key also to open the 1934 fair.

Bibliography

Manuscripts and Other Collections

George R. Adams and Fred M. Smith diaries. Found in a barrel of papers in a San Francisco junk shop by the late Charles S. Hubbell of Seattle, and acquired by the University of Washington, they provide an insight nowhere else available into the Russian-American Expedition's Scientific Corps and Robert Kennicott's last days.

Charles H. Brown's diary. Building the first transcontinental telegraph line. Provided by his niece, Mrs. Margaret Brown Burgess Ward.

John D. Caton papers. Library of Congress. Thirty boxes, 9,000 items, describing Caton's pioneer telegraph activities.

Ezra Cornell papers. Cornell University Archives. Twenty boxes covering period from 1844 to 1860, and other related material. In chronological order, well preserved. Other Cornell papers at DeWitt Historical Society Museum, Ithaca, New York moved to Cornell University Archives. Other Cornell letters in New York Historical Society Library. Mrs. Edith M. Fox, curator and archivist at Cornell.

Department of State records on microfilm at National Archives. Dispatches between Secretary of State William H. Seward and and Russian minister to United States. Other letters scattered among thousands of communications with foreign legations.

Thomas A. Edison papers. More than 100,000 letters, documents, patents, notebooks, and ledgers in 55x90-foot underground vault at Edison National Historic Site, West Orange, New Jersey. Also about 10,000 volumes in Edison's three-story library at the site, which includes a museum and Edison's laboratory. Much material not sorted. Many pages of manuscript are based on letters and documents made available by Norman R. Spei-

den, Harold S. Anderson, and Mrs. Kathleen McGuirk.

The Esquimaux. First publication in what is now Alaska. Issued by members of the Russian-American Expedition at Port Clarence.

Cyrus W. Field papers. Manuscript Division, New York Public Library. Twenty-one boxes. Box 1 contains transatlantic cable correspondence.

Fred L. Hester. "Atlanta's Telegraph History." A booklet of mimeographed pages and clippings issued in 1963 by the Atlanta chapter of the Morse Telegraph Club. Provided by Mr. Hester.

Amos Kendall papers. Library of Congress. Two boxes. Some items concern his pioneer telegraph activities.

Samuel F. B. Morse papers. Library of Congress. 101 volumes and boxes containing diaries, scrapbook, letters, clippings, newspapers, drawings, agreements, in chronological order.

Henry O'Reilly papers. New York Historical Society. Largest and most useful telegraph papers. Thirty-one very large boxes of letters. Also, in 50 volumes, thousands of early documents, annual reports, pamphlets, and clippings. Period: 1845–1870. Also 33 boxes of O'Reilly letters at Rochester Public Library, and others among Smith, Wade, and Morse Papers.

Helen C. Phillips, "History of the U.S. Army Signal Center and School, Fort Monmouth, N.J.," manuscript. Phillips is retired museum director and U.S. Signal Corps historian.

Charles M. Scammon, private journal. One hundred twenty-nine pages on the Russian-American Expedition. Provided by Joel W. Hedgepeth of Rockport, Texas, in 1946.

Hiram Sibley biographies and papers. Princeton thesis on his life by his great-grandson,

Harper Sibley, Jr., January 1949. Borrowed from Harper Sibley. Also, at the Hiram Sibley mansion, Rochester, New York, a biography believed written by Hiram Sibley's son Hiram W. Sibley. Numerous Sibley letters are included among papers of other pioneers.

F. O. J. Smith papers. Maine Historical Society: twenty-seven boxes of letters and other papers, five scrapbooks, two notebooks, five books, and ten pamphlets covering the period 1818–1888. Also Smith's book-length manuscript on the invention of telegraph, lost for 100 years. Treatment of Smith in *Maine Political History* by Elizabeth Ring, director and vice president of the Maine Historical Society at Portland. Also two boxes of Smith letters at New York Public Library.

Alvah Strong's autobiography, part of it relating to the telegraph. Provided by his grandson, Alvah G. Strong of Alton, New York.

D. T. Tillotson scrapbook. Cornell University Archives. Clippings and correspondence with telegraph pioneers.

Alfred Vail papers. Smithsonian Institution Archives. Fifty volumes of letters. Handwritten records of his building Morse instruments and operating the first line. Diaries, notebooks, scrapbooks of clippings, and documents. Also, at Morristown, New Jersey Public Library, 745-page abstract (1918) from the papers by S. Ward Righter of East Orange, and clippings, pamphlets, and documents. Also, at New York Historical Society, hundreds of pages of handwritten extracts from Vail papers.

Jeptha H. Wade papers. Western Reserve Historical Society, Cleveland. Large boxes contain Wade letters, notebooks, ledger, and lengthy autobiographical letter to grandson Jeptha H. Wade, a Boston lawyer in charge of family papers, who in 1966 added helpful information.

J. W. Wanless diary. Describes work and hardships of Russian-American Expedition in Siberia. Diary presented to University of Washington Libraries, Seattle, by Frederic Bronner of Victoria, British Columbia, and his late wife, daughter of Lieut. Wanless. Photostatic copy of entire diary and maps provided by Isabella Sims, Gifts and Exchanges librarian, University of Washington Libraries at Bronner's request.

Ferdinand Westdahl diary. Diary of first officer of the bark *Golden Gate,* and later astronomer and surveyor of Alaskan construction party of Russian-American Expedition. Made available by Dr. Philip R. Westdahl of San Francisco.

Charles F. West. "Personal Reminiscences of Telephone and Telegraph Experiences from 1876 to 1926." Manuscript in author's possession.

"Western Union Telegraph Company Corporate History." Western Union Library. A mimeographed list of more than 400 companies merged with Western Union, with facts of organization, leases, and mergers.

Western Union records. Minutes of board meetings for a century. Vault in secretary's office.

"Technical Progress (1915–1940)." Western Union Engineering Department. Three mimeographed volumes.

Copies of numerous speeches by various industry leaders.

Testimony at hearings of Federal Communications Commission and Senate and House committees. Original transcripts and printed copies.

Numerous annual reports and news releases of companies.

Books, Articles, Papers, Speeches

American Telegraph Magazine (October 1852). Announcing coinage of the word "Telegram."

Appleyard, Rollo. Article on Hans Christian Oersted. *Electrical Communication Journal* (April 1928).

Avery, Elroy McKendree. *Cleveland and Its Environs*. 2 volumes. Chicago and New York: Lewis Publishing Co., 1918.

Bancroft, Frederic. *The Life of William H. Seward*. New York and London: Harper & Brothers, 1900.

Bancroft, Hubert Howe. *History of Alaska, 1730–1885*. Volume 28. San Francisco: A. L. Bancroft & Co., 1886.

____. *Chronicles of the Builders*. 7 volumes. San Francisco: The History Company, 1892.

Barlow, William. Article reporting his tests. *Edinburgh Philosophical Journal* (1825).

Barnard, Henry. *Aarmsmear. A Memorial*. Commissioned by Samuel Colt's widow and privately printed, 1866.

Barnett, Lincoln. "The Voice Heard Round the World." *American Heritage Magazine* (April 1965).

Bates, Alice L. "History of the Telegraph in California." *Southern California Historical Society* 9/3 (1914): 181-87.

Bates, David Homer. *Lincoln in the Telegraph Office*. New York: Century, 1907.

Becker, Carl. *Cornell University: Founders and the Founding Fathers*. Ithaca NY: Cornell University Press, 1943.

"Behn Brothers." *Fortune Magazine* (December 1930).

Berger, Meyer. *The Story of the New York Times, 1851–1951*. New York: Simon and Schuster, 1951.

Bishop, Morris. *A History of Cornell*. Ithaca NY: Cornell University Press, 1962.

Bok, Edward. *The Americanization of Edward Bok*. New York: Charles Scribner's Sons, 1923. An autobiography.

Brewer, A. R. *A Retrospect*. New York: James Kempster, 1901. 36-page pamphlet.

Briggs, Charles F., and Augustus Maverick. *The Story of the Telegraph and a History of the Great Atlantic Cable*. New York: Rudd & Carleton, 1858.

Bright, Charles. "The Evolution of the Submarine Telegraph." *Minutes of Proceedings, Institution of Civil Engineers*. Volume 157. London, 1904. 40 pages.

____. *The Life Story of Sir Charles Tilston Bright*. London: Constable & Co., 1910.

Bright, Edward Brailsford. *The Life Story of the Late Sir Charles Tilston Bright*. Westminster, England: A Constable & Co., 1899.

Brooks, John. *Telephone*. New York: Harper & Row, 1975.

Bryan, George S. *Edison, the Man and His Work*. Garden City NY: Garden City Publishing Co., 1926.

Carnegie, Andrew. *Autobiography of Andrew Carnegie*. Boston and New York: Houghton Mifflin Company, 1924.

Casson, Herbert N. *The History of the Telephone*. Chicago: A. C. McClurg & Co., 1910.

Chevigny, Hector. *Russian America*. New York: Viking, 1965.

Clampitt, John Wesley. *Echoes From the Rocky Mountains*. Chicago and New York: Belford, Clarke & Co., 1889.

Coll, Steve. *The Break Up of AT&T*. New York: Atheneum, 1986.

Colles, Julia Keese. *Authors and Writers Associated with Morristown*. Morristown NJ: Vogt Bros., 1895.

Collier, R. J. "An Up To Date Look at Holography." *Bell Telephone Laboratories Record* (April 1967).

Coon, Horace. *American Tel & Tel*. New York and Toronto: Longmans, Green and Co., 1939.

Cortissoz, Royal. *The Life of Whitelaw Reid*. New York: Charles Scribner's Sons, 1921.

Cousins, Margaret. *The Story of Thomas Alva Edison*. New York: Random House, 1965.

Dall, William H. "Alaska as It Was and Is, 1865–1895." *Bulletin of the Philosophical Society of Washington* 13 (n.d.): 123-61.

Dall, William H. *Alaska and Its Resources*. Boston: Lee and Shephard, 1870.

Deshler, Charles D. Article on the News of July 4, 1776. *Harper's New Monthly Magazine* (New York, July 1892).

Directory of American Bibliography. Volume 19. New York: Scribners, n.d.

Dorf, Philip. *The Builder.* New York: Macmillan, 1952.

Driggs, Howard R. *Westward America.* New York: G. P. Putnam's Sons, 1942.

Dunbar, E. E. *Romance of the Age.* New York: D. Appleton & Co., 1867.

Dunlap, Orrin E., Jr. *Marconi, the Man and His Wireless.* New York: Macmillan, 1938.

Dyer, Frank Lewis, and Thomas Commerford Martin, in collaboration with William Henry Meadowcroft. *Edison, His Life and Inventions.* 2 volumes. New York: Harper & Bros., 1910 and 1929.

Engineering News, The (April 14, 1886). Article honoring Vail. Also other issues of the journal on related subjects as cited in notes.

Fahie, J. J. *A History of the Electric Telegraph to the Year 1837.* London: n.p., 1884.

Farrar, Hector J. "Joseph Lane McDonald and the Purchase of Alaska." *Washington Historical Quarterly* 12 (Seattle, April 1921): 88ff.

Field, Henry M. *The Story of the Atlantic Telegraph.* New York: Charles Scribner's Sons, 1903.

Flick, Alexander C. *Samuel Jones Tilden.* New York: Dodd-Mead, 1939.

Geneen, Harold S. Remarks at ICD Rehabilitation and Research Center awards dinner, October 13, 1976.

Gramling, Oliver. *A.P., The Story of News.* New York and Toronto: Farrar and Rinehart, 1940.

Golder, Frank A. "The Purchase of Alaska." *American Historical Review* (April 1920).

Greely, A. W. "The Photographic History of the Civil War." *Review of Reviews* (1911).

Griffen, Joseph, editor. *History of the Press of Maine.* Brunswick ME, 1872.

Graves, Charles. *The Thin Red Lines.* London: Standard Art Book Co. Ltd., n.d.

Half Century of Cable Service to the Three Americas, A. New York: All America Cables, Inc., 1928. A booklet.

Hall, Arthur D. *A Methodology for Systems Engineering.* Princeton NJ: D. Van Nostrand Co., 1962.

Harlow, Alvin F. *Old Wires and New Waves.* New York: Appleton Century, 1936.

Harrower, Charles S. *Address in Memoriam of Harrison Gray Dyar.* New York: C. H. Jones & Co., 1875.

Harvey, O. J., and E. G. Smith. *A History of Wilkes-Barre, Luzerne County, Pa.* 1929.

Hayes, Jeff W. *Tales of the Sierras.* Portland OR: F. W. Baltes & Co., 1900.

Heilbroner, Robert L. "Epitaph for the Steel Master." *American Heritage Magazine* (August 1960).

Hendrick, Burton J. *The Life of Andrew Carnegie.* Volume 1. Garden City NY: Doubleday, Doran & Company, Inc., 1932.

Herron, Edward A. *First Scientist of Alaska: William Healey Dall.* New York: Julian Messner, Inc., 1958.

Hillenthal, J. A. *The Alaskan Melodrama.* Liveright NY, 1936.

"History of Speedwell Iron Works." *The Telegraph and Telephone Age* (April 1, 1937).

Holmes, E. T. *A Wonderful Fifty Years.* New York: privately published, 1917.

Hoyt, Edwin P. *The Vanderbilts and Their Fortunes.* Garden City NY: Doubleday & Company, Inc., 1962.

Hudson, Frederic. *Journalism in the United States from 1690 to 1872.* New York: Harper & Brothers, 1873.

"Indian Raid on Julesburg." *The Telegraph World* (October 1937).

Jones, Alexander. *Historical Sketch of the Electric Telegraph, including Its Rise and Progress in the United States.* New York: G. P. Putnam's Sons, 1852.

Jones, Francis Arthur. *The Life History of Thomas A. Edison.* New York: Grosset & Dunlap, 1931.

Josephson, Matthew. *The Robber Barons.* New York: Harcourt, Brace and Company, 1934.

Josephson, Matthew. *Edison.* New York: McGraw-Hill, 1959.

Judson, Isabella Field. *Cyrus W. Field, His Life and Work.* New York: Harper & Bros., 1896.

Kappel, Frederick R. Talk before American Bar

Association, August 11, 1964.

Kendall, Amos. *Morse's Telegraph and the O'Reilly Contracting.* Louisville, 1848. Pamphlet.

Kennan, George. *Tent Life in Siberia.* New York: G. P. Putnam's Sons, 1910.

Krock, Arthur. *Memoirs. Sixty Years on the Firing Line.* New York: Funk & Wagnals, 1967.

Lake, George B. Article on Ampere. *Clinical Medicine and Surgery* (1937).

Larkin, Oliver. *Samuel F. B. Morse and American Democratic Art.* Boston: Little Brown, 1954.

Laut, Agnes. *Vikings of the Pacific.* New York: Macmillan, 1904.

Lee, James Melvin. *History of American Journalism.* Boston and New York: Houghton Mifflin Co., 1917 and 1923.

Lessing, Lawrence. *Man of High Fidelity.* Philadelphia and New York: J. B. Lippincott Co., 1956. A biography of Edwin Howard Armstrong.

Lyman, George Dunlap. *Saga of the Comstock Lode. Boom Days in Virginia City.* New York: Scribner's, 1937.

Lehman, Milton. *This High Man.* New York: Farrar Straus, 1963. A biography of Goddard.

Mabee, Carleton. *The American Leonardo. A Life of S. F. B. Morse.* New York: A. A. Knopf, 1943.

Mabon, Prescott C. *A Personal Perspective on Bell System Public Relations.* New York: AT&T, 1972.

Mack, Effie Mona. "Telegraphing the State Constitution." In *History of Nevada.* Glendale CA: Arthur H. Clark Co., 1936.

Mackay, Corday. "Russian-American Line in British Columbia." *The British Columbia Historical Quarterly* (July 1946).

McDonald, Philip B. *A Saga of the Seas. The Story of Cyrus W. Field and the Laying of the First Atlantic Cable.* New York: Wilson-Erickson, 1937.

McKelvey, Blake. *Rochester. The Flower City, 1855–1890.* Cambridge MA: Harvard University Press, 1949.

Miller, Francis Trevelyan. *Thomas A. Edison, Benefactor of Mankind.* Philadelphia: John C. Winston Co., 1931.

Morse, Edward Lind. *Samuel F. B. Morse, His Letters and Journals.* 2 volumes. Boston: Houghton Mifflin, 1914.

Mott, Frank Luther. *American Journalism. A History: 1690–1960.* New York: Macmillan, 1962.

Myers, Gustavus C. *History of the Great American Fortunes.* New York: Modern Library, 1937.

Nerney, Mary Childs, Harison Smith, and Robert Hoas. *Thomas A. Edison.* New York: n.p., 1934.

Neuberger, Richard. "The Telegraph Trail." *Harper's Magazine* (October 1946): 369.

Nevada State Journal. February 14, 1927. Obituary of Governor Frank Bell.

Newhall, Beaumont. *The Daguerreotype in America.* New York: Duell, Sloan & Pearce, 1961.

"New Telephone Industry, The." *Business Week* (February 13, 1978).

Northrop, H. D. *Life and Achievements of Jay Gould.* Philadelphia: National Publishing Co., 1892.

O'Brien, J. Emmet. *Telegraphing in Battle; Reminiscenses of the Civil War.* Scranton PA: Raeder Press, 1910.

O'Brien, J. Emmet. "Military Telegraph Stories." *Century Magazine* (September 1889).

O'Connor, Richard. *Gould's Millions.* Garden City NY: Doubleday, 1962.

Orr, J. W. "The Lightning World." *The American Odd Fellow Magazine* (1867).

Page, Arthur W. *The Bell Telephone System.* New York and London: Harper & Brothers, 1941.

Paine, Albert Bigelow. *In One Man's Life (Theodore N. Vail).* New York: Harper & Brothers, 1921.

Paris, Comte de (Louis-Philippe-Albert d'Orleans). *History of the Civil War in America.* 4

volumes. Philadelphia: Joseph H. Coates & Co., 1875–1888.

Parker, Jane Marsh. "How Men of Rochester Saved the Telegraph." Publication Fund Series. Rochester NY: Rochester Historical Society, 1926.

Petroleum Today (Winter 1971).

Plum, William R. *The Military Telegraph During the Civil War in the United States.* 2 volumes. Chicago: Jansen, McClurg & Co., 1882.

Portland City Guide. Portland ME: n.p., 1940.

Prescott, George B. *History, Theory, and Practice of the Electric Telegraph.* Boston: Ticknor and Fields, 1860.

Prescott, George B. *Bell's Electric Speaking Telephone.* New York: D. Appleton & Co., 1884.

Prescott, George B. *Electricity and the Electric Telegraph.* 2 volumes. New York: D. Appleton and Company, 1885.

Prime, Samuel I. *The Life of Samuel F. B. Morse.* New York: D. Appleton and Company, 1875.

Reid, James. D. *The Telegraph in America.* New York: John Polhemus, 1879, 1886.

Rhodes, Frederick Leland. *Beginnings of Telephony.* New York: Harper & Brothers, 1929.

Rhodes, Frederick Leland. *(Outside plant development engineer, AT&T) John J. Carty, an Appreciation.* New York: privately printed, 1932.

Riedman, Sarah R. *Trailblazer of American Science. A Life of Joseph Henry.* Chicago, New York, San Francisco: Rand McNally & Co., 1961.

Rose, William Ganson. *Cleveland. The Making of a City.* Cleveland and New York: World Publishing Co., 1950.

Rosewater, Victor. *History of Cooperative News Gathering in the United States.* New York and London: D. Appleton and Co., 1930.

Russell, R. W. *History of the Invention of the Electric Telegraph.* New York: n.p., 1853.

Russell, Sir William Howard. *The Atlantic Telegraph.* London: Day & Son, Ltd., 1865.

Rohan, Jack. *Yankee Arms Maker: The Incredible Career of Samuel Colt.* New York: Harper Brothers, 1935.

Salmon, Lucy Maynard. *The Newspaper and the Historian.* New York: Oxford University Press, 1923.

Sarnoff, Paul. *Russell Sage: The Money King.* New York: Ivan Obolensky, Inc., 1965.

Scientific American. Various issues as cited.

Scrugham, James G. *Nevada.* 2 volumes. Chicago and New York: American Historical Society, 1935.

Seward, Frederick W. *Seward at Washington as Senator and Secretary of State. 1861–1872.* Volume 3. New York: Derby & Miller, 1891.

Shaffner, Taliaferro P. *Telegraph Manual.* New York: Pudney and Russell, 1859.

Shiels, Archie W. *Seward's Ice Box.* Private Edition. Bellingham WA: n.p., 1933.

Sibley, Hiram W. "Memories of Hiram Sibley." Publication Fund Series. Rochester NY: Rochester Historical Society, n.d. 2:127-34.

Simonds, William Adams. *Edison—His Life, His Work, His Genius.* Indainapolis and New York: Bobbs Merrill, 1934.

Swan, William Upham. "Early Visual Telegraphs." *Massachusetts Proceedings of the Bostonian Society for 1933.* Boston: n.p., 1933.

Taylor, William B. *A Historical Sketch of Henry's Contribution to the Electro-Magnetic Telegraph.* Washington: Government Printing Office. 1879.

Telcon Story, 1850–1950, The. London: Telegraph Construction & Maintenance Co. Ltd., 1950. Small memorial volume.

Telecommunications Reports. A privately published communications industry newsletter. Washington, D.C. Many years.

The Telegrapher. Article on reception by sound (November 28, 1864).

Thompson, Robert Luther. *Wiring a Continent.* Princeton NJ: Princeton University Press, 1947.

Thompson, Silvanus P. *Philipp Reis, Inventor of*

the Telephone. London and New York: E. & F. N. Spon, 1883.

Thorwald, Jurgen. *The Century of the Surgeon.* New York: Pantheon Books, 1956.

Turnbull, Lawrence. *The Electro-Magnetic Telegraph.* Philadelphia: A. Hart, 1853.

Twain, Mark. *Roughing It.* Hartford CT: American Publishing Co., 1873.

U.S. News & World Report, Interview with Nobel Prize Winner Charles H. Townes (December 23, 1968).

U.S. News & World Report (November 28, 1977): 66.

Van Pelt, Daniel. *Leslie's History of the Greater New York.* New York: Arkell Publishing Co., 1898.

Visscher, William Lightfoot. *The Pony Express.* Chicago: Rand, McNally & Co., 1908.

Ware, Eugene F. *Indian War of 1864.* Topeka KS: Crane & Co., 1911.

"Who Invented the Quadruplex Telegraph?" *The Electrician* (January 1886).

Whymper, Frederick. *Travel and Adventure in the Territory of Alaska.* New York: Harper, 1869.

Willis, William. *The History of Portland from 1632 to 1864.* Portland ME: Bailey and Hoyes, 1865.

Wilson, Ben Hur. "Telegraph Pioneering." *The Palimpsest* (State Historical Society of Iowa, November 1925).

_____. "Across the Prairies of Iowa." *The Palimpsest* (August 1926).

Wilson, James Grant. *The Memorial History of the City of New York.* New York: New York History Co., 1893.

Wilson, William Bender. *A Few Acts and Actors in the Tragedy of the Civil War.* Philadelphia: published by the author, 1892.

Wolff, Leon. *Lockout.* New York: Harper & Row, n.d.

Wright, Albert Hazen. *Pre-Cornell and Early Cornell X.* Ithaca NY: n.p., 1965.

Periodicals, Documents

American Heritage Magazine.

Bell Laboratories Record. April 1967 and other numbers.

Bell Telephone Magazine. Winter 1964–1965 and other numbers.

Documents of U.S. Senate and House of Representatives concerning purchase of Alaska. Fortieth Congress, and others. Library of Congress, Washington.

Forbes Magazine. Issues of September 15, 1976 and September 15, 1968.

New York Times and other newspapers, as indicated in notes and text.

Public Relations Journal. October 1966 and other numbers.

Radio Amateur News 1 (1919–1920). Experimentor Publishing Co., New York.

The Telegraph and Telephone Age. Various issues in addition to those specifically cited above.

The Telegrapher. All issues for a number of years, including those specifically cited above.

The Telegraph World. Various numbers in addition to those cited above.

The Telegraphic Journal and Electrical Review 6 (London: Haughton & Co., 1878).

Telecommunications Reports. All issues of this weekly news service covering telephone, telegraph, and radio since 1934. Washington, D.C. Fred W. Henck, editor.

Television Digest. Various issues as cited in notes.

Wall Street Journal. Daily for many years. Source of facts about numerous companies, especially in data processing.

Western Union Technical Review. All issues since 1947. Mary Killilea Malone, editor

Index